ANNUAL REVIEW OF HEAT TRANSFER

ANNUAL REVIEW OF HEAT TRANSFER

Formerly Annual Review of Numerical Fluid Mechanics and Heat Transfer

Editor
C. L. Tien, *University of California, Berkeley, California*

Founding Editor
T. C. Chawla, *Argonne National Laboratory, Argonne, Illinois*

Titles in the Series

Annual Review of Numerical Fluid Mechanics and Heat Transfer, Vol. 1

Editor
T. C. Chawla, *Argonne National Laboratory, Argonne, Illinois*

Annual Review of Numerical Fluid Mechanics and Heat Transfer, Vol. 2

Editors
C. L. Tien, *University of California, Berkeley, California*
T. C. Chawla, *Argonne National Laboratory, Argonne, Illinois*

Annual Review of Heat Transfer, Vol. 3

Editor
C. L. Tien, *University of California, Irvine, California*

Annual Review of Heat Transfer, Vol. 4

Editor
C. L. Tien, *University of California, Berkeley, California*

ANNUAL REVIEW OF HEAT TRANSFER

Volume 4

Edited by

C. L. Tien

Chancellor and A. Martin Berlin Professor
University of California
Berkeley, California

HEMISPHERE PUBLISHING CORPORATION

A member of the Taylor & Francis Group

Washington Philadelphia London

ANNUAL REVIEW OF HEAT TRANSFER, Volume 4

1 2 3 4 5 6 7 8 9 0 B R B R 9 8 7 6 5 4 3 2 1

The editors were Joyce Duncan and Lynne Lackenbach. This book was set in Times Roman by Harper Graphics. Braun-Brumfield, Inc. was the printer and binder.

A CIP catalog record for this book is available from the British Library.

Library of Congress Cataloging-in-Publication Data

ISBN 1-56032-176-8
ISSN 1049-0787

CONTENTS

CONTRIBUTORS

HAIM H. BAU
Department of Mechanical Engineering
University of Pennsylvania
Philadelphia, Pennsylvania

M. QUINN BREWSTER
Department of Mechanical and Industrial Engineering
University of Illinois
Urbana, Illinois

S. H. CHAN
Mechanical Engineering Department
University of Wisconsin
Milwaukee, Wisconsin

YOGESH JALURIA
Mechanical Engineering Department
Rutgers University
Piscataway, New Jersey

A. MAJUMDAR
Department of Mechanical and Energy Systems Engineering
Arizona State University
Tempe, Arizona

PER F. PETERSON
Nuclear Engineering Department
University of California
Berkeley, California

ROGER H. RANGEL
Mechanical Engineering Department
University of California
Irvine, California

DA YU TZOU
Mechanical Engineering Department
University of New Mexico
Albuquerque, New Mexico

YUZHOU WANG
Department of Mechanical Engineering
University of Pennsylvania
Philadelphia, Pennsylvania

PREFACE

The serial publication, *Annual Review of Heat Transfer,* is designed to provide expository, but penetrating, state-of-the-art reviews of important new developments in the general field of heat transfer. These special characteristics were clearly exhibited in the previous volume (Volume 3), which covered such topics as biological freezing, binary vapor condensation, film condensation interface, high-T_c superconductors, multiphase transport in porous media, convection in packed beds and granular flow, microscales of turbulence, and scaling and asymptotic analysis in convection. Many of the contributing authors are already leaders in these emerging fields.

The present volume again contains eight chapters of new, challenging topics including chaos, fractal analysis, thermal shock in solids, thermal processing of materials, species-controlled condensation, heterogeneous propellant combustion, mixing of vaporizing droplet sprays, and fouling heat transfer. Again, the newness of these topics necessitated the choice of up and coming researchers, whose talents are vividly demonstrated in their personal backgrounds. All the participating authors deserve praise for their contributions.

The editor expresses his thanks to the heat transfer community for its continuing support and cooperation. Any suggestions and input from members of the scholarly community will be greatly appreciated.

C. L. Tien

CHAPTER
ONE

CHAOS: A HEAT TRANSFER PERSPECTIVE

Haim H. Bau and YuZhou Wang

1 INTRODUCTION

During the last two decades, the topic of chaos has attracted a considerable amount of attention in both the scientific and popular literatures (see, for example, a best-selling book by Gleick [1]). The topic has even been referred to as revolutionary. Its significance is not restricted to any particular scientific discipline. Chaos is of interest to fluid mechanicians, heat transfer practitioners, electrical engineers, chemists, biologists, physiologists, ecologists, and economists, among others. The purpose of this chapter, however, is to introduce the topic of chaos to heat transfer practitioners who may not be well versed in chaos theory. The chapter is self-contained. Various concepts and the associated technical jargon are introduced as part of the exposition. We limit our discussion to dissipative systems. These are systems in which energy is being dissipated by viscous or friction action, in contrast to conservative systems in which energy is being conserved.

Deterministic chaos is a complicated, randomlike behavior exhibited by deterministic systems. For fluid mechanics and heat transfer practitioners, chaos is a familiar phenomenon. After all, turbulence involves randomlike behavior generated by a deterministic system modeled by the Navier-Stokes equations. The excitement of the last few years has been derived not so much from the "discovery" of chaos, but rather from the realization that a very complicated, turbulentlike behavior may be exhibited by systems with a relatively small number of degrees of freedom. This realization has given rise to the hope that complicated phenomena such as turbulence, which are modeled by systems with an infinite number of degrees of freedom (mathematically speaking, partial differential equations are considered to have an infinite number of degrees of freedom), may possibly be approximated by relatively low-dimensional systems (i.e., a finite number of ordinary differential equations). The practical implications are, of course, immense. Although some progress toward this goal has been made in recent years, the objective of modeling turbulence, for example, with a low-dimensional system has not yet been attained. Nevertheless, the study of chaos and, more generally, the study of dynamical systems has yielded a number of byproducts which are quite useful in their own right.

First and foremost, studying the evolution of chaos in low-dimensional systems has enhanced our understanding of the generation of turbulence. The issue of the relationship between chaos and turbulence is no longer controversial. Even Landau and Lifshitz [2], in a new edition of their book, have revised the celebrated Landau-Hopf conjecture on the evolution of turbulence by adopting modern concepts of dynamical systems theory.

More generally, a fundamental understanding of the evolution of chaotic behavior is of great importance to both scientists and engineers, since such knowledge may assist us in enhancing or retarding chaotic behavior depending on the particular application at hand (whether chaos is a desired or an undesired phenomena). For example, chaos may enhance mixing [3], and it may be desired in processes involving heat transport, chemical reactions, and combustion. On the other hand, one would like to avoid chaotic behavior by a car's suspension, which might lead to an uncomfortable ride and early failure of components, and by an emergency cooling system in nuclear reactors (where, as we shall see below, chaotic behavior may imply large temperature fluctuations which, in turn, may lead to structural failure).

Another surprising aspect of the theory of dynamical systems is the universality of the transition routes to chaos. Diverse physical systems ranging from solid-state physics to convective systems exhibit similar behavior. For example, a frequent route to chaos consists of a period doubling sequence. It turns out that all period doubling sequences exhibit similar behavior.

One useful aspect of the theory of chaotic systems is its tenet that the full dynamics of such a system are embedded in the history of a single variable. Thus we should be able to obtain information on the dynamics of a chaotic system from the time series of a single quantity. For example, by examining the behavior of temperature as a function of time at a certain point in a chemical reactor, we can determine whether the system is deterministic or stochastic. If the system is deterministic, we can, in principle, estimate how many equations (or degrees of freedom) are needed to model the system. Moreover, the time series should enable us to construct an object in phase space (an attractor) around which the trajectories of suitable variables will evolve. The availability of such an attractor may allow us to obtain short-term predictions for the future behavior of the system even if the mathematical model is not known. Such predictions may be extremely useful for control purposes as well as for early warning of an impeding disaster. For example, if we assume that the economy as a whole is a deterministic system (there is no assurance that this is, indeed, the case), then by monitoring an economic measure such as the Dow-Jones, we can (in principle) obtain short-term predictions for future behavior of the stock market and even estimate the range of validity (timewise) of these predictions.

Another important feature of chaotic systems is unpredictability. Such systems are very sensitive to variations in initial data and to small disturbances. Since we usually do not have precise knowledge of the initial conditions for a real system we are attempting to model and since all real systems are subject to disturbances, this implies, in practical terms, that long-term predictions for the behavior of chaotic systems are not possible even when the mathematical model is known precisely. For example, no matter how large the computers we acquire, we will not be able to forecast weather for an extended period of time.

In short, from both practical and historical perspectives, chaos should be of special interest to heat transfer practitioners. Many of the problems encountered in heat transfer involve chaotic (or turbulent) processes. Also, since efficient heat transfer requires efficient mixing, it is often desirable to operate under chaotic conditions. Finally, many of the early examples of chaotic behavior were discovered and much of the current research is conducted in the area of thermal convection.

To facilitate the introduction to and understanding of chaos, we will be discussing in this chapter a simple example of a thermal convection loop. Unfortunately, because of space and time limitations, we will not be able to cover the topic exhaustively, nor will we cite all the pertinent literature. However, we will provide a short description of a sample of the available literature for those readers whose appetite has been wetted but not quite satisfied by this chapter. Rather than simply write a survey paper, we have chosen to develop a kind of tutorial on the subject of chaos. In the next section we discuss some experimental and theoretical aspects of fluid motion in convection loops. Then we briefly point out various general consequences of chaos theory. Concrete examples are borrowed conveniently from the convection loop problem. Although much of the material presented here is based on contributions of many researchers (appropriate citations are provided as often as possible), the chapter also includes some original contributions by the authors.

2 EXAMPLE OF A SYSTEM EXHIBITING CHAOTIC BEHAVIOR

There are many examples of low-dimensional physical and mathematical systems which exhibit chaotic behavior. A frequent and celebrated example used to demonstrate complicated dynamical behavior is the logistic difference equation, $x_{n+1} = 4\lambda x_n (1 - x_n)$, where $0 < \lambda < 1$ and $0 < x_n < 1$ [4, 5]. We urge the reader to compute values of x_n for various choices of λ and, in particular, values of $\lambda > 0.75$.

Since this chapter is aimed at the heat transfer practitioner, we will discuss at greater length a somewhat more difficult example than the aforementioned one. Appropriately enough, the example is borrowed from the area of heat transfer and, more specifically, from thermal convection in loops (thermosyphons).

2.1 Thermal Convection Loop: Experimental Observations

Thermal convection loops provide a means for circulating fluid without the use of pumps. Such loops are of interest for solar heaters, emergency reactor-core cooling, and process industries. They also are of interest for understanding warm springs, seawater circulation in the oceanic crust, and formation of ore deposits. For a general review of applications and analyses of these loops (excluding their chaotic behavior), see an article by Metrol and Greif [6] and the literature cited therein. Here we shall use the thermal loop as a simple example of a system which exhibits complicated dynamical behavior.

Consider, for example, a thermal convection loop constructed from a pipe bent into a torus and standing in the vertical plane as depicted in Fig. 1. The diameter of the pipe is d; the radius of the torus is R; and θ is the angular location of a point on the torus. The wall temperature of the pipe, $T_w(\theta, t)$, may vary with both the angular location θ and time t. Variations in the wall temperature may cause a temperature distribution inside the fluid, which, in turn, under appropriate conditions, may induce fluid motion in the loop.

Thermal convection loops of this type have been constructed at a number of laboratories. For examples, see Creveling et al. [7], Gorman et al. [8–10], Erhard and Muller [11], and Greif et al. [12]. For the last few years, we have been operating a similar loop at the University of Pennsylvania, as part of our undergraduate laboratory course, to demonstrate experimentally chaotic dynamics [13].

Below we describe a typical experiment which our students carry out in the laboratory. The experiment consists of observations of fluid motion in the loop as the heating rate is gradually increased. The measured quantities include the temperature differences between locations 9 and 3 o'clock and between locations 6 and 12 o'clock around the loop, which we denote by $y(t)$ and $z(t)$, respectively. The aforementioned quantities are continuously monitored as functions of time. We use a thymol blue solution as a working fluid. The solution is slightly acidic and has a yellow color. By subjecting electrodes submerged in the fluid to a DC voltage, we generate blue tracers next to the anodes, which allow us to observe the streak lines. At the beginning of

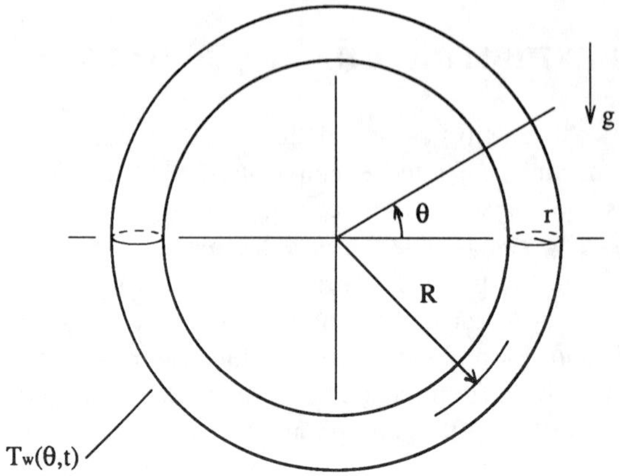

Figure 1 Schematic description of the thermal convection loop. A pipe of diameter d is bent into a torus of radius R. The torus is standing in the vertical plane.

the experiment, the fluid in the loop is isothermal. The temperature difference between the top and bottom is increased in small steps and maintained constant in time between subsequent increases. For very low temperature differences between bottom and top, no fluid motion is detected in the loop. As the temperature difference is increased beyond a certain threshold, motion starts. Due to the symmetry of the heating and cooling, the flow may start either in the clockwise or the counterclockwise direction. For low heating rates, the motion initially is relatively steady, though its intensity increases as the temperature difference between bottom and top increases. When the temperature difference increases beyond a certain point, the flow ceases to be steady and one observes oscillations in the temperature difference y. The amplitude of these oscillations increases until a reversal in the flow direction occurs. The oscillatory behavior continues indefinitely with occasional reversals in the flow direction. The experimentally observed time series is similar to the one depicted in Fig. 2. (The figure we use here was actually generated by a mathematical model we present later. Similar time series have been observed in experiments). The time series appears to have complicated, nonperiodic, turbulentlike time dependence. Analysis of the signal in the frequency domain (Fig. 3) demonstrates a continuous spectrum with gradually decaying amplitude as the frequency increases. The absence of dominating peaks in the power spectrum (Fig. 3) suggests that the phenomena is neither periodic nor quasi-periodic. The power spectrum depicted in Fig. 3 is typical for chaotic systems. In order to emphasize further the complexity of the observed phenomenon and its randomlike nature, we depict in Fig. 4 the positive and negative oscillations from Fig. 2 as positive

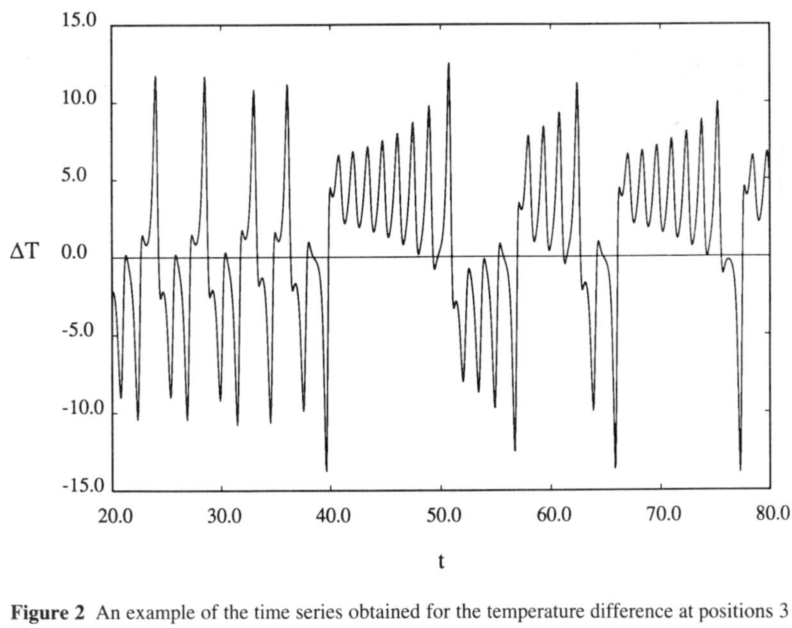

Figure 2 An example of the time series obtained for the temperature difference at positions 3 and 9 o'clock around the loop as a function of time (Ra = 20, p = 4). Sufficient time is allowed for initial transients to die out. The oscillatory, aperiodic behavior continues indefinitely.

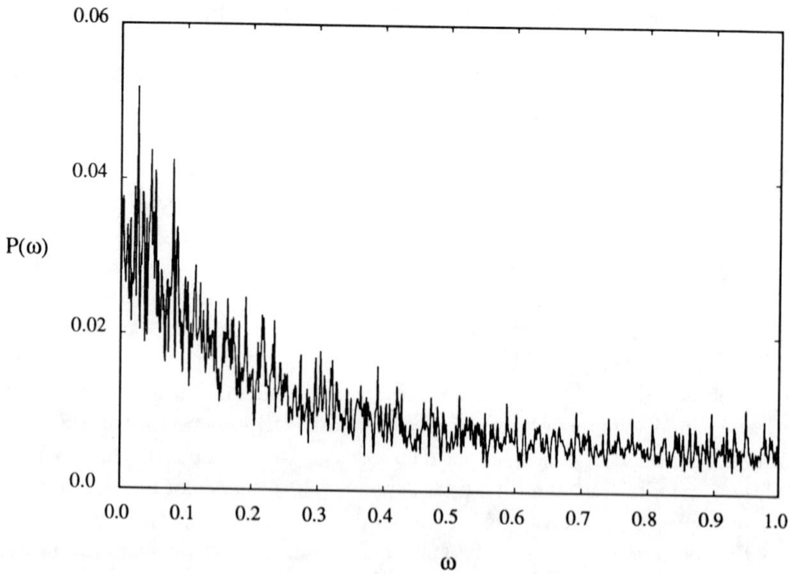

Figure 3 The power spectrum of the time series depicted in Fig. 2.

and negative ticks, respectively. We denote the positive and negative ticks with the letters P and N, respectively, to generate the sequence $N^2PN^2PN^2PNPN^2P^8 \ldots$, where the exponents indicate the number of adjacent oscillations in the positive/negative directions of the flow. Such a sequence is referred to as *symbolic dynamics* (similarly, we could have denoted the positive and negative oscillations by 1 and 0, respectively, to obtain a binary sequence). This sequence immediately reminds us of a similar random sequence which one would obtain by tossing a coin and counting the number of heads and tails. Yet, as we shall point out shortly, the signal we are dealing with (Figs. 2–4) is fully deterministic.

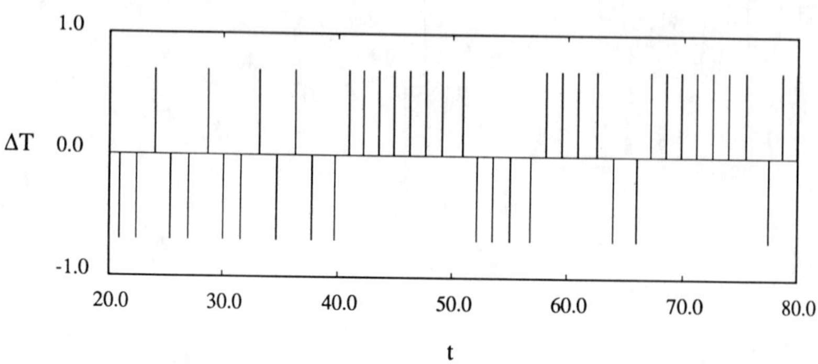

Figure 4 The counterclockwise and the clockwise oscillations shown in Fig. 2 are depicted as positive (P) and negative (N) ticks.

A plausible physical explanation for the complicated dynamical behavior described above has been provided by Welander [14]. Briefly, consider steadystate motion in the loop in the counterclockwise direction. Suppose that, due to some random thermal disturbance, the temperature of an ascending parcel of fluid is greater than is normally the case. As a result, the buoyancy force will increase and the fluid will move in the loop faster than usual. The faster-moving fluid will spend less time in the heater and cooler, causing the average temperatures in the ascender/descender to be lower/higher than usual with a net reduction in the buoyancy force and a corresponding slowdown in the fluid's velocity. This process can continue indefinitely, with the oscillations building up to such a degree that at a certain time the direction of the buoyancy force is reversed and the flow changes direction.

2.2 Thermal Convection Loop: Mathematical Model

In order to gain some additional insight into the aforedescribed phenomenon, we shall construct a simple mathematical model and study it in some detail. Often loops like the one described here are analyzed using a one-dimensional model consisting of mass, momentum, and energy balances [15]. We shall work within the framework of Boussinesq's approximation. The fluid is assumed to be incompressible and Newtonian. The continuity, momentum, and energy equations are

$$u = u(t) \tag{1}$$

$$\dot{u} = \frac{1}{\pi} \operatorname{Ra} P \oint T \cos(\theta)\, d\theta - \operatorname{Pu} \tag{2}$$

and

$$\dot{T} = -u \frac{\partial T}{\partial \theta} + B \frac{\partial^2 T}{\partial \theta^2} + [T_w(\theta, t) - T] \tag{3}$$

In the above, all quantities are nondimensional; $\mathrm{Ra} = g\beta\, \Delta T \tau^2/(2RP)$ is the loop's Rayleigh number; β is the thermal expansion coefficient; g is the gravitational acceleration; and ΔT is the time-averaged, wall temperature difference between the loop's bottom and top. The time scale is $\tau = \rho C_p d/(4h)$, where ρ is the average fluid's density, C_p is the thermal capacity, and h (which we assume to be constant) is the heat transfer coefficient between the fluid and the pipe's wall. $P = 32\nu\tau/d^2 = 8$ Pr/Nu is the loop's Prandtl number, where ν is the kinematic viscosity. $\mathrm{Pr} = \nu/\alpha$ and $\mathrm{Nu} = hd/k$ are the conventional Prandtl and Nusselt numbers, respectively. α and k are the fluid's thermal diffusivity and conductivity; and $B = 1/\mathrm{Nu}(d/2R)^2$ is the Biot number. The length scale is the torus's radius R.

In addition to the aforementioned Boussinesq approximation, the mathematical model presented here assumes implicitly that the friction and heat transfer laws are similar to those of laminar, fully developed, Poisuelle pipe flow. One would expect and we did, in fact, observe in the experiments the development of secondary circulation which may significantly modify both the friction and heat transfer laws (but has the positive effect of improving temperature uniformity at each cross section of the loop).

Such more realistic friction and heat transfer laws are not known a priori. To obtain these correlations, one may need to solve a spatially three-dimensional model or conduct experiments. We justify the use of the simpler correlations on the grounds that the model still provides a qualitatively correct picture, as has been confirmed by our own experiments and those of others (i.e., [8–10, 16], as well as by theoretical studies by Hart [17, 18] and by Yorke et al. [19], in which more complicated heat transfer and friction factor correlations were used.

Equations (1)–(3) constitute a set of nonlinear differential-integral equations for which an exact solution is not available. As is often done in the solution of problems in mathematical physics, we employ a spectral technique. Briefly, we expand the wall and fluid temperatures in Fourier series in terms of the angle θ:

$$T_w(\theta, t) = W_0(t) + \sum_{n=1}^{\infty} W_n(t) \sin(n\theta) \tag{4}$$

and

$$T(\theta, t) = \sum_{n=0}^{\infty} S_n(t) \sin(n\theta) + C_n(t) \cos(n\theta) \tag{5}$$

Next, we substitute the series (4)–(5) into the governing equations (1)–(3) and require these equations to be satisfied in the sense of weighted residuals to obtain an infinite set of ordinary differential equations. Typically, one would truncate the set and work with a finite set of equations. In this particular case, three equations in the set decouple from the rest of the set (with exact closure) and can be solved independently of the other equations without need of truncation [20]. Thus the full dynamics of the problem are embedded in three ordinary differential equations. Once these three equations are integrated, the remaining set of (slaved) equations can be solved as well.

In summary, by proper choice of a projection space (in this case, a space whose coordinates consist of trigonometric functions), we succeed in reducing the problem in question from an infinite number of dimensions (degrees of freedom) to three dimensions. Of course, this reduction constitutes a major simplification from both the analytical and numerical points of view.

The set of three equations is

$$\frac{\dot{u}}{P} = c - u \tag{6}$$

and

$$\dot{c} = -us - c \tag{7}$$

$$\dot{s} = uc - s - \mathrm{Ra}[1 + \varepsilon f(t)] \tag{8}$$

In the above, we removed the dependence on the Biot number, B, via the simple algebraic transformation $\{u, c, s, \mathrm{Ra}, P, t\} \leftarrow 1/(1 + B)\ \{u, \mathrm{Ra}\ C_1, \mathrm{Ra}\ S_1, \mathrm{Ra}/(1 + B), P, t(1 + B)^2\}$. Also, without loss of generality, we set $W_1(t) = -[1 + \varepsilon f(t)]$.

Equations (6)–(8) are invariant under the transformation $(u, c, s, t) \rightarrow (-u, -c, s, t)$. Thus all solutions of the equations appear either as symmetric objects or asymmetric pairs which are mirror images of each other under the aforementioned invariance. Any symmetry-breaking bifurcation will result, at the very least, in a pair of solution branches, one being the image of the other. For example, the no-motion solution bifurcates into two asymmetric solutions, one corresponding to clockwise and the other to counterclockwise motion. We mention this since often when we plot bifurcation diagrams, we shall depict only one of the solution branches and not its mirror image.

In the special case of $\varepsilon = 0$, the wall temperature is time-independent and Eqs. (6)–(8) reduce to the celebrated Lorenz [21] equations. It is interesting to note that Lorenz derived his equations as a crude approximation for atmospheric convection (essentially the Benard problem). It turns out that his equations may not represent the Benard problem adequately but they give a fairly good approximation for the loop problem which we discuss here. The Lorenz equations have been investigated extensively in the scientific literature. See, for example, Sparrow [22] and the references cited therein. We shall consider below the cases of $\varepsilon = 0$ (*time-independent* wall temperature) and the case of $\varepsilon > 0$ and $f(t) = \sin(\omega t)$ (time-periodic wall temperature). Clearly, there are many other interesting possibilities, which for lack of space we shall not explore here.

Equations (6)–(8) are nonlinear and admit a number of possible solutions. Depending on the magnitude of the system's parameters, these are no motion, periodic motion, and chaotic motion.

2.3 Thermal Convection Loop with Time-Independent Wall Temperature

We start our theoretical investigation of the loop's equations (6)–(8) by first considering the *time-independent* wall temperature case ($\varepsilon = 0$). First we find the steady solutions and examine their stability. Subsequently, we show that as the Rayleigh number increases, the steady-state solutions are replaced by complicated time-dependent flows.

2.3.1 Time-Independent Solutions and Their Stability. The time-independent solutions can readily be obtained by solving Eqs. (6)–(8) with their left-handsides set to zero. For the purposes of the following discussion, it is convenient to define a new coordinate system, which we obtain by the shift $(x, y, z) = (u, c, s + \mathrm{Ra})$. Equations (6)–(8) admit both the no-motion solution,

$$x = y = z = 0 \tag{9}$$

which corresponds to a fixed point at the origin of the phase space (point C_0 in Fig. 5), and the motion solutions,

$$x = y = \pm\sqrt{\mathrm{Ra} - 1} \qquad z = \mathrm{Ra} - 1 \tag{10}$$

which correspond to steady motions either in the counterclockwise $(+)$ or clockwise $(-)$ direction. The corresponding fixed points are denoted, respectively, by C^+ and C^- in phase space (Fig. 5). Thus the system admits three different possible steady solutions.

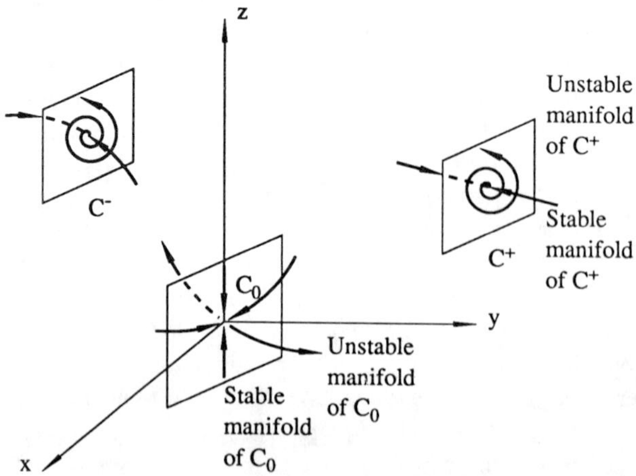

Figure 5 The fixed points corresponding to the steady-state solutions of the loop equations are depicted in phase space. Also shown schematically are the eigenvectors associated with the linearized stability problem and the stable and unstable manifolds.

In the phase-space representation (Fig. 5), we view Eqs. (6)–(8) as if they were describing the motion of a particle in an n-dimensional space (here $n = 3$). Accordingly, the solutions (x, y, z) can be viewed as coordinates of location and their time derivatives as velocities. Steady-state and time-dependent solutions will be represented by fixed points and trajectories (of the imaginary particle) in phase space. The phase-space portrait is very useful in obtaining qualitative information on the nature of the solutions of differential equations. We also note that for any given set of initial conditions, the solution of Eqs. (6)–(8) is unique. This implies that trajectories cannot cross each other in phase space.

The stability of the solutions (with respect to small disturbances) can be investigated using linear theory. That is, we can introduce a small disturbance around the various steady solutions [Eqs. (9) and (10)] which we denote, in turn, by (x_0, y_0, z_0). Upon linearizing the equations for the disturbance, we obtain the coefficients' matrix of the linearized system (or the system's Jacobian):

$$\begin{pmatrix} -P & P & 0 \\ -(\mathrm{Ra} - z_0) & -1 & -x_0 \\ y_0 & x_0 & -1 \end{pmatrix}. \tag{11}$$

The stability of the solutions is determined by the sign of the real part of the eigenvalues of the Jacobian matrix, Eq. (11). For the no-motion solution, the matrix has the three real eigenvalues:

$$\sigma_{1,2} = \frac{1}{2}[-(P + 1) \pm \sqrt{(P - 1)^2 + 4P\,\mathrm{Ra}}] \qquad \sigma_3 = -1 \tag{12}$$

of which two (σ_2 and σ_3) are always negative (for $\mathrm{Ra} > 0$), while the third (σ_1) is negative for $\mathrm{Ra} < 1$ and positive for $\mathrm{Ra} > 1$. Thus the no-motion solution is stable for $\mathrm{Ra} < 1$ and nonstable for $\mathrm{Ra} > 1$. The linear theory tells us that for $\mathrm{Ra} < 1$, any trajectories starting in the vicinity of C_0 will end up at C_0. The region in which this phenomenon happens is called the basin of attraction of C_0. In general, the basin of attraction occupies a portion of the phase space. In this particular case, it is easy to verify that for Ra < 1, the no-motion solution is globally stable. We also call such a solution globally attracting, since no matter what initial conditions we introduce, we eventually end up with the no-motion solution. In other words, for Ra < 1, the basin of attraction of the no-motion solution includes the entire phase space. At Ra $= 1$, the no-motion solution loses stability in favor of one of the motion solutions [Eq. (10)]. That is, when we increase Ra beyond 1, the solution bifurcates into either counterclockwise or clockwise motion. We use the term *bifurcation* to indicate a qualitative change in the behavior of the solution. This result is depicted in a bifurcation diagram (Figs. 6*a* and 6*b*), where stable and nonstable solutions are denoted by solid and dashed lines, respectively. Figure 6*a* is a schematic description of the bifurcation sequence which is not plotted to scale, while Fig. 6*b* represents actual data obtained by numerical integration. The term *nonstable* (or saddle, fixed point) indicates that some, but not all, of the eigenvalues of the Jacobian matrix have a positive real part. This is in contrast to the term *unstable* solution, which indicates that *all* the eigenvalues have a positive real part. The distinction is made since unstable solutions will become stable when we reverse the direction of the time axis (go backwards in time) while nonstable solutions remain nonstable under such a time reversal. Thus it is easy to compute unstable (repelling) solutions simply by integrating backwards in time. We shall show later that all solutions of dissipative systems are either stable or nonstable.

There are three eigenvectors (X_i) in phase space associated with the three eigenvalues of the no-motion solution. We can project the linearized equations on the eigenspace to obtain a set of three decoupled linear equations of the form $dX_i/dt = \sigma_i X_i$ ($i = 1, 2, 3$). The eigenvectors are depicted schematically in Fig. 5 for Ra $>$ 1. Clearly, a trajectory which starts along eigenvectors corresponding to eigenvalues with negative real parts will end up at the origin (C_0) as we integrate the linearized equations forward in time ($t \rightarrow \infty$), while trajectories started on eigenvectors corresponding to eigenvalues with positive real parts will end up at C_0 as we integrate the equations of motion backward in time ($t \rightarrow -\infty$). We refer to the spaces spanned by the former and latter eigenvectors as the stable (L^s) and unstable (L^u) linear manifolds of C_0, respectively. Similarly, we can define the stable (W^s) and unstable (W^u) manifolds of the nonlinear system. These will be spanned by trajectories which go asymptotically to C_0 as we integrate the nonlinear equations forward ($t \rightarrow \infty$) and backward ($t \rightarrow -\infty$) in time, respectively. The nonlinear manifolds may be curved surfaces with a rather complicated topological structure, which are tangent to the corresponding linear manifolds at C_0. Any trajectory which starts on one of the manifolds will remain on it forever. We note that C_0 possesses a two-dimensional stable and a one-dimensional unstable manifold. Any trajectory which is not on the stable manifold will eventually land on the unstable manifold and, for sufficiently small Ra > 1, will end up at one of the fixed points (C^+ or C^- in Fig. 5). For not too large Ra > 1, trajectories starting to the right/left of W^s will end up at C^+/C^-. Thus the stable manifold acts as a separatrix between the basins of attraction of C^+ and C^-. For illustration purposes, we depict in Figs. 7 and 8 the results of a numerical integration of the nonlinear equations for $P = 4$ and Ra $= 7$ and 8 with initial data very close

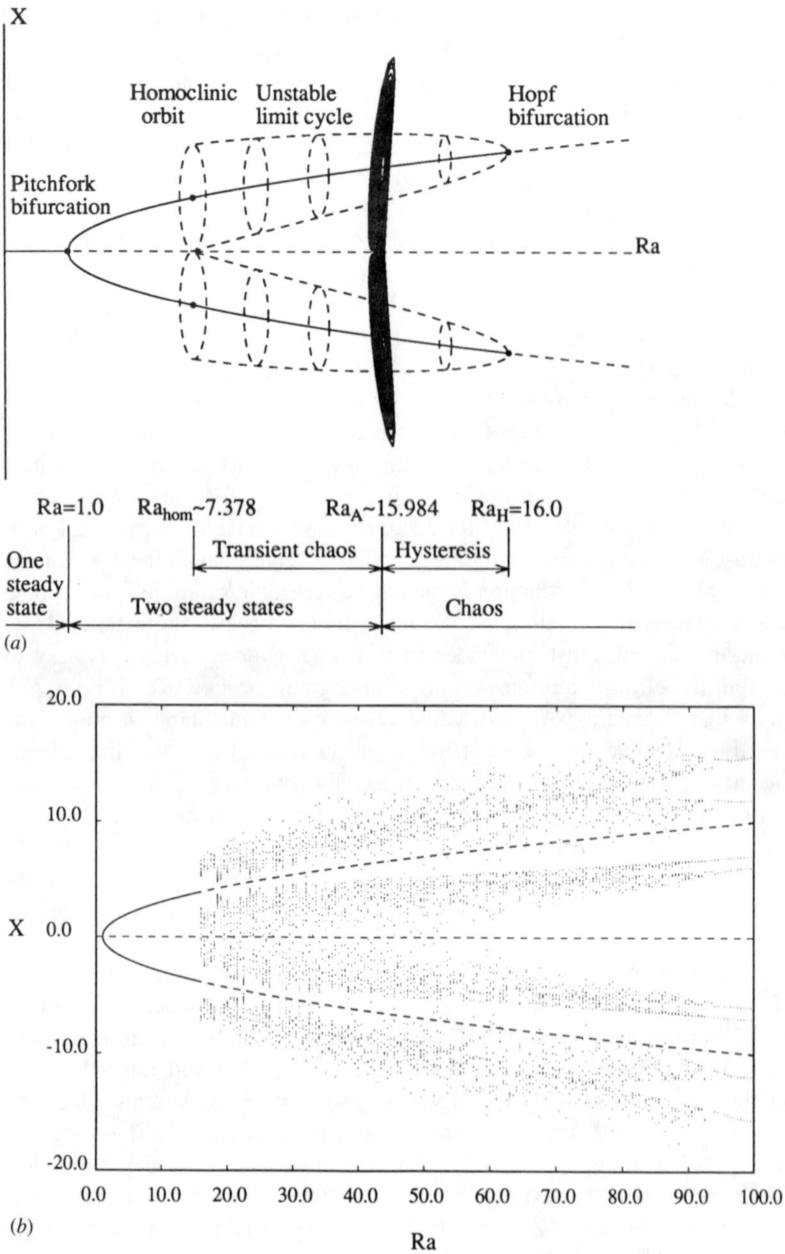

Figure 6 A bifurcation diagram depicting the various possible solutions as the Rayleigh number is being increased ($p = 4$) is shown in two different ways. (*a*) An unscaled schematic description illuminating the various regions in the bifurcation sequence. The dark region represents the appearance of the strange attractor. Stable and nonstable solutions are denoted by solid and dashed lines, respectively. (*b*) A bifurcation diagram obtained from numerical experiments. The x value of the penetration points through the plane $z = \text{Ra} - 1$ is shown as a function of the Rayleigh number. The dense distribution of points indicates chaotic behavior. A few points at any vertical cross section of the graph indicates a periodic window (see, for example, the right-hand side of the figure).

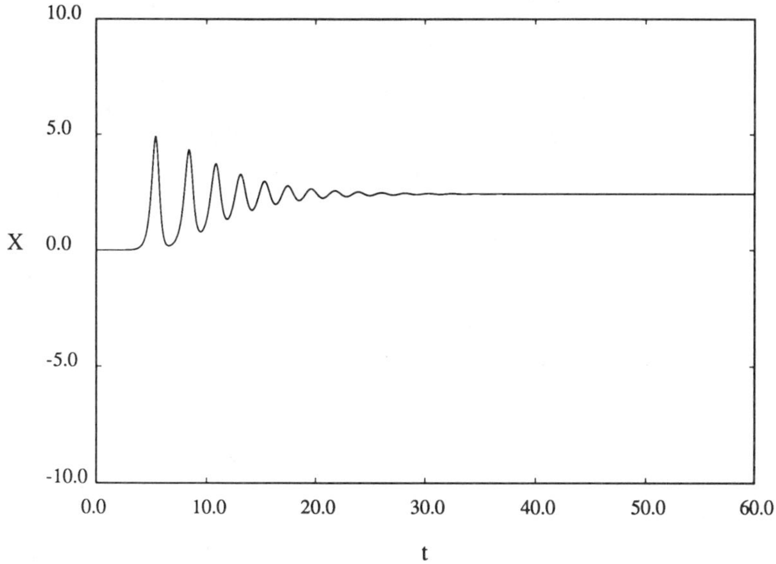

Figure 7 Time series exhibiting the approach to the steady-state solution (C^+) for Ra = 7 and initial conditions on the right-hand-side unstable manifold of C_0.

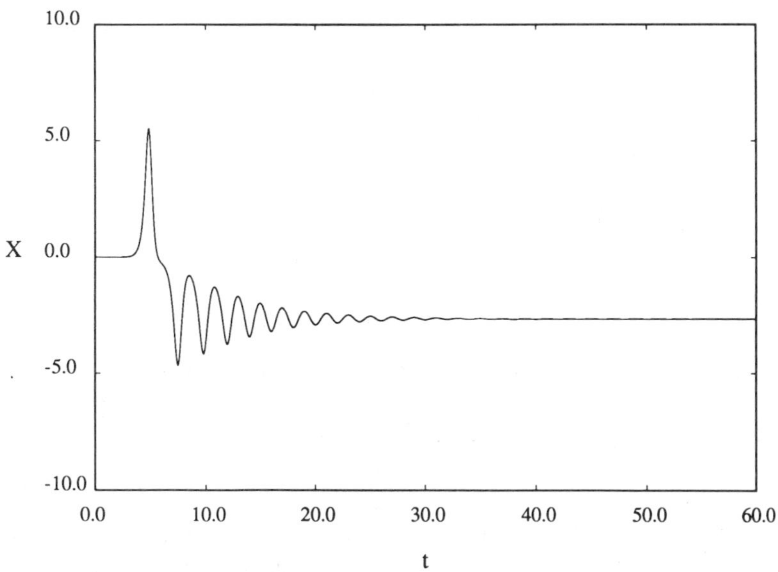

Figure 8 Time series exhibiting the approach to the steady-state solution (C^-) for Ra = 8 with initial conditions similar to those in Fig. 7. Note that the change in Ra number causes trajectories with similar initial data to end up at different fixed points.

to the origin (C_0) and, on its right-hand side, unstable manifold. As time goes by, for Ra = 7/Ra = 8, the trajectory approaches the fixed point C^+/C^-. This suggests that for Ra = 8, W^s must bend in a rather complicated fashion.

The loss of stability of C_0 occurs supercritically via a pitchfork bifurcation as depicted in Fig. 6. Thus, for Ra > 1, there is motion in the loop in either the counterclockwise or clockwise direction. The linear stability of this motion can be determined by the sign of the real part of the eigenvalues of the matrix [Eq. (11)]. These eigenvalues satisfy the cubic equation:

$$\sigma^3 + \sigma^2(2 + P) + \sigma(P + \mathrm{Ra}) + 2P(\mathrm{Ra} - 1) = 0 \tag{13}$$

Although the explicit form of the roots of Eq. (13) is fairly lengthy, it is a straightforward procedure to determine the conditions when the real part of the eigenvalues is positive or negative. It turns out that all eigenvalues have a negative real part for $1 < \mathrm{Ra} < \mathrm{Ra}_H$, where $\mathrm{Ra}_H = P(P + 4)/(P - 2)$; i.e., for $P = 4$, $\mathrm{Ra}_H = 16$. For $\mathrm{Ra} > \mathrm{Ra}_H$, the real part of two complex conjugate eigenvalues is positive. The loss of stability is strict. That is, as Ra is increased past Ra_H, the real part of the complex eigenvalues increases monotonically from a negative to a positive value. Such a loss of stability is known as a *Hopf bifurcation*. The complexity of the eigenvalues indicates that the loss of stability occurs in an oscillatory fashion. This is emphasized by the spiraling nature of the eigenvectors of the linearized problem (Fig. 5). (The oscillatory approach of the trajectory to C^+/C^- in Figs. 7 and 8 indicates that two of the eigenvalues of C^+/C^- are already complex for some $\mathrm{Ra} < \mathrm{Ra}_H$; but, for $\mathrm{Ra} < \mathrm{Ra}_H$, the corresponding eigenvectors belong to the stable manifold of C^+/C^-. Only for $\mathrm{Ra} > \mathrm{Ra}_H$ do these eigenvectors become a part of the unstable manifold). As Fig. 5 illustrates, the stable/unstable manifolds of C^+ and C^- are one-two-dimensional. The conditions of the Hopf bifurcation guarantee the existence of a limit cycle which emerges at the bifurcation point. We calculate the limit cycle bifurcating from the counterclockwise solution using the method of projection [23] for $P = 4$ and $|\mathrm{Ra} - 16| << 1$ to obtain the periodic solution:

$$\begin{pmatrix} x \\ y \\ z \end{pmatrix} = \begin{pmatrix} \sqrt{\mathrm{Ra} - 1} \\ \sqrt{\mathrm{Ra} - 1} \\ \mathrm{Ra} - 1 \end{pmatrix} + \sqrt{\frac{29}{62}(16 - \mathrm{Ra})} \begin{pmatrix} 2\cos(2\sqrt{5}t) \\ 2\cos(2\sqrt{5}t) - \sqrt{5}\sin(2\sqrt{5}t) \\ \frac{-5}{\sqrt{3}}\sin(2\sqrt{5}t) - \frac{2\sqrt{5}}{\sqrt{3}}\cos(2\sqrt{5}t) \end{pmatrix} \tag{14}$$

This limit cycle exists only for Ra < 16 and therefore it is subcritical and nonstable. That is, this limit cycle will never be observed in an experiment. The limit cycle is shown schematically by a dashed circle in Fig. 6*a*. We can compute the limit cycle numerically using special routines for the determination of periodic solutions [22]. It turns out that the period of the limit cycle increases as the Ra number decreases.

Numerical experiments indicate that for $Ra > Ra_H$, the solution to Eqs. (6)–(8) behaves in a very complicated, aperiodic way like that depicted in Fig. 2. The general form of the time series does not depend on the choice of initial conditions (after initial transients die out) or the choice of the integration procedure. However, the details of the time series depend crucially both on initial conditions and the integration routine. We shall discuss this issue below.

To gain a better appreciation of the complexity of the flow, we depict in Figs. 9–11 the phase portrait of the trajectories which correspond to the time series shown in Fig. 2. Figures 9–11 were generated by integrating Eqs. (6)–(8) for a period of time and disregarding initial transients. Figure 9 depicts the structure of the flow in the three-dimensional phase space, while Figs. 10 and 11 show two-dimensional projections of the same trajectories. Because of the uniqueness of the solutions, the trajectories cannot cross each other in phase space (Fig. 9). The same is not true, however, with respect to the two-dimensional projections (Figs. 10 and 11). In Figs. 9–11, we see that the trajectories wind around the fixed points C^+/C^- and pass close to C_0 without ever settling down. The trajectories occupy a finite subspace of the phase space. They appear to be bounded and lie *almost* on a two-dimensional sheet which includes the points C^+, C^- and the z axis (Fig. 9). Notice also the symmetry between the trajectories around points C^+ and C^-. Regardless of the initial conditions we choose, our solution will eventually end up as part of the set depicted in Figs. 9–11; hence the set is considered attracting. Since the attracting set is much more complicated than previously encountered attractors (such as fixed points and limit cycles), it is entitled a *strange* or *chaotic attractor*. This particular attractor is known as the *Lorenz attractor*. As we shall see later, the Lorenz attractor is just one example of a strange attractor. There is a rigorous mathematical definition of chaotic (strange) attractors (i.e., [4]), which we shall not reproduce here. For our purposes, we shall adopt the

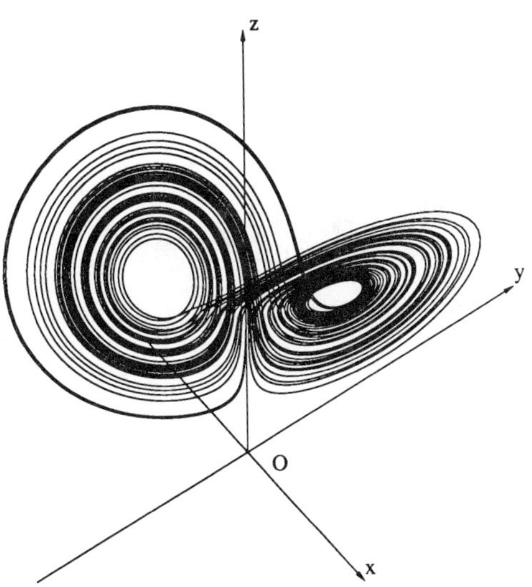

Figure 9 The trajectories depicted in phase space for Ra = 20 and p = 4. Sufficient time is allowed for the initial transient to die out.

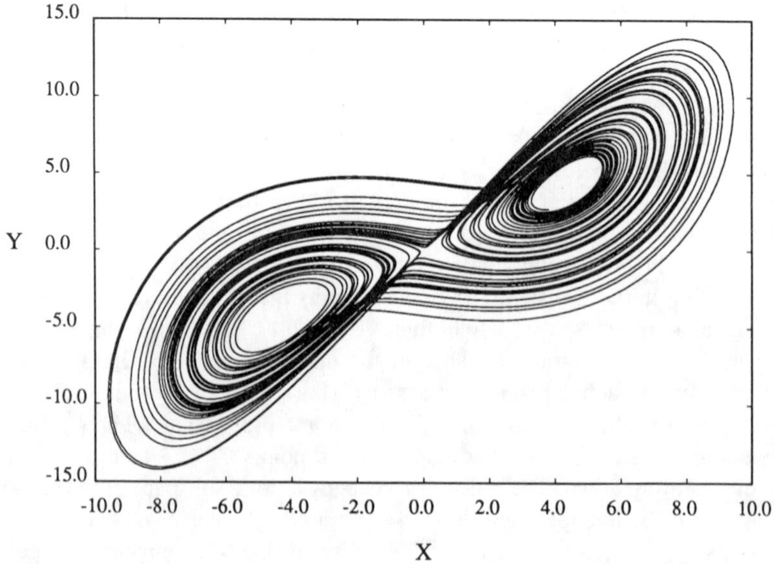

Figure 10 Two-dimensional projection of the trajectories depicted in Fig. 9 on the plane x–y.

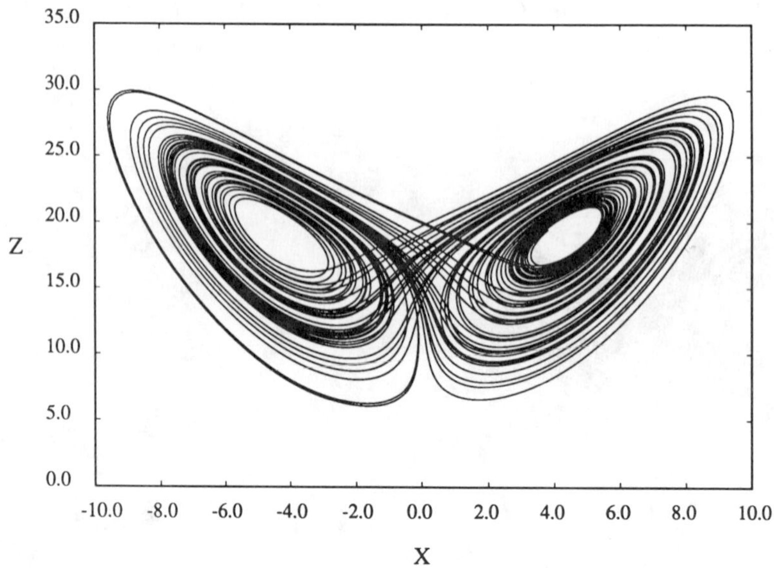

Figure 11 Two-dimensional projection of the trajectories depicted in Fig. 9 on the plane x–z.

less rigorous and more practical approach in which we refer to attractors which are not fixed points, limit cycles, or quasi-periodic as strange or chaotic attractors.

It is somewhat surprising that relatively innocent-looking differential equations such as Eqs. (6)–(8) lead to such complicated dynamical behavior (Figs. 2 and 9–11). Below, we shall attempt to obtain some insight with respect to the mathematical mechanism responsible for this complexity.

2.3.2 The Homoclinic Explosion. In Figs. 7 and 8 we have demonstrated that for Ra = 7 and Ra = 8, trajectories starting on the RHS/LHS unstable manifolds approach asymptotically the points C^+/C^- and C^-/C^+, respectively. Note that as we increase the Rayleigh number, initial conditions similar to those which caused us to end at C^+ (Fig. 7) now lead us to C^- (Fig. 8). The stable manifold W^s still acts as a separatrix between the basins of attraction of C^-/C^+. Thus W^s must have a rather complicated topological shape. These results suggest and numerical experiments confirm that there is a Ra = $\mathrm{Ra}_{\mathrm{hom}}$ (~7.378 . . . for $P = 4$) for which a trajectory starting on the unstable manifold of C_0 does not approach asymptotically either C^+ or C^-. Such a trajectory starts on the unstable manifold of C_0 and ends on the stable manifold of C_0. This situation is further illustrated in Fig. 12, where we show trajectories corresponding to $\mathrm{Ra} < \mathrm{Ra}_{\mathrm{hom}}$, $\mathrm{Ra} \approx \mathrm{Ra}_{\mathrm{hom}}$, and $\mathrm{Ra} > \mathrm{Ra}_{\mathrm{hom}}$. A trajectory which starts and ends at the same fixed point is called a *homoclinic orbit*. In contrast, trajectories which connect two different fixed points are called *heteroclinic orbits*. The homoclinic orbit is a periodic solution of Eqs. (6)–(8) with an infinite period. The infinite period results from the fact that as we approach the origin (C_0) of phase space, the velocity of the trajectory goes to zero. Numerical experiments suggest that the limit cycle obtained at the Hopf bifurcation point becomes a homoclinic orbit as $\mathrm{Ra} \rightarrow \mathrm{Ra}_{\mathrm{hom}}{}^+$. We shall demonstrate below that the homoclinic orbit is responsible for the complicated dynamical behavior we observed in Fig. 12.

We shall focus on trajectories starting in the vicinity of the unstable manifold of C_0 for Ra values slightly below and slightly above $\mathrm{Ra}_{\mathrm{hom}}$. Our objective is to investigate whether there are any other trajectories (in addition to the homoclinic orbit itself) which pass in the vicinity of C_0 without ever settling down at either C^+ or C^-. The existence of such trajectories is suggested by our phase portrait (Figs. 9–11). We follow here the exposition given by Sparrow [22], to which we refer the reader for additional details and references. We use a coordinate system which coincides with the eigenvectors (X_i) of the linear stability problem for C_0. We construct a small cubic box B centered on C_0 with edges of size 2ζ and enclose the RHS/LHS unstable manifold of C_0 within the tube S^+/S^- (Fig. 13). The tube S^+/S^- leaves the box through a sector D^+/D^- on its face ($|X_1| = \zeta$) and reenters the top of the box ($X_3 = \zeta$) through sector E^+/E^-. Our objective is to find the relationship between subsequent reentry points in E. In other words, we wish to establish a two-dimensional mapping (or return map) from E to E. Such a return map (which is an example of a *Poincaré map*) can provide a two-dimensional description of the system's dynamics instead of the three-dimensional phase portrait. Thus the return map allows us to deal with a lower-dimensional system. Also, note that the relationships between subsequent penetration points in the return (Poincaré map) are provided by difference equations rather than

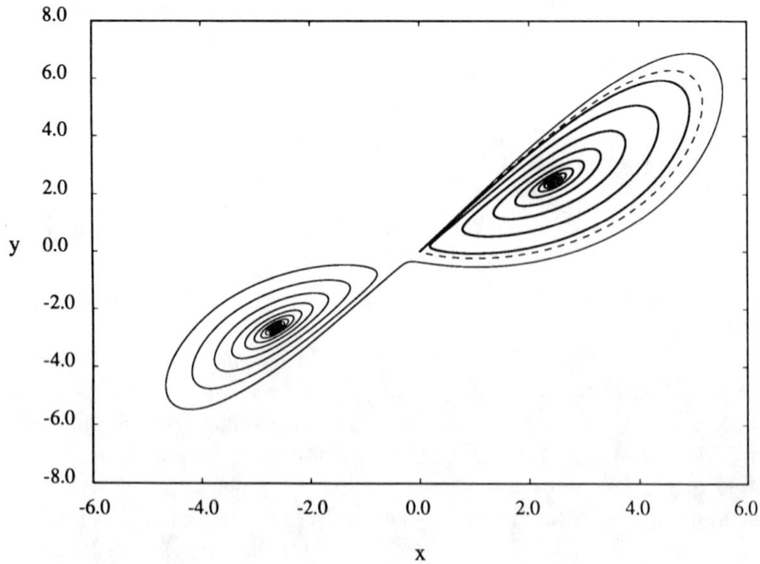

Figure 12 Three trajectories, starting on the right-hand side: unstable manifold of (C_0) for Ra = 7 (light solid line), Ra = Ra_{hom} ≈ 7.378 (dashed line), and Ra = 8 (heavy solid line). The trajectories end up, respectively, at the fixed points C^+, C_0, and C^-.

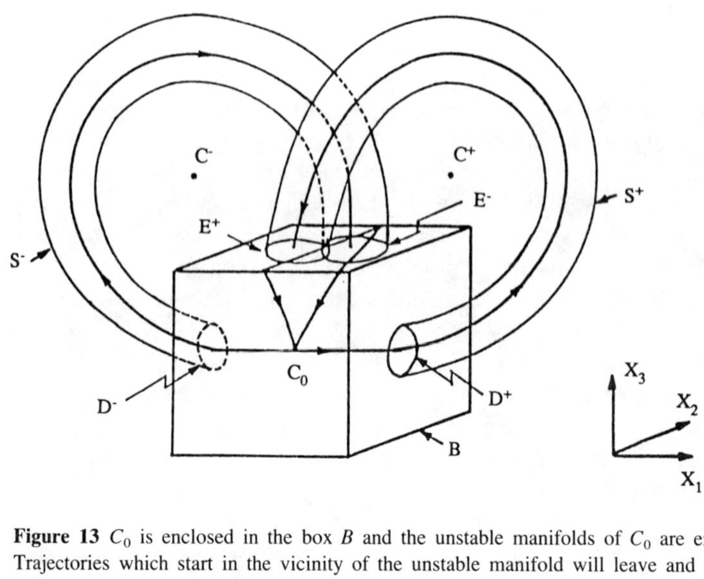

Figure 13 C_0 is enclosed in the box B and the unstable manifolds of C_0 are enclosed by the tubes S. Trajectories which start in the vicinity of the unstable manifold will leave and reenter the box through segments D and E, respectively.

differential ones. To construct the return map (or $E \rightarrow E$ map), we shall first calculate the trajectory within the box to establish the mapping of E to D and then follow the trajectory as it traverses through the tube S to establish the mapping of D to E. Combining these results will yield the desired return map of E to E.

If the box is sufficiently small, we can linearize Eqs. (6)–(8) inside the box. It is convenient to project the equations on a coordinate system defined by the corresponding eigenvectors (X_i, $i = 1, 2, 3$) to obtain equations of the form $dX_i/dt = \sigma_i X_i$, where σ_i are the eigenvalues determined in Eq. (12). These equations can be solved exactly to determine the relationship between points at which the trajectory enters and exits the box (i.e., the mapping from E to D). Accordingly, a trajectory entering the top of the box at a point (X_1, X_2, ζ) in section E will exit the box at the point $\{\zeta, X_2(\zeta/X_1)^{\sigma_2/\sigma_1}, \zeta(\zeta/X_1)^{\sigma_3/\sigma_1}\}$ in section D. For example, for Ra $= 9$ and $P = 4$, the exit point will be given by $\{\zeta, X_2(X_1/\zeta)^{2.3}, \zeta(X_1/\zeta)^{0.27}\}$. Recall that $X_1/\zeta < 1$. Thus we have contraction and stretching in the X_2 and X_3 directions, respectively. In other words, two trajectories entering the box at two separate points in E will emerge closer together in the X_2 direction and farther apart in the X_3 direction. This stretching is typical of chaotic systems, and it explains why these systems are so sensitive to initial conditions. That is, when we integrate the differential equations (6)–(8) with slightly different initial conditions, we end up with vastly different solutions (see Fig. 14 for an example where we depict two trajectories with slightly different initial data). The contraction in the X_2 direction will cause the trajectories to shrink in volume. We shall return later to this issue.

Next, we consider the path of a trajectory within the tube S. To obtain a map from D to E, we integrate numerically a trajectory starting on the unstable manifold of C_0 at a point $r_D^{\pm} = (\zeta, X_2^{\pm}, X_3^{\pm})$ and find the location of the point $r_E^{\pm} = (X_1^{\pm}, X_2^{\pm}, \zeta)$ at which this trajectory pierces the top of the box. We use the superscript $+/-$ to distinguish between the right and left unstable manifolds. Next, using a Taylor series

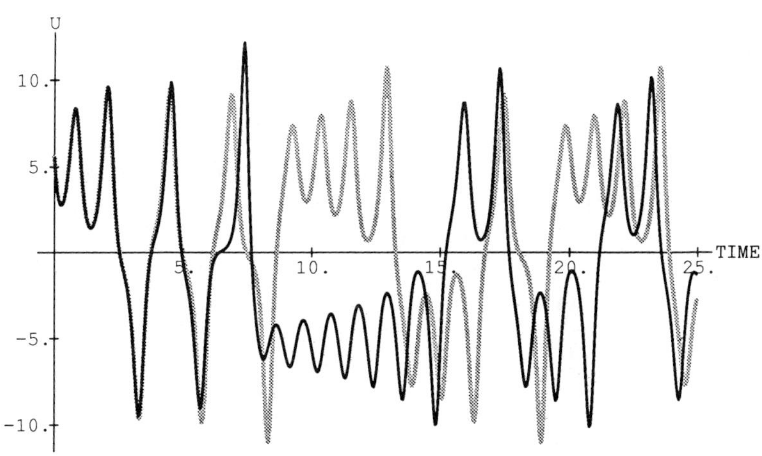

Figure 14 Sensitivity to initial conditions. Two solutions with slightly different initial conditions (black and gray lines) exhibit vastly different behavior.

expansion, we calculate the penetration points $r'_E = (X'_1, X'_2, \zeta)$ of nearby trajectories starting at some location r'_D in D. That is, $r'_E = r^*_E + A(r'_D - r^*_D)$, where A is a two-dimensional matrix consisting of the derivatives of r'_E with respect to X_i ($i = 2$, 3) in D. Combining this result and the mapping of $E \rightarrow D$ we obtained earlier, we get a two-dimensional return map for $E \rightarrow E$.

In our particular example, we can introduce a further simplification. Due to the strong contraction in the X_2 direction, trajectories farther apart in this direction are drawn together. In other words, most of the action occurs in the X_1 direction. Accordingly, we shall depict only the X_1 coordinate of subsequent penetration points in E, thereby reducing our two-dimensional return map to a one-dimensional map. The $(i + 1)$ return point can be calculated from the ith return point using the approximate formula

$$X_{1,i+1} = X_1^* + a\zeta\left(\frac{X_{1,i}}{\zeta}\right)^{\sigma_1/\sigma_3} \tag{15}$$

where a is the first entry in the matrix A and X_1^* is the point at which the unstable manifold of C_0 penetrates the top of B in E. The resulting first return maps are depicted schematically in Figs. 15*a* and 15*b* for Ra $<$ Ra$_{\text{hom}}$ and Ra $>$ Ra$_{\text{hom}}$, respectively, where the solid lines indicate the $X_{1,i+1}$ location of a penetration point as a function of the previous penetration point $(X_{1,i+1})$. It turns out that $X_1^* > 0$, $=0$, and <0 for Ra$<$, $=$, and $>$ Ra$_{\text{hom}}$, respectively.

The one-dimensional maps depicted in Fig. 15 provide a convenient means of following the history of a trajectory. To this end, we draw 45° lines in Figure 15. We can find the subsequent penetration point by erecting a vertical line from the present penetration point up to the solid line. The next penetration point is found by running a horizontal line from the intersection of the vertical line with the solid curve to the 45° line, and subsequently erecting another vertical line and finding the intersection with the solid curve. The procedure is depicted by light, solid lines in Figs. 15*a* and 15*b*. Figure 15*a* (Ra $<$ Ra$_{\text{hom}}$) demonstrates that as we continue to iterate, the penetration point moves away from the center of the box (the magnitude of the coordinate X_1 keeps increasing) until eventually the trajectory will not penetrate B anymore. This trajectory leaves BS (the box B and the tube S) and is likely to end up at one of the fixed points C^+ or C^-. Thus, for Ra $<$ Ra$_{\text{hom}}$, the only solutions for the equations are the steady fixed points. The situation is different for Ra $>$ Ra$_{\text{hom}}$ (Fig. 15*b*). In Fig. 15*b*, we see that the trajectories starting in regions $|X_1| > A$ and $0 < |X_1| < B$ will eventually get stretched out and leave BS. On the other hand, trajectories starting in the interval $B < |X_1| < A$ will be mapped back on the interval $|X_1| < A$. Some of these trajectories will be mapped on $B < |X_1| < A$ and stay within BS, while others will map on $0 < |X_1| < B$ and leave BS. As we continue this process, it becomes clear that there are trajectories which stay in BS for all future times. These trajectories form a Cantor-like set on the interval $B < |X_1| < A$ (see Fig. 30, below, for an example of a Cantor set). These trajectories, which include a countable infinity of periodic orbits and an uncountable infinity of nonperiodic orbits, are titled a strange, *nonwandering* set. The nonwandering set is a time-dependent, nontransient solution

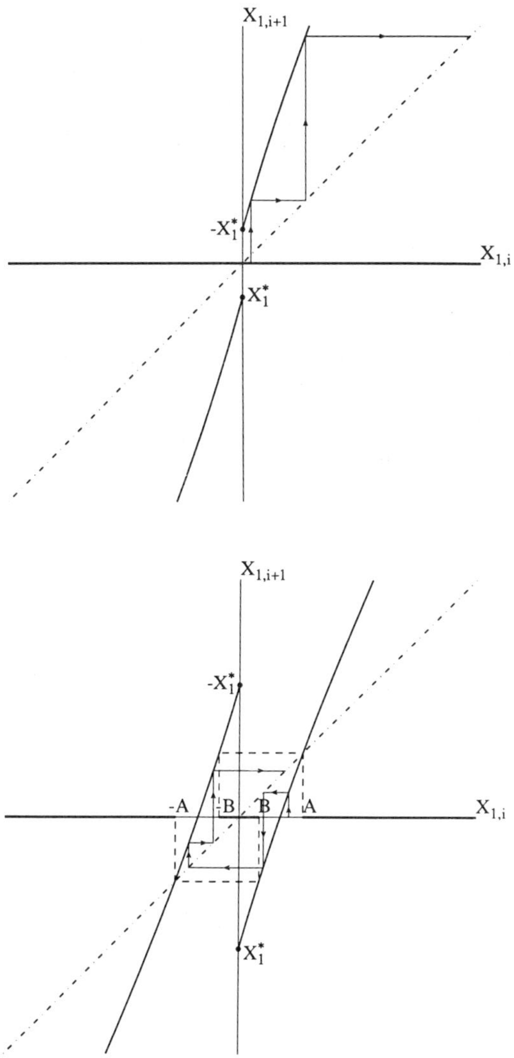

Figure 15 The return ($E \rightarrow E$) map depicted schematically (heavy solid line) for (*a*) Ra $<$ Ra$_{\text{hom}}$ and for (*b*) Ra $>$ Ra$_{\text{hom}}$. The subsequent penetration points can be traced conveniently by erecting a vertical line from the X_i axis to the solid line, followed by a horizontal line from the intersection with the solid line to the 45° diagonal line (see light, solid lines), etc. For Ra $<$ Ra$_{\text{hom}}$ (Fig. 15*a*), the X_i value of subsequent penetration points increases monotonically. For Ra $>$ Ra$_{\text{hom}}$ (Fig. 15*b*), there exists a region $B < |X| < A$, in which some of the trajectories will remain forever.

of Eqs. (6)–(8). That is, trajectories within the nonwandering set will continue to wander around C^+ and C^- without ever settling down (in a similar fashion to what we see in Figs. 9–11). The bifurcation at Ra = Ra$_{\text{hom}}$ which led to the creation of the nonwandering set is called a *homoclinic explosion*. For values of Ra slightly above Ra$_{\text{hom}}$, the nonwandering set is not attracting. That is, these orbits are nonstable. Any small disturbance will move the solution onto an orbit which does not belong to the aforementioned Cantor set, and the trajectory will eventually leave the nonwandering set and end up at one of the attracting fixed points (C^+ or C^-). Thus we cannot observe the nonwandering set in an experiment. The only experimental evidence of the existence of the nonwandering set is the existence of chaotic transients. That is, if we disturb the steady state of the system for $\mathrm{Ra} > \mathrm{Ra}_{\mathrm{hom}}$, the transient is likely to exhibit a complicated timewise behavior before the solution settles at one of the steady states (see Fig. 16 for an example). The length of the transient will increase as the magnitude of Ra increases beyond $\mathrm{Ra}_{\mathrm{hom}}$. The time length of the transient depends on initial conditions. Averaged over all initial conditions, the length of the transient scales like $(\mathrm{Ra}_A - \mathrm{Ra})^{-1/2}$, where Ra_A is the point at which the nonwandering set becomes an attractor.

As Ra is further increased, another bifurcation occurs at $\mathrm{Ra} = \mathrm{Ra}_A > \mathrm{Ra}_{\mathrm{hom}}$ (for $P = 4$, $\mathrm{Ra}_A \approx 15.98$). This time the bifurcation is caused by a heteroclinic connection generated by the unstable manifold of C_0 becoming tangent to the stable manifold of the limit cycle which was born at the Hopf bifurcation point. At Ra_A, the nonwandering set becomes attracting (strange attractor). The modifier ''strange'' is used to emphasize the complicated structure of the attracting set. Thus, for $\mathrm{Ra} > \mathrm{Ra}_A$, trajectories may wind forever around the attracting set. This structure is rugged and insensitive to small disturbances.

Since the solutions to Eqs. (6)–(8) are bounded, we need $\mathrm{Ra}_A < \mathrm{Ra}_H$. This implies the existence of a region $\mathrm{Ra}_A < \mathrm{Ra} < \mathrm{Ra}_H$, in which, in addition to two attracting steady solutions, we have an attracting, time-dependent solution—the strange attractor. The stable manifold of the limit cycle (generated at the Hopf bifurcation point) forms the boundary between the basins of attraction of the steady and chaotic solutions. Consequently, we may expect to see hysteresis phenomena. For example, when we conduct an experiment and gradually increase the magnitude of the Ra number, we will observe the transition to time-dependent, chaotic flow at $\mathrm{Ra} = \mathrm{Ra}_H$. On the other

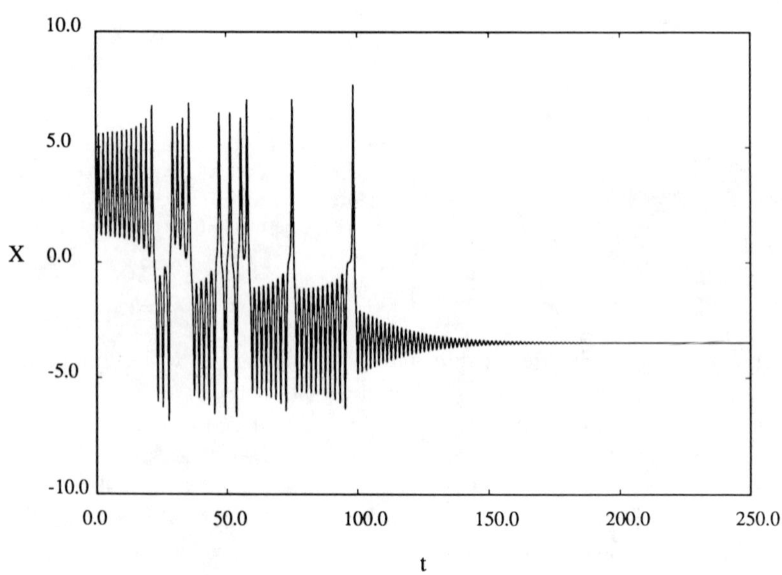

Figure 16 Transient chaotic behavior for $\mathrm{Ra}_A > \mathrm{Ra} > \mathrm{Ra}_{\mathrm{hom}}$.

hand, if we start the experiment at a Rayleigh number larger than Ra_H and gradually decrease the magnitude of the Rayleigh number, the chaotic phenomena will disappear at $Ra = Ra_A < Ra_H$.

2.3.3 Periodic Windows. As we increase the Rayleigh number beyond Ra_A, we observe chaotic, time-dependent flow. The chaotic behavior does not prevail, however, for all values of the Rayleigh number $Ra > Ra_A$. As we increase the Rayleigh number, the chaotic behavior is interrupted occasionally by periodic windows. There are probably a few periodic windows, only one of which is visible in Fig. 6*b*. The periodic windows are represented by a small number of points in the vertical cross section of Fig. 6*b* (see, for example, the region $Ra > 93$), while the chaotic flow appears as a high-density collection of points. The periodic windows disappear typically through a sequence of period doubling bifurcations which eventually lead to chaos. As period doubling is a common route to chaos, we shall discuss the period doubling cascade in more detail later in the chapter. A detailed investigation of some of the periodic windows of the Lorenz system is available in Sparrow [22].

It is also interesting to note that in the limit of $Ra \to \infty$, the solution exhibits periodic behavior. This explicit form of the solution can be derived analytically in terms of elliptic functions [24]. An example of the periodic trajectory for a large Ra number is depicted in Fig. 17.

2.3.4 Volume Contraction in Phase Space and Poincaré Cross Sections. The trajectories of a dissipative system such as the one modeled by Eqs. (6)–(8) contract volume in phase space. Here, we use the term ''volume'' in a generalized form. That is, volume in N-dimensional phase space will have N dimensions. Suppose that the initial conditions of the system include all possible points in phase space. That is, V is initially three-dimensional. By simple manipulation, we obtain

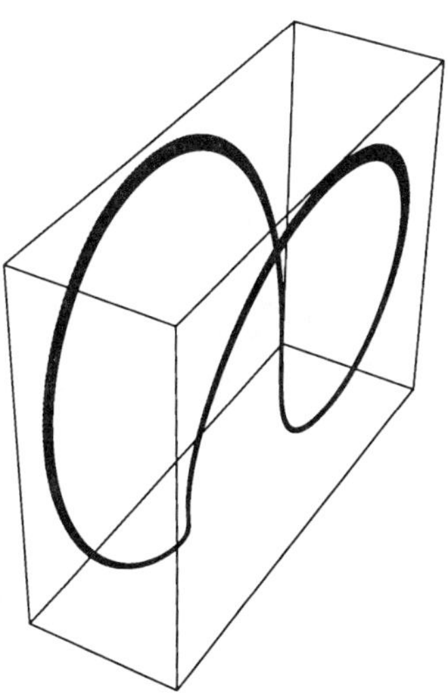

Figure 17 For large Ra number, the system exhibits periodic behavior, as shown by the trajectory depicted in phase space for $Ra = 500$ and $p = 4$.

$$\frac{\dot{V}}{V} = \frac{\partial \dot{x}}{\partial x} + \frac{\partial \dot{y}}{\partial y} + \frac{\partial \dot{z}}{\partial z} = -P - 2 \tag{16}$$

That is, the element V_0 is contracted by the flow into a volume element $V(t) = V_0 \exp[-(P + 2)t]$. For example, for $P = 4$, the volume will shrink by a factor of ~400 after one unit time. Thus, the system is highly dissipative. This implies that after initial transients die out, the volume occupied by the trajectories in phase space is smaller than three dimensions.

The volume-shrinkage effect is exhibited in a dramatic way in the *Poincaré cross section*. In situations where the equations are affected by T periodic forcing, the Poincaré map is generated by a stroboscopic sampling of the system's variables at times nT, with n being an integer (we shall see an example of such a map in Sec. 2.4.2). In the absence of periodic forcing, the method of constructing the map is less obvious. It can be done by drawing an $(N - 1)$ hypersurface in the N-dimensional phase space and documenting the penetration points of trajectories as they pierce through that surface in a given direction. The choice of the surface is a judgment call. For example, in Fig. 18 we depict the penetration points of trajectories through the plane $z = \text{Ra} - 1$. This plane contains the fixed points C^+/C^-. At first glance, the Poincaré section seems to consist of two line segments, which would imply that the attractor is confined to a two-dimensional surface. In fact, if we zoom in on the line, we will discover that it consists of a large number of closely packed sheets. Zooming

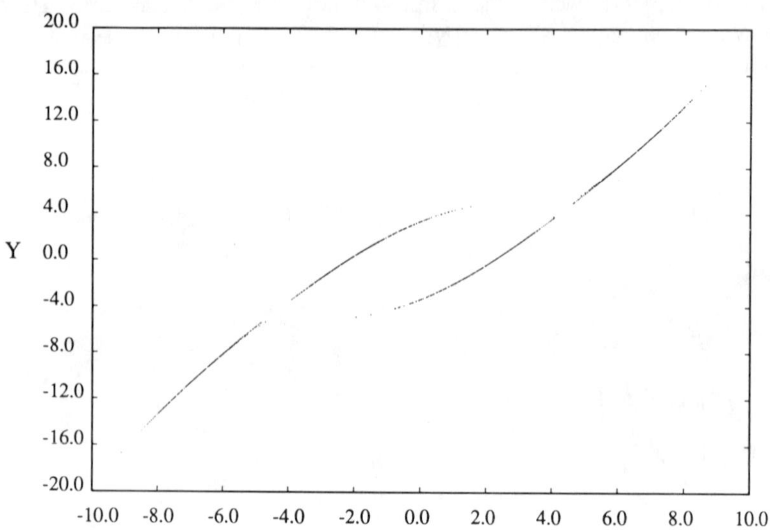

Figure 18 Poincaré cross section through the plane $z = \text{Ra} - 1$. The penetration points appeared to form a single curve because of the strong contraction and poor resolution. If the figure were magnified, one would observe that each of the curves consists of numerous sheets.

in on any individual sheet reveals that each individual sheet consists of a large number of sheets. This structure appears to be self-similar in the sense that each additional magnification reveals a structure similar to the one we saw in the previous magnification. The almost single line structure is a result of the shrinkage we observed in the X_2 direction (Sec. 2.3.2). Due to the volume shrinkage, the attractor must have the dimension $D < 3$. The structure we see in Fig. 18 suggests that the attractor does not quite lie on a surface. Also, the complicated dynamical behavior exhibited in Figs. 9–11 excludes two-dimensionality, since in two-dimensional space, trajectories cannot cross and we can have at most periodic behavior. In short, the attractor's dimension is a fractional (*fractal*) dimension $2 < D < 3$. In fact, for the Lorenz attractor, $D \approx 2.06$. We shall define the fractal dimension in Sec. 3.

The contraction phenomenon is common to dissipative systems. Consequently, their Poincaré cross sections typically consist of a well-defined, often esthetically appealing structure rather than an unorganized collection of points.

Due to the relatively simple structure of the Poincaré map, it can be used to generate low-dimensional iterative maps. We saw an example of such a construction in Fig. 15, and we shall encounter another example later on. Such iterative maps provide a simple means of studying the system's dynamics.

2.3.5 Summary of Results for the Loop with a Time-Independent Wall Condition

1. Trajectories in three-dimensional phase space are volume-contracting. The attractor has a fractal dimension between 2 and 3.
2. For $\mathrm{Ra} < 1$, only the no-motion solution is attracting. There is an attractor (fixed point) of dimension 0.
3. For $1 < \mathrm{Ra} < \mathrm{Ra}_A$, the clockwise and counterclockwise motion solutions are attracting. Their basins of attraction are separated by the stable manifold of C_0.
4. For $\mathrm{Ra}_A < \mathrm{Ra} < \mathrm{Ra}_H$, there are three attracting solutions, the two steady motion solutions and the strange attractor. The stable manifold of the unstable limit cycle generated by the Hopf bifurcation at Ra_H acts as a separatrix between the basins of attractions of the steady solutions and the strange attractor.
5. For $\mathrm{Ra} > \mathrm{Ra}_H$, the steady motion solutions lose stability. As far as we know, the strange attractor is the only attracting solution, with the basin of attraction encompassing the entire phase space. The possibility of the existence of additional attractors (yet undiscovered) cannot be excluded. The chaotic motion exhibits high sensitivity to initial conditions.
6. As the Ra number increases, the chaotic motion may be replaced by periodic motions for limited ranges of Rayleigh numbers. These are called periodic windows.
7. For $\mathrm{Ra} \to \infty$, the motion is periodic in time.

We have demonstrated above that the homoclinic orbit leads to the birth of a strange, nonwandering set. A similar process (of homoclinic or heteroclinic explosion) can also lead to the destruction of the chaotic attractor.

2.4 Thermal Convection Loop with Time-Periodic Wall Modulation

Next we shall consider a thermal convection loop with a time-periodic wall temperature. This example has some interesting features. First, we shall see that time modulation of the wall temperature advances the transition point to chaotic behavior. Second, we shall observe that under sufficiently large temperature modulation, chaos does not occur suddenly via a homoclinic explosion, as is the case when the wall temperature is time-independent, but appears at the end of a sequence of period doubling bifurcations. Period doubling is a frequent route to chaos that is observed in diverse physical systems. Also, the structure of the strange attractor at the onset of chaos is considerably simpler than that of the Lorenz attractor. This fact will enable us to gain some further insight into the dynamics of chaos.

2.4.1 The No-Motion Solution and Its Stability. We investigate Eqs. (6)–(8) with $f(t) = \sin(\omega t)$. Equations (6)–(8) admit a no-motion, T-periodic (herein we refer to solutions having the forcing period $T = 2\pi/\omega$ as T-periodic) solution of the form:

$$\{u, c, s\} = \{0, 0, -\varepsilon/\varepsilon_0 \sin(\omega t + \phi)\} \tag{17}$$

where $\varepsilon_0 = \sqrt{(1 + \omega^2)}$ and $\phi = \arcsin(-\omega/\varepsilon_0)$. Physically, this solution corresponds to no fluid motion in the loop. The thermal interaction between the fluid and the pipe's walls causes the fluid's temperature to oscillate at the same period as the wall temperature, albeit with a certain phase lag. In the special cases of $\varepsilon \rightarrow 0$ and/or $\omega \rightarrow \infty$, the fluid's temperature is fixed in time and the solution is identical to the one presented in Eq. (9).

We use Floquet theory to determine the stability of the no-motion solution. The interested reader is referred to Wang and Bau [25] and Wang [26] for details. The results of the stability analysis for loop Prandtl number $P = 4$ are depicted in Fig. 19 as a function of the modulation amplitude ε. Note that the no-motion solution can bifurcate into either T (harmonic) or $2T$ (subharmonic) motion solutions. It also appears that the modulation has a stabilizing effect in the sense that the Rayleigh number at the onset of motion is slightly larger than it is in the unmodulated case.

2.4.2 Beyond Linear Stability Analysis. Next we investigate the behavior of the flow as the Rayleigh number increases beyond the critical value for the onset of convection. We shall present our results in terms of the variables $\{x, y, z, t\} = \{u, c, s + \varepsilon/\varepsilon_0 \sin(\omega t + \phi)\}$. Thus, the no-motion solution is equivalent to $\{x, y, z\} = \{0, 0, 0\}$. We use a fourth-order-accurate Runge-Kutta scheme [27] to integrate the nonlinear differential equations (6)–(8) with $f(t) = \sin(\omega t)$. We also employ standard procedures (i.e., [22]) to locate periodic solutions and study their stability. The solutions are presented (after initial transients decay) as trajectories or orbits in a four-dimensional space $\Sigma = \{(x, y, z, t) \in R^3 * S^1 | \, t = t_0\}$, where t is defined on the cylindrical surface S as Mod(t, T). We define Poincaré cross sections in Σ as R^3 hypersurfaces with Mod$(t, T) = m/nT$, where $0 \leq m < n$, and n are integers. Thus, by using Poincaré sections, we transform the continuous time system to a discrete system. Since it is inconvenient to depict higher-order-dimensional surfaces, we shall resort to projections onto lower-dimensional hypersurfaces.

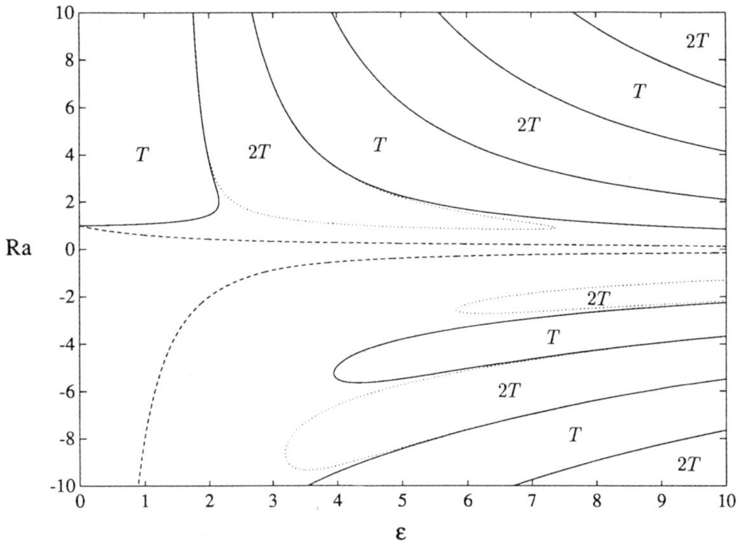

Figure 19 Stability curve for the modulated loop. The Rayleigh number at the onset of motion is depicted as a function of the modulation's amplitude (ε). The symbols T and $2T$ indicate harmonic and subharmonic solutions, respectively.

Next, we follow the chain of events as the Rayleigh number increases. To this end, we fix the magnitude of $\varepsilon = 1$ in Fig. 19 and gradually increase the magnitude of the Rayleigh number. Our results are summarized in a bifurcation diagram (Fig. 20), where we depict the x values associated with the Poincaré map ($m = n = 1$) as a function of the Ra number. In other words, we probe the velocity stroboscopically using the period T. Thus, in any vertical cross section of Fig. 20, a single point at a given Ra value indicates a T-periodic motion; two points indicate a subharmonic with period $2T$; N points indicate a subharmonic with period $N * T$; and a smear of points indicates chaotic behavior.

At the boundary of the stability region in Fig. 19, at Ra ≈ 1.0562, the no-motion solution loses stability and bifurcates into a pair of T-periodic solutions via a symmetry-breaking bifurcation. These two nonsymmetric orbits are mirror images of each other (only the counterclockwise solution is shown in Fig. 20). As the Ra number is further increased, the system undergoes a sequence of period doubling. For example, at $\mathrm{Ra}_1 \approx 2.0513818\ldots$, the T-periodic solution loses stability via a flip bifurcation in favor of a $2T$-periodic solution. That is, for $\mathrm{Ra} > \mathrm{Ra}_1$, we have a stable $2T$-periodic solution and a nonstable T-periodic solution. A sample of the projections of the T-, $2T$-, and $4T$-periodic orbits is shown in Fig. 21. In these periodic solutions (Fig. 21), the velocity is always positive. A mirror image of an orbit with only a negative velocity also exists but is not shown here. The magnitudes of the Rayleigh numbers (Ra_n) at each of the period doubling points up to period 2^9T are given in Table 1 together with the ratio $\delta_n = (\mathrm{Ra}_n - \mathrm{Ra}_{n-1})/(\mathrm{Ra}_{n+1} - \mathrm{Ra}_n)$. Note that as the Rayleigh number increases, the interval between bifurcation points decreases rather rapidly. It appears that as

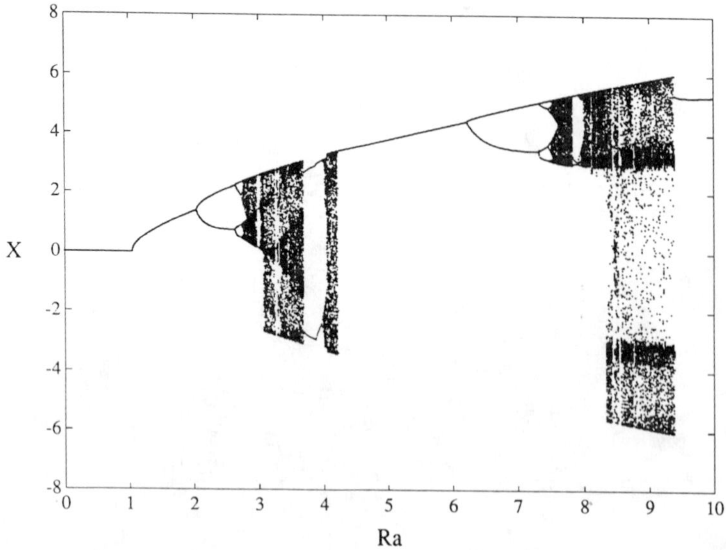

Figure 20 The upper half of a bifurcation diagram for a thermal convection loop with modulated wall temperature. The x values are stroboscopically probed every period T. In any vertical cross section, a single point indicates a T-periodic solution, two points indicate a $2T$-periodic solution, etc. A smear of points indicates chaotic behavior.

$n \to \infty$, $\delta_n \to 4.669201609\ldots$, which is the celebrated Feigenbaum number [28]. The Feigenbaum number is a universal constant which appears in many situations when the route to chaos involves a period doubling sequence. The period doubling sequence is terminated at an accumulation point $Ra_\infty \approx 2.75794\ldots$, a value we have estimated by extrapolation with the aid of the aforementioned Feigenbaum number.

Beyond the accumulation point, the periodic motion is replaced by an asymmetric, chaotic attractor. Note that the transition to chaos in the loop with periodic wall temperature modulation occurs at a substantially lower Rayleigh number than it does in the loop with a timewise fixed wall temperature. Immediately after the onset of chaos, we observe a reversed, semiperiodic (noisy) doubling sequence. We use the term semiperiodic, as did Lorenz [29], to describe a scenario when the nonperiodic orbit keeps penetrating the Poincaré section within a well-defined region. In other words, the semiperiodic trajectory is confined within a tube surrounding a nonstable periodic orbit. These nonstable periodic orbits are possibly the ones which lost stability in the period doubling sequence prior to the onset of chaos. N semiperiodic attractors will consist of N such nonoverlapping regions or tubes. The spaces between the tubes are in the basin of attraction of the various tubes. Immediately beyond the accumulation point there is a $2^N T$-semiperiodic attractor, with N being large (Fig. 20). As the Rayleigh number increases, semiperiodic regions merge together to reduce the semiperiodicity. In Fig. 22 we depict, for example, a semiperiodic attractor with semi-

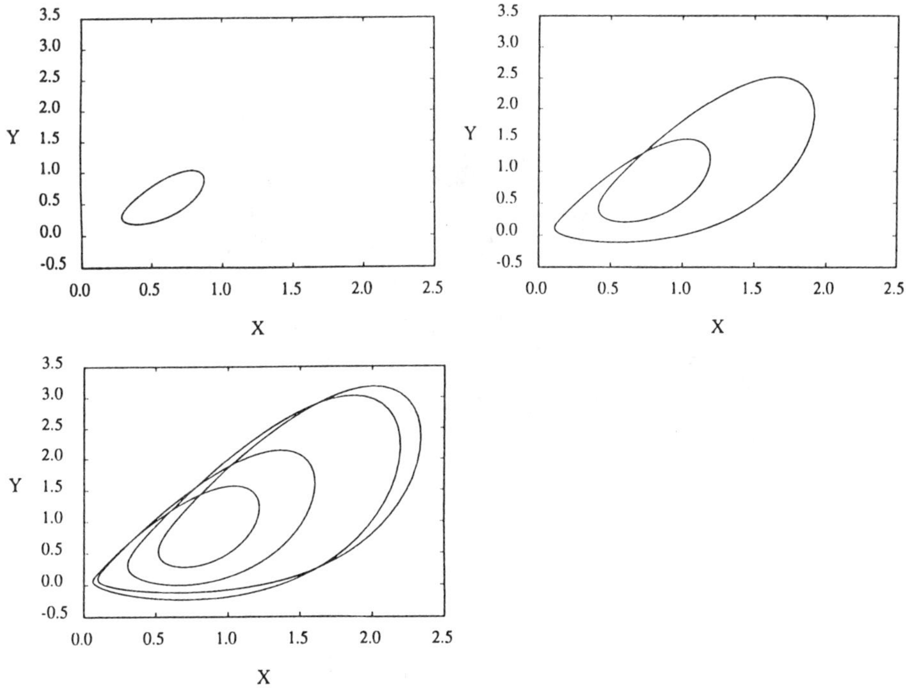

Figure 21 A sample of T-, $2T$-, and $4T$-periodic orbits in the period doubling sequence, at Ra = 1.5, 2.3, and 2.7. P = 1.

periodicities of 16 (Ra = 2.76. . .) and 4 (Ra = 2.77. . .). The ratios of Rayleigh numbers of the reversed bifurcation sequence typically obey the Feigenbaum sequence.

In Fig. 23 we depict the two-dimensional projection of the attractor observed soon after the termination of the reversed, semiperiodic doubling sequence (Ra = 2.92). The attractor (Fig. 23) resembles the Rossler attractor, which has the basic structure of a Möbius disk with some folding action at the edge of the strip. To gain a better understanding of the nature of the attractor, we construct a sequence of Poincaré cross

Table 1 Period doubling points and Feigenbaum number
$\varepsilon = 1, P = 1, \omega = 1$

	Ra_n	$Ra_n - Ra_{n-1}$	δ_n
Ra_1 $2^0T \rightarrow 2^1T$	2.0513818		
Ra_2 $2^1T \rightarrow 2^2T$	2.6431927	0.591811	6.3624
Ra_3 $2^2T \rightarrow 2^3T$	2.7362112	0.0930173	5.4005
Ra_4 $2^3T \rightarrow 2^4T$	2.7534340	0.0172240	4.8542
Ra_5 $2^4T \rightarrow 2^5T$	2.75698227	0.00354827	4.7064
Ra_6 $2^5T \rightarrow 2^6T$	2.75773619	0.00075392	4.6770
Ra_7 $2^6T \rightarrow 2^7T$	2.757897384	0.000161194	4.6711
Ra_8 $2^7T \rightarrow 2^8T$	2.757931891	0.000034507	

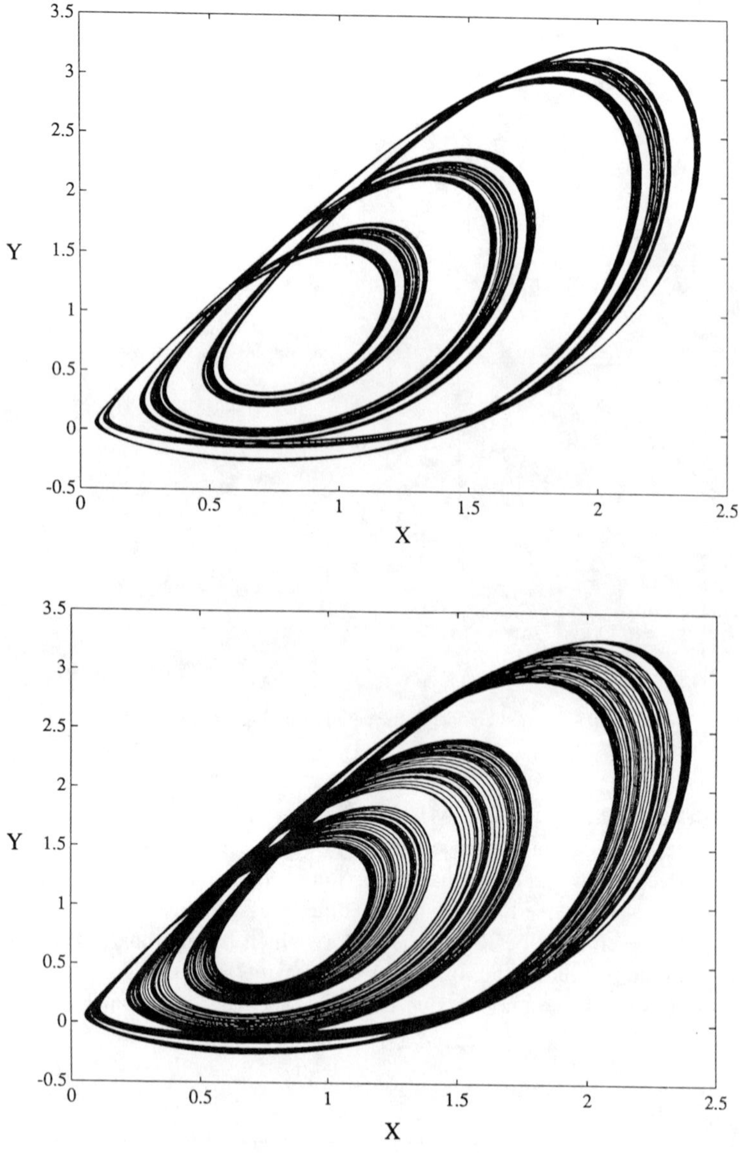

Figure 22 Semi-(noisy) periodicity: (*a*) at Ra = 2.76 with semiperiodicity 16; (*b*) at Ra = 2.77 with semiperiodicity 4.

sections within a single period (Fig. 24). Due to the strong dissipative nature of the loop equations, the Poincaré cross section appears almost as a single curve. We shall see, however, that this impression is misleading and the curve does have ''thickness.'' We select a set of points on a Poincaré cross section forming a curve with the mid- and endpoints denoted, respectively, by A_0 and B_0. Next, we take six subsequent

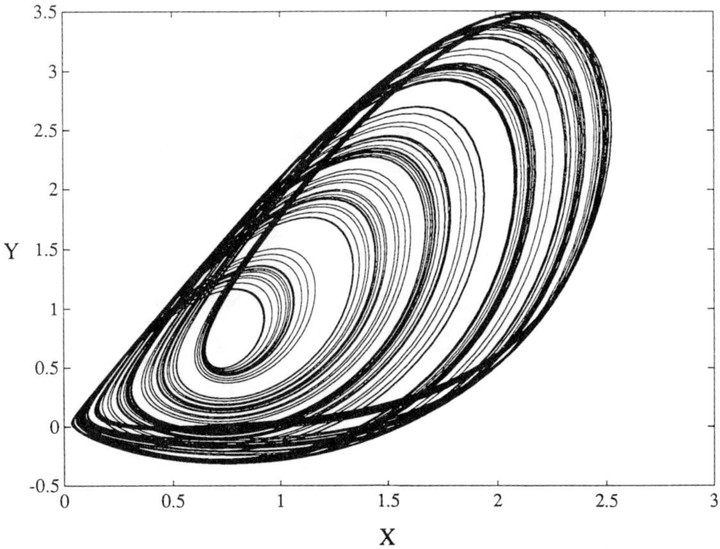

Figure 23 Rossler-like, chaotic attractor; Ra = 2.92, $\varepsilon = 1$, $P = 1$, and $\omega = 1$.

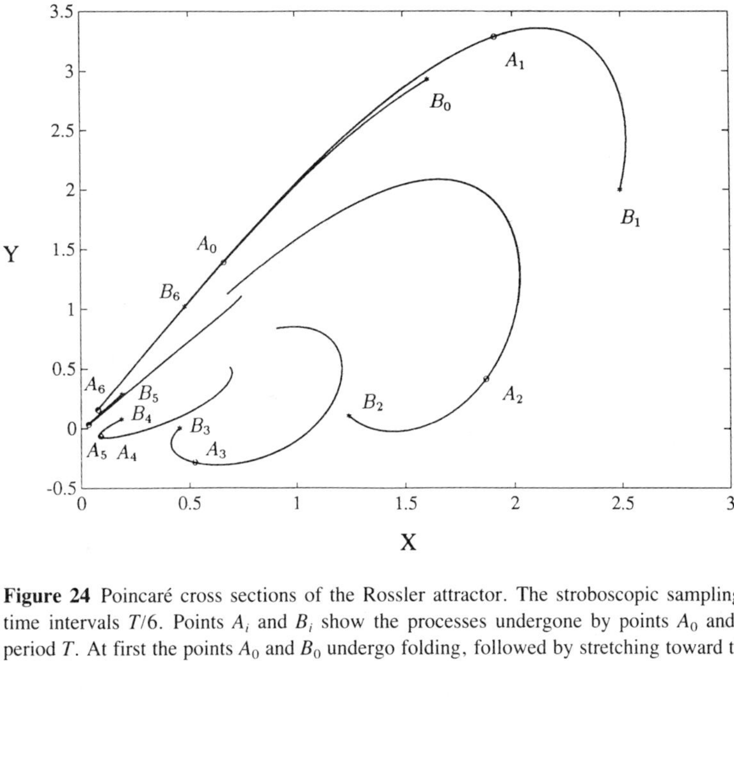

Figure 24 Poincaré cross sections of the Rossler attractor. The stroboscopic sampling is performed with time intervals $T/6$. Points A_i and B_i show the processes undergone by points A_0 and B_0 during a single period T. At first the points A_0 and B_0 undergo folding, followed by stretching toward the end of the cycle.

Poincaré sections at times $t = i/6T$, with $i = 1, \ldots, 6$. The points corresponding to A_0 and B_0 are denoted by A_i and B_i in each cross section. We observe at first the folding action of the attractor. Toward the end of the period T, the folding is nearly complete. The point B_0 is folded to point B_6, almost at the middle of the original interval. The point B_6 is not quite located on the original curve (A_0–B_0); but, due to the strong contraction, it appears to be so. Since the solution is unique, two different points cannot occupy the same position in space. Thus the attractor must have a fractal structure. That is, any line within the attractor will reveal, upon magnification, multiple lines, and so on ad infinitum. Toward the end of the folding action, we also observe stretching as, for example, the interval A_4–B_4 increases in length to A_6–B_6. This stretching, which we have encountered earlier, is responsible to the high sensitivity of the attractor to initial conditions. The stretching is always accompanied by folding, since the folding assures that the solution of the equations (the attractor) remains bounded. Figure 24 also demonstrates the mixing action of the attractor. Points farther apart are brought close together by the folding action, while points initially close together are pulled apart by the stretching action.

The simple structure of the Poincaré cross section (Fig. 24) allows us to construct a one-dimensional iterative map for the attractor, i.e., $X_{n+1} = F(X_n; \text{Ra})$. This is done by recording the X_{n+1} crossing of the Poincaré plane as a function of the X_n crossing. Alternatively, we can designate as X_n all the points in a given cross section and find their counterparts in the subsequent T-advanced cross section. The corresponding one-dimensional iterative map is depicted in Fig. 25. We include the argument Ra in the function $F(\cdot;\cdot)$, to emphasize that the iterative map as well as the Poincaré cross section from which it was generated depend on the Rayleigh number and that their structures will change as we vary the Rayleigh number. As we have seen before, the iterative map allows us to study the dynamics of the attractor. Once we set the initial conditions, the future of the trajectory is determined for all times. By erecting a 45° diagonal line as we have done in Fig. 15, we can follow the history of trajectories and observe the stretching and folding action of the attractor. For example, a fixed point on the map (the intersection of the 45° line with the curve) represents a T-periodic solution. The stability of the solution can be determined from the slope of the curve in the vicinity of the point of interest. That is, $|dX_{j+1}/dX_j| < 1$, (>1) correspond to stable (unstable) fixed points. nT-periodic solutions are fixed points of the map $F^n(X)$, which implies n successive applications of the map F, that is, $F(F(\ldots(X)))$. In the map in Fig. 25, all periodic solutions are nonstable.

It is interesting to note that the one-dimensional map we obtain (Fig. 25) is quite similar to the one representing the logistic difference equation we mentioned at the beginning of Sec. 2. This similarity is not incidental. First, since the chaotic flow requires stretching, we must have $|dF^n(X; \text{Ra})/dX| > 1$ for all n at the corresponding fixed points. Second, since the solution is bounded, we must have folding or the map has to bring trajectories back to a given range. Both requirements can be satisfied only by a nonmonotonic function with extremum points, such as the one depicted in Fig. 25. The simplest analytic expression for such a function is $F(X) = 1 - \alpha X^2$. Note that the map is not invertible. Thus there is no one-to-one correspondence between the differential equation and the one-dimensional iterative map. As the differential

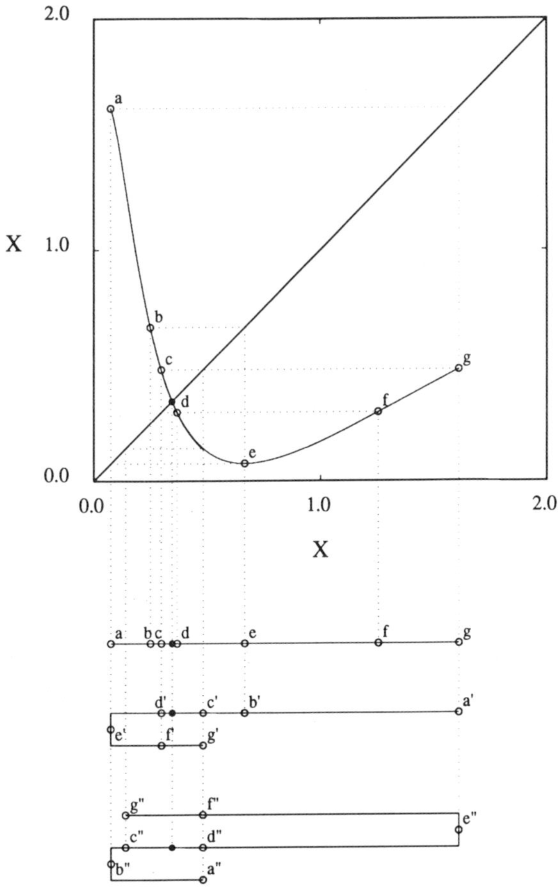

Figure 25 One-dimensional iterative map constructed from the Poincaré cross section in Fig. 24. The folding and stretching is described schematically by the process undergone by the points a–g. In the iterative map (which is noninvertible), the points are mapped back on the same curve. In the differential equations (which are invertible), in contrast, points cannot occupy positions already occupied by other points. In order to reflect this fact, we show schematically the interval folding back on itself. After the first iteration (corresponding to period T), the interval a–g is stretched into the interval a'–g' and folded back upon itself. Additional stretching and folding occurs in the second iteration (a''–g''), and so on ad infinitum. The process describes schematically the formation of the fractal structure.

equations are invertible, from any given point in phase space, we can go backward in time and recover the initial conditions. The same is not true with respect to the one-dimensional map. The same point X_{n+1} can correspond to two different X_n points in Fig. 25.

Next, in Fig. 26, we depict the time series for the velocity given the same conditions as in Fig. 23. The velocity oscillates in time, but there are no reversals in the direction of the flow.

As the Rayleigh number increases, the width of the chaotic attractor increases (Fig. 20) until, at Ra $\approx$ 3.073 . . . , the two Rossler-type attractors (the one shown in Fig. 23 and its mirror image, which is not shown) and the nonstable, no-motion solution intersect. At this point, the width of the attractor doubles. The new attractor, whose projection is shown in Fig. 27, resembles the Lorenz attractor. The sudden change in the size of the attractor at the point of intersection of a nonstable solution with the attractor (Fig. 20) is known as the boundary crisis [30]. The time series for the velocity is depicted in Fig. 28 for the same conditions as those of Fig. 27.

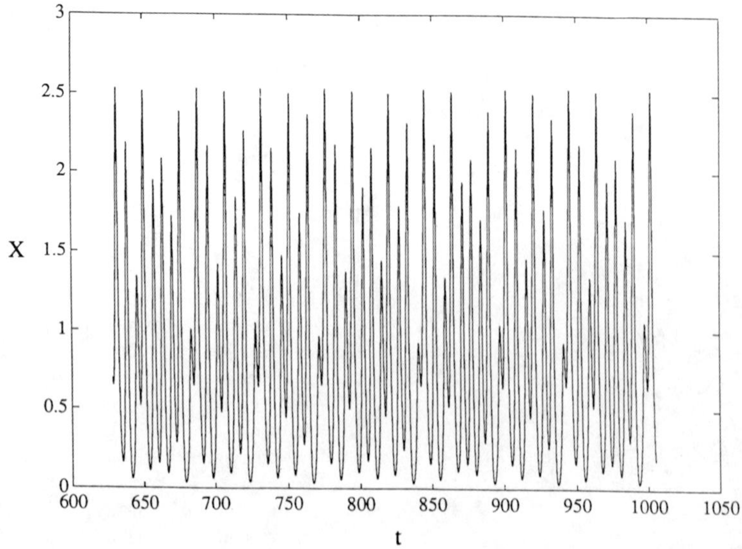

Figure 26 The time series for the velocity x depicted for Ra = 2.92, $\varepsilon = 1$, $P = 1$, and $\omega = 1$.

As the Rayleigh number is further increased, the chaotic flow is interrupted by periodic windows, each of which undergoes a sequence of period doubling into chaos. As in the unmodulated case, the system exhibits periodic behavior for very large Rayleigh numbers. A sample of the T-periodic trajectory is depicted in Fig. 29.

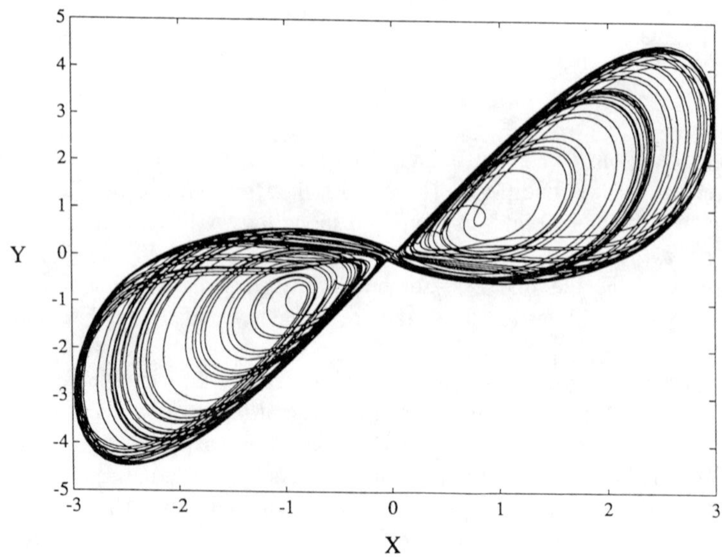

Figure 27 Chaotic attractor at Ra = 3.56, $\varepsilon = 1$, $P = 1$, and $\omega = 1$.

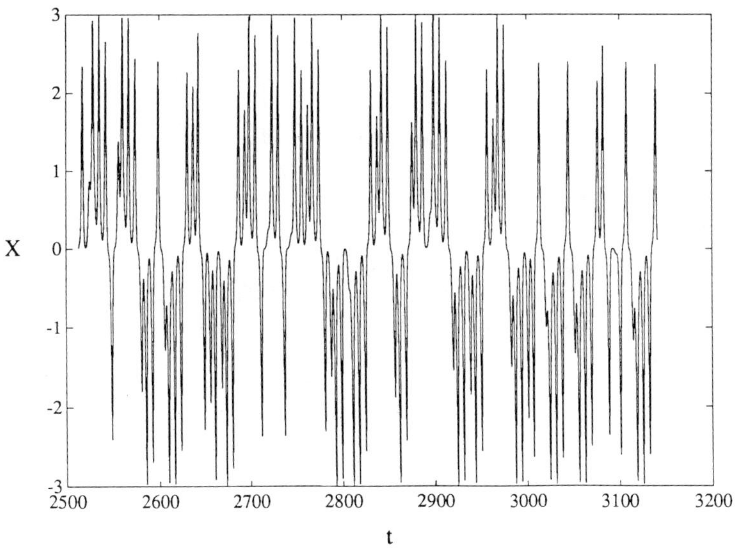

Figure 28 Time series of the velocity for Ra = 3.56, ε = 1, P = 1, and ω = 1.

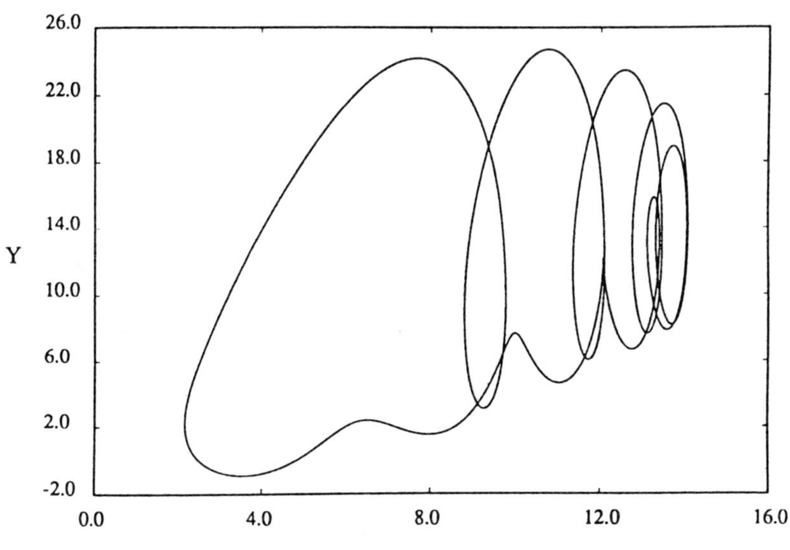

Figure 29 Trajectory of T-periodic solution for Ra = 100, ε = 1, P = 1, and ω = 1.

3 FRACTAL DIMENSIONS

All of us are familiar with the concept of the Euclidean dimension, which always admits an integer value. For example, a point, line, area, and volume have, respectively, dimensions of 0, 1, 2, and 3. When we deal with strange (chaotic) attractors, we need to generalize the concept of a dimension since, on the one hand, due to the contraction property of the attractor, the attractor's dimension will be smaller than that of the corresponding phase space, while, on the other hand, due to the folding action typical to chaotic flows, strange (chaotic) attractors will have a fractal structure and usually possess a noninteger dimension. For example, we have argued above that the Lorenz attractor has dimension (D) smaller than 3 and larger than 2. Knowledge of the attractor's dimension is useful in order to classify the attractor. Also, the attractor's dimension can provide a clue to the number of degrees of freedom (equations) needed to model the attractor (we shall address this issue below).

The concept of a noninteger dimension is a new one and requires a definition. This concept was formalized by Mandelbrot [31] and entitled *fractal dimension*. There are a number of definitions and means of calculating the fractal dimension. We shall introduce two of them in this section: the Hausdorff (or coverage) dimension (D_H) and the correlation dimension (D_c). A third, the Lyapunov dimension (D_L), will be discussed in Sec. 4. The Hausdorff (or coverage) dimension is easy to understand conceptually but is not convenient to use in practice for numerically or experimentally obtained data, while the correlation dimension allows for simpler numerical algorithms. The advantage of the Lyapunov dimension is that it can be computed from the Lyapunov exponents, which we often calculate independently. The three definitions, though they frequently lead to similar results, are not equivalent. Typically, $D_H \leq D_c \leq D_L$.

3.1 Hausdorff (Coverage) Dimension (D_H)

Consider a set of points in an M-dimensional phase space. We cover the points with $N(\zeta)$ hyper (M-dimensional)-cubes of edge ζ. The Hausdorff dimension is defined as

$$D_H = \lim_{\zeta \to 0} \left[\frac{\log N(\zeta)}{-\log(\zeta)} \right] \tag{18}$$

For example, for attractors consisting of a single (fixed) point, a line of length L, a surface of area S, and a volume V, $N(\zeta) = 1, L\zeta^{-1}, S\zeta^{-2}$, and $V\zeta^{-3}$, respectively; and the above definition [Eq. (18)] leads to $D_H = 0, 1, 2$, or 3, which coincides with the Euclidean concept of dimension. Thus, for Euclidean spaces, the Hausdorff dimension reproduces the Euclidean dimension and can be considered as a generalization of the latter.

Next, let us consider the Hausdorff dimension of a non-Euclidean structure. A frequent example used for this purpose is the (one-third) Cantor set. The Cantor set is constructed by taking a unit interval (Fig. 30), dividing it into three segments, and removing the middle one. The process continues ad infinitum, with us removing during each operation the center third of each of the remaining segments. This is quite

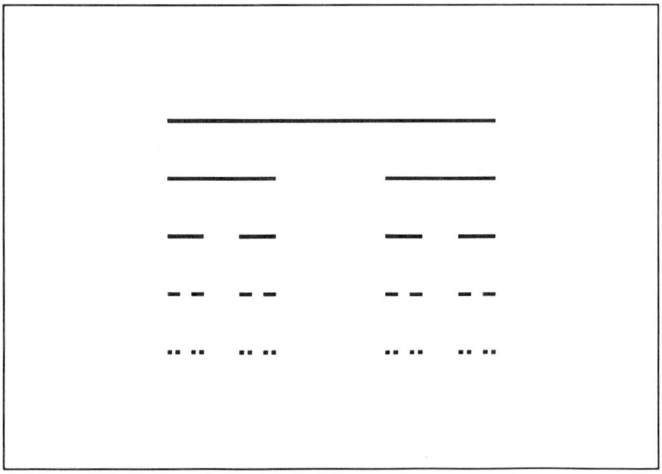

Figure 30 The one-third Cantor set.

similar to the process we observed in the first return map in Fig. 15. From a classical mathematical point of view, the Cantor set has zero measure. The same is not true with respect to its fractal dimension. Here, at the mth stage of the process we have $N(\zeta) = 2^m$ elements in the set, each requiring a cube of edge $\zeta = (1/3)^m$ to cover. Upon substitution in Eq. (18), we obtain $D_H = \log 2/\log 3 \approx 0.63$.

The Hausdorff dimension is useful when the analytic structure of the attractor is known. When the attractor is known only empirically (which is usually the case), Eq. (18) leads to slow convergence and misleading results. To overcome this difficulty, other definitions of a fractal dimension are often used. One of the more frequently used definitions is the correlation dimension.

3.2 Correlation Dimension (D_c)

Suppose that our attractor consists of N data points obtained either numerically or experimentally (the construction of the attractor from experimental data is discussed in Sec. 5). We define the correlation sum:

$$C(r) = \lim_{N\to\infty} \left(\frac{\textit{number} \text{ of pairs of points within a sphere of radius } r}{N^2} \right) \tag{19}$$

In practice, we increase N until $C(r)$ is independent of N. The correlation sum typically scales as

$$C(r) \approx \alpha r^{Dc} \tag{20}$$

where D_c is the correlation dimension and r is chosen to be smaller than the size of the attractor [as r approaches the size of the attractor, $C(r)$ saturates] and larger than the smallest spacing between points. In practice, we plot $C(r)$ as a function of r on a log-log plot and obtain the dimension D_c from the slope of the curve (see Fig. 31 for

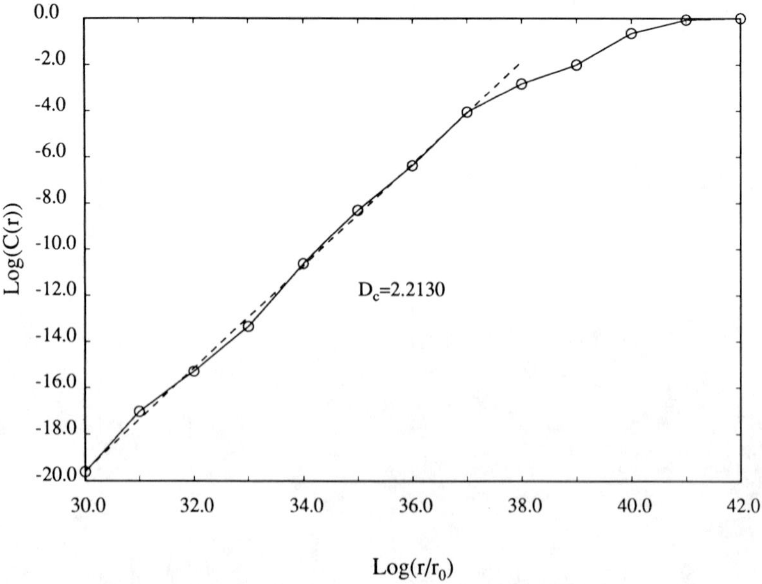

Figure 31 The correlation dimension for the Lorenz attractor shown in Fig. 27 ($P = 1$, Ra $= 3.56$, $\varepsilon = 1$, and $\omega = 1$). D_c is determined from the slope of the solid line described by the dashed line.

an example). We can verify, as we did in the case of the Hausdorff dimension, that for Euclidean objects such as point, line, surface, and volume, the correlation dimension reproduces the integer, Euclidean dimension. For example, for an attractor consisting of a single curve (line), surface, and volume, $C(r) \approx r$, r^2, r^3 and $D_c = 1$, 2, and 3, respectively. For the Cantor set, the correlation dimension $D_c \approx D_H$.

Many variants of the definition in Eq. (19) have been introduced to reduce the computational effort in calculating $C(r)$. For example [32], one can choose a single point, say, i, and calculate the number of points inside a sphere of radius r centered at that point i. This is called a pointwise dimension. Typically, one evaluates the pointwise dimension at a number of points and verifies that the scatter of the results is acceptable. In Fig. 31, we show $C(r)$ as a function of r for the attractor shown in Fig. 27. As expected, $2 < D_c < 3$.

4 THE LYAPUNOV EXPONENT

As we mentioned earlier, sensitivity to initial conditions (i.e., Fig. 14) is a common feature of strange (chaotic) attractors. This sensitivity to initial conditions is a result of the stretching action of the attractor. The magnitude of the stretching, or the rate at which two adjacent trajectories diverge, can be measured by the Lyapunov exponent. The Lyapunov exponent is an extension of the concept of eigenvalues (for linear systems with constant coefficients) and Floquet exponents (for linear systems with time-periodic coefficients). Like the fractal dimension, the Lyapunov exponent can be used as a means of classifying the attractor.

Consider the flow,

$$\frac{dX}{dt} = f(X, t) \tag{21}$$

where X is an N-dimensional vector and $f(\cdot,\cdot)$ is a nonlinear operator. In order to investigate the dependence of the solution on initial conditions, we consider the $N*N$ Jacobian matrix $Df_X[X(t), t]$ which consists of partial derivatives of the system variables $[X_i(X^0, t)]$ with respect to the initial conditions (X_j^0). That is, $Df_X[X(t), t] = [\partial X_i(X^0, t)/\partial X_j^0]$. For the loop problem, $Df_X[X(t), t]$ is identical with the matrix Eq. (11). The time evolution of two nearby trajectories initially separated by small distance $\delta X(0)$ in phase space is governed by the variational equation:

$$\frac{d\delta X}{dt} = Df_X[X(t), t]\, \delta X \tag{22}$$

In general, analytic integration of Eq. (22) is not possible. The equation can be integrated numerically together with Eq. (21). We formally express the solution as

$$\delta X(t) = \Pi(t)\, \delta X(0) \tag{23}$$

where $\Pi(t)$ is the $N*N$ matrix with N eigenvalues, denoted by $\gamma_i(t)$. The Lyapunov exponents λ_i are defined as

$$\lambda_i = \lim_{t\to\infty} \left\{ \frac{1}{t} \log[|\gamma_i(t)|] \right\} \qquad i = 1, \ldots, N \tag{24}$$

where we arrange the exponents in descending order $\lambda_1 > \lambda_2 > \cdots > \lambda_N$. It can be verified that for linear systems with constant coefficients, the Lyapunov exponents are identical with the eigenvalues of the matrix $Df_X[X(t), t]$. The magnitude of the Lyapunov exponent may depend on the point in phase space for which we carry out the calculation.

To make the ideas expressed above more concrete, we shall calculate the Lyapunov exponents for a simple atypical example in which the linear system (22) is diagonizable, and, by appropriate transformation, can assume the form

$$\frac{d\delta X_i}{dt} = \beta_i[X(t), t]\, \delta X_i \tag{25}$$

which upon integration yields

$$\delta X_i(t) = \delta X_i(0) \exp\{\textstyle\int \beta_i[X(\tau), \tau]\, d\tau\} \tag{26}$$

The Lyapunov exponents are given by

$$\lambda_i = \lim_{t\to\infty} \left[\frac{1}{t} \log\left(\left| \frac{\delta X_i(t)}{\delta X_i(0)} \right| \right) \right] \tag{27}$$

Alternatively, the Lyapunov exponent can be defined in a more geometric fashion as the growth rate of an infinitesimal hypersphere centered initially at point X in phase space. As time goes by, the sphere will move with the flow while being deformed into an ellipsoid. We denote the principal axis of the ellipsoid with χ_i. The Lyapunov exponent can be defined as

$$\lambda_i = \lim_{t \to \infty} \left[\frac{1}{t} \log \left(\left| \frac{\chi_i(t)}{\chi_i(0)} \right| \right) \right] \tag{28}$$

Trajectories converge/diverge in the direction associated with negative/positive Lyapunov exponents. For stationary solutions (fixed points), all the Lyapunov exponents will be negative. For a limit cycle, $\lambda_1 = 0$ and $\lambda_i < 0$ for $i > 1$. For chaotic attractors, which always exhibit sensitivity to initial conditions, *at least* one of the Lyapunov exponents will be positive (i.e., $\lambda_1 > 0$). Also, since dissipative systems are volume contracting in phase space, we always have $\Sigma_{i=1}^{N} \lambda_i < 0$.

As the number of positive Lyapunov exponents determine the number of directions in phase space in which expansion occurs, these exponents should relate to the dimension of the attractor. That is, if we have K positive Lyapunov exponents, we would expect that the Lyapunov dimension is larger than K but smaller than $(K + 1)$. Accordingly, one would define the Lyapunov dimension as

$$D_L = K - \frac{1}{\lambda_{K+1}} \sum_{j=1}^{K} \lambda_j \tag{29}$$

where $K \leq N$ is the largest integer for which $\lambda_K > 0$, and since $\lambda_{K+1} < 0$, $D_L > K$. The definition (29) is based on heuristic arguments which we do not go into here. The Lyapunov dimension is not equivalent to D_H and/or D_c and its magnitude may not always coincide with that of D_H and/or D_c. Typically, D_L provides an upper bound for the attractor's dimension.

5 EXTRACTING THE PHASE PORTRAIT FROM A TIME SERIES

One of the rather surprising aspects of dynamical system's theory is a procedure for reconstructing the system's phase portrait from information embedded in a single time series without detailed knowledge of the underlying physics or the corresponding mathematical model. The procedure is based on a set of rigorous mathematical theorems derived by F. Takens (see, for example, [33]). The practical implementation is described in a review article by Gershenfeld [32]. By reconstructing the attractor, we can gain some knowledge about the system's dynamics and possibly use this information to see into the future.

5.1 Reconstruction

Briefly, the technique allows us to generate any chosen number of scalar variables $x_i(t)$, $(i = 1, 2, \ldots, n)$ from a single time series $x(t)$. This can be done by assigning time derivatives to each of the scalars $x_i(t)$ [that is, $x_i(t) = d^i x/dt^i$] or by introducing time delays τ_i so that $x_i(t) = x(t + \tau_i)$, where the τ_i may be all the same ($\tau_i = \tau$) or different. The latter technique is typically used, as it is difficult to obtain accurate data for the time derivatives of the signal. Employing this procedure, one can construct the attractor which corresponds to the given time series. For illustration purposes, we reconstruct for $n = 3$ the Lorenz attractor (Fig. 32) from the time series (Fig. 2). Note that the reconstructed attractor resembles closely the one constructed directly by integrating the differential equations (Fig. 10). Indeed, the reconstructed variables are related to the original ones through a smooth transformation. That is, the topological structure of the attractor is preserved.

Of course, the power of the technique is really revealed when we attempt to reconstruct the attractor of a time series for which we know neither the mathematical model nor the number of degrees of freedom. In such a case, we need to guess the number of degrees of freedom. Typically, one starts with a small number of degrees of freedom, say, $n = 3$. Once the attractor has been constructed, one can estimate various properties of the attractor such as its fractal dimension (D_c). Subsequently, one gradually increases the dimension of the phase space (n) and recalculates the dimension of the attractor (D_c). It is likely that initially D_c will increase monotonically as a function of n. However, once n is sufficiently large, D_c will not vary any further. The smallest value of n beyond which no further variations of D_c occur will be the dimension of the phase space. The situation is depicted schematically in Fig. 33, where we show the variation of the fractal dimension of the attractor (D_c) as a function of the phase-space dimension (n) for deterministic and stochastic systems.

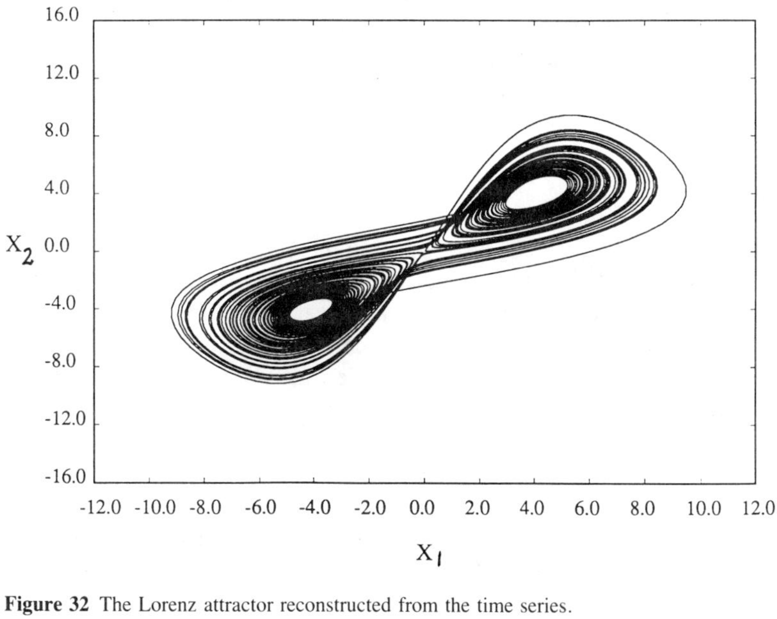

Figure 32 The Lorenz attractor reconstructed from the time series.

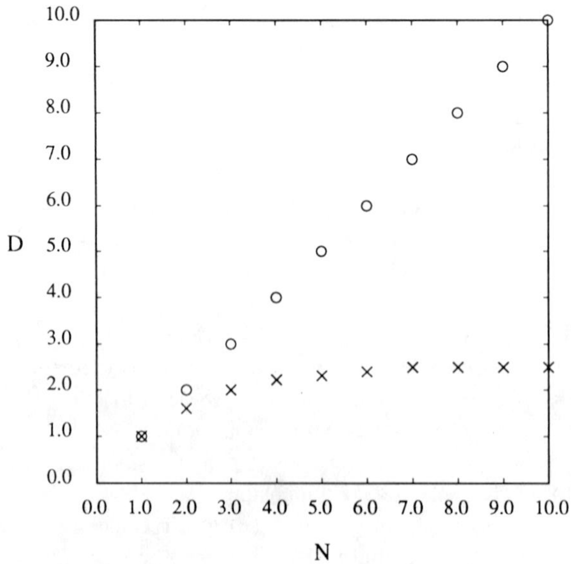

Figure 33 The variation of the fractal dimension (D_c) is shown as a function of the phase space's dimension (n) for a deterministic, chaotic system (×) and for a stochastic system (○).

Since such a construction is possible only if the system is deterministic, the above procedure allows us to test whether the system is deterministic or stochastic. Typically, in a stochastic system, D_c will not reach an asymptotic value. That is, if D_c keeps increasing without limit as n increases (Fig. 33), we can conclude that either the system is stochastic or that the required number of degrees of freedom is exceedingly large. Another test of determinism is based on examining the power spectrum. In deterministic systems, the amplitude of the power spectrum typically decays as the frequency increases (see Fig. 3, which depicts the power spectrum for the time series generated from the Lorenz equations), while in stochastic systems the power spectrum is relatively flat.

5.2 Predicting the Future

Once the attractor is available, we can also construct the vector field associated with the attractor. Such a vector field can be used for predictive purposes. For example, suppose that we constructed the attractor for a dynamical system for which only the time series is available (but not the mathematical model). We wish to attempt to predict the future given a certain set of initial conditions. We can find an approximate point on the attractor which corresponds to the given initial conditions. We can integrate forward from that point and construct the future behavior of the time series. An example of such a construction is provided in Fig. 34, where the solid line represents the prediction obtained from the reconstructed attractor (Fig. 34), while the dashed line was obtained from direct integration of the differential equations. Of course, due to the sensitivity of the system to small perturbations, the prediction is valid only for a limited amount of time, the magnitude of which we can estimate by calculating the

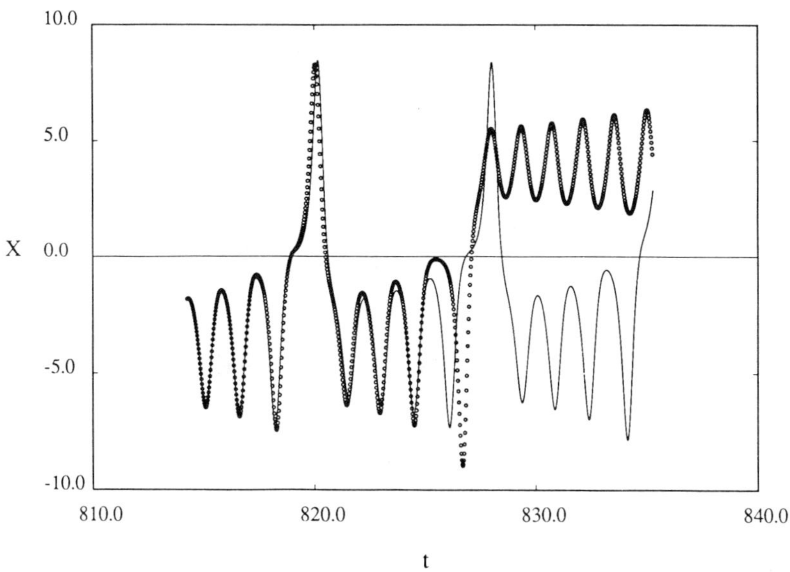

Figure 34 Prediction of future behavior generated from the reconstructed attractor (solid line) compared with data generated by direct integration of the mathematical model (dashed line).

Lyapunov exponent. Such a predictive tool is of course of great value. Potentially, it could be used in conjunction with a feedback control system or as an alarm for impending disaster. Such a tool could even be put to use for making money by predicting stock market fluctuations (assuming, of course, that the economy is a deterministic system).

6 RELATIONSHIP BETWEEN CHAOS AND TURBULENCE

Until a few years ago, the issue of whether there is any relationship between chaos and turbulence was the subject of heated debate. For many years, the prevailing theory of transition to turbulence was based on the Landau-Hopf conjecture, which envisioned turbulence resulting from a sequence of bifurcations in which steady flow is replaced by a periodic one, which subsequently undergoes additional bifurcations, each adding additional incommensurate frequencies to the flow. That is, the time-periodic flow bifurcates into a quasi-periodic flow (two-dimensional torus), which, as the control parameter (i.e., Reynolds or Rayleigh number) increases, further bifurcates into a three-dimensional torus, and so on ad infinitum. Thus, according to the Landau-Hopf conjecture, turbulence is a superposition of a large number of solutions of the Navier-Stokes equations having incommensurate frequencies. It turns out that the Landau-Hopf scenario is highly improbable, since oscillations with independent frequencies tend to influence one another in such a way as to produce synchronization of the independent oscillations into periodic oscillations with a longer period. Thus, if we conduct an experiment, we are likely to see the eventual disappearance of the quasi-periodic flow and the establishment of a new periodic flow. This is known as *frequency locking*.

Dynamical system's theory identifies turbulence with the existence of a strange attractor. From the examples we presented earlier in the chapter, it is evident that the strange attractor may appear after a few bifurcations or as a result of a period doubling sequence. The examples we provided as well as much of the available research literature focus on low-dimensional systems. Thus the question remains whether there is a connection between infinite-dimensional systems (Navier-Stokes equations) and chaos theory. Below, we provide some evidence that such a connection indeed exists. Most of the evidence linking turbulence and chaos pertains to confined flows; but there is also some suggestive evidence pertaining to open flows. We describe some of this in Sec. 6.1.

It is probably fair to say that only a few scientists doubt any more that chaos theory explains appropriately the transition to turbulence. The prevailing (and open) question is whether chaos theory can help us model turbulence.

6.1 Transition to Turbulence

A relationship between chaos theory and turbulence has been established for confined flows such as the Benard problem of a horizontal fluid layer heated from below and cooled from above (i.e., [34]) and the Taylor-Couette flow between rotating cylinders (i.e., [35]). In these experiments the controlling parameter (i.e., the Rayleigh number) was gradually increased while the state of the system was monitored. It was observed that turbulence occurs after a limited (small) number of bifurcations, in contrast to the scenario envisioned by Landau, which requires a large number of bifurcations before the appearance of turbulence. Also, Landau's theory suggests that only systems with a large number of degrees of freedom will exhibit chaotic behavior, while we have seen in the previous section that low-dimensional systems can exhibit turbulentlike behavior. In fact, some of the routes to turbulence observed in closed systems are quite similar to what we have seen in low-dimensional systems. For example, Libchaber and Maurer [34] conducted careful, delicate experiments with liquid helium confined in a small box and heated from below. As they increased the Rayleigh number, they observed a sequence of four period doublings before the appearance of weak turbulence (or chaos). We use the term "weak turbulence" to indicate loss of time but not spatial coherence. The reason why only four period doublings were observed before detection of turbulence is probably due to the fact that additional period-doubling bifurcations occur at fast-decreasing increments of the Rayleigh number (i.e., Table 1) and thus may be hard to detect experimentally. Also, as the period increases, the amplitude of the low-frequency signals in the power spectrum reduces, possibly below the noise level.

For open flows, no direct experimental evidence exists to link chaos and turbulence. This may be partially explained by the fact that external flow experiments are more difficult than ones in closed systems, since coherent structures propagate in both space and time, on the one hand, and it is harder to control boundary conditions with the same precision as can be done in closed flows, on the other hand. In a recent theoretical work, however, Aubry and co-workers [36] modeled the wall region of the turbulent boundary layer with a low-dimensional dynamical system which undergoes transition to chaos via a homoclinic explosion. Aubry et al. [36] were successful in capturing the bursting events associated with streamwise vortex pairs which were observed in prior experimental work by others. Thus their work provides a link between low-dimensional chaotic dynamics and open-flow turbulence.

Thus far in this chapter we have explored the evolution of turbulence via a sequence of (perfect) bifurcations. That is, at critical (bifurcation) points, two solution branches intersected and one branch lost and the other gained stability. There are, however, a number of examples in which the transition to turbulence is observed experimentally at much lower values of the control parameter (i.e., Reynolds or Rayleigh number) than is predicted by theory. For example, theory predicts that Poiseuille flow in a circular duct will be stable for all Reynolds numbers, yet from experimental observations we know that turbulence will occur at a Reynolds number of an order of a few thousands. Clearly, the mechanism of subsequent (perfect) bifurcations fails to explain the route to turbulence in these cases. It is plausible that turbulence in such cases is attributable to random ''jumping'' of the solution among a (possibly large) number of isolated attractors. It is quite common that nonlinear systems may have a number of coexisting, isolated attractors. If the basins of attraction are sufficiently narrow, jumping from one basin of attraction to another may be caused by a random noise. There is also some evidence that, under such conditions, the boundaries of the basins of attraction may be fractal in nature, with one basin penetrating into the other in a complicated way so that any point within the attractor is close to the attractor's boundary. Some recent numerical experiments with low-dimensional difference and differential equations provide credence to this scenario [37]. It is also possible that, in the presence of noise, the multiple attractors merge into a single one.

We have seen that chaos theory is quite successful in providing an explanation for the onset of turbulence. The other important question is whether the theory can assist us in modeling turbulent flows. Modeling turbulent flows is an expensive proposition at best and often an impossible task altogether. In the next section we shall explore a few ideas of how chaos theory can assist us in this task.

6.2 Turbulent Modeling

By now, we know that dissipative systems, of which the Navier-Stokes equations are an example, are attracted to a low-dimensional attractor in phase space. We even have a mean of estimating the dimension (or the number of degrees of freedom) of the attractor. Thus, at least theoretically, we should be able to project the Navier-Stokes equations on a relatively low-dimensional space. Such a capability would be of enormous utility, as it would allow us to model turbulence with a relatively small number of equations. The major difficulty is that we do not know the coordinates (or eigenfunctions) on which we should project the Navier-Stokes equations. Clearly, the selection of the eigenfunctions will have a crucial impact on our success or failure. For example, earlier in this chapter we succeeded in reducing the infinite-dimensional problem of the thermal convection loop into a three-dimensional problem by electing to project on trigonometric functions. The results would not have been as spectacular had we selected any other eigenfunctions (i.e., Chebychev polynomials). There are a number of procedures which allow us to construct an "optimal" eigenspace in some sense. Below, we shall explore briefly one such technique and comment on its shortcomings.

The Karhunen-Loeve (KL) procedure, which is also known as the proper orthogonal decomposition theorem [38–40], allows us to construct "optimal" eigenfunctions from experimental or numerical data. The eigenfunctions are optimal in the sense that the corresponding series of eigenmodes converges more rapidly than any other representation. The same procedure was employed by Aubry et al. [36] to reduce the Navier-Stokes equations for the turbulent boundary-layer flow into a small number (<10) of ordinary differential equations. Briefly, suppose that the velocity vector field $\mathbf{v}(\mathbf{x}, t) = (v_1, v_2, v_3)(\mathbf{x}, t)$ or another suitable variable is available to us as a function of the location vector $\mathbf{x}$ and time t in domain D. We seek a set of orthonormal functions $\phi^{(n)}(x) = (\phi_1{}^{(n)}, \phi_2{}^{(n)}, \phi_3{}^{(n)}(\mathbf{x})$ that maximizes the projection

$$\lambda^{(n)} = \langle \overline{(\boldsymbol{\phi}^{(n)}(x), v(x, t))^2} \rangle \tag{30}$$

subject to the constraint that $(\boldsymbol{\phi}^{(m)}, \boldsymbol{\phi}^{(n)}) = \delta_{mn}$, where $(a, b) = \int_D ab\, dx$ is the inner product on the domain D; the overbar indicates a time average over a prechosen time interval; $\langle\ \rangle$ denotes a probalistic expectation or an average over a number of observations; and δ_{mn} is the Kronecker delta. Since, as the value of n increases, the field of admissible functions $\boldsymbol{\phi}$ decreases, we have $\lambda^{(1)} > \lambda^{(2)} > \lambda^{(3)}$. . . . From Eq. (21), it is clear that the function $\boldsymbol{\phi}^{(1)}$ carries the dominant part of the Fourier expansion of $\mathbf{v}(\mathbf{x}, t)$ with respect to the double average.

Using calculus of variations, the problem can be reduced to a Fredholm integral equation of the first kind:

$$\int_D R_{i,j}(x, x')\phi_j(x')\, dx' = \lambda\phi_i(x) \tag{31}$$

where we use Einstein's summation rule with respect to repeating indices. The kernel (the correlation tensor) $R_{i,j}(x, x') = \langle \overline{v_i(x, t)v_j(x', t)} \rangle$ which can be calculated from experimental or numerical data is symmetric. Thus the eigenfunctions $(\boldsymbol{\phi}^{(m)})$ can be calculated by solving the integral Eq. (31). The Hilbert-Schmidt theory of integral equations guarantees that real eigenvalues $\lambda^{(1)} > \lambda^{(2)} > \lambda^{(3)} > \ldots > 0$ exist and that the eigenfunctions $(\boldsymbol{\phi}^{(m)})$ are real, orthonormal to each other $(\boldsymbol{\phi}^{(m)}, \boldsymbol{\phi}^{(n)}) = \delta_{mn}$, and form a complete set. In practice, since the tensor $R_{i,j}(\mathbf{x}, \mathbf{x}')$ is known only in finite form (a matrix of size K^2, say), we can calculate at most K eigenvectors and eigenfunctions. Consequently, the field $\mathbf{v}(\mathbf{x}, t)$ can be reconstructed as

$$v_i(x, t) = \sum_{n=1}^{N} a^{(n)}(t)\phi_i^{(n)}(x) \tag{32}$$

where $N < K$. By substituting the series into the time-dependent Navier-Stokes equations and requiring that the equations are satisfied in the sense of weighted residuals, one obtains a finite set of N ordinary differential equations for the coefficients $a^{(n)}(t)$.

As there is no apparent relationship between the eigenspace generated by the KL procedure and the eigenspace mentioned in the Ruelle and Taken's theorem (specifying the number of degrees of freedom needed to model a dynamical system, Sec. 4.1), a question remains with respect to the magnitude of the truncation N. One way of determining N is based on the magnitude of the turbulent "energy" captured by the decomposition. It is an easy matter to show that the average turbulent energy,

$$E = \int_D \langle \overline{v_i(x, t) v_i(x, t)} \rangle \, dx = \sum_{n=1}^{\infty} \lambda_n \tag{33}$$

Thus the fraction $(1/E) \Sigma_{n=1}^{N} \lambda_n$ gives us an indication of the fraction of turbulent energy captured by the model. Typically, one would like to have this fraction larger than 90%.

The KL procedure gives us a means of modeling a *known* turbulent flow with a low-dimensional dynamical system. Such an exercise can be quite useful in gaining insight into the behavior of the flow and establishing links between chaos theory and turbulence. The usefulness of the procedure for predicting behavior of unknown flows is less clear. In order to construct the eigenfunctions, we need to know the flow field $\mathbf{v}(\mathbf{x}, t)$. Moreover, the eigenfunctions are likely to change as we vary the control parameter (i.e., the Reynolds number). Also, note that the procedure is computationally expensive.

7 NUMERICAL CHAOS—A NOTE OF CAUTION

Much of our knowledge about chaos and its existence is generated by numerical studies. There are only a very few, relatively simple examples in which the existence of chaos can be proved rigorously. Thus, one may and should wonder whether the complicated behavior observed in numerical experiments is an artifact of the computer simulation or a true reflection of the dynamics of a system. With respect to the thermal convection loop described earlier in the chapter, we have a fairly high level of confidence that the numerically generated complicated dynamics are a true representation of the differential equations, at least in a qualitative sense. This confidence results on the one hand from the fact that this system has been studied by many researchers utilizing a variety of numerical and quasi-analytical techniques to achieve similar results, and on the other hand, from the qualitative agreement between the theoretical results and experiments. There are numerous examples (i.e., [41]), however, in which one can demonstrate that chaotic behavior is a numerical artifact, merely a consequence of numerical instability. This is often the case when one uses finite-difference schemes with an exceedingly large time step to simulate differential equations. A celebrated example is the logistic differential equation,

$$\frac{dx}{dt} = x(1 - x) \tag{34}$$

whose solution is the fixed point $x = e^t/(C + e^t)$, but its finite-difference counterpart, $x_{n+1} = (1 + \Delta t)x_n - \Delta t x_n^2$, is known to exhibit period doubling sequence and chaos as the time Δt increases above the value 2. However, numerical chaos is not inevitable, and it can be avoided if sufficient care is exercised.

8 SOME RELEVANT LITERATURE

This section lists some additional literature on chaos (beyond that already cited) for those readers who wish to explore this topic further. As the number of relevant publications is rapidly growing, this list is far from exhaustive, and no judgment as to a work's merits should be inferred from its inclusion in or omission from this list.

Chaos, by Gleick [1], is a popular, nonmathematical, entertaining book describing the general theory of chaos and the personalities which contributed to its development. The book makes good reading, and one of its main virtues is its emphasis on the universality of chaos theory and its applicability across disciplinary lines.

Nonlinear Dynamics and Chaos, by Thompson and Stewart [42], *Chaotic Vibrations*, by Moon [43], *Order Within Chaos*, by Berge, Pomeau, and Vidal [44], and *Deterministic Chaos*, by Schuster [45], are general survey books which do not require much mathematical background. Each of these books can serve as a general introduction to the theory and a good starting point.

Practical Numerical Algorithms for Chaotic Systems, by Parker and Chua [46], is a how-to-do-it book. It is on a similar level to those sources mentioned above. The authors also wrote a computer program (which we have not tried) to simulate and diagnose chaotic systems. A simple, inexpensive program to model low-dimensional dynamical systems on the IBM PC is available in *Differential and Difference Equations* through computer experiments by Kocak [47].

For the more mathematically inclined, *An Introduction to Chaotic Dynamical Systems*, by Devaney [4], introduces the topic through finite-difference equations. This is a true math text with proofs but also with a few examples. *Nonlinear Oscillations, Dynamical Systems, and Bifurcations of Vector Fields*, by Guckenheimer and Holmes [48], and *Global Bifurcations and Chaos: Analytical Methods*, by Wiggins [49], include a discussion of differential equations. Both books require some mathematical sophistication. *The Lorenz Equations: Bifurcations, Chaos, and Strange Attractors*, by Sparrow [22], consists of a detailed investigation of the Lorenz equations. This book is self-contained, clearly written, and can serve as an excellent introduction to chaos.

In addition, numerous research articles spread among diverse journals and conference proceedings deal with various aspects of chaos theory. For obvious reasons, it will not be possible to go into any detail here. The largest concentration of related articles probably appears in *Physica D*. A collection of some of the earlier articles pertaining to chaos, many of which are now "classics," was assembled into the bound volume *Universality in Chaos*, by Cvitanovic [50].

ACKNOWLEDGMENT

This work was supported, in part, by the National Science Foundation (Grant CBT 8351658).

REFERENCES

1. J. Gleick, *Chaos*, Viking, N.Y., 1987.
2. L. D. Landau and E. M. Lifshitz, *Fluid Mechanics*, 2nd ed., Pergamon Press, 1987.
3. J. M. Ottino, *The Kinematics of Mixing: Stretching, Chaos and Transport*, Cambridge University Press, 1989.
4. R. L. Devaney, *An Introduction to Chaotic Dynamical Systems*, 2nd ed., Addison-Wesley, Reading, Mass., 1989.
5. R. M. May, Simple Mathematical Models with Very Complicated Dynamics, *Nature*, vol. 261, pp. 459–467, 1976.
6. A. Metrol and R. Greif, A Review of Natural Circulation Loops, in *Natural Convection: Fundamentals and Applications* (W. Aung, S. Kakac, and R. Viskanta, eds.), Hemisphere, New York, 1985.
7. H. F. Creveling, J. F. De Paz, J. Y. Baladi, and R. J. Schoenhals, Stability Characteristics of a Single Phase Thermal Convection Loop, *J. Fluid Mech.*, vol. 67, pp. 65–84, 1975.
8. M. Gorman, P. J. Widmann, and K. A. Robins, Chaotic Flow Regimes in a Convection Loop, *Phys. Rev. Lett.*, vol. 52, pp. 2241–2244, 1984.
9. M. Gorman, P. J. Widmann, and K. A. Robins, Nonlinear Dynamics of a Convection Loop: A Quantitative Comparison of Experiment with Theory, *Physica 19D*, pp. 255–267, 1986.
10. P. J. Widmann, M. Gorman, and K. A. Robins, Nonlinear Dynamics of a Convection Loop II: Chaos in Laminar and Turbulent Flows, *Physica D, 36*, 157–166, 1989.
11. P. Ehrhard and U. Muller, Dynamical Behavior of Natural Convection in a Single-Phase Loop, *J. Fluid Mech.*, 217, 487–518, 1990.
12. R. Greif, Y. Zvirin, and A. Metrol, The Transient Behavior of a Natural Convection Loop, *J. Heat Transfer*, vol. 101, pp. 684–688, 1979.
13. H. H. Bau, New, Improved Formula for an ME Laboratory, *ASME Mech. Eng.*, pp. 84–87, February 1989.
14. P. Welander, On the Oscillatory Instability of Differentially Heated Fluid Loops, *J. Fluid Mech.*, vol. 29, pp. 17–30, 1967.
15. H. H. Bau and K. E. Torrance, Transient and Steady Behavior of an Open, Symmetrically-Heated, Free Convection Loop, *Int. J. Heat Mass Transfer*, vol. 24, pp. 597–609, 1981.
16. P. Ehrhard, Dynamisches Verhalten der Naturkonvektion in geschlossenen Kreislaufen, Ph.D. thesis, Kfk 4373, Kernforschungszentrum Karlsruhe, 1988.
17. J. E. Hart, A New Analysis of the Closed Loop Thermosyphon, *Int. J. Heat Mass Transfer*, vol. 27, pp. 125–136, 1984.
18. J. E. Hart, A Note on the Loop Thermosyphon with Mixed Boundary Conditions, *Int. J. Heat Mass Transfer*, vol. 28, pp. 939–947, 1985.
19. A. Yorke, E. D. Yorke, and J. Mallet-Paret, Lorenz-like Chaos in Partial Differential Equations, *Physica 24D*, pp. 279–291, 1987.
20. W. V. R. Malkus, Non-periodic Convection at High and Low Prandtl Number, *Mem. Soc. Roy. Sci. Liege*, 6^{e} ser., vol. IV, pp. 125–128, 1972.
21. E. N. Lorenz, Deterministic Nonperiodic Flow, *J. Atmospheric Sci.*, vol. 20, pp. 130–141, 1963.

22. C. Sparrow, *The Lorenz Equations: Bifurcations, Chaos, and Strange Attractors*, Springer-Verlag, 1982.
23. G. Iooss and D. D. Joseph, *Elementary Stability and Bifurcation Theory*, 2nd ed., Springer-Verlag, 1989.
24. K. A. Robins, Periodic Solutions and Bifurcation Structure at High R in the Lorenz Model, *SIAM J. Appl. Math.*, vol. 36, pp. 457–472, 1979.
25. Y.-Z. Wang and H. H. Bau, Period Doubling and Chaos in a Thermal Convection Loop with Time Periodic Wall Temperature Variation, *Proc. Int. Heat Transfer Conf.*, vol. 2, pp. 357–362, 1989.
26. Y.-Z. Wang, Problems in Thermal Convection, Ph.D. thesis, University of Pennsylvania, Philadelphia, 1991.
27. W. H. Press, B. P. Flannery, S. A. Teukolsky, and W. T. Vetterling, *Numerical Recipes*, Cambridge, 1986.
28. M. J. Feigenbaum, Universal Behavior in Nonlinear Systems, *Los Almos Science*, 1, pp. 4–27, 1980.
29. E. N. Lorenz, Noisy Periodicity and Reverse Bifurcation, in *Nonlinear Dynamics* (R. H. G. Helleman, ed.), New York Academy of Sciences, New York, pp. 287–291, 1980.
30. C. Grebogi, E. Ott, and J. A. Yorke, Crises, Sudden Changes in Chaotic Attractors, and Transient Chaos, *Physica 7D*, pp. 181–200, 1983.
31. B. B. Mandelbrot, *Fractals—Form, Chance and Dimension*, W. H. Freeman, San Francisco, 1977.
32. N. Gershenfeld, An Experimentalist's Introduction to the Observation of Dynamical Systems, in *Directions in Chaos*, Vol. 2 (Hao Bai-lin, ed.), World Scientific, Singapore, pp. 310–382, 1988.
33. D. Ruelle, *Chaotic Evolution and Strange Attractors*, Cambridge University Press, 1989.
34. A. Libchaber and I. Maurer, A Rayleigh-Benard Experiment: Helium in a Small Box, in *Nonlinear Phenomena at Phase Transitions and Instabilities* (T. Riste, editor), 259–286, 1982.
35. J. P. Gollub and H. L. Swinn of Turbulence in a Rotating Fluid, *Physics Rev. Lett.* 35, pp. 927–930, 1975.
36. N. Aubry, P. Holmes, J. L. Lumley, and E. Stone, The Dynamics of Coherent Structures in the Wall Region of a Turbulent Boundary Layer, *J. Fluid Mech.*, vol. 192, pp. 115–173, 1988.
37. F. T. Arecchi, R. Badii, and A. Politi, Generalized Multistability and Noise Induced Jumps in a Nonlinear Dynamical System, *Phys. Rev. A*, vol. 32, pp. 402–408, 1985.
38. J. L. Lumley, *Stochastic Tools in Turbulence*, Academic Press, 1970.
39. L. Sirovich, Turbulence and the Dynamics of Coherent Structures, Parts I–II, *Quart. Appl. Math.*, vol. XLV, pp. 561–590, 1987.
40. L. Sirovich, Chaotic Dynamics of Coherent Structures, *Physica D*, vol. 37, pp. 126–145, 1989.
41. E. N. Lorenz, Computational Chaos—A Prelude to Computational Instability, *Physica D*, vol. 35, pp. 299–317, 1989.
42. J. M. T. Thompson and H. B. Stewart, *Nonlinear Dynamics and Chaos*, John Wiley, New York, 1986.
43. F. Moon, *Chaotic Vibrations*, John Wiley, New York, 1987.
44. P. Berge, Y. Pomeau, and C. Vidal, *Order Within Chaos*, John Wiley, New York, 1984.
45. H. G. Schuster, *Deterministic Chaos*, Physik Verlag, 1984.
46. T. S. Parker and L. O. Chua, *Practical Numerical Algorithms for Chaotic Systems*, Springer-Verlag, 1989.
47. H. Kocak, *Differential and Difference Equations*, Springer-Verlag, 1988.
48. J. Guckenheimer and P. Holmes, *Nonlinear Oscillations, Dynamical Systems, and Bifurcations of Vectors Fields*, Springer-Verlag, 1983.
49. S. Wiggins, *Global Bifurcations and Chaos: Analytical Methods*, Springer-Verlag, 1988.
50. P. Cvitanovic, *Universality in Chaos*, Adam Hilger, Bristol, 1984.

CHAPTER

TWO

ROLE OF FRACTAL GEOMETRY IN THE STUDY OF THERMAL PHENOMENA

A. Majumdar

ABSTRACT

The ability of fractal geometry to quantify the disorder and chaos in structures and shapes offers promising prospects of understanding the physical mechanisms of disordered thermal phenomena. A fundamental analytical approach of modeling disorder can be formulated based on these physical insights. This chapter reviews the relevant literature on fractals and explores the possibilities of applying this simple and powerful technique to several areas in thermal sciences. The areas of surface thermophysics, combustion, transport phenomena in porous media, and thermal problems in biology are discussed in this study.

NOMENCLATURE

a	area, m^2
A	area, m^2
b	pore size, m
B	applied magnetic field, T
$B(t)$	self-affine function of time t
$\overline{B}(t)$	probability distribution of $B(t)$
c	empirical load exponent in Eq. (15)
C	specific heat, J kg^{-1} K^{-1}
C_{sf}	empirical constant in Eq. (24)

Ca	capillary number ($= u\mu/\sigma$)
$C(\mathbf{r})$	statistical correlation function with lag r, defined in Eq. (6)
d	Euclidean dimension of space in which fractal is embedded; diameter, m
D	fractal dimension
D_s	fractal dimension of surface, $2 < D_s < 3$
$\mathcal{D}$	diffusion coefficient, $m^2\ s^{-1}$
$E(k)$	energy spectrum of turbulence, $m^3\ s^{-2}$
F	load in contact conductance N
g	gravity, $m\ s^{-2}$
h	contact conductance or heat transfer coefficient, $W\ m^{-2}\ K^{-1}$
h_{fg}	heat of vaporization, $J\ kg^{-1}$
H	scaling exponent in Eq. (10); hardness of a material, $N\ m^{-2}$; induced magnetic field, $A\ m^{-1}$
I	intensity, $W\ m^{-2}\ sr^{-1}$
J_c	critical current density, $A\ m^{-2}$
k	thermal conductivity, $W\ m^{-1}\ K^{-1}$; wavenumber of turbulent eddy, m^{-1}
K	permeability, m^2
K_{ij}	proportionality constant for i cluster and j cluster to combine, s^{-1}
l	length scale, m; direction cosine
ℓ	integral length scale of turbulence or length of a blood vessel, m
ℓ_g	Gibson length scale, m
ℓ_k	Kolmogorov length scale, m
L	length, m: L_f, actual path length; L_s, straight-line distance
m	magnification
M	measure, m^D; total mass, kg
n	refractive index
n_i	number of clusters of mass i
N	number of units
p	probability
P	N; pressure, $N\ m^{-2}$; power of q spectrum; m^3 for spatial frequency
Pe	Peclet number ($= ul\rho C/k$)
q	heat flux, $W\ m^{-2}$; variance of velocity fluctuation, $m^2\ s^{-2}$; wavevector in Eq. (30), m^{-1}
Q	energy flux, $W\ m^{-2}$
r	radius and distance, m
$\mathbf{r}$	position vector
R_1	Reynolds number based on integral scale ($= u\ell\rho/\mu$)
R_g	radius of gyration defined in Eq. (27)
s	number of particles in a cluster
S	structure function defined in Eq. (10) m^2
t	time, s
T	temperature, K; tortuosity
u	velocity, $m\ s^{-1}$
$\mathbf{u}$	velocity vector
v_0	laminar flame velocity, $m\ s^{-1}$

v_t	turbulent flame velocity, m s^{-1}
w	perfusion rate, m s^{-1}
x	distance or location, m
X_e	thermal equilibration length, m
z	height of a surface, m; generation of branching
α	nondimensional optical roughness factor defined in Eq. (19)
α_d	dispersion diffusivity, m^2 s^{-1}
β	length scale for dispersivity, m, defined in Eq. (35)
γ	scaling constant determining the frequency spectrum in Eq. (12); exponent for diffusion coefficient
δ	thickness of flame front, m; Kronecker delta function
Δ	period of microgrooves, m; diameter exponent in Eq. (38)
ε	unit of measurement, m; porosity
ε_c	smallest pore size, m
η	exponent for probability defined in Eq. (26)
θ	angle, sr
λ	wavelength, m
μ	dynamic viscosity, kg m^{-1} s^{-1}
ξ	ratio of gas to solid thermal conductivities
ρ	reflectivity; density, kg m^{-3}
σ	standard deviation of height distribution, m; surface tension, N m^{-1}; shape factor of blood vessels in Eqs. (41) and (42)
σ'	mean slope
τ	lag in structure and correlation function, defined in Eqs. (10) and (31), respectively
ϕ	scalar function—electric potential, temperature, or concentration
ψ	empirical constant in Eq. (15)
ω	frequency, m^{-1} or s^{-1}

Subscripts

a	apparent area
b	blood
c	cavity
e	eddy
eff	effective
g	gas
i	inner cutoff
j	jth generation or node
ij	referring to i and j directions in a coordinate system
l	liquid
m	maximum likelihood of instability growth; metabolism
max	maximum
min	minimum
o	outer cutoff
sat	saturation

t tissue
w wall

Superscripts

$'$ deviation

1 INTRODUCTION

The last two decades have witnessed the development of a new field of mathematics called *fractal geometry*, which has found numerous and diverse applications in the physical and biological sciences [1]. Fractal geometry is a mathematical language which quantifies the disorder and chaos of structures and shapes. The importance of fractals lies in their description and characterization of nature which appears disordered over a large number of length scales. The most apparent examples of natural shapes include coastlines, clouds, earth's topography, and rivers. Since clouds are not spheres and mountains are not cones, classical Euclidean geometry fails to characterize these disordered structures using its mathematical tools of ordered shapes such as a point, a smooth curve, a smooth surface, or a smooth sphere. The reason for such a failure is because natural objects are seldom smooth over several length scales. Fractal geometry finds the order behind such disorder and can therefore effectively quantify the disorder of nature. Since science and engineering involve the study and, perhaps, control of nature, it is not surprising to find the increasing importance of fractals in these fields.

In thermal sciences the occurrence of disorder in either a phenomenon or a medium is quite frequent. Common examples include the roughness of surfaces, which influences contact conductance, reflection of radiation, and boiling; the transport of mass, momentum, and energy in a turbulent flow or in a porous medium; the agglomeration of soot particles and its effects on energy transfer in a combustion system; and the multiscale branched network of the lung or of blood vessels, which governs the distribution of heat and mass in the human body. The lack of a general technique to characterize the disorder in such systems has proved to be an obstacle for researchers seeking insight of the basic mechanisms of disordered phenomena. In addition, it has prevented the development of a general methodology with which to study the effects of a disordered medium on heat transfer. The ability of fractal geometry to quantify the disorder in nature provides an opportunity to confront and tackle these problems. This chapter emphasizes the role of fractals in developing a formal methodology of characterizing disorder in thermal phenomena and in providing clues to understand the physical mechanisms which result in disorder. It is encouraging to note that since fractal geometry forms a link between shapes and structures of different length scales, it could be used to understand a microscale thermal phenomenon by making observations on a macroscale.

The idea of applying fractals to thermal sciences is supported by the fact that, in physics and chemistry [2], fractal geometry has cast new light on several phenomena in solid-state physics [3], nonlinear dynamics [4], and polymer science [5] and heterogeneous chemistry [6]. The application of fractal geometry in biological sciences is quite recent and promising indeed [7]. The nonlinear behavior and electrical stability of the heart can be well described by fractals. In addition, fractal geometry can characterize the structure of the distribution networks of the blood circulatory, the pulmonary, and the nervous systems [8]. Fractal geometry is gradually being recognized as a useful mathematical tool in some engineering fields such as fluid mechanics [9, 10], combustion [11, 12], surface contact mechanics and friction [13], surface thermophysics [14], and materials science [15].

The applications of fractal geometry in thermal sciences is in its nascent stages of development. From the few studies reported so far, it appears that the fractals will have a strong impact on research involving disordered phenomena in thermal sciences. With this motivation, the purpose of this chapter has been designed to be twofold: first to discuss the current status of this new field in thermal sciences and second to explore new areas of research. The chapter is divided into several sections. Section 2 gives a mathematical background of the theory of fractal geometry. Section 3 discusses its role in understanding interfacial and surface thermophysics. The applications of fractals in combustion are covered in Sec. 4. Section 5 focuses on transport phenomena in porous media. The applications of fractals in bioheat transfer are explored in Sec. 6.

2 MATHEMATICAL BACKGROUND

Fractal geometry is a generalization of Euclidean geometry. Therefore, before presenting the theory of fractals it is first necessary to understand the mathematics of the more conventional Euclidean geometry. Euclidean geometry describes ordered and smooth objects such as points, curves, surfaces, and spheres using dimensions of 0, 1, 2, and 3, respectively. Associated with each dimension is a measure of the object such as the length of a line, the area of a surface, or the volume of a cube. These measures are invariant with respect to the unit of measurement, implying that the length of a line remains independent of whether a centimeter or a micrometer scale is used. This arises from the smoothness of the object, which makes a smooth curve linear or a smooth curved surface planar at some small length scale. However, a multitude of objects found in nature appear disordered and wrinkled at a large number of length scales for which the measures of length, area, and volume are scale-dependent. This suggests that the dimensions of such objects cannot be integers. A generalized concept of dimension and the origins of fractal geometry are now discussed.

Fractal geometry was founded by Mandelbrot [16] when he showed that for decreasing units of measurement, the length of a natural coastline does not converge but, instead, increases monotonically. On plotting the length L as a function of the unit of measurement ε on a log-log plot, he found a simple relation of the form $L \sim \varepsilon^{1-D}$, as shown in Fig. 1. He finally made an interesting conclusion that the real number D associated with every coastline is the dimension of the coastline. This simple observation has deep implications on the concepts of order and disorder and marked the origins of fractal geometry. In order to understand the relevance of this new field of mathematics to thermal sciences, it is necessary to provide a formal background of

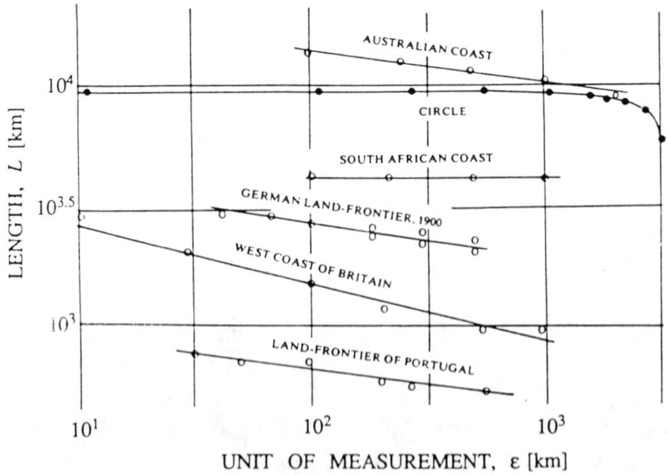

Figure 1 Increase in the lengths of coastlines for decreasing units of measurement [1].

fractal geometry. More details on the theory of fractals and its applications in physical sciences can be found in texts [1, 17–19].

2.1 Hausdorff Dimension

The length of a line is measured by breaking the line into small units of length ε and then adding the number of units in the form

$$L = \sum \varepsilon^1 \tag{1}$$

Similarly, the area of surface is measured by breaking up the surface into small squares of size $\varepsilon \times \varepsilon$ and then adding the number of units as

$$A = \sum \varepsilon^2 \tag{2}$$

It should be noted that in Eqs. (1) and (2) the exponents 1 and 2 correspond to the dimensions of the objects. Therefore the length of a line has units of meters, whereas the area of a surface has units of square meters or meter^2. The unique property of these measures, length and area, is that they are independent of the unit of measurement ε, and in the limit $\varepsilon \rightarrow 0$ these measures remain finite and nonzero. This concept of Euclidean measure and dimension can be generalized to the form

$$M = \sum \varepsilon^D \tag{3}$$

where M is the measure and D is a real number. The exponent D is the dimension of an object if it makes the measure M independent of the unit of measurement ε in the limit $\varepsilon \rightarrow 0$. The unit of the measure M is therefore meter^D. This is a simple definition of the *Hausdorff dimension*, whereas a more rigorous one can be found in Mandelbrot [1]. Contrary to common understanding, this generalization allows the dimension of an object to take noninteger values.

If, in the above argument, it is assumed that an object is broken into N equal parts, then Eq. (3) can be written as $M = N\varepsilon^D$. Since the measure is invariant with the unit of measurement, it follows that $N \sim \varepsilon^{-D}$. Now, if the length of an object of dimension D is evaluated, then the length would vary as $L = N\varepsilon^1 \sim \varepsilon^{1-D}$, as was observed for the lengths of the coastlines. It can easily be shown that only when $D = 1$ will the length be independent of ε. Similarly, for a smooth surface of dimension equal to 2, the volume, which corresponds to an exponent of 3, is zero, whereas the length, corresponding to an exponent of 1, is infinite.

2.2 Self-Similarity

The generalized concepts of measure and dimension are fundamental to the issue of self-similarity. Consider a one-dimensional line of unit length as shown in Fig. 2*a*. Each segment of the line, of length $1/m$, is *similar* to the whole line and needs a magnification of m to be an exact replica of the whole line. Since the length of the line remains independent of $1/m$, it follows that the number of units is $N = m$. Now consider the square in Fig. 2*b*, which has a side of unit length. Each small square of side $1/m$ is similar to the whole square and needs a magnification of m to be an exact replica of the whole square. However, the number of small squares in the whole is $N = m^2$. In general, for an object of dimension D, it follows from the above argument that

$$N = m^D \tag{4}$$

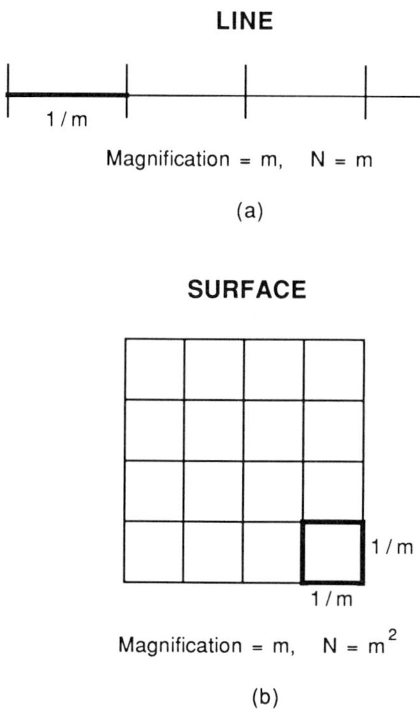

Figure 2 Self-similarity in Euclidean objects of integer dimensions: (*a*) line; (*b*) surface.

Hence the dimension of the object can be obtained as

$$D = \frac{\log N}{\log m} \tag{5}$$

This definition of dimension, which is based on the self-similarity of an object, is called the *similarity dimension* [1]. To perceive what an object of a noninteger dimension looks like, follow the recursive construction of Fig. 3*a*, which yields the Koch curve of dimension 1.26. The first step in this construction breaks a straight line into three parts and replaces the middle portion by two segments of equal lengths. In the subsequent stages each straight segment is further broken into three parts and the middle segment of each is replaced by two parts. If this recursion is done infinite times, then the Koch curve is obtained. This curve has mathematical properties some

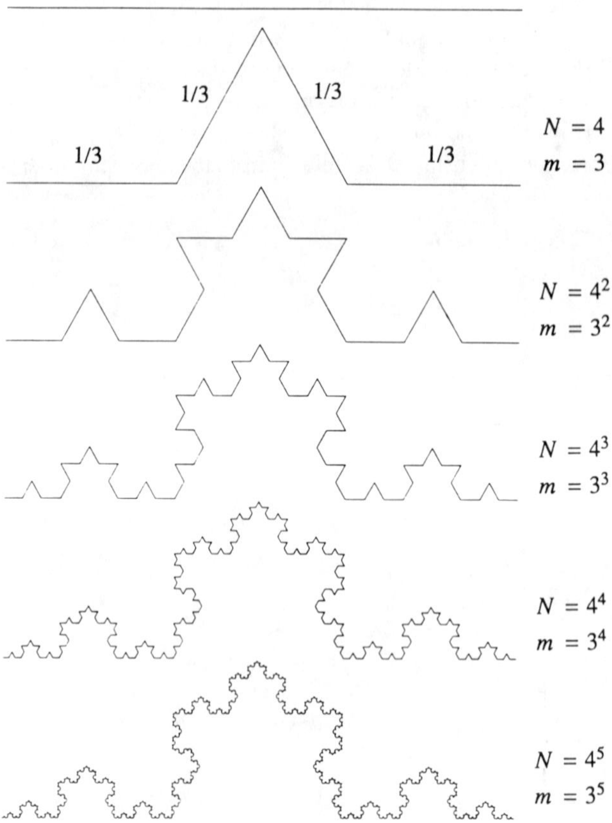

$$D = \frac{\log N}{\log m} = 1.26$$

Figure 3 Recursive construction of (*a*) Koch curve, $D = \log 4/\log 3 = 1.26$;

of which are unique to fractal objects. The curve is continuous but not differentiable anywhere. The nondifferentiability arises from the fact that if the curve is repeatedly magnified, more and more details of the curve keep appearing. This implies that a tangent cannot be drawn at any point and therefore the curve cannot be differentiated. The Koch curve is infinitely wrinkled and is not smooth at any length scale. Second, the curve is exactly self-similar. That is, if a small portion of the curve is appropriately magnified, it will be an exact replica of the whole Koch curve. Third, although the curve contains details at a large number of length scales, the dimension remains constant at all scales. Other examples of recursive construction of self-similar fractal objects such as the Cantor set and Sierpinski carpet are shown in Figs. 3*b* and 3*c*. The Cantor set is formed by repeatedly dividing a line segment into three parts and omitting the middle segment. Infinite recursions of this process result in a set of invariant points which are individually of zero dimension but collectively of dimension $D = \ln 2/\ln 3 = 0.63$. The Sierpinski carpet is formed by dividing a square into nine parts and omitting the middle part. It is of dimension $D = \ln 8/\ln 3 = 1.89$ and looks similar to a cross section of a deterministic porous medium. These examples of self-similarity are constructed by deterministic algorithms. Most objects found in nature, however, are not deterministic but random. For example, consider the coastline of an island. Although a coastline is not *exactly* self-similar, it is *statistically* self-similar. Statistical self-similarity implies that the probability distribution of a small part of an object will be the same as the probability distribution of the whole object if the small part is magnified appropriately.

To obtain the fractal dimension of statistically self-similar objects, it is necessary to look at the pair correlation function $C(\mathbf{r})$, defined as

$$C(\mathbf{r}) = \frac{1}{N} \sum \rho(\mathbf{r} + \mathbf{r}')\, \rho(\mathbf{r}') \tag{6}$$

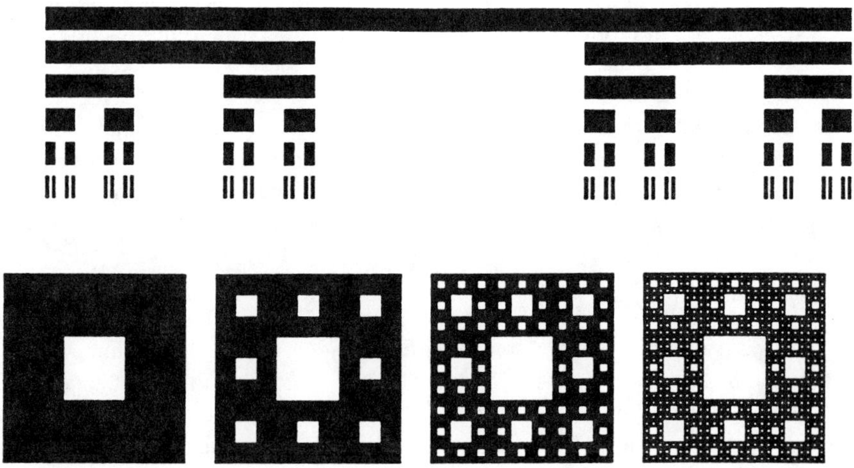

Figure 3 (*b*) Cantor set, $D = \log 2/\log 3 = 0.63$; (*c*) Sierpinski carpet, $D = \log 8/\log 3 = 1.89$.

where $\rho(\mathbf{r}') = 1$ if the point defined by vector $\mathbf{r}'$ is within the object and zero otherwise. Therefore $C(\mathbf{r})$ is the expectation value of the event that two points separated by $\mathbf{r}$ belongs to an object. If the object is isotropic, then the correlation function depends on the scalar distance r. For a self-similar fractal object, the correlation function follows a power-law behavior of the form

$$C(r) \sim r^{-\alpha} \tag{7}$$

which satisfies a scaling relation $C(\beta r) \sim \beta^{-\alpha} C(r)$. The fractal dimension D of the object is related to the exponent α by the relation $D = d - \alpha$, where d is the dimension of the Euclidean space in which the fractal object is embedded. Consider the situation where particles of a particular size are aggregating to form a self-similar fractal structure such that the correlation function follows Eq. (7). The number of particles N of the object within a sphere of radius L can be determined from the correlation function as [19]

$$N(L) \approx \int_0^L C(r)\, d^d r \tag{8}$$

Using the power-law behavior of the correlation function in Eq. (8), it can be shown that $N(L) \sim L^{d-\alpha}$. For Euclidean objects the number of parts in a whole object varies as $N \sim L^n$, where $n = 1$ for a straight line, $n = 2$ for a surface, and $n = 3$ for a cube. This suggests that the exponent $d - \alpha$ obtained from Eq. (8) is the dimension of the aggregate of particles. The fractal dimension of statistically self-similar objects can be found by either evaluating the correlation function or counting the number of parts within a distance L of the object.

2.3 Self-Affinity

The definition of self-similarity is based on the property of *equal* magnification in all directions as described in Fig. 2*b*. However, there are many objects in nature which have unequal scaling in different directions. For example, consider the function $B(t)$, which represents the x location of a particle in Brownian motion as a function of time, t. If this curve is repeatedly magnified, more and more details keep appearing, which suggests that it is in some sense similar to the Koch curve. However, since the time t and the location B are two different physical entities, they usually scale differently. Therefore, the function $B(t)$ is not self-similar but *self-affine*. This can be shown quantitatively by the statistics of the Brownian function, which follows the relation [20]

$$\overline{B}(\gamma t) = \gamma^{1/2}\overline{B}(t) \qquad \text{assuming that } B(0) = 0 \tag{9}$$

where $\overline{B}$ is the probability distribution of B. Equation (9) shows that if time t scales by γ, then the position of the particle B scales by $\gamma^{1/2}$. Since the time and the position scale differently, the Brownian function $B(t)$ is statistically self-affine.

One important example of self-affine fractals is the generalization of Brownian motion to the whole class of functions called *fractional Brownian motions* or fBm [21]. The statistics of this generalized Brownian motion follow the relation

$$S(\tau) = \langle (B(t + \tau) - B(t))^2 \rangle \approx \tau^{2H} \tag{10}$$

where $\langle \ldots \rangle$ implies temporal average. The function in Eq. (10) is often called the *structure function* of the curve $B(t)$. Brownian motion is special case of fBm and corresponds to $H = \frac{1}{2}$. The dimension of self-affine fractals cannot be obtained from Eq. (5), which is based on the self-similarity of an object. Instead, Voss [22] showed that the dimension is related to the parameter H by the relation $D = E + 1 - H$, where E is equal to the number of arguments of the function B. In the above example, since t is the only argument of B, the dimension of the function is $D = 2 - H$. Mandelbrot [20] showed that the length of self-affine fractal curves do not follow the relation $L \sim \varepsilon^{1-D}$ for all values of ε, and therefore the dimension of self-affine curves cannot be obtained by measuring their lengths. Instead, the dimension of self-affine functions can be obtained either from the structure function $S(\tau)$ in Eq. (10) or from their power spectrum $P(\omega)$, which follows the relation [22]

$$P(\omega) \approx \frac{1}{\omega^{5-2D}} \tag{11}$$

where ω is the frequency. If the power spectrum is plotted against the frequency on a log-log plot, the fractal dimension can be obtained from the slope of the spectrum.

Fractional Brownian motions and self-affine fractals are very important for studying physical phenomena, since most involve different physical entities which scale differently in time and space. An example is the velocity field of a turbulent flow.

With this background of fractal geometry, its applications to thermal sciences are now discussed.

3 THERMAL PHENOMENA AT SOLID SURFACES

Surfaces and interfaces of solids play a crucial role in a broad range of physical phenomena. The importance of surfaces in thermal sciences lies in the fact that a surface forms the boundary between two bulk regions and therefore strongly influences the transport of mass, momentum, and energy across the two regions. In addition, the nature of atomic bonding is different at the surface than in the bulk. This often leads to forces on the surfaces which are quite different from those of the bulk. Owing to the ubiquity of surfaces in nature and their influence on basic physical mechanisms, the physics and chemistry of surfaces involve a vast spectrum of phenomena, which include: conduction between two contacting bodies [23]; thermal boundary resistance at cryogenic temperatures [24]; interaction of surfaces with radiation [25]; phase change at a surface [26]; energy losses on superconducting surfaces [27]; as well as friction, wear, and lubrication [28].

The two important properties of a surface which govern a surface phenomenon are the surface material and the surface structure. This study will focus on the structure of surfaces, which is usually found to be disordered at several length scales. Before discussing the applications of fractal geometry to study surface structure, it is first necessary to identify some unique features of the structure. Smooth surfaces are rare in nature, whereas rough surfaces are abundant. The roughness of surfaces is usually multiscale, for which the range of roughness can span from large length scale like that of the earth, right down to atomic scales [29]. Another unique feature of surface roughness is that, unless artificially textured, most natural surfaces are random. The random and multiscale behavior is seen qualitatively in Fig. 4, which shows the appearance of roughness under repeated magnification, suggesting a self-affinity relation between the length scales.

Although the idea of using fractal geometry to describe surface roughness appears lucrative, it is first necessary to confirm whether rough surfaces in nature do follow such a geometry. Therefore one must first focus on the various methods of surface formation. This section will then discuss the use of fractal geometry to characterize roughness of both natural and machined surfaces. Subsequently, the fractal characterization will be used to gain insights into surface-related thermal phenomena.

3.1 Surface Formation

There are three mechanisms of surface formation which are commonly encountered in science and engineering—fracture of solids, deposition of solids, and solidification of liquids. Mandelbrot et al. [30] showed experimentally that when metals are fractured by tensile or impact loading, the surfaces created are rough and irregular and can be described by fractals. The variation of fractal dimension for differently heat-treated stainless steel samples was attributed to the difference in the microstructure of the grains in the material. This study prompted Termonia and Meakin [31] to develop a phenomenological model of crack propagation at atomic scales which showed that the cracks generated were fractal. The second method of surface formation is deposition of materials, such as thick and thin films. Commercial films are generally deposited by one of two methods—diffusion-limited deposition and ballistic deposition. A recent review by Meakin [32] identifies the fractal-nature thin film surfaces created by both deposition processes. Solidification of a liquid is the third mode of surface formation. Vicsek [19, 33] has shown that when the surface growth is represented by the Laplace equation $\nabla^2 T = 0$, where T is temperature, the surface structure is fractal. The Laplace equation is a special case of the heat equation $\partial T/\partial t = \alpha \nabla^2 T$, in which the time

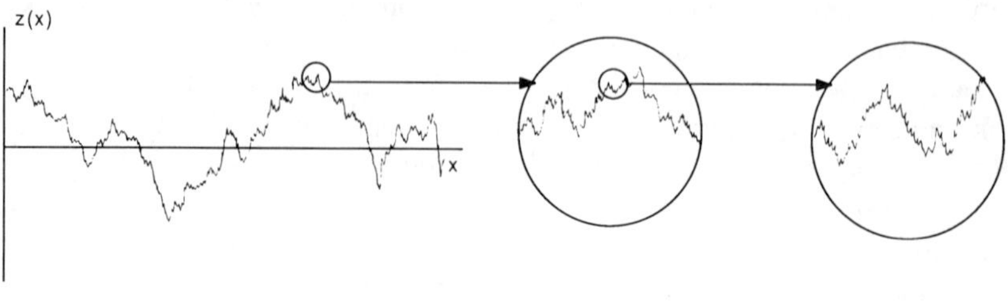

Figure 4 Qualitative description of self-affinity.

variation is negligible. It can be concluded that most natural modes of surface formation yield fractal surfaces. The reasons for surfaces to be fractal were also explored by Kardar et al. [34], who found that the evolution of growing surfaces during deposition and solidification follows scaling laws which are typically represented by self-affinity.

3.2 Fractal Characterization and Simulation of Surface Roughness

The random and multiscale nature of surfaces warrants a scale-independent method of characterization. Conventionally, the deviation of a surface from its mean plane is assumed to be a random phenomenon, for which statistical parameters such as the standard deviations of the surface height, slope, and curvature are used for characterization [35]. However, it has been found that the surface topography is a nonstationary random process for which the variance of the height distribution is related to the length of the sample [36]. It is also known that the variances of the slope and curvature depend strongly on the resolution of the roughness-measuring instrument or any other form of filter and are hence not unique [37, 38]. Consequently, instruments with different resolutions and scan lengths yield different values of these statistical parameters for the same surface [13]. The conventional methods of characterization are therefore fraught with inconsistencies, which give rise to the term ''parameter rash'' [39] commonly used in contemporary literature. The underlying problem with the conventional methods is that although rough surfaces contain roughness at a large number of length scales, the characterization parameters depend only on a few particular length scales such as the instrument resolution or the sample length. A logical solution to this problem is to use scale-invariant parameters to characterize the surface. The scale invariance of the fractal dimension to characterize the Koch curve or the self-affine record of Brownian motion suggests that fractal geometry can be used to characterize rough surfaces.

3.2.1 Roughness Characterization. A multiscale surface profile $z(x)$, as shown in Fig. 4, can be considered to be composed of a superposition of waves of all wavelengths and random phases. To characterize such a profile, it is necessary to determine the amplitude of roughness at each wavelength. This is obtained by finding the power spectrum $P(\omega)$ of the profile by the relation

$$P(\omega) = \frac{1}{L}\left|\int_0^L z(x)e^{i\omega x}\,dx\right|^2 \tag{12}$$

Here frequency ω is the reciprocal of the wavelength and has units of inverse length. The integral of Eq. (8) is the Fourier transform of $z(x)$ which provides the average amplitude of waves of frequency ω.

For a self-affine fractal surface profile of dimension D, the power spectrum follows the relation [13, 14, 40]

$$P(\omega) = \frac{G^{2(D-1)}}{\omega^{(5-2D)}} \tag{13}$$

The parameter G is a scaling constant and has units of length. The structure function $S(\tau)$ can be found as

$$S(\tau) = \left\langle \left(z(x) - z(x+\tau)\right)^2 \right\rangle = \int_{-\infty}^{\infty} P(\omega)\left(e^{i\omega x} - 1\right) d\omega$$
$$= \vartheta G^{2(D-1)} \tau^{(4-2D)} \tag{14}$$

where $\vartheta = \Gamma(2D - 3)\sin\left((2D - 3)\pi/2\right)/(2 - D)$ and Γ is the gamma function. Comparison of Eq. (14) with Eq. (10) shows that $S(\tau)$ satisfies the criterion of a fractional Brownian motion of dimension D.

It should be noted that the parameters G and D characterize the power spectrum and the structure function. For a particular surface, these parameters can be found by either plotting the power spectrum or the structure function on a log-log plot. The dimension D can be found from their slopes and G can be obtained from the intercept of the $P(\omega)$ or the $S(\tau)$ axis, respectively. The important aspect of the fractal characterization is that G and D are independent of ω or τ and therefore scale-independent. Before investigating whether real surfaces do follow the power-law spectrum or structure function, it is instructive to understand the interpretations of the parameters G and D. Figure 5 shows four fractal profiles that were generated artificially. The method of generating surface roughness is discussed later. The first three profiles show that as D is increased from 1 to 2, the high-frequency components become comparable in amplitude to the low-frequency ones. Therefore D decides the ratio of the amplitude of different frequency modes. Comparison of the third and fourth profile shows that as G is reduced, the amplitude of roughness reduces over all frequencies while keeping the ratio unaffected.

3.2.2 Roughness Measurements. Figure 6 shows the structure function of lapped, ground, and shape-turned stainless steel surfaces. The profiles $z(x)$ were measured by traversing a diamond stylus of radius 2.5 μm in contact with the surface [40]. The solid line $S(\tau) \sim \tau^1$ corresponds to the structure function of a Brownian surface profile with $D = 1.5$. It is evident from Fig. 6 that for $\tau < 40$ μm, the profiles tend to be Brownian. The coincidence of the surface statistics at small length scales or high frequencies has been observed not only for stainless steel but also for other materials such as glass and sandstone [41]. However, for $\tau > 40$ μm, the structure functions of surface profiles diverge. The reason for the divergence lies in the type of machining used to polish the surface. Polishing reduces the amplitude of roughness and makes it seemingly smoother. However, polishing can reduce the amplitude only to a particular length scale, smaller than which the surface remains unaffected. This critical length scale depends on the machining tool; examples are the grain size of a grinding wheel or the tool radius for turning. Smaller than this length scale, the unaffected surface roughness behaves like an unprocessed one. In this unprocessed region, the dimensions of all the profiles seem to coincide. Also evident in Fig. 6 is that for $\tau > 40$ μm, the smoother the surface the lower the value of the structure function. Since the value of G is decided by the position of the structure function along the vertical axis, polishing the surface reduces the value of G.

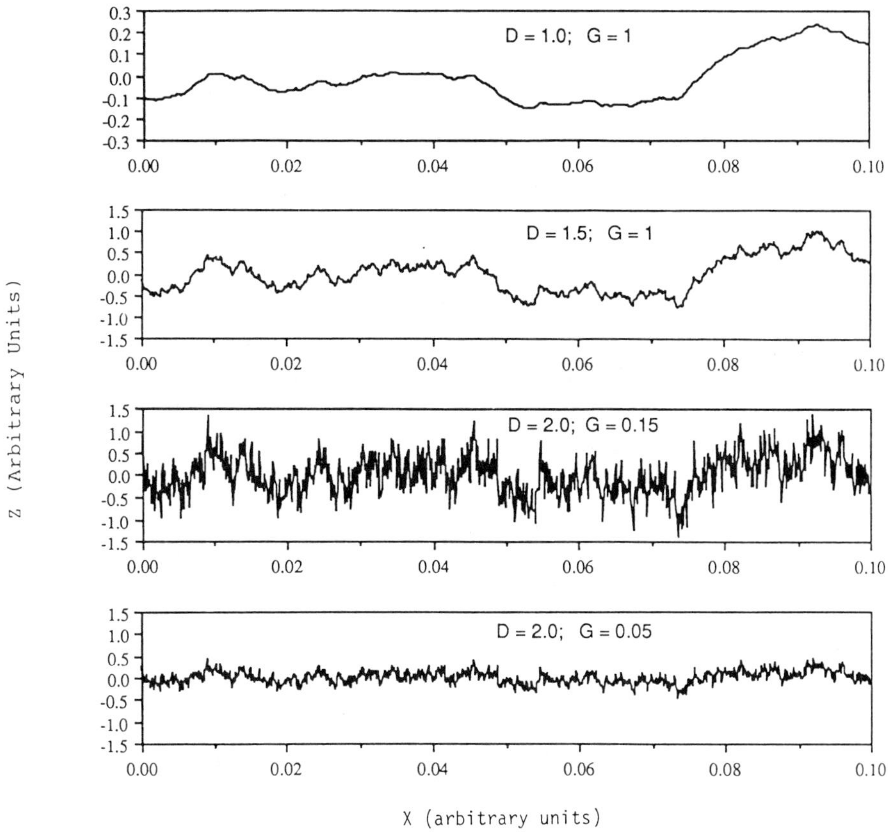

Figure 5 Comparison of stimulated fractal profiles for different values of D and G.

Figure 7 shows the topographies and averaged power spectra of an untextured magnetic thin-film rigid disk surface [13]. The spectra follow a power law of $\omega^{-2.24}$ ($D = 1.38$). Figure 8 shows the topography and spectra of a textured magnetic thin film disk. The x direction corresponds to the circumferential direction in which the grooved texture is made. It is observed that the y spectrum shows a nonfractal power-law behavior, whereas the x spectrum corresponds to $D = 1.31$. Since the spectra of Figs. 7 and 8 show both fractal and nonfractal behavior, it is worth asking when are surfaces fractal and when are they nonfractal. The different modes of surface formation indicate that natural processes yield fractal surfaces. However, when a surface is processed, in some cases it produces a fractal structure whereas in others it does not. Based on the experiments, it was speculated [13] that when the processing technique is random and directionally unbiased within a range of length scales, the surfaces remain fractal in that range. However, when the processing produces deterministically biased texture on the surface, the surface is observed to be nonfractal within the range of length scales affected by the texturing.

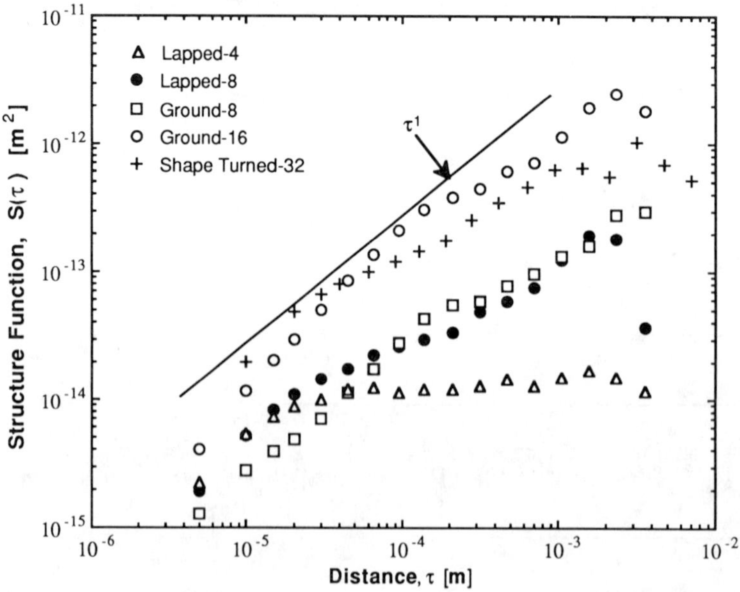

Figure 6 Structure function of machined stainless steel surfaces. Numbers associated with each machining process correspond to average roughness (μin) as specified for General Electric Roughness Specimen (Cat. No. 8651831G1).

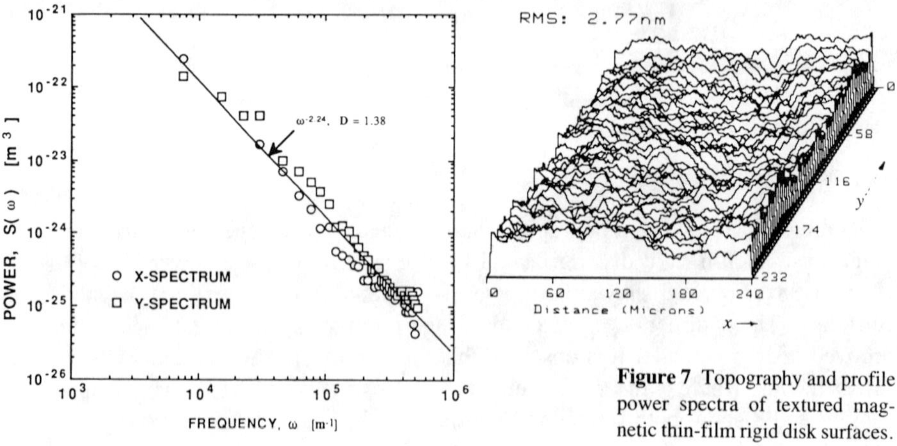

Figure 7 Topography and profile power spectra of textured magnetic thin-film rigid disk surfaces.

The structure and surfaces of deposited thin films are known to be fractal [32]. Chin et al. [42] measured the surface topographies of sputter-deposited thin films of high-temperature superconductors by a scanning tunneling microscope. The films were made of Y-Ba-Cu-O, Er-Ba-Cu-O, and Bi-Sr-Ca-Cu-O and were deposited on single-crystal substrates of MgO, $SrTiO_3$, and Al_2O_3. Both in-situ and annealed samples, with and without gold films on the superconductor surfaces, were considered. The

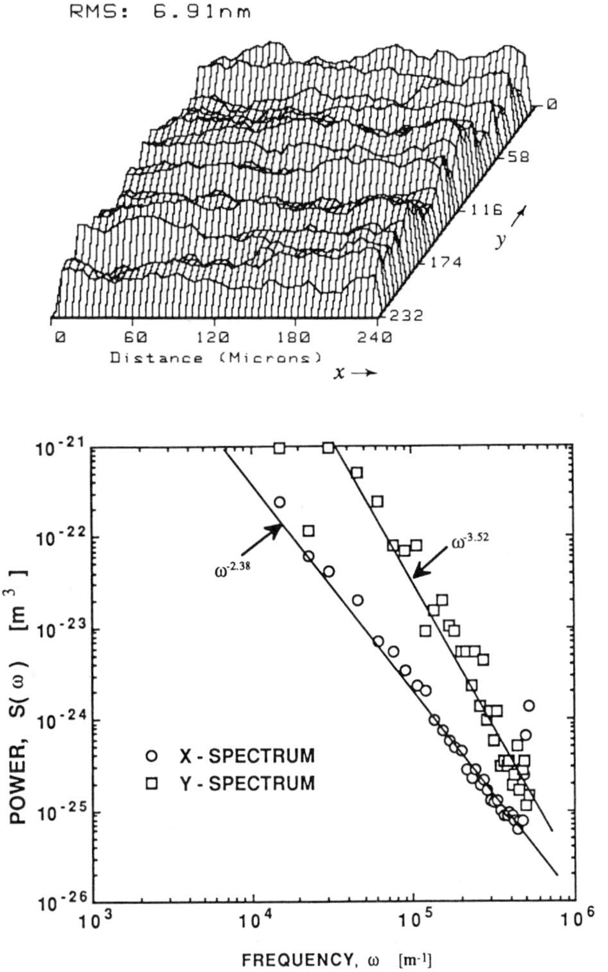

Figure 8 Topography and profile power spectra of textured magnetic thin-film rigid disk surfaces.

averaged structure function of all surface profiles showed a fractal behavior within the length scale 1.5 μm and 10 nm. Figure 9 shows the structure function of annealed and in-situ samples of Y-Ba-Cu-O on a $SrTiO_3$ substrate. It is evident that annealing increases the amplitude of roughness thereby increasing the value of G.

The above examples show that the roughness of a variety of surfaces, processed and unprocessed, follows a self-affine fractal structure in a wide range of length scales. Several examples of the fractal behavior of deposited thin film and polymeric rough surfaces have been discussed in a recent review by Majumdar et al. [43]. The fractal structure can be characterized by scale-independent parameters such as G and D. In Sec. 3.3, the effects of fractal surfaces on thermal phenomena at surfaces are explored.

3.2.3 Simulation of Surface Profiles. In addition to roughness characterization, fractals have also been used deterministically to simulate a seemingly random surface

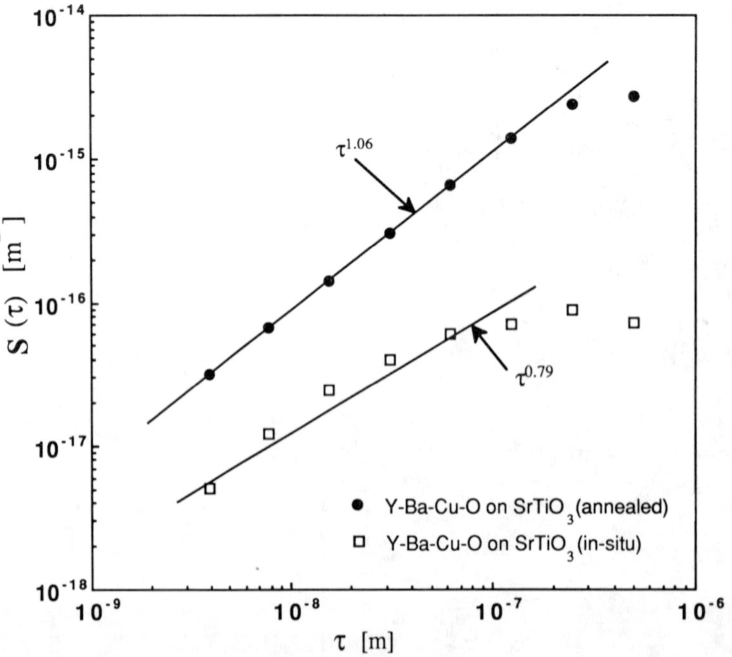

Figure 9 Power spectra of surface profiles of high-T_c thin-film superconductors: ●, Y-Ba-Cu-O on $SrTiO_3$ annealed sample; □, Y-Ba-Cu-O on $SrTiO_3$ in-situ sample.

profile, $z(x)$ [18]. A simple technique is to use the Weierstrass-Mandelbrot (W-M) function [40] given as

$$z(x) = G^{(D-1)} \sum_{n=n_1}^{\infty} \frac{\cos 2\pi\gamma^n x}{\gamma^{(2-D)n}}; \qquad 1 < D < 2; \qquad \gamma > 1 \tag{15}$$

where γ^n are the discrete frequency modes. Although the power spectrum of the W-M function is discrete in frequency, the average spectrum follows that of equation (13). The mathematical properties of the W-M function have been discussed in detail by Majumdar and Tien [40]. Once the parameters D and G are found from the power spectrum or the structure function of a real surface profile, they can be used in equation (15) to artificially generate a rough surface. With $\gamma = 1.5$ in equation (15), the W-M function was used to create the artificial profiles of Fig. 5.

Figure 10 shows a simulation of a lapped stainless surface profile, the structure function of which is given as Lapped-4 in Fig. 6. It can be observed that the structure function does not follow a single power law but has two power laws for $\tau < 40$ μm and for $\tau > 40$ μm. Each power-law region has corresponding values of D and G. To simulate such a profile, the W-M function was split into two parts corresponding to frequencies γ^n less than or greater than 1/40 μm^{-1}. Finally, in each part of the W-M function, the respective values of D and G were used. It is evident that there is a striking resemblance between the real and the fractal profiles. The advantage of such

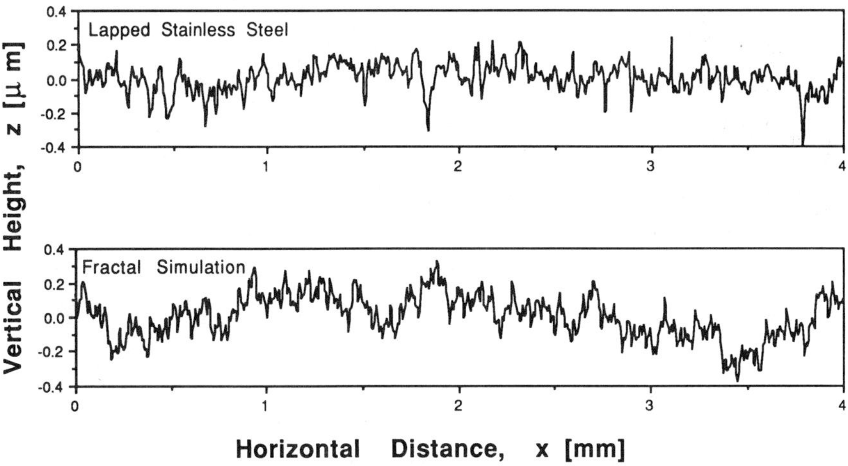

Figure 10 Fractal simulation of a lapped surface profile.

a simulation is that it can be used to simulate a random surface phenomenon that may be difficult to observe or control experimentally.

3.3 Surface Thermophysics

The roughness structure of surfaces influences a wide range of thermal phenomena such as contact conductance, thermal boundary resistance, reflection and emission of radiation, energy losses in superconductors, boiling, and freezing. The applications of fractal characterization of surface roughness to these phenomena are now discussed.

3.3.1 Contact Conductance. When a compressive load is applied between two nominally flat surfaces, the presence of surface roughness produces imperfect contact at the interface. An imperfect contact imposes a resistance to the flow of heat and electricity across the interface. For surfaces in thermal contact this results in a sharp temperature rise across the interface. Such a temperature rise plays a significant role in the thermal performance of microelectronic chips, composite materials, porous insulations, and biomedical and aerospace instrumentation [44]. Previous studies on thermal contact resistance [23, 45–46] have concluded that the nondimensional conductance and the load, under vacuum environment, are related as

$$\frac{h\sigma}{k} = \psi\sigma'\left(\frac{F}{HA_a}\right)^c \tag{16}$$

where h is the contact conductance, σ is the effective standard deviation of the surface heights, k is the effective thermal conductivities, σ' is the effective mean slope of the surfaces, F is the compressive load, H is the surface hardness of the softer material, and A_a is the apparent area of the surface. The parameters ψ and c are empirical constants obtained from experimental observations. From the background on fractal

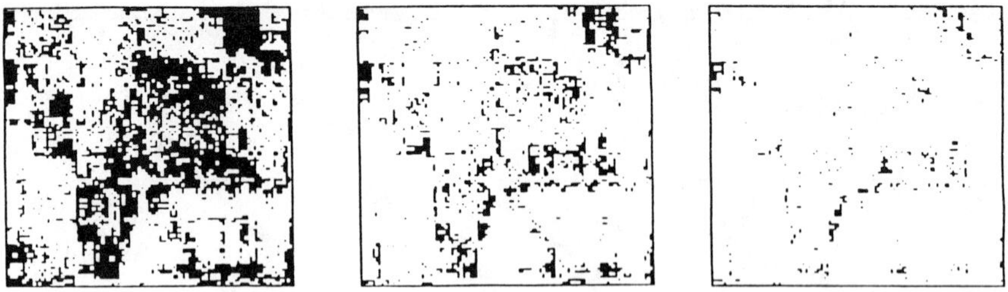

Figure 11 Horizontal cross sections in the x–y plane of a simulated rough surface at various heights above the mean plane $z = 0$.

surfaces, it can be concluded that the mean slope depends on the resolution of the roughness-measuring instrument and therefore should not be used in a contact conductance relation. The value of c is a source of uncertainty, since it varied between 0.85 [45] and 0.99 [46] depending on the surfaces considered. Although empirical studies were made for two decades, the effect of surface topography on contact conductance was not properly understood.

To question the validity of Eq. (16), Majumdar and Tien [47] recently developed a fractal network model which studied the effect of a fractal surface on contact conductance. They visualized the contact interface to be made of a distribution of contact spots spread randomly over the interface, a simulation of which is shown in Fig. 11. The distribution of these contact spots was assumed to be $N(A > a) \sim a^{-D/2}$, where N is the total number of spots of area larger than a and D is the fractal dimension of the profile of a surface. This size distribution is obtained from geomorphology [48], where the islands on the surface of the earth follow the same distribution. Due to its self-affinity, the earth's surface can be visualized as a magnification of an engineering surface. The number-area distribution was verified experimentally by Majumdar and Bhushan [13]. The network model assumed that the contact spots at the interface form a series-parallel network of resistances. The basic resistance element of the network is a roughness of size $z \sim l^{(2-D)} \sim a^{(2-D)/2}$, where l is the length scale of a contact spot of area a. The complete series-parallel network was then solved to obtain a relation of the form

$$\frac{h\sigma}{k} = f(D)\left(\frac{F}{HA_a}\right)^{D/2} \tag{17}$$

where $f(D)$ is a function of the fractal dimension D. Comparison of Eqs. (16) and (17) shows that the network model proves the empirical exponent c to be related to the spectrum of the rough surface profile and equal to $D/2$. Since D varies between 1 and 2, it is expected that c would vary between 0.5 and 1. However, since machining increases the dimension of a surface, the lower limit was observed to be 0.85.

Figure 12 shows a comparison between the experimental observations and the theoretical predictions. It was observed that the theory tended to overpredict the experiments possibly due to several inherent assumptions in the theory. First, the heat conduction was assumed to be one-dimensional and heat spreading in the lateral direction was ignored. This results in the overprediction of conductance, as was observed [40]. Second, most metal surfaces react with the atmosphere to produce an oxide film covering the surface. Since the conductivities of oxides are usually lower than those of metals, these films impose high thermal resistance. Third, the shapes of the cavities were assumed to be parallelopiped, whereas they are actually more conical in shape. A more refined study of contact conductance based on a fractal model is therefore necessary.

3.3.2 Thermal Boundary Resistance. At cryogenic temperatures, the interface between a solid and a liquid or two dissimilar solids poses a resistance to heat transport. The thermal boundary resistance between a solid and liquid helium was first observed by Kapitza [49] and this phenomenon is now called Kapitza resistance. The occurrence of such a resistance is explained by the mismatch of acoustic impedances. Khalatnikov [50] developed the theory of acoustic mismatch to predict the boundary resistance R between a solid and a liquid to vary as $R \sim T^{-3}$. Although the acoustic mismatch theory (AMT) predicts reasonably correct trends at temperatures below 1 K, experiments indicate that the actual resistance is much less than the predictions, and usually show trends different from the T^{-3} behavior. This is evident in Fig. 13, which shows some measurements of the thermal resistance between copper and liquid ^{3}He at different pressures of the liquid [51]. The results are plotted as RT^3 versus T. The predictions

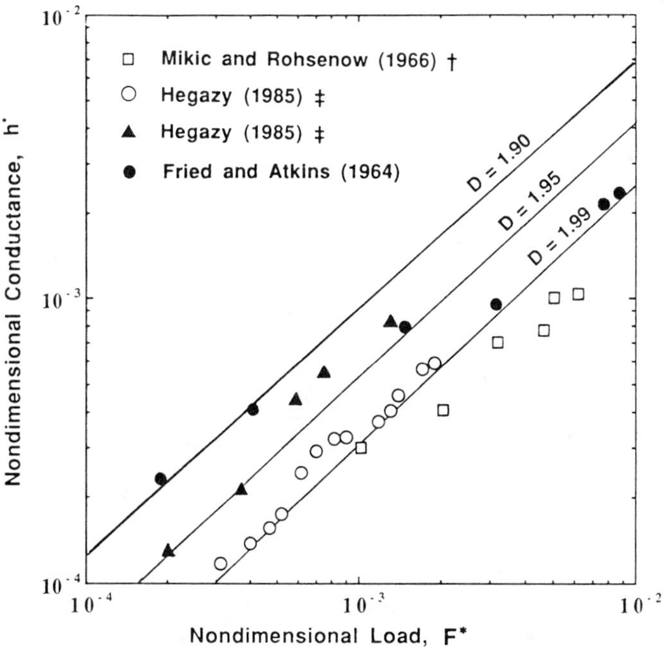

Figure 12 Comparison of network model for contact conductance with previous experimental data [47].

of the AMT would be a horizontal line on Fig. 13. The experiments suggest that first, although RT^3 is a constant below 0.1 K, its value is much less than that predicted by the AMT. For instance, the experiments show that for liquid ^{3}He at a pressure of 0 atm and temperatures below 0.1 K, $RT^3 \sim 200$ [$K^4 - cm^2/W$] whereas the AMT predicts it to be close to 1500 [$K^4 - cm^2/W$]. Secondly, for $T > 0.1$ K the resistance behaves as $R \sim T^{-n}$, where $n > 3$. Several explanations have been proposed for this anomalous behavior and are discussed in an excellent review by Swartz and Pohl [24]. Although it is not certain which effects are dominant, it is clear that surface roughness plays a vital role in reducing the thermal boundary resistance below the predictions of the AMT where such an effect is neglected.

Analogous to the reflection and transmission of electromagnetic waves or photons by a boundary between two media of different refractive indices, lattice vibrations or phonons that are responsible for heat transport also undergo similar wave phenomena at boundaries between two media with different acoustic impedances. The AMT assumes a flat boundary to specularly reflect phonons. However, it is known that the microscale surface roughness can seriously affect phonon transport across the interface.

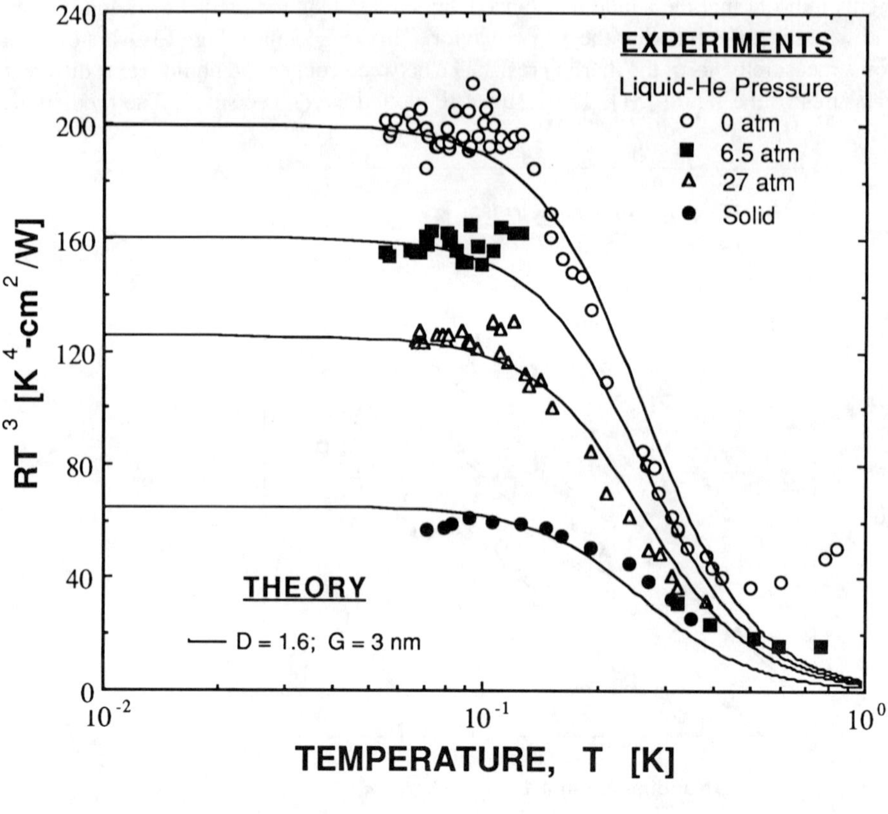

Figure 13 Comparison between theoretical predictions [55] and experimental observations [51] for the product of the thermal boundary resistance and the third power of temperature as a function of temperature.

Most previous studies [52–54] on roughness contribution have focused on scattering of phonons by Rayleigh surface waves coupled through surface roughness. Shiren [54] assumed

$$P(\omega) = \frac{2\pi\sigma^2\ell^2}{(1 + \omega^2\ell^2)^{3/2}} \tag{18}$$

as the power spectrum of a rough surface to model the coupling between the phonons and Rayleigh waves. In Eq. (18), ℓ is the correlation length and the σ is the standard deviation of the surface roughness. By manipulating the values of σ and ℓ, he was able to curve-fit the experimental results for $T > 0.2$ K. However, for $T < 0.2$ K the theory showed trends quite different from the experimental observations in Fig. 13. Therefore, the effect of surface roughness on phonon transmission was not properly understood.

To explain the experimental observations in the whole temperature range, Majumdar [55] recently developed a theory based on the fractal characterization of surface roughness. It was shown that when phonons are incident on an interface, the surface area exposed to the phonons depended on the frequency or wavelength of the phonon. For phonons of longer wavelength a surface appears smoother than for phonons of shorter wavelength. Therefore, short-wavelength phonons are exposed to more surface area than long-wavelength phonons. Since at low temperatures the phonon spectrum follows the Planck blackbody distribution, the wavelength of phonons that are dominant in energy follows the Wien's law. Therefore, as the temperature is increased the dominant phonons have shorter wavelengths. Consequently, for increasing temperature, the surface area exposed to phonon transport increases and the resistance to phonon transport decreases. Majumdar [55] used the fractal characterization to find a relation between the exposed area and the phonon frequency in terms of the roughness parameters G and D. Finally, the simple relation $RT^3 \sim (1 + \eta T^{\beta})^{-1}$ was derived where η is a function of G and β is a function of D. The theoretical predictions are shown as the solid lines in Fig. 13 where it is evident that for $D = 1.6$ and $G = 3$ nm, the predictions are in excellent agreement with the experiments for the whole temperature range. Roughness measurements have shown that these values of D and G are typically observed at microscales for polished surfaces [40].

Although it may seem that thermal boundary resistances are significant only at extremely low temperatures, they play a crucial role in the heat transport in thin films [56, 57] at temperatures of the order of 50 K. When the thickness of the film is of the order of the phonon mean free path, phonon scattering at the boundaries determines thin film conductivities. Flik and Tien [56] assumed diffused scattering at the interface to show the effect of film thickness and temperature on the thin-film thermal conductivity. Since the structure of a boundary can be described by fractal geometry, it is worth exploring the effects of the degree of specular and diffuse reflection of phonons on thermal conductivities of thin films. The operating temperatures of high-temperature superconductors, such as Y-Ba-Cu-O, have been speculated to be around 50 K. Since the use of thin-film oxide superconductors is predicted to have a significant impact on technology [58], it is expected that thermal phenomena such as phonon scattering by boundaries in thin film superconductors [59] will be important.

3.3.3 Effects of Roughness on Surface Radiation. Surface roughness has a strong influence on surface radiative properties such as emissivity, absorptivity, and reflectivity [25]. It is well known that when the roughness of a surface is increased, the reflection becomes more diffused. To analyze the scattering of electromagnetic waves by a random rough surface, Davies [60] proposed the first model which used the Helmholtz-Kirchhoff diffraction theory [61] in conjunction with Kirchhoff's approximation of tangent planes [62]. The randomness of the surface, which poses the main difficulty in modeling this phenomenon, was characterized by statistical parameters such as the standard deviation and correlation of the surface heights. It was assumed that the surface was a stationary random process. However, it was later shown that surface topography is a nonstationary random process [36]. The fractal nature of rough screens and their effect on electromagnetic wave scattering was first analyzed by Berry [63] and by Berry and Blackwell [64], who coined the word "diffractals." The progress in research in this field is discussed in detail in a recent review by Ogilvy [65].

The reflection of radiation by fractal rough surfaces was modeled by Majumdar and Tien [66], who showed that the averaged reflectivity in the specular direction is given by

$$\rho = 2 \sum_{k=0}^{\infty} \frac{(-2^{(4-2D)}\alpha)^k}{k![(4 - 2D)k + 1][(4 - 2D)k + 2]} \tag{19}$$

where α is given by

$$\alpha = \left(4\pi \frac{\sigma}{\lambda} \cos \theta\right)^2 \Gamma(2D - 3) \sin \frac{(2D - 3)\pi}{2} \tag{20}$$

and Γ is the gamma function. Note that σ depends on the fractal parameters G and D and the length of the sample considered, which in this case is the square root of the illuminated area. A comparison with previous experiments [67, 68] in Fig. 14 shows that reflection in the specular direction increases with decreasing values of dimension. The limiting case of $D = 2$ corresponds to the prediction of coherent scattering by the Davies model. Therefore, it was concluded that a decrease in the fractal dimension resulted in more incoherent or diffuse reflection in the specular direction. The experiments were in good agreement for values of $D = 1.5$ and 1.8, which are expected values for machined rough surfaces.

The fractal model of Majumdar and Tien [66] neglected the effects of shadowing, masking, and multiple reflection, which are significant when surface roughness is large compared to the wavelength of radiation. They also give rise to off-specular peaks for reflection at high angles of incidence [69]. More research in this area is therefore required. The effect of shadowing can be modeled by the structure function

$$\langle (z(x_1) - z(x_2))^2 \rangle \approx |x_1 - x_2|^{(4-2D)} \tag{21}$$

where $z(x_1)$ and $z(x_2)$ are two locations on the rough surface. The variance of the slope joining two points separated by x_1 and x_2 is given by

$$\left\langle \frac{(z(x_1) - z(x_2))^2}{(x_1 - x_2)^2} \right\rangle \approx |x_1 - x_2|^{-2(D-1)} \tag{22}$$

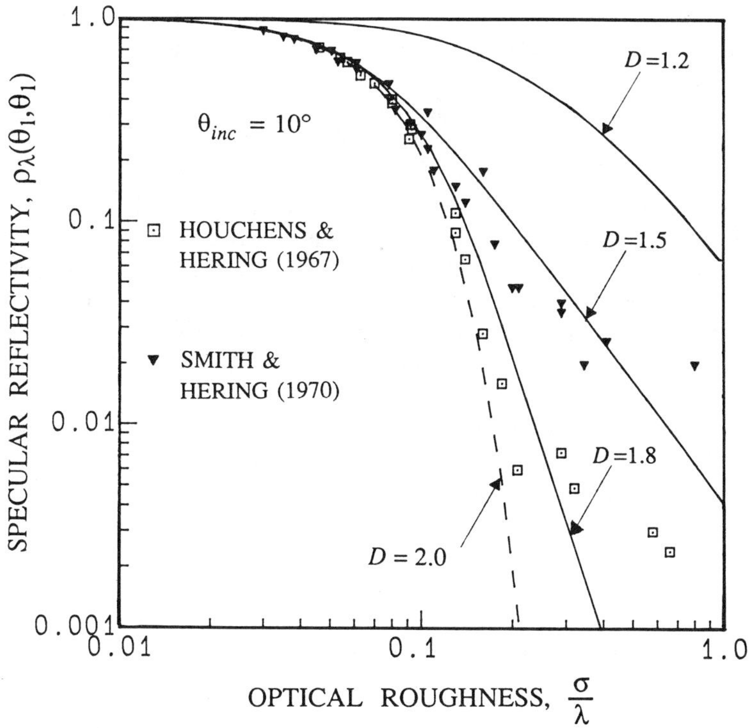

Figure 14 Comparison of experiments [67, 68] with predictions [66] of reflection in the specular direction as functions of optical roughness and fractal dimension.

If it is assumed that the slope of a cord joining two surface points follows a Gaussian distribution with a variance given by Eq. (22), the probability of a cord facing or being in a shadow of the incident beam can be evaluated [70]. The previous fractal model [66] can therefore be appended by a shadow factor which will depend on the angle of incidence and the fractal characterization parameters G and D.

Experiments show that surface roughness increases the directional spectral emissivity of surfaces [71]. It is also observed that the emissivities of rough surfaces increase more rapidly with the angle from the normal to the mean plane than for smooth surfaces. The theoretical explanations for such a phenomenon are still inconclusive. The only analysis of emission by a random rough surface was developed by Birkebak and Abdulkadir [72], who suggested that it is due to the random blockage of surface peaks by adjacent peaks that results in an increase in emissivity. However, their theory is based on the standard deviation of the surface slope, which was experimentally shown to be a scale-dependent parameter theory [13, 38, 40]. Therefore, the prediction of the random blockage theory would depend on the resolution of the roughness-measuring instrument and thus would be inconclusive. The question still remains as to why the emissivity increases with surface roughness. Recent experiments on microgrooved surfaces have shown rather interesting results which may explain such an increase. Lofving [73] measured the absorption of a surface having periodic grooves in the shape of a triangular wave of period Δ. He found sharp peaks in absorption for angles θ satisfying the relation $\sin\theta = 1 - n\lambda/\Delta$, where $n = 1, 2, \ldots$. He explained this phenomenon as possibly caused by a coupling between thermally excited surface plasmons and free propagating waves on the rough surface. More recently, Hesketh et al. [74, 75] showed a similar "organ pipe" mode of emission for a heavily doped microgrooved semiconductor surface. They explained this behavior by a classical treatment of radiative exchange in enclosures, in this case periodic grooves, with either diffuse or specular walls. Although the predictions of this model match quite well with the theory, the possibility of noncontinuum or non-Planckian radiation still exists. Since both experimental studies on periodic microgrooved surfaces show increase in emissivity at particular angles, it may give clues as to why the emissivity of a random rough surface is more than that of a smooth surface. A random rough surface is multiscale in structure with an infinite number of wavelengths. Although roughness of each wavelength produces peaks at different angles, emission in a particular angle will consist of the cumulative effect of emission from all wavelengths. The fractal characterization of random rough surfaces provides the structure of roughness or microgrooves of any wavelength and can therefore provide an opportunity to model the effect of random rough surfaces on surface emissivities. However, the physics of the phenomenon—coupling of surface plasmons with surface waves, classical radiative exchange, or non-Planckian radiation—still remains unresolved.

3.3.4 AC Losses in Superconductors. Alternating magnetic fields have been observed to produce energy losses in low-temperature superconductors such as Nb and Nb_3Sn [27]. It was shown experimentally that losses were caused mainly by surface roughness [76] and grain sizes [77] of the superconducting material. In superconducting systems these losses are manifested as the power needed to run the cooling system. It is predicted that these losses will be more drastic in the high-temperature oxide superconductors due to the anisotropy in their critical current density J_c [27].

The questions which one may ask are what are the mechanisms for such energy losses and what is the effect of surface roughness. When an applied magnetic field H penetrates a superconductor, it induces a magnetic field B inside the superconductor which results from an opposing current flow on the surface and, for type II superconductors in the bulk. For a time-varying magnetic field H, the inherent magnetic hysteresis of the superconductor results in a loss of energy Q (W/m^2) given by

$$Q = \int H \, dB \tag{23}$$

and is a source of heat. Beam [78] assumed that the induced current inside a superconductor with a smooth surface is the maximum corresponding to its critical current density J_c. He developed a simple model to show that the energy losses were proportional to H_p^n, where $n = 3$ and H_p is the peak magnetic field. However, it is observed experimentally that surface roughness tends to decrease n and influence energy losses [77]. An elaborate study on the effects of surface roughness was performed by Bussiere and Suenaga [76], who showed that mechanical polishing, elec-

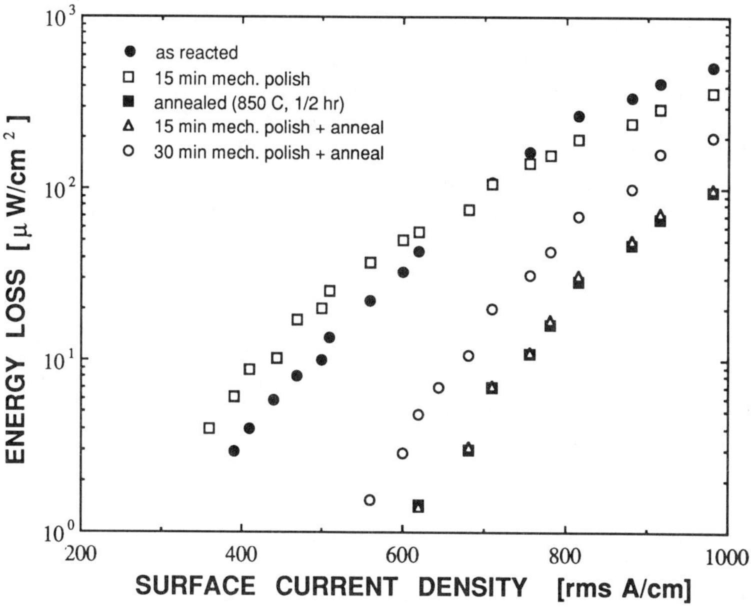

Figure 15 AC losses at 60 Hz of Nb_3Sn as-reacted sample for various surface treatments [76].

tropolishing, and annealing can reduce losses by about two orders of magnitude, as observed in Fig. 15. They concluded that the increase in losses for rough surfaces is due to the porosity of the surface, which reduces J_c. Mechanical polishing also damaged the bulk material beneath the surface but could be removed by annealing. This study showed qualitatively the dependence of energy losses on surface roughness, which was measured by a profilometer or shown by scanning electron micrographs. However, there is no theory that quantitatively predicts the effect of roughness on energy losses. The main difficulty lies in quantifying the structure of surface porosity or roughness and finding the dependence of J_c on the roughness-characterizing parameters. Since the fractal characterization has been able to quantify the structure, it could be used to find the roughness dependence of J_c.

Although the decrease in J_c on a rough surface is one explanation for an increase in losses, there is a second possibility. The losses in Fig. 14 are measured as watts per unit area. The area that is used in this case is the apparent area of the surface. However, it is known that since rough surfaces are fractal, the area will depend on the scale of measurement. It is found that the losses increase with surface roughness. What may be happening is that currents on the superconductor surface follow the topography of the surface, that is, the mountains and valleys. The area covered by the current path will be more for a rough surface than for a smooth one. Therefore, instead of measuring the losses in terms of watts per unit apparent area, what could be used is the actual area that the current has to flow across. Therefore, if the losses are normalized by the actual area, then the curves in Fig. 14 could possibly collapse into one curve. The role of fractals in modeling this phenomenon is to determine the actual area as a function of the roughness of the surface and can therefore be used to collapse the data. If the data do collapse, then it proves that it is the increase in actual area that is responsible for the increase in losses.

The discovery of high-temperature superconductors is expected to have considerable impact on power transmission and electronic devices such as SQUIDS and antennas, in which alternating magnetic fields are present. The prediction of energy losses due to surface roughness will greatly enhance the design of cooling systems for thermal stability. Fractal geometry can be used to quantify the surface porosity or to measure the actual surface area of the current flow which are expected to be responsible for energy losses.

3.3.5 Boiling. Boiling heat transfer has been studied for nearly five decades, and the literature on various aspects of boiling is quite overwhelming [79]. It is well known that the presence of a surface decreases the superheat requirement for nucleation [80] and gives rise to heterogeneous nucleation. In pool and forced boiling, the cavities in a surface trap vapor in them and provide nucleation sites for bubbles to form. Although extensive research has been conducted in boiling, the correlations among the heat flux, the superheat, and other parameters are nearly always empirical [81]. They are listed in texts [26, 82] and in a review paper [79] and therefore will not be discussed here. As an example, consider one of the earliest correlations, given as [83]

$$\frac{C_l(T_w - T_{\text{sat}})}{h_{fg}} = C_{\text{sf}}\left\{\frac{q}{\mu_l h_{fg}}\left[\frac{\sigma}{g(\rho_l - \rho_g)}\right]^{1/2}\right\}^r \left(\frac{C_l \mu_l}{k_l}\right)^s \tag{24}$$

where C_l, μ_l, k_l, and ρ_l are the specific heat, dynamic viscosity, thermal conductivity, and density, respectively, of the liquid; T_w and T_{sat} are the wall and saturation temperatures; h_{fg} is the latent heat of vaporization; q is the applied heat flux; σ and ρ_g are the liquid surface tension and vapor density; and g is the acceleration due to gravity. The empirical constants C_{sf}, and the exponents r and s, depend on the fluid and surface combination. Although the fluid properties are well represented in Eq. (24), surface parameters are all absorbed in the empirical constants. What are the effects of surface roughness on pool boiling? At present, there is no satisfactory way to quantitatively include the effects of solid surfaces in any correlation [79]. The reason for such a dilemma is that there is no way to quantify the roughness of surface such that the shape and distribution of active nucleation sites can be predicted. Most investigations usually provide the surface machining process, such as sandblasting, finishing by emery paper, grinding, or electropolishing. However, the effects of these processes on the sizes and the number of surface cavities are not known. Therefore there seems to be a lack of quantitative knowledge of the influence of machining process on boiling phenomena. This is probably the biggest obstacle preventing the development of a single theory for nucleate pool boiling. The other uncertainties which a solid surface poses are the size of a bubble at breakoff and the frequency of bubble generation at a nucleation site. The possible applications of a quantitative fractal characterization of rough surfaces to the analysis of boiling heat transfer are now explored.

Consider nucleate pool boiling of a pure liquid from a horizontal solid surface. The range of cavity sizes, of mouth radius r_c, that will actively generate bubbles can be found from the model of Hsu [82]. However, the number of cavities of a particular size on a given rough surface is still an unknown factor. In the analysis of contact conductance, Majumdar and Tien [40] showed that the distribution of contact spots followed a power-law relation of the form $N(A > a) \sim a^{-D/2}$, where N is the total number of spots greater than area a and D is the fractal dimension of the surface profile and is related to the surface dimension D_s by $D_s = D + 1$. This is the same as the distribution of islands on the earth's surface, where the islands are formed by taking a horizontal cross section of a rough surface at a particular height. If, however, the rough surface is turned upside down, the mountains which are cut to form islands become cavities on the surface. Therefore the cavities on a surface should also follow the distribution

$$N(R > r) = \left(\frac{r_{\max}}{r}\right)^D \qquad r_{\min} \leq r \leq r_{\max} \tag{25}$$

where $r_{\max}$ and $r_{\min}$ can be predicted by the Hsu model. Experiments show that such a power-law behavior of nucleation site distribution was observed [84, 85], and is shown in Fig. 16. Such distributions were conventionally characterized by an exponent m, which was determined by empirical correlations. Mikic and Rohsenow [86] used this to develop a correlation for nucleate pool boiling in which they showed that the relation between heat flux and wall superheat is $q \sim \Delta T^{m+1}$. This type of power-law behavior was observed in previous experiments [83, 87]. Since Eq. (25) can now be

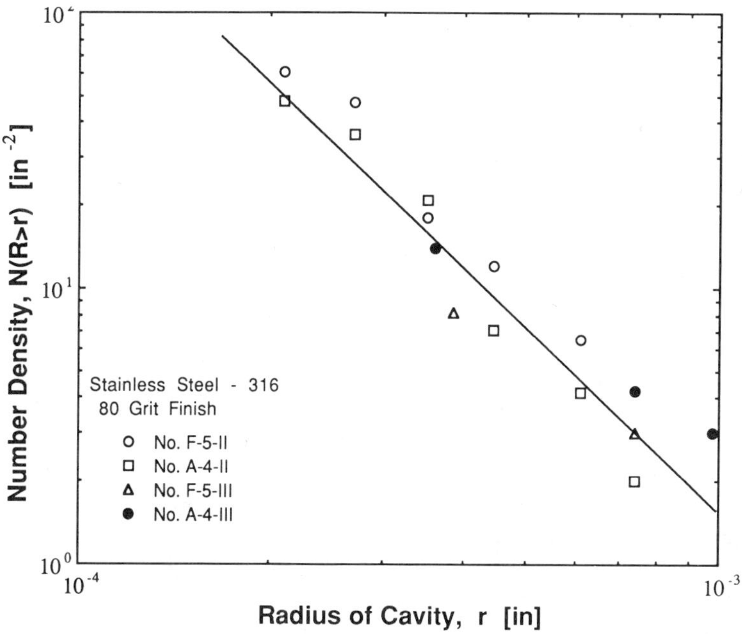

Figure 16 Power-law behavior of nucleation site distribution in pool boiling [84, 85].

used for the size distribution, the fractal dimension of a surface could have a direct effect on nucleate boiling heat transfer. In addition, if the fractal parameters G and D of a surface are known, the depth z of a cavity of size r can be determined by the relation $z = G^{D-1}r^{2-D}$, which follows directly from the fractal characterization. From the depth and the radius of a cavity, the angle of a cavity can easily be obtained. Thus the fractal characterization provides details not only about the size distribution of the active cavities but also about the geometry of a cavity of any size. Since there are no experiments available which present results in terms of fractal surface parameters, it is difficult to verify this speculation. However, the possibility of quantifying surface roughness at any length scale appears to have promising applications in nucleate boiling.

Another aspect of nucleate boiling phenomena in which fractal geometry could find applications is the problem of bubble emission frequency. Conventionally, the frequency of bubble emission is related to the waiting time t_w for the conduction layer to grow in the liquid adjacent to the surface, the growth time t_g for the bubble to grow from vapor entrapped in a cavity to the point of departure, and bubble departure diameter D_d [79]. In all investigations of bubble emission frequency it is assumed that for a particular heat flux or wall superheat, the frequency remains constant. This suggests that bubble emission is a periodic behavior. It should be noted, however, that the phenomenon of bubble emission from a nucleation site is analogous to the "dripping faucet" problem, which is known to show chaotic behavior [88]. The time between successive drops in a leaking faucet is quite regular when the flow rate is small. However, as the flow rate is increased, there exists a regime where the drops still fall separately but in a never-repeating pattern. The change between periodicity and seemingly random patterns is an example of transition to chaos. The reason for such a transition lies in the nonlinear behavior of surface tension forces involved in droplet formation and breakoff from a faucet. When the flow rate is low, formation of a droplet is not affected by the breakoff of the previous droplet. The liquid meniscus gets enough time to relax into a stable position and all the influence of the previous droplet is lost. However, when the flow rate is increased, the receding meniscus after a droplet breakoff does not get enough time to relax before the next droplet starts to form. The periods of two or even three droplets start to overlap, giving rise to chaotic behavior. Although it may seem that if the period of the nth droplet t_n is plotted against the period of the previous droplet t_{n-1}, the points will be distributed randomly, it is interesting to note that the points form a regular horseshoelike strange attractor [88]. In nucleate boiling, such a situation could possibly be encountered by increasing the heat flux or wall superheat before it reaches the film boiling regime. Here again, if the bubble growth rate overlaps with the breakoff of the previous bubble, it might lead to nonperiodic behavior of bubble emission. This type of phenomenon will be more pronounced when the waiting period is much smaller than the growth period, since the latter involves surface tension forces. A strange attractor of a chaotic phenomenon is characterized by the fractal dimension of the attractor [89, 90]. Therefore, fractal geometry can be applied to determine the distribution of bubble emission frequencies during nucleate boiling.

4 COMBUSTION

An understanding and control of combustion in fires was one of the earliest scientific endeavors of mankind. Although combustion is a complex physical phenomenon involving chemical reaction, heat transfer, and fluid dynamics, there are many types of combustion—laminar flames, for example—that are now quite well understood and easily modeled by a set of tractable equations [91]. However, the complexity of combustion in a *disordered* medium poses severe challenges for scientists and engineers to understand the basic physical mechanisms involved. Some examples include premixed and nonpremixed turbulent combustion and combustion in a porous medium. In turbulent combustion, the disorder of fluid turbulence plays a major role in the structure of a flame and the mixing of chemical species. The disordered structure of a solid porous medium also enhances dispersion of species and provides more surface area for surface catalysis. Although these combustion processes are of great technological importance, their analysis is invariably limited by the lack of methods to quantify disordered structures. In the last few years, however, fractal geometry has cast new light on turbulent premixed combustion [11, 92]. This section reviews some of the studies on applications of fractals in combustion and explores new areas where fractals may give insight into the physics or provide mathematical models with which to analyze complex problems. The applications of fractals in modeling flow in porous media are described in Sec. 5.

4.1 Turbulent Combustion

Turbulent combustion can be classified into two categories—diffusion flames and premixed flames [91]. In the first category, the fuel and oxidizer do not come in contact with each other before the combustion takes place. For such a case, turbulence in the fluid flow helps in diffusing the chemical species. For turbulent premixed combustion, the fuel and the oxidizer are mixed before the combustion occurs. Most applications of fractal geometry have been in this type of combustion where the flame front becomes wrinkled and multiscale in structure due to turbulence. To understand the role of turbulence in combustion, it is first necessary to identify some unique characteristics of turbulent fluid flow.

4.1.1 Turbulence Scales and Fractals. If a velocity measurement is taken in a turbulent flow, it is found that the velocity v is a fluctuating quantity in time and space. Such an oscillating velocity field is usually characterized by the energy spectrum $E(k)$, where E is the turbulent energy of an eddy of size $1/k$, and k is the wavenumber of an eddy. The energy spectrum is obtained by taking a power spectrum of the velocity field. The span of the spectrum over a range of wavenumber shows that turbulence consists of a multiscale structure of eddies. The energy spectrum of a turbulent flow possesses some unique characteristics which suggest some scaling relation between the energies of eddies of different sizes. The length scale of eddies where most of the turbulent energy is concentrated is called the integral length scale ℓ. The Kolmogorov length scale ℓ_k represents the smallest eddy size. In eddies smaller than the Kolmogorov scale, viscous forces are dominant over inertia forces. The energy spectrum between the integral and Kolmogorov length scales, that is, $\ell < 1/k < \ell_k$, follows a power-law behavior of the form $E(k) \sim k^{-5/3}$, which was analytically derived by Kolmogorov and is often called the inertial subrange. In this region, there is a cascade of energy from larger eddies which break up due to instabilities into smaller eddies. The turbulence in the inertial subrange is isotropic and homogeneous and is therefore the simplest form of turbulence. Anisotropy in turbulence, caused by boundaries, occurs near the integral scale.

The power-law behavior of the energy spectrum in the inertial subrange suggests that the velocity field is fractal in structure. Mandelbrot [93] showed that iso-surfaces of scalar quantities, such as temperature or concentration, are fractal in structure and suggested that the fractal dimension of Gauss-Kolmogorov turbulence is 7/3 whereas for Gauss-Bergers turbulence it is 5/2. In addition, he suggested that the Kolmogorov exponent of 5/3 should be changed to $5/3 + (3 - D)/3$, which takes into account the intermittency of turbulence [94]. However, in recent years, Chorin [95, 96] has shown that the Kolmogorov exponent of 5/3 already takes intermittency into account and therefore should not be changed. He showed that the shape of a vortex tube in a homogeneous and isotropic turbulence is analogous to the structure of a polymer and can be modeled by a self-avoiding walk of fractal dimension 5/3 as opposed to a Brownian random walk of dimension 3/2. He suggested that the Kolmogorov exponent of 5/3 is the fractal dimension of the velocity trace $v(x)$. A comparison of the energy spectrum $E(k) \sim k^{-5/3}$ with the power spectrum of Eq. (13) indicates that $D = 5/3$, confirming Chorin's hypothesis.

Not only has theoretical modeling shown that turbulence is fractal, experiments on atmospheric turbulence have also confirmed the fractal nature. The most famous one is that of Lovejoy [97], who found the perimeters of clouds to be fractal of dimension $D = 1.35$ over about five decades of length scales. If the turbulence is isotropic, which it is in the inertial subrange, then the dimension of iso-surfaces is 2.35. Lovejoy suggested that the fractal dimension of clouds is a consequence of turbulent diffusion in the atmosphere. Subsequently, Hentschel and Procaccia [98] presented a theory of relative turbulent diffusion which proved that the dimension of 2.35 is universal and is therefore manifested in the shape of clouds. Later, Sreenivasan and Meneveau [10] conducted laboratory experiments to show the fractal dimension of a turbulent interface was 2.40 for a boundary layer, 2.32 for an axisymmetric jet, 2.37 for a plane wake, and 2.40 for a mixing layer. These developments in the recent past paved the way for applications of fractals in turbulent combustion.

4.1.2 Turbulent Premixed Flames. To discuss the role of fractal geometry in turbulent combustion, it is first necessary to understand a regime map concerning length and time scales of turbulence and chemical reaction. Figure 17 shows a map of this kind [91]. The horizontal axis is the ratio of the Kolmogorov scale to the thickness of the flame δ. The vertical axis is the ratio of the mean velocity fluctuation $\sqrt{2q}$, where q is the variance of the velocity fluctuation, and the laminar flame speed v_o.

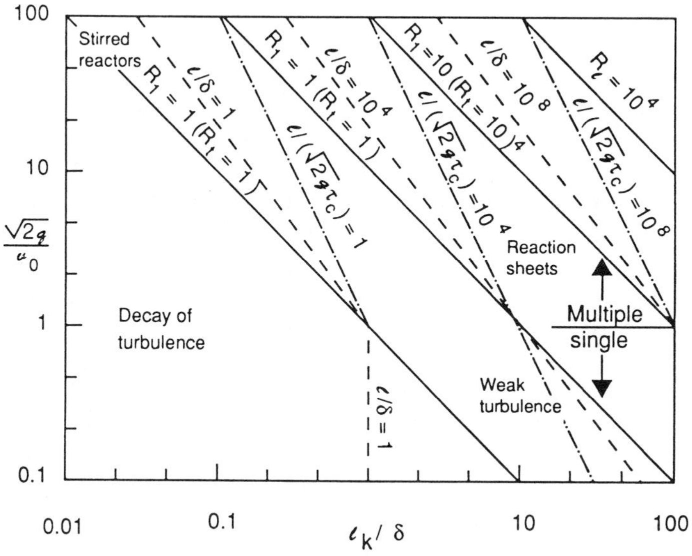

Figure 17 Regime map of premixed turbulent combustion [91].

The flame thickness and the flame speed are related by $v_o = \delta/\tau_c$, where τ_c is the chemical conversion time. The turbulence Reynolds number based on the integral scale is $R_1 = [(\sqrt{2q}/v_o)(\ell_k/\delta)]^4$. The Damköhler number, which is the ratio of time scales of transport properties and of chemical reaction, is equal to $(\ell_k/\delta)^2$. In Fig. 17, high values of ℓ_k/δ suggest that the flame thickness is smaller than the smallest scale of turbulence. Therefore, the flame front exists within the smallest eddy and the transport phenomena are controlled by a laminar diffusion process. In this regime of combustion, the role of turbulence is to distort and wrinkle the flame front into a convoluted reaction sheet. The structure of this sheet follows the iso-surface structure of turbulence and is therefore fractal in nature. For low values of turbulent fluctuation, or q, the flame front appears as a simply connected single sheet, whereas for higher fluctuations the flame sheet may collide at certain locations to form multiply connected sheets. The other extreme regime occurs when the flame thickness is much larger than the integral scale $\delta \gg \ell$ and the combustion is a stirred reaction process. In this case, turbulent diffusion plays a major role in enhancing the mixing process. Since turbulent diffusion is known to have the universal fractal dimension of 2.35 [97, 98], the application of fractal geometry in this type of combustion seems possible. In the intermediate region, $\ell_k \ll \delta \ll \ell$, a combination of both turbulent diffusion and flame wrinkling influences the combustion process.

The application of fractals to premixed turbulent flames was recently initiated by Gouldin [11] and has spawned several new investigations. All these studies have focused their attention on the reaction sheet regime of turbulent combustion. Gouldin's study is based on the observation of Damköhler that the turbulent flame speed v_t is related to the laminar flame speed v_o by $v_t/v_o = A_i/A_o$, where A_i is the average area of the wrinkled flame sheet and A_o is the cross-sectional area of the turbulent flow. The fractal nature of the flame sheet suggests that the area of the surface depends on the scale of measurement ε, by $A \sim \varepsilon^{D-2}$. It was assumed that the inner or smallest scale cutoff was the Kolmogorov scale, whereas the outer or largest scale cutoff was the integral scale. Therefore, the ratio of the areas was found to be $A_i/A_o = (\ell/\ell_k)^{D-2}$, where $\ell/\ell_k \sim R_1^{3/4}$. It was speculated that for small levels of turbulent fluctuation, that is, small $\sqrt{2q}/v_o$, the inner cutoff is increased. The dimension in this study was assumed to be 2.37 based on the average of cloud dimension [97] and that of jet boundaries of free shear flows [10]. Experiments on methane-air V-flames [99] have shown that although the inner and outer cutoffs were in agreement with the theory, the dimension was found to be 2.1, well below that assumed for turbulence. This disagreement was conjectured to be due to the influence of turbulence intensity. Similar experimental results were reported by Murayama and Takeno [100], who found the fractal dimension to be lower than that of unreacting flows but increasing with turbulent intensity. This was also observed in premixed combustion in an internal combustion engine, where an increase in $\sqrt{2q}/v_o$ from 0.5 to 4.0 increased the fractal dimension of stoichiometric flames from 2.12 to a limiting value of 2.37 [92]. For a lean condition with equivalence ratio equal to 0.59, the fluctuations satisfied the condition $\sqrt{2q}/v_o \gg 1$, for which the dimension was found to be 2.36. Their observations on the structure of flame front for three different engine speeds and two equivalence ratios are shown in Fig. 18. It is evident that for increasing engine speeds and decreasing equivalence ratio, the flame structure appears more wrinkled, producing multiple reaction sheets. The limiting values of the dimension are quite close to those observed by Lovejoy [97] and by Sreenivasan and Meneveau [10]. However, the influence of $\sqrt{2q}/v_o$ on the fractal dimension of flame surfaces still remains unknown. It is possible that the Gibson scale $\ell_g = \ell(v_o/\sqrt{2q})^3$, and not the Kolmogorov scale, determines the inner cutoff of the flame structure, as suggested by Peters [101]. An investigation on the

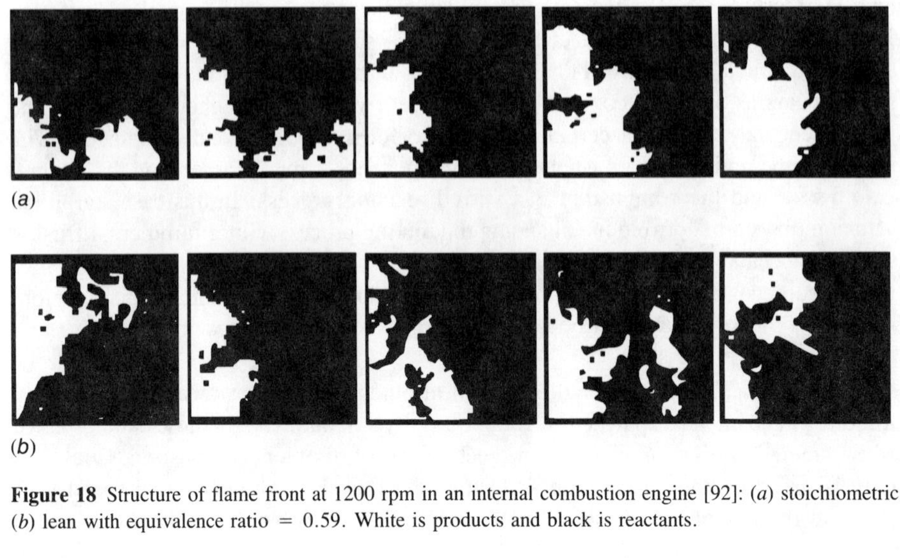

Figure 18 Structure of flame front at 1200 rpm in an internal combustion engine [92]: (*a*) stoichiometric; (*b*) lean with equivalence ratio = 0.59. White is products and black is reactants.

interactions of turbulent fluctuations, chemical reaction, and heat generation, with the fractal dimension and cutoff length scales of the flame structure, is warranted.

4.1.3 Turbulent Diffusion Flames. To the author's knowledge, applications of fractals in turbulent diffusion flames have not been reported so far. Most conventional analyses in the literature of turbulent diffusion flames deal with probability distributions of the temperature and mass fractions of the fuel and oxidizer [91]. These probability distributions are then used in differential equations of mass, energy, and species conservations to obtain Favre averages of the relevant quantities. Depending on the type of turbulent flows—mixing layers, jets, wakes, and boundary layers—the probability distributions have unique signatures. Although the distributions are flow-dependent, there could be a fundamental universality underlying these differences. The experiments of Sreenivasan and Meneveau [10] on boundary layers, jets, wakes, and mixing layers show that the interface between the turbulent and nonturbulent regions, as well as the structures of material lines, are fractal in nature. The dimensions of the interfaces in these structures lie between 2.32 and 2.4, which possibly suggests such a universality. The applications of fractals in quantifying the structure of turbulent diffusion flames therefore seems logical.

4.2 Soot Agglomeration

The combustion of fossil fuels produces ultrafine soot particles which evolve out of a complex growth process involving nucleation, surface growth, and agglomeration [102]. Agglomerated soot particles play a significant role in emission, absorption, and scattering of radiation during combustion. In addition to the optical properties of soot [103], the structure of soot agglomerates influences scattering and absorption. For determining the optical properties of soot agglomerates, conventional theories have modeled them as spheres forming either a closely packed structure with rotational symmetry or chains oriented randomly in space [104, 105]. However, transmission electron microscopy of minute soot agglomerates has shown that the structure is neither closely packed nor a long chain but an open-branched structure [12, 106]. Such structures follow a power-law pair correlation of the form of Eq. (7) and are therefore self-similar fractals. It is observed that under certain conditions, soot agglomerates have a sparse structure containing branches of long chains, whereas sometimes they are more compact [106]. The ability of fractal geometry to quantify the compactness or spareness of an agglomerate is a powerful tool for studying the dynamics of agglomeration and its interactions with the combustion process. Recent developments in computer simulation of growth and aggregation have given key insight into the physics governing the process of soot agglomeration. The following discussion will briefly explain these models and then explore the possibilities of applying the concepts to study combustion.

4.2.1 Aggregation Models. The first model, developed by Witten and Sander [107], is called the diffusion-limited aggregation (DLA) model. In this, a two- or three-dimensional lattice is considered in which a seed particle is placed at the lattice point

Figure 19 Computer simulation of two-dimensional DLA structure producing an aggregate of $D = 1.71$ [17].

in the center. A second particle is then allowed to wander randomly in this lattice, thus representing a diffusion process. If this particle touches the seed particle, it forms a cluster and does not move. Then a third particle is left at the boundary of the lattice and allowed to walk randomly. If a particle reaches the lattice boundary instead of the cluster, then it is removed and a new particle is introduced. If this process is continued it produces a cluster as shown in Fig. 19, of fractal dimension $D = 1.70$ in two-dimensional space and $D = 2.53$ in a three-dimensional space. The fractal dimension can be determined either by the power-law correlation function method or by counting the number of particles within a radius L, as described in Sec. 2. A variant of the DLA model considers the probability of a particle sticking to the cluster to be less than 1 [108]. It was observed that as this probability is decreased, the structure becomes more compact.

A generalization of the DLA model is the dielectric-breakdown model (DBM), which considers the electrostatic potential of the cluster to be 0 and that of the lattice boundary to be 1 [109]. At every step, the model evaluates the potential of all the lattice points by solving the Laplace equation, $\nabla^2\phi = 0$ in conjunction with the boundary conditions of the cluster and the boundary. If there are N lattice points adjacent to the cluster, the electric field at the jth adjacent point can be determined by $\mathbf{E}_j = -\nabla\phi_j$. The probability p_j for the cluster to form the jth bond is given by

$$p_j = \frac{|\nabla\phi_j|^\eta}{\sum_{j=1}^{N} |\nabla\phi_j|^\eta} \tag{26}$$

where the parameter η allows for nonlinear dependencies between the probability and the local field. For $\eta = 0$, the probabilities are independent of the previous cluster shape and therefore a compact cluster of $D = 2$ is produced in a two-dimensional space. The other limiting case of $\eta \rightarrow \infty$ produces a straight chain of $D = 1$. Therefore, it should be noted that compact clusters and straight chains considered in previous radiation scattering and absorption models [104, 105] correspond to these limiting cases. The value of $\eta = 1$ corresponds to the DLA model. The relation between D and η is presented in a recent review by Pietronero et al. [110], which discusses the theory of Laplacian fractals underlying the DBM model.

The model which seems to be most applicable to the soot agglomeration is the cluster-cluster aggregation (CCA) model [111, 112]. A two-dimensional simulation of CCA considers a square lattice with particles filling only a fraction ρ of lattice points. At each time step, a particle or a cluster is selected randomly and moved by one lattice unit in a random direction. When two clusters touch each other, they become attached. Figure 20 shows the evolution of clusters joining other clusters to form larger agglomerates. As time proceeds, the number of clusters decreases, and large open-branched structures appear. The dimension of the clusters have been found to be $D = 1.44$ in 2-D simulation and $D = 1.78$ in 3-D simulation. The mobility of a cluster can be varied by defining a size-dependent diffusion coefficient as $\mathcal{D} \sim s^{\gamma}$, where s is the number of particles in a cluster and $\gamma < 0$. Thus smaller clusters are allowed to move faster than larger ones. Details on the relation between γ and D can be found in Jullien [113] and Vicsek [19].

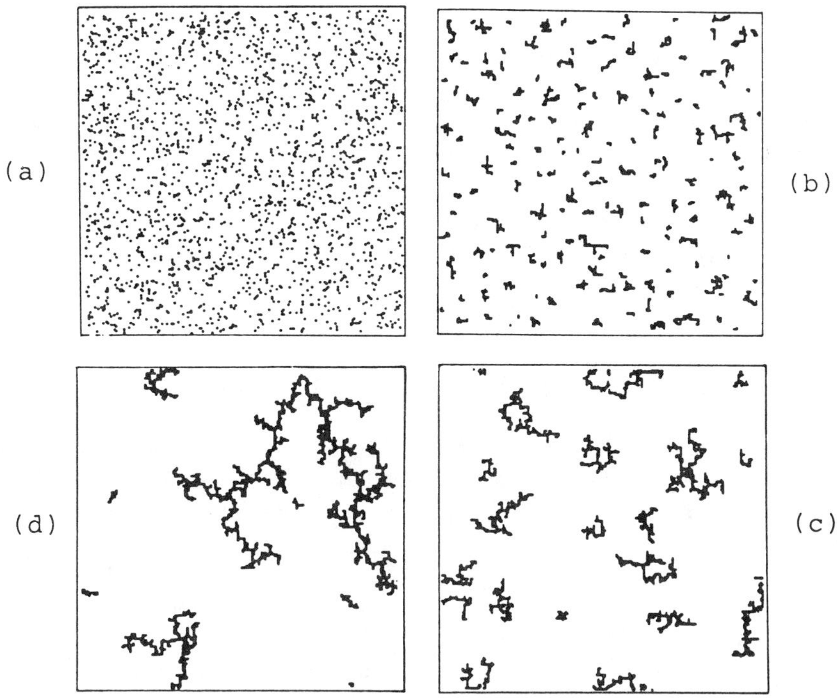

Figure 20 Computer simulation of CCA showing evolution of diffusion-limited clusters at increasing times (*a*)–(*d*) [19].

Whereas the fractal dimension provides information regarding the compactness of an agglomerate, its characteristic size is represented by the radius of gyration R_g, defined as

$$MR_g^2 = \sum_{i=1}^{N} m_i r_i^2 \tag{27}$$

where M is the total mass of the cluster and m_i is the mass of particle i whose radius is m_i from the centroid.

4.2.2 Application to Combustion. The fractal dimension determines the compactness of a soot agglomerate—a value close to d, the dimension of the space, suggests a compact structure, whereas a value close to 1 implies a straight-chainlike structure. The dimension can provide some insight into the physical process of agglomeration. As suggested by the DBM model, a chainlike structure is formed when the growth occurs mainly at the tip of a branch. This is likely to occur when the attractive forces between the clusters are strong. The strength of these forces depends on the surface energies or the thermodynamic and chemical state of the cluster. When the sticking probability is low, that is, when the attractive forces are small, then the structure is more compact. Therefore, the chemistry of the combustion process should have some influence on the structure of soot agglomerates. Since the fractal dimension is a good indication of the strength of attractive forces, there is possibly a connection between the chemistry of combustion and the fractal dimension.

It was found experimentally that soot agglomerates produced by combustion of acetylene in a coannular diffusion burner have a fractal dimension between 1.76 and 1.90 [12]. The dimension of soot agglomerates in a 3-D space appears to be close to that obtained by the CCA model. Physically, the CCA model best represents the scenario in a combustion process where small carbon particles, first formed by nucleation and condensation, move by diffusion and then form small clusters with other carbon particles. These small clusters further diffuse to form larger clusters, and the process goes on. The dynamics of the agglomeration of a cluster of mass i and one of mass j is given by the Smoluchowski equation [114]:

$$\frac{dn_s}{dt} = \frac{1}{2} \sum_{i+j=s} K_{ij} n_i(t) n_j(t) - n_s(t) \sum_{j=1}^{\infty} K_{sj} n_j(t) \tag{28}$$

where $n_i(t)$ is the number of clusters of mass i at time t. The probability of an i cluster combining with a j cluster is proportional to K_{ij}. If the cluster diffusivity depends on the mass as i^{γ}, then one can expect the combination rate to follow the relation [19]

$$K_{ij} \sim (i^{1/D} + j^{1/D})^{d-2}(i^{\gamma} + j^{\gamma}) \tag{29}$$

Since experiments show that the fractal dimension of soot agglomerates lie close to 1.78, the size distribution of agglomerates can be found as a function of time from Eq. (28). Further, if the velocities of the combustion products are known, the spatial distribution of soot agglomerates can also be determined. The knowledge of the spatial distribution allows one to evaluate the radiative energy transfer from any location of a flame.

The optical properties of fractal soot agglomerates, which are needed for evaluating radiative transfer, were obtained by Berry and Percival [115], who found that if $D > 2$, the scattering cross section grows as $N^{1-2/D}$. However, for $D < 2$ and $kR_g \gg 1$ the cross section grows to a saturation value which exceeds the cross section of an individual sphere by a factor $(kr)^{(3-D)}$. Here $k = 2\pi/\lambda$, λ is the wavelength of incident radiation, and r is the radius of a single spherule. The absorption cross sections were found to be independent of D and were merely the sum of cross sections of individual particles.

The structure of soot agglomerates can be optically diagnosed by static light scattering, which was developed by Schaefer et al. [116] and Weitz and Oliveria [117]. They showed that the intensity of scattered radiation behaved as

$$\begin{aligned} I(q) &\sim 1 && \text{for } qR_g \ll 1 \\ &\sim (qR_g)^{-D} && \text{for } qR_g \ll 1 \end{aligned} \tag{30}$$

where $q = (4\pi n/\lambda)\sin(\theta/2)$, n is the refractive index, and θ is the angle between the scattered and incident light. Whereas the fractal dimension of soot agglomerates can be found by static light scattering, the aggregation kinetics and the growth of clusters can be detected by dynamic light scattering [118].

4.3 Radiative Transfer in Turbulent Flames

The temporal and spatial fluctuations of temperature and radiative properties in a turbulent flame have strong effects on the radiative transfer within a flame [119, 120]. Both mean radiation levels and fluctuating radiation intensities [120] are influenced by the interactions of turbulence and radiation. Recent studies have shown premixed flames to be fractal [92, 99, 100] in structure. Based on the fluid mechanics [10], it is speculated that diffusion flames are also fractal. Since the fractal structure seems to be common in turbulent combustion, a new model of radiative transfer, incorporating this feature, needs to be developed.

Current models of radiative transfer in nonhomogeneous media [120, 121] determine the total transmittance by breaking up the radiation path length into regions with homogeneous properties. The scattering of radiation is neglected in these models. The size of each region is determined by the characteristic eddy lengths. The temporal fluctuation is modeled by assuming the correlation function of property $f(t)$ to decay exponentially as [120]

$$C(\tau) = \frac{\langle f(t+\tau)f(t)\rangle}{\langle f^2(t)\rangle} = \exp\left(\frac{-\tau}{t_e}\right) \tag{31}$$

where t_e is the temporal integral scale. Such a correlation function corresponds to a power spectrum of the form $P(\omega) \sim (1 + \omega^2 t_e^2)^{-1}$ [122]. Note that for high-frequency fluctuation this corresponds to a spectrum of the form $P(\omega) \sim \omega^{-2}$, which implies a fractal dimension of 1.5 corresponding to a Brownian process. However, it is known that the dimension of turbulent premixed flame can vary from 2.1 to 2.35, which suggests that the fluctuating property will have a dimension of 1.1 or 1.35. The current models therefore need to be generalized to accommodate various fractal dimensions encountered in turbulent flames. The structure function of Eq. (14) suggests that for a function $f(t)$ of dimension D, the correlation function would vary as $C(\tau) \sim 1 - \tau^{(4-2D)}$.

In the classical treatise of Tatarski [122], the Maxwell equations were solved to obtain the electromagnetic field for a fluctuating refractive index. He found that the mean scattered intensity is a function of a volume integral of the spatial correlation function of the refractive indices, $\langle n(\mathbf{r}' + \mathbf{r})n(\mathbf{r}')\rangle$, where n is the refractive index and $\mathbf{r}'$ and $\mathbf{r}$ are position vectors. Here again, the correlation function can be expressed in terms of the fractal dimension.

In addition to generalizing these models, radiative transfer can be modeled by simulating the temperature and property fluctuation by the Weierstrass-Mandelbrot function. These fluctuations can be incorporated in the equation of radiative transfer [25] to obtain the mean and fluctuating intensities.

5 TRANSPORT PHENOMENA IN POROUS MEDIA

The study of transport phenomena in porous media is essential for understanding the physical mechanisms involved in geothermal heat transfer [123], porous heat pipes [124], porous insulation, drying, and packed-bed combustors. A porous medium can be defined as a structure consisting of a solid phase which partially occupies a volume of space. The empty spaces consist of small pores which are connected to neighboring pores, thus allowing fluid flow. The structure of such a network of connected paths is usually random in direction and size and forms a complicated geometric structure. A fluid parcel thus follows a tortuous path through a porous medium. This randomness in fluid motion due to the structure of a porous medium is called dispersion [125]. Dispersion in flow acts in conjunction with molecular diffusion and is an additional transport mechanism. The following section will discuss the applications of fractal geometry in modeling dispersion.

When two immiscible fluids flow in a porous matrix, surface tension or capillary forces compete with viscous forces. A typical example of this phenomenon is the displacement of oil by water, which is commonly used in oil recovery from porous rocks. When viscous forces are dominant, invasion of low-viscosity fluid into a high-viscosity one creates viscous fingering [17]. In contrast, for dominating capillary forces, the process of invasion of a wetting fluid into a nonwetting fluid is called invasion percolation [126]. The structure of both viscous fingers and percolation fronts are known to be fractal. Section 5.2 discusses the physical mechanisms involved in growing such fractal structures in multiphase flow in porous media.

5.1 Dispersion

Due to the randomness in the flow field, transport phenomena in porous media are usually analyzed by averaging the conservation equations over a small control volume to obtain volume-averaged equations [127]. A conservation equation of a property ϕ can be written as

$$\frac{d\langle\phi\rangle}{dt} + \nabla\cdot\mathbf{u}\langle\phi\rangle = \nabla\cdot[\mathcal{D}\nabla\phi - \varepsilon\langle\mathbf{u}'\phi'\rangle] \tag{32}$$

where ϕ could be a scalar such as temperature or species concentration, or a vector such as velocity; $\mathbf{u}$ is the volume-averaged velocity; $\mathcal{D}$ is the diffusion coefficient; ε is the porosity; and $\mathbf{u}'$, ϕ' are the spatial deviations of the respective quantities. The terms within $\langle\ \rangle$ imply volume-averaged quantities. The term $-\varepsilon\langle\mathbf{u}'\phi'\rangle$ is the dispersion term, which is the product of the spatial deviations of the velocity and the property. The spatial deviations, of course, depend on the influence of the porous microstructure on the flow field. Therefore, although the process of volume averaging seems to solve the problem of randomness of the medium, the residual dispersion term creates the well-known "closure problem" commonly encountered in modeling turbulent flows.

The most common method of modeling dispersion is to relate the dispersion term to the gradient of the property as [128]

$$-\varepsilon\langle\mathbf{u}'\phi'\rangle = \alpha_d\nabla\langle\phi\rangle \tag{33}$$

where α_d is the dispersion diffusivity. Therefore, when Eq. (33) is substituted in Eq. (32), dispersion acts as an enhancement to molecular diffusion. However, the problem lies in finding the relation between the dispersion diffusivity and the microstructure of the porous medium. Previous investigations have used mixing length theories of turbulence [129–131] to relate the dispersion diffusivity to the velocity as $\alpha_d = \gamma u\ell$, where ℓ is the mixing length. The mixing length is then obtained from the characteristic length scale of the porous microstructure such as the average diameter of particles. However, it has been shown that the sizes of particles in a porous medium can span a wide range of length scales [132, 133] and follow a fractal structure. Therefore, the multiscale nature of porous media suggests that dispersion can be scale-dependent. This is exactly what has been observed in experiments [134], which can now be predicted by fractal dispersion models [135, 136].

Wheatcraft and Tyler [136] used simple scaling arguments to characterize the structure of a dispersed streamline by fractal geometry. Consider a path of particle to be wrinkled and multiscaled. Let L_f be the actual length of the path and let L_s be the straight-line distance traveled from the origin to the final destination. The smallest scale of movement is ε_c, which is of the order of the pore size. Assuming the streamline to be a self-similar fractal of dimension $D \geq 1$, the actual length and the straight-line distance are related as

$$L_f = \varepsilon_c^{1-D} L_s^D \tag{34}$$

The tortuosity T is defined as L_f/L_s and is therefore equal to $\varepsilon_c^{1-D} L_s^{D-1}$. Therefore the tortuosity depends on the scale of measurement L_s. Wheatcraft and Tyler [136] defined dispersivity as $\beta = \alpha_d/u$, which has units of length and is analogous to the mixing length, where u is the average scalar velocity. Using their fractal model, they found that the dispersivity varied with the scale of measurement L_s as

$$\beta = \frac{1}{2}\,\mathrm{var}(\varepsilon_c^{1-D})L_s^{(2D-1)} \tag{35}$$

where var( ) implies variance of a quantity. This model allowed for a random spatial distribution of minimum pore sizes. The scale dependence of the dispersivity was compared with experimental data [135] of a wide variety of porous media to obtain a fractal dimension of 1.3 for the fluid path. The results of the fractal model were verified by numerical simulation in which the fluid paths of different dimensions were constructed stochastically as shown in Fig. 21. It was observed that the dispersivity of these curves did follow the scale dependence of Eq. (35).

Several conclusions can be made from this fractal model. First, the dispersion diffusivity is scale-dependent and therefore cannot be related to a single length scale of mixing length. Second, dispersion is also a function of the fractal dimension of a streamline, which determines the degree of wrinkling. Lastly, the dispersivity is related to the microstructure of the porous medium through the variance of the smallest pore-size distribution and the fractal dimension of the connected pore spaces.

The fractal model of dispersion raises some fundamental issues about characterization of a porous medium. Porosity and permeability are conventionally used for characterization. The permeability K is obtained from the Darcy equation $\mathbf{u} = -(K/\mu)\nabla P$ where μ is the viscosity and P is the pressure. Both porosity and permeability are macroscopic parameters which do not characterize the microstructure of a porous medium. It seems that tortuosity is a more relevant microstructural parameter than permeability, since it is known that permeability K is related to tortuosity as $K = (\varepsilon r/$

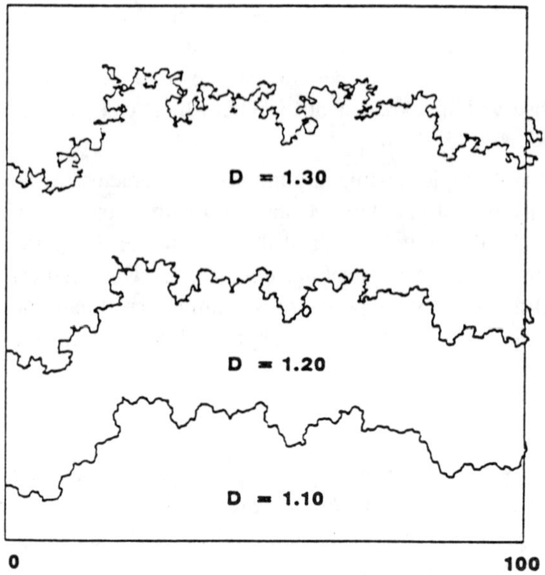

Figure 21 Fractal model of the path of a dispersed fluid particle in a porous medium [136].

$T)^2$ [137], where ε is the porosity and r is the characteristic radius of particles comprising the porous medium. Since tortuosity depends on the scale of measurement and the fractal dimension of a streamline, it appears that the fractal dimension is more fundamental. A fractal characterization of a porous medium will therefore be able to provide the tortuosity and permeability. However, a method of determining the fractal dimension of a streamline a priori has not yet been developed. It is also possible that the dimension of a streamline may be different near a wall (close to 1) than in the core region. Therefore, although the idea of using fractal geometry to model dispersion seems attractive, there are still some fundamental problems that have to be resolved.

5.2 Fractal Patterns of Fluid Interfaces

The interface between two fluid phases present in a porous medium often becomes wrinkled and distorted due to hydrodynamic instabilities and surface tension forces. These issues are of primary importance in oil recovery by water or steam and in waste removal within the earth. Two nondimensional parameters which characterize different regimes of such flows [138] are the ratio of the fluid viscosities, $M = \mu_2/\mu_1$, where fluid 2 invades fluid 1, and the capillary number, $\mathrm{Ca} = u\mu_2/\sigma$, where u is the scalar velocity obtained by the Darcy equation and σ is the surface tension. The structure of interfaces for low values of M and Ca have been shown to be fractal and are discussed in recent texts [17, 19] and by Wong [139]. The following sections present the main features of the study of interfacial fractal patterns.

5.2.1 Viscous Fingering. The first example is that of viscous fingering [17], which occurs when the viscous forces are larger than capillary forces. Flow in a porous medium is often simulated by a Hele-Shaw cell, since the flow field in the cell satisfies the Darcy equation. Consider a Hele-Shaw cell consisting of parallel plates with a small gap of thickness b in between them. If the fluid is incompressible, mass conservation of fluid within the medium, $\nabla \cdot \mathbf{u} = 0$, suggests that when $\mathbf{u}$ is given by the Darcy equation, the pressure field satisfies the Laplace equation $\nabla^2 P = 0$. The gap thickness is related to the permeability by $K = b^2/12$ [17]. In Sec. 4.2.1 it was discussed that when a field satisfies the Laplace equation, the structure generated by the field is fractal. These structures are often called Laplacian fractals [110]. In the Hele-Shaw cell, when a low-viscosity fluid invades a high-viscosity fluid, hydrodynamic instabilities cause the front to break up into branched fingers. These fingers are created when the pressure gradient or the Darcy velocity is large enough to overcome the surface tension forces which stretch the interface and prevent instabilities to grow. The wavelength of instability which is most likely to grow is given by

$$\lambda_m = \frac{\pi b}{\sqrt{\mathrm{Ca}}} \tag{36}$$

In a porous medium, the length scale b is the pore size and the capillary number corresponds to the local velocity inside a pore. Since the distribution of pore size is random, the local values of λ_m are also random. Therefore, a porous medium induces randomness in the structure of the branched fingers, which have been shown to be fractal in structure [140]. If a low-viscosity fluid is injected into high-viscosity fluid in a porous medium, the viscous fingers created under a sufficient pressure gradient appear very similar to an agglomerate formed by diffusion-limited aggregation [141] as shown in Fig. 19. Such a structure can be characterized by counting the number of elements $N(r)$ of the structure within the radius r as

$$N(r) \sim N_o \left(\frac{r}{R_g} \right)^D \tag{37}$$

where R_g is the radius of gyration defined in Sec. 4.2.1. Maloy et al. [140] found that for different varieties of packed-bed porous medium and different capillary numbers, the plots of $N(r)$ versus r/R_g coincided and were best fitted by a fractal dimension of $D = 1.62$.

In oil recovery from rocks and waste removal in soil, it is necessary to have a stable interface between the invading and the receding fluids. However, since water has lower viscosity than oil or most waste products, viscous fingering at the water interface is natural although detrimental. The idea of using fractal geometry to characterize the structure of the viscous fingers will allow one to predict the efficiency of an oil recovery process.

5.2.2 Invasion Percolation. In this regime of flow, the capillary forces are dominant over the viscous forces which are caused by pressure gradients. The capillary force is characterized by a capillary pressure P_c, which is equal to $\sigma \cos \theta / b$, where b is the pore radius and θ is the contact angle between the interface and the pore wall. The capillary forces increase for decreasing sizes of pores or, more appropriately, the size of pore throats. When a wetting fluid such as water invades a nonwetting fluid such as oil, the water will spread spontaneously through pores of small radii. When the interface comes across a large pore, the capillary pressure may not be larger than an externally applied pressure and therefore water will not permeate through such a pore. Therefore, the spreading of a wetting fluid through a nonwetting fluid in a porous

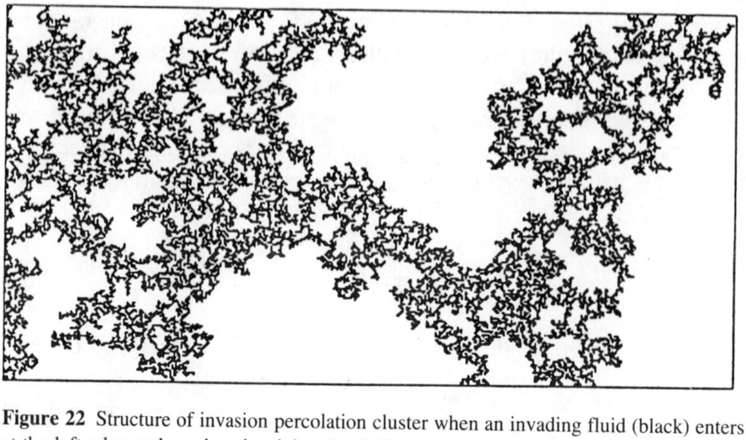

Figure 22 Structure of invasion percolation cluster when an invading fluid (black) enters a porous medium at the left edge and reaches the right edge [17].

medium depends on the link between pores' sizes less than a critical radius. Since the distribution of pores of different sizes is random, the structure of the interface is random and can be characterized by fractal geometry [126].

Wilkinson and Willemsen [126] simulated such a phenomenon by considering a random network of pores on a lattice. The nodes of this lattice were the pore sites, and the bonds in the lattice were the pore throats. By assigning random numbers to sites and bonds to represent the likelihood of a wetting fluid invading a pore, the percolating front was easily established. Invasion of a wetting fluid will be more likely in smaller pores, whereas invasion of nonwetting fluids will be more likely for larger pores. Figure 22 shows the structure of the invasion percolation where an invading fluid enters at the left edge and percolates to the right edge. If the length of the whole lattice is L, then the number of sites invaded by the wetting fluid is given as $N(L) = AL^D$, where D is the fractal dimension of the percolation cluster and A is a constant. In a two-dimensional simulation, the value of D was found to be 1.89. When water percolates through oil, the regions of trapped oil surrounded by water never reduce in size. Under this condition it was found that the fractal dimension of the cluster was 1.82. Experiments at very low capillary numbers, $\mathrm{Ca} = 10^{-5}$, have shown that the fractal dimension of the cluster of air invading glycerol is 1.84 [142].

During condensation or evaporation of wetting fluids in a porous medium, surface tension forces are quite dominant [143, 144]. Majumdar and Tien [143] showed that

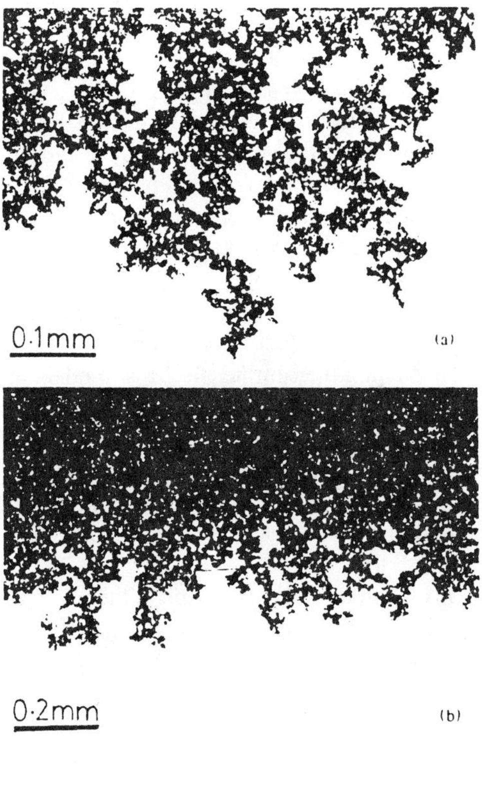

Figure 23 Fractal structure of a liquid front at different magnifications (*a*) and (*b*) during drying [144].

during film condensation in a porous medium, a region of coexisting liquid and vapor phases is present between the liquid film adjacent to the wall and the pure vapor zone. In this two-phase zone, vapor and liquid are transported in opposite directions through capillary forces and is therefore an invasion percolation phenomenon. A similar scenario is found in drying, where the interface between the liquid and the vapor is fractal, as shown in Fig. 23 [144]. Conventionally, the liquid concentration at any location is measured by the liquid saturation, which is the fraction of liquid volume to the vapor volume in a large control volume of pore spaces [124]. The saturation in the film condensation problem was found to follow a boundary-layer profile, with saturation of unity at the interface between the liquid zone and zero at the vapor interface. This behavior can be seen qualitatively by the microstructure of the percolation front in Fig. 23.

6 HEAT AND MASS TRANSFER IN FRACTAL ANATOMY

The regulation of body temperature and the exchange of oxygen, carbon dioxide, and nutrients in biological tissues are examples of heat and mass transfer within a mammalian body. One of the unique features of the transport of heat and mass within an organ or tissue is that they are distributed by a system of branching networks. The finely branched structure of the bronchial tree [145] and the architecture of the vascular network surrounding an ovary [146] are some examples shown in Fig. 24. The repeated branching of the bronchial system serves as a paradigm of anatomical complexity found in most distribution networks. In the vascular system, the internal diameter of blood vessels varies by a factor of 2500 proceeding from the aorta to capillary. This results in a 10^7-fold decrease in cross-sectional area of a single vessel. However, due to the multiple branching, there is actually a 300-fold increase in total cross-sectional area at the final stage of branching [147]. The complexity of the architecture can be further emphasized by the fact that branching continues through about 30 generations until billions of capillaries are formed. In fact, about 10^6 capillaries can be found in 1 cm^3 of tissue, occupying only about 0.5% of the volume [147] which allows the blood to reach the vicinity of nearly every cell in a tissue but occupying negligible volume. Although the treelike structure of these networks has considerable effect on the heat and mass transfer, the immense complexity of such networks is a major obstacle to developing a suitable methodology for understanding their influence. Most models of bioheat transfer employ volume-averaging techniques to overcome this problem [148, 149]. However, the lack of information of the transport properties in these models raises a "closure" problem which requires the knowledge of the microstructure. Therefore, there is a need to develop a method of characterizing the structure of any branching network so that the influence of the structure on heat and mass transfer can be studied at all levels of branching.

Although the treelike architecture of distribution networks is complex, there seems to be some pattern underlying the structural chaos. The branching at any stage appears to be similar to that of the previous stages, the difference being only in the sizes of vessels. Such similitude gives rise to self-similar scaling relations between the sizes of mother and daughter branches at every stage of bifurcations. The applications of

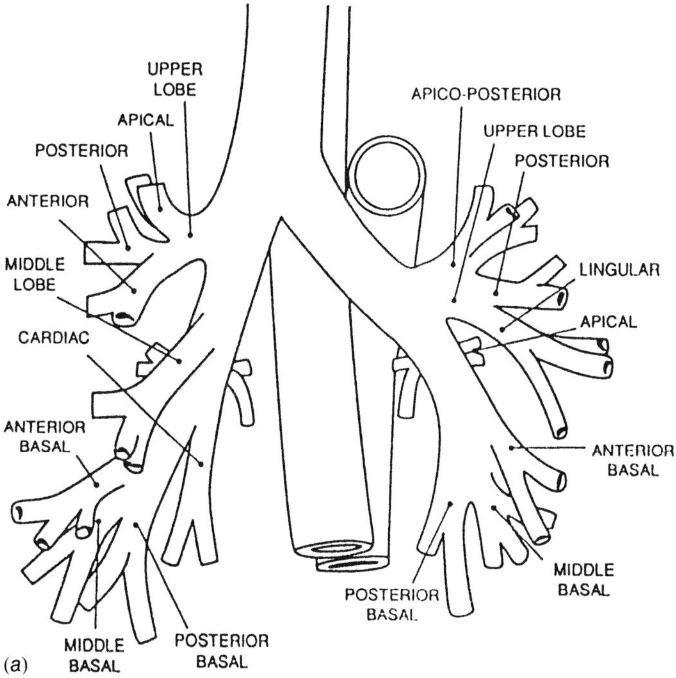

(a)

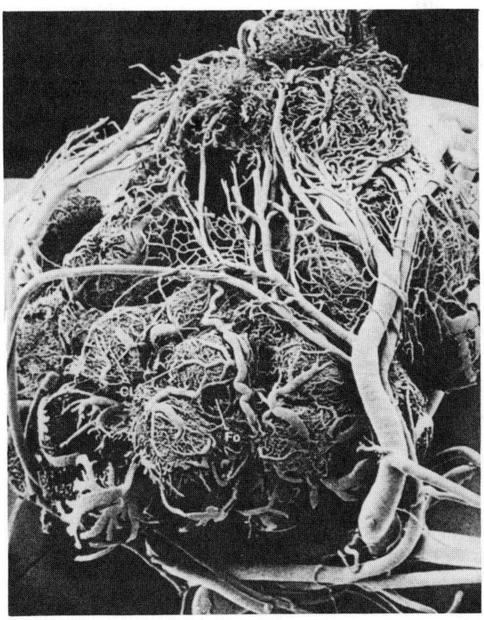

(b)

Figure 24 Branched structure in (*a*) the bronchial tree [145] and (*b*) vascular network surrounding a rat ovary [146].

fractal geometry in characterizing the self-similarity of the branching are relatively recent [1, 7]. The following section discusses the role of fractal geometry in characterizing the structure of distribution networks. The possibility of using the fractal features in modeling heat and mass transfer in tissues seems very appealing and will be explored subsequently.

6.1 Self-Similarity in Distribution Networks

Although the walls of the bronchial vessels are two-dimensional, the bifurcations of a distribution network produces billions of small vessels which seem to nearly fill a three-dimensional volume of tissue. Therefore, if a two-dimensional object appears to be volume-filling, then the whole collection of branched vessels will have a dimension between 2 and 3. It therefore seems possible to apply fractal geometry to characterize the near-space-filling branching networks.

To understand the self-similarity in treelike networks it is first necessary to investigate the Cantor set in Fig. 3*b*. The Cantor set is formed by dividing a line segment into three parts and omitting the middle portion. The number of parts remaining is two, each of which has to be magnified three times to be similar to the original line segment. The dimension of the Cantor set is therefore equal to $D = \ln 2/\ln 3$. If, however, the number of parts is N set equal to 2, or $N = 2$, and the magnification m required for each part is chosen to be $2^{1/2} < m < 2$, then the dimension of the object $D = \ln N/\ln m$ will be $2 > D > 1$. Now consider the bifurcation as shown in Fig. 25. The bifurcation at generation 1 (mother) produces two similar branches at generation 2 (daughters). Further bifurcations will produce branches at generation 3, 4, . . . , and so on. In this figure the number of parts $N = 2$. If the magnification $m = 2^{1/2}$ and the branches are not overlapping, then the network is likely to be space-

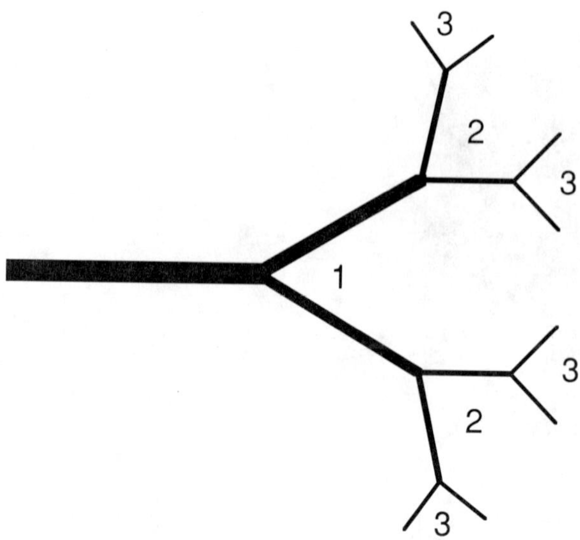

Figure 25 Two stages of self-similar branching.

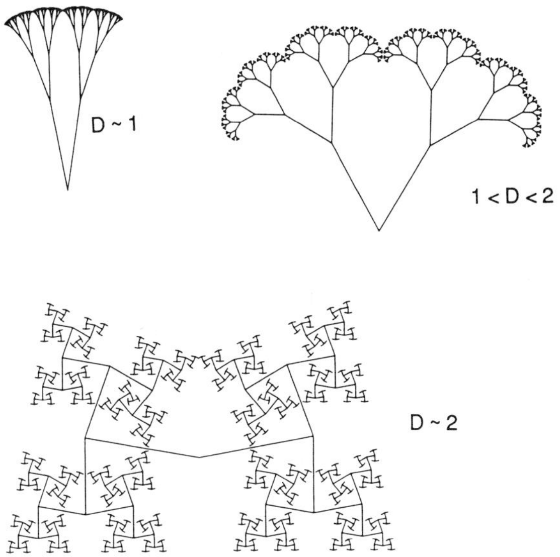

Figure 26 Self-similar branching in trees of different dimensions.

filling in a two-dimensional plane. In a three-dimensional space, the minimum scaling ratio required for space filling is $m = 2^{1/3}$. Note that for the branches to be nonoverlapping the angle between the branches should be greater than a critical angle θ_{crit} [1]. Depending on the scaling ratio between the mother and daughter branches, the dimension of the tree can be obtained. For different values of the dimensions, self-similar branching can produce trees of various shapes as shown in Fig. 26 [1]. When the dimension is barely greater than 1, the tree appears to be nearly straight and broomlike. As the dimension is increased, the canopy of branches opens out and becomes more lush. It has been observed that the ratio of the lengths of the mother and daughter branches in the bronchial tree is close to 1.62 [8]. An interesting observation about this number is that the ratio of the nth term to the $(n - 1)$th term in the Fibonacci series is equal to 1.618 for large values of n. The Fibonacci scaling has been observed in a number of botanical and physiological structures [8]. Poppel [150] reports that the scaling ratio for the vascular network varies from 1.36 to 1.92 depending on the type of tissue.

The relation between the diameters of the mother and daughter branches is also of concern for determining fluid flow rate in branches of different generations. West and Goldberger [8] noted that for the zth generation of the bronchial tree, the diameter d of the branch followed a power-law scaling relation of the form $d = A(z)z^{-n}$, where $A(z)$ is a periodic function in z. This relation was obtained from the scaling behavior of the Weierstrass function. Figure 27 shows the power-law decrease in diameter of the bronchial branch of a human lung. West and Goldberger [8] suggested that the exponent n is the fractal dimension of the tree structure. However, for the Weierstrass function the dimension is related to the exponent as $D = 2 - n$, as discussed earlier for Eq. (15). Therefore, one should be cautious in using any exponent as the fractal dimension of an object.

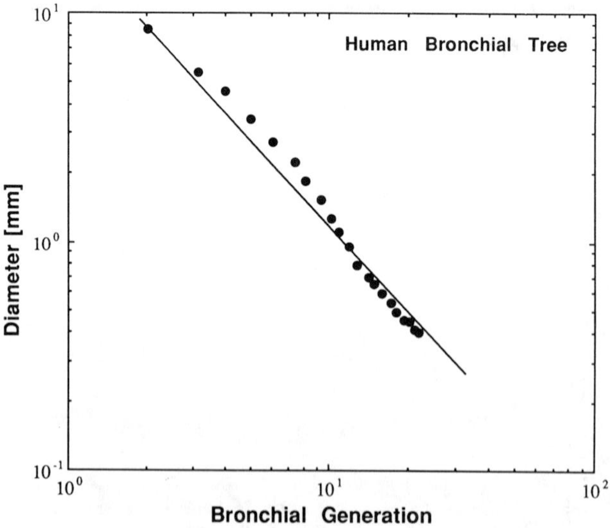

Figure 27 Power-law decrease of bronchial diameter with generation of branching in the human lung [8].

Mandelbrot [1] suggested that the diameter of the mother branch d and those of the two daughter branches d_1 and d_2 are related by the equation

$$d^\Delta = d_1^\Delta + d_2^\Delta \tag{38}$$

where Δ was called the diameter exponent. For botanical trees $\Delta = 2$, whereas for the bronchi $\Delta = 3$ and for the arteries $\Delta = 2.7$. The value of $\Delta = 3$ has also been found experimentally for the pulmonary arterial tree of dogs [151]. Mandelbrot noted that self-similarity implies that $\Delta = D$, although in some cases $\Delta < D$, for which the self-similarity breaks down. The diameter exponent of $\Delta = 3$ is of particular interest to heat and mass transfer. The air flow rate in a bronchus is equal to the velocity multiplied by the cross-sectional area, which varies as d^2. Since the diameter varies as d^Δ, the velocity varies as $d^{\Delta-2}$. Therefore, as the diameters of the bronchial tree decrease, the air slows down. This suggests that the mass transfer at the alveoli will tend to be a purely diffusion process with marginal convection effects. It turns out that besides slowing the fluid down as it moves through the bronchial tree, the diameter exponent of $\Delta = 3$ also yields the least resistance to air flow [1] and is therefore a fascinating optimization of the human body.

The fractal structure in the anatomy, in essence, provides a mechanism for converting a volume of dimension 3 (blood in large vessels and air in the upper respiratory tract) into something approaching a two-dimensional structure such as the alveolar-capillary surface, thereby facilitating heat and mass transfer. The architecture of the tracheobronchial tree, the vascular network, the neural network, the biliary network, the His-Purkinje system in the heart, the urinary collecting systems, and the placenta all have the common feature of self-similarity [7], although the nature of self-similarity is different in each case. It is therefore useful to characterize the structure of branching networks by fractal geometry.

6.2 Heat and Mass Transfer

In the human body, heat is generated at the cell level by metabolism and is transported by conduction through tissues and by convection through flow of blood. Heat is finally lost to the environment by conduction, convection, and evaporation at the skin. The most complex of all these thermal phenomena is the transport of heat due to convection. Consider blood to be entering a vessel at temperature T_b within a tissue at temperature T_t. The heat balance between the blood and tissue yields the relation [148]

$$X_e \frac{dT_b}{dx} = T_t - T_b \tag{39}$$

where X_e is the equilibration length defined as $X_e = d\rho_b C_b u/4h$, where d is the vessel diameter, ρ_b is the blood density, C_b is the blood heat capacity, u is the mean blood velocity, and h is the heat transfer coefficient. This equation suggests that the blood temperature decreases as $\exp(-x/X_e)$ while flowing through the vessel. When the length of the vessel ℓ is larger than X_e, the blood temperature equilibrates with the local tissue temperature. Table 1 presents details of the radii r, X_e, and lengths of different vessels of the vascular network [152]. It is evident that in precapillary arterioles, capillaries, and venules, blood is essentially at the same temperature as the tissue. However, larger vessels are more thermally significant, since the blood temperature does not equilibrate with the tissue. Chen [148] used a continuum approach of volume averaging and derived a bioheat equation of the form

$$\rho C \frac{\partial T}{\partial t} = \nabla \cdot k\, \nabla T + \nabla \cdot k_e\, \nabla T + w_j \rho_b C_b (T_j - T) - \rho_b C_b w_b \cdot \nabla T + q_m \tag{40}$$

The second term on the right-hand side corresponds to an eddy conduction term due to randomness in blood vessel direction with respect to the temperature gradient in the tissue. The third term represents the blood acting as a heat source with w_j as the perfusion rate per unit volume at the jth generation of the vascular network, and T_j is

Table 1 Physical and Thermal Length Scales of Blood Vessels [152]

Vessel	Percent vascular volume	r (μm)	X_e (m)	ℓ/X_e
Aorta	3.30	5000	190	0.002
Large artery	6.59	1500	4	0.05
Arterial branch	5.49	500	0.3	0.3
Terminal branch	0.55	300	0.08	0.1
*	1.00	175	0.009	1
Arteriole	2.75	10	5×10^{-6}	400
Capillary	6.59	4	2×10^{-7}	6000
Venules	12.09	15	2×10^{-6}	800
Terminal vein	3.30	150	0.1	0.1
Venous branch	29.67	1200	0.3	0.3
Large vein	24.18	3000	5	0.04
Vena cava	5.49	6250	190	0.002

the temperature of the blood at the jth generation. The fourth term is the convection term to account for the mean blood flow, and the last term is the heat generation due to metabolism.

Although the temperature field in a tissue can ideally be determined by Eq. (40), the problem lies in evaluating the transport coefficients k_e, w_j, and $|w_b|$. This is reminiscent of the closure problem in turbulence modeling, where the eddy diffusivity is an unknown quantity. The reason for such a difficulty is that in volume averaging the detailed structural information of the vascular architecture at the microscopic level is lost. Consequently, the transport coefficients, which depend strongly on the microscopic structure, cannot be readily determined. If, however, the vascular microarchitecture can be modeled as self-similar fractals, the transport coefficients can possibly be obtained.

Weinbaum et al. [153] and Jiji et al. [154] noted an interesting feature about the vascular network which has a strong effect on the heat transfer. The main arteries and veins of a peripheral tissue are initially 200–300 μm in diameter and are placed in close proximity to each other. Even after branching, they remain in pairs until they reach the size of capillaries, where they branch out in the cutaneous layer. The close proximity allows countercurrent heat exchange between the warm blood in the arteries and cooler blood in the veins. They showed that this heat exchange is much larger than the heat source term in Eq. (40). Subsequently, a new bioheat equation was developed which included this effect in thermally significant vessels [149], and is given as

$$\rho C \frac{\partial T}{\partial t} = \frac{\partial}{\partial x_i}\left[(k_{ij})_{\text{eff}} \frac{\partial T}{\partial x_j}\right] - \frac{\pi^2 n r^2 k_b^2}{4\sigma k} \text{Pe}^2\, l_j \frac{\partial l_i}{\partial x_i}\frac{\partial T}{\partial x_j} + q_m \tag{41}$$

Here l_i is the direction cosine of the vessel pair with respect to the coordinate axis x_i, Pe is the Peclet number, k is the conductivity of the tissue, σ is the shape factor for the radii and the distance between the vessels, k_b is the conductivity of the blood, n is the number density of vessels per unit area enclosing the control volume, and $(k_{ij})_{\text{eff}}$ is the effective conductivity tensor, given as

$$(k_{ij})_{\text{eff}} = k\left(\delta_{ij} + \frac{\pi^2}{4\sigma k^2}\, n r^2 k_b^2\, \text{Pe}^2\, l_i l_j\right) \tag{42}$$

The second term in the conductivity tensor is the enhancement of heat transfer due to countercurrent heat exchange.

The Weinbaum-Jiji bioheat equation relies on a number of microvascular information such as the number density of vessels, the velocity of blood in a vessel, the shape factor, and the direction of the vessels. Here again, the volume-averaging approach has created parameters which invariably depend on the microstructure of the vascular network and therefore raise the closure problem.

The conclusion that can be made from both these models of bioheat transfer is that although volume averaging allows one to develop differential equations representing energy balance, the transport properties in these equations depend on the microstructure of the vascular network. The lack of a proper method of characterizing the microstructure disallows the evaluation of these properties. This problem can possibly be circumvented by characterizing the microstructure by fractal geometry and then determining the transport properties of the above-mentioned models. Another option is to adopt a completely new approach of developing a fractal network model of bioheat transfer which, instead of determining the transport properties, would study the heat transfer in the series-parallel network of vessels at both the micro and macro levels. Table 1 suggests that the branching in such a network would be terminated for vessels smaller than 10 μm, which are thermally insignificant. Recently, Baish [155] developed a fractal-based bioheat transfer theory which showed that while measuring the temperature within a tissue, the probability of encountering a fluctuation of size ε would follow a power-law distribution of the form ε^{-n}. Such power-law distributions suggest a self-similar or self-affine scaling.

In contrast with heat transfer, efficient mass transfer is known to occur mainly in the capillaries [156]. The effectiveness of the mass transfer in capillaries is enhanced due to the low flow rates and increase in total surface area. A convenient method of modeling the mass transfer in a single capillary is to consider a unit volume of tissue, called the Krogh cylinder [157], surrounding the capillary. However, the total effect of the billions of capillaries on the mass transfer would require the information about the vascular structure at both the arterial and venous side of the capillary, which may be different in some cases [158]. It is here that the fractal characterization of the network will play a role in studying the influence of the vascular structure on the mass transfer. The fractal approach seems to be more suited to mass transfer, since the vessel sizes involved in the transport span the entire range, in contrast with the thermally significant vessels in heat transfer.

7 SUMMARY AND CONCLUSIONS

This chapter reviews the literature on applications of fractal geometry in thermal sciences and explores the possibilities of using fractals in studying a large variety of thermal phenomena. The mathematical language of fractal geometry involving the concepts of fractional dimension, self-similarity, and self-affinity, which are used for quantifying disorder and chaos in shapes, is first described. These ideas are then used to characterize the roughness of various solid surfaces. It is shown that the fractal representation of surface roughness is a powerful technique to get physical insight into surface thermal phenomena such as contact conductance, thermal boundary or Kapitza resistance, reflection and emission of radiation, hysteretic losses in superconductors, and boiling. The effect of turbulence on premixed and diffusion flames is then studied by first inspecting the fractal nature of turbulent flows. It was found that although fractals have been used to quantify premixed flame structures, the application of fractals in diffusion flames is still an open area of research. The disordered structure of soot agglomerates, which influences the radiative transfer during combustion, can also be quantified by fractal geometry and simulated by several aggregation models. It was identified that the fractals can be applied to study the interactions between turbulence and radiative transfer in a nonhomogeneous medium. Some recent studies on flows in porous media have shown that the phenomenon of dispersion, which is an additional transport mechanism, can be modeled by characterizing a distorted streamline as a fractal curve. The interface between two fluids in a porous medium is often distorted and wrinkled due to surface tension and viscous forces in conjunction with the randomness of the porous structure. The fractal nature of these interfaces is well known and can be used to study interfacial heat and mass transfer in porous media. The applications of fractals in bioheat transfer is very recent and is an open field of research. The self-similar branching of distribution networks such as the vascular network and the bronchial tree in the lung can be modeled as fractals so that the microstructure and distribution of these vessels in a tissue can be quantified. This will facilitate the study of micro- and macroscale heat and mass transfer in biology and medicine.

Before using the concepts of fractal geometry in studying any disordered physical phenomenon, one might be inclined to ask the question: As opposed to conventional methods, what is the advantage of using fractals? By definition, disorder is something which is not order. Conventionally, order is associated with Euclidean shapes and structures which are smooth and regular. Disorder is characterized by the lack of smoothness over a large range of length scales. How does one characterize disorder? Fractal geometry generalizes the concept of Euclidean dimension and therefore finds the order in disordered structures. It forms the link between the structures at *micro-* and *macroscales*. It is now well known that a large variety of shapes and structures found in nature are disordered at some length scales and follow fractal geometry.

How can fractals help in understanding thermal phenomena? In the study of surface thermal phenomena, most previous investigations have resorted to empirical studies. This lack of understanding of the interaction between a rough surface and a physical phenomenon lies in the inability to characterize the disorder in the surface structure. Since every physical phenomenon has one or more length scales, it is influenced by the surface structure of all scales larger than that of the phenomenon. Fractal geometry quantifies the structure at all length scales and therefore forms a fundamental approach to understanding the mechanisms of surface thermal phenomena.

It is also quite common to find that the method of volume or time averaging is used as a continuum approach to overcome the structural disorder at microscales. These techniques are often used in turbulence, flows through porous media, and bioheat transfer. Although it may seem that averaging a set of conservation equations circumvents the problem of analyzing a disordered medium, the notorious closure problem invariably appears in these situations. The reason is that the disordered fluctuations lead to additional transport mechanisms which depend on the microstructure on the chaotic medium. If the disorder can be quantified at all scales by fractal geometry, these transport mechanisms can be properly understood. The idea of finding order in disorder by fractal geometry is therefore a new concept and seems to be a promising approach in the study of thermal phenomena.

ACKNOWLEDGMENTS

I gratefully thank Professor C. L. Tien and his research group for their helpful support and constant encouragement. My thanks to Dr. A. Joshi for her help in proofreading.

The support of the National Science Foundation through grant CTS-9010311 is greatly appreciated.

REFERENCES

1. B. B. Mandelbrot, *The Fractal Geometry of Nature*, W. H. Freeman, New York, 1982.
2. L. Pietronero and E. Tosatti, *Fractals in Physics*, Elsevier, New York, 1986.
3. S. H. Liu, Fractals and Their Applications in Condensed Matter Physics, *Solid State Phys.*, vol. 39, pp. 207–273, 1986.
4. R. L. Devaney, *Introduction to Chaotic Dynamical Systems*, Benjamin Cummings, Menlo Park, Calif., 1986.
5. D. W. Schaefer, Polymers, Fractals and Ceramic Materials, *Science*, vol. 243, pp. 1023–1027, 1989.
6. D. Avnir, *The Fractal Approach to Heterogeneous Chemistry*, John Wiley, New York, 1989.
7. A. L. Goldberger and B. J. West, Fractals in Physiology and Medicine, *Yale J. Biol. Med.*, vol. 60, pp. 421–435, 1987.
8. B. J. West and A. L. Goldberger, Physiology in Fractal Dimensions, Am. Sci., vol. 75, pp. 354–365, 1987.
9. D. L. Turcotte, Fractals in Fluid Mechanics, *Ann. Rev. Fluid Mech.*, vol. 20, pp. 5–16, 1988.
10. K. R. Sreenivasan and C. Meneveau, The Fractal Facets of Turbulence, *J. Fluid Mech.*, vol. 173, pp. 357–386, 1986.
11. F. C. Gouldin, An Application of Fractals to Modeling Premixed Turbulent Flames, *Combustion and Flame*, vol. 68, pp. 249–266, 1987.
12. R. J. Samson, G. W. Mulholland, and J. W. Gentry, Structural Analysis of Soot Agglomerates, *Langmuir*, vol. 3, pp. 272–281, 1987.
13. A. Majumdar and B. Bhushan, Role of Fractal Geometry in Roughness Characterization and Contact Mechanics of Surfaces, *J. Tribology*, vol. 112, pp. 205–216, 1990.
14. A. Majumdar, Fractal Surfaces and Their Applications to Surface Phenomena, Ph.D. thesis, University of California, Berkeley, 1989.
15. B. B. Mandelbrot and other eds., *Fractal Aspects of Materials*, Extended Abstracts of Symposiums, Materials Research Society, Pittsburgh, 1984, 1985, 1986, 1987, 1988.
16. B. B. Mandelbrot, How Long Is the Coast of Britain? Statistical Self-Similarity and Fractional Dimension, *Science*, vol. 155, pp. 636–638, 1967.
17. J. Feder, *Fractals*, Plenum Press, New York, 1988.
18. H. O. Peitgen and D. Saupe, *The Science of Fractal Images*, Springer-Verlag, New York, 1988.
19. T. Vicsek, *Fractal Growth Phenomena*, World Scientific, N.J., 1989.
20. B. B. Mandelbrot, Self-Affine Fractals and Fractal Dimension, *Physica Scripta*, vol. 32, pp. 257–260, 1985.
21. B. B. Mandelbrot and J. W. van Ness, Fractional Brownian Motions, Fractional Noises and Applications, *SIAM Rev.*, vol. 10, pp. 422–437, 1968.
22. R. F. Voss, Random Factal Forgeries, in *Fundamental Algorithms for Computer Graphics*, (R. A. Earnshaw, ed.), Springer-Verlag, Berlin, pp. 805–835, 1985.
23. M. G. Cooper, B. B. Mikic, and M. M. Yovanovich, Thermal Contact Conductance, *Int. J. Heat Mass Transfer*, vol. 12, pp. 279–300, 1969.
24. E. T. Swartz and R. O. Pohl, Thermal Boundary Resistance, *Rev. Modern Phys.*, vol. 61, pp. 605–668, 1989.
25. R. Siegel and J. R. Howell, *Thermal Radiation Heat Transfer*, 2nd ed., Hemisphere, New York, 1981.
26. W. M. Rohsenow and H. Choi, *Heat, Mass and Momentum Transfer*, Prentice Hall, Englewood Cliffs, N.J., 1961.
27. E. B. Forsyth, Energy Loss Mechanisms of Superconductors Used in Alternating-Current Power Transmission Systems, *Science*, vol. 242, pp. 391–399, 1988.
28. N. P. Suh, *Tribophysics*, Prentice Hall, Englewood Cliffs, N.J., 1986.
29. E. D. Williams and N. C. Bartelt, Thermodynamics of Surface Morphology, *Science*, vol. 251, pp. 393–400, 1991.

30. B. B. Mandelbrot, D. E. Passoja, and A. J. Paullay, Fractal Character of Fracture Surfaces of Metals, *Nature*, vol. 308, pp. 1571–1572, 1984.
31. Y. Termonia and P. Meakin, Formation of Fractal Cracks in a Kinetic Fracture Model, *Nature*, vol. 320, pp. 429–431, 1986.
32. P. Meakin, Fractal Scaling in Thin Film Condensation and Material Surfaces, *CRC Crit. Rev. Solid State and Materials Sci.*, vol. 13, pp. 143–189, 1987.
33. T. Vicsek, Formation of Solidification Patterns in Aggregation Models, in *Fractals in Physics*, (L. Pietronero and E. Tosatti, eds.), Elsevier, New York, pp. 247–250, 1986.
34. M. Kardar, G. Parisi, and Y. C. Zhang, Dynamic Scaling of Growing Interfaces, *Phys. Rev. Lett.*, vol. 56, pp. 889–892, 1986.
35. P. R. Nayak, Random Process Model of Rough Surfaces, *J. Lubrication Technol.*, vol. 93, pp. 398–407, 1971.
36. R. S. Sayles and T. R. Thomas, Surface Topography as a Nonstationary Random Process, *Nature*, vol. 271, pp. 431–434, 1978.
37. T. R. Thomas, Defining the Microtopography of Surfaces in Thermal Contact, *Wear*, vol. 79, pp. 73–82, 1982.
38. B. Bhushan, J. C. Wyant, and J. Meiling, A New Three-Dimensional Noncontact Digital Optical Profiler, *Wear*, vol. 122, pp. 301–312, 1988.
39. J. I. McCool, Relating Profile Instrument Measurements to the Functional Performances of Rough Surfaces, *J. Tribology*, vol. 109, pp. 264–270, 1987.
40. A. Majumdar and C. L. Tien, Fractal Characterization and Simulation of Rough Surfaces, *Wear*, vol. 136, pp. 313–327.
41. S. R. Brown and C. H. Scholz, Closure of Random Elastic Surfaces in Contact, *J. Geophys. Res.*, vol. 90, no. B7, pp. 5531–5545, 1985.
42. D. K. Chin, T. van Duzer, S. Whitfield, R. E. Thompson, J. Clarke, A. Majumdar, M. I. Flik, and C. L. Tien, Surface Topography of High-T_c Thin Film Superconductors, in preparation.
43. A. Majumdar, B. Bhushan, and C. L. Tien, *Role of Fractal Geometry in Tribology, Advances in Information Storage Systems*, ASME Series vol. 1, pp. 231–266, 1991.
44. L. S. Fletcher, Recent Developments in Contact Conductance Heat Transfer, *J. Heat Transfer*, vol. 110, pp. 1059–1070, 1988.
45. C. L. Tien, A Correlation for Thermal Contact Conductance of Nominally-Flat Surfaces in Vacuum, *Proc. 7th Thermal Conductivity Conf.*, U.S. Bureau of Standards, Washington, D.C., pp. 755–759, 1968.
46. B. B. Mikic, Thermal Contact Conductance; Theoretical Considerations, *Int. J. Heat Mass Transfer*, vol. 17, pp. 205–214, 1974.
47. A. Majumdar and C. L. Tien, Fractal Network Model of Contact Conductance, *J. Heat Transfer* (in press).
48. B. B. Mandelbrot, Stochastic Models for the Earth's Relief, the Shape and the Fractal Dimension of the Coastlines, and the Number-Area Rule for Islands, *Proc. Natl. Acad. Sci., USA*, vol. 72, pp. 3825–3828, 1975.
49. P. L. Kapitza, The Study of Heat Transfer in Helium II, *J. Phys.* (USSR), vol. 4, p. 181, 1941.
50. I. M. Khalatnikov, *Journal of Experimental and Theoretical Physics (USSR)*, vol. 22, p. 687, 1952.
51. A. C. Anderson, J. I. Connolly and J. C. Wheatley, Thermal Boundary Resistance Between Solids and Helium below 1 K, *Physical Review*, vol. 135, pp. A910–A921, 1964.
52. T. Nakayama, Scattering of High-Energy Phonons at Irregular Surfaces Without and With Liquid Helium, *Phys. Rev. B*, vol. 33, pp. 8664–8676, 1986.
53. A. D. Lapin, Scattering of Surface Waves Propagating over an Uneven Liquid-Solid Interface, *Sov. Phys. Acoustics*, vol. 15, p. 387, 1970.
54. N. S. Shiren, Surface Roughness Contribution to Kapitza Conductance, *Phys. Rev. Lett.*, vol. 47, 1466–1469, 1981.
55. A. Majumdar, Effect of Interfacial Roughness on Phonon Radiative Heat Conduction, *J. Heat Transfer* (in press).
56. M. I. Flik and C. L. Tien, Thermal Conductivity of High-T_c Thin-Film Superconductors, presented at *ASME Winter Annual Meeting*, San Francisco, December 10–15, 1989.

57. M. I. Flik, Thermal Phenomena in High-T_c Thin-Film Superconductors, Ph.D. thesis, University of California, Berkeley, 1989.
58. T. H. Geballe and J. K. Hulm, Superconductivity—The State That Came in from the Cold, *Science*, vol. 239, pp. 367–375, 1988.
59. M. I. Flik and C. L. Tien, Thermal Phenomena in High-T_c Thin-Film Superconductors, *Ann. Rev. Heat Transfer*, Vol. 3, Hemisphere, New York, 1990.
60. H. Davies, The Reflection of Electromagnetic Waves from a Rough Surface, *Proc. IEE*, vol. 101, part IV, pp. 209–214, 1954.
61. M. Born and E. Wolf, *Principles of Optics*, 5th ed., Pergamon Press, Oxford, 1975.
62. P. Beckmann and A. Spizzichino, *The Scattering of Electromagnetic Waves from Rough Surfaces*, Macmillan, New York, 1963.
63. M. V. Berry, Diffractals, *J. Phys. A*, vol. 12, pp. 781–797, 1979.
64. M. V. Berry and T. M. Blackwell, Diffractal Echoes, *J. Phys. A*, vol. 14, pp. 3101–3110, 1981.
65. J. A. Ogilvy, Wave Scattering from Rough Surfaces, *Rep. Prog. Phys.*, vol. 50, pp. 1553–1608, 1987.
66. A. Majumdar and C. L. Tien, Reflection of Radiation by Rough Fractal Surfaces, to be presented at *AIAA Thermophysics and Heat Transfer Conference*, Seattle, June 18–20, 1990.
67. A. F. Houchens and R. G. Hering, Bidirectional Reflectance of Rough Metal Surfaces, in *Thermophysics of Spacecrafts and Planetary Bodies*, (G. B. Heller, ed.), Vol. 20, pp. 65–89, 1967.
68. T. F. Smith and R. G. Hering, Comparison of Bidirectional Reflectance Measurements and Model for Rough Metallic Surfaces, *Proc. 5th Symp. Thermophysical Properties*, pp. 429–435, 1970.
69. K. E. Torrance and E. M. Sparrow, Theory of Off-Specular Reflection from Roughened Surfaces, *J. Optical Soc. Am.*, vol. 57, pp. 1105–1114, 1967.
70. P. Beckmann, Shadowing of Random Rough Surfaces, *IEEE Trans. Antennas and Propagation*, vol. AP-13, pp. 384–388, 1965.
71. D. K. Edwards and I. Catton, Radiation Characteristics of Rough and Oxidized Metals, in *Advances in Thermophysical Properties at Extreme Temperatures and Pressures* (S. Gratch, ed.), ASME, New York, pp. 189–199, 1965.
72. R. C. Birkebak and A. Abdulkadir, Random Rough Surface Model for Spectral Directional Emittance of Rough Metal Surface, *Int. J. Heat Mass Transfer*, vol. 19, pp. 1039–1043, 1976.
73. S. Lofving, Hemispherical Emittance of Rough Metal Surfaces, *Appl. Phys. Lett.*, vol. 36, pp. 632–633, 1980.
74. P. J. Hesketh, J. N. Zemel, and B. Gebhart, Organ Pipe Radiant Modes of Periodic Micromachined Silicon Surfaces, *Nature*, vol. 324, pp. 549–551, 1986.
75. P. J. Hesketh, B. Gebhart, and J. N. Zemel, Measurements of the Spectral and Directional Emission from Microgrooved Silicon Surfaces, *J. Heat Transfer*, vol. 110, pp. 680–686, 1988.
76. J. F. Bussiere and M. Suenaga, Reduction of AC Losses of Nb_3Sn by Surface Treatment, *J. Appl. Phys.*, vol. 47, pp. 707–714, 1976.
77. M. Suenaga and M. Garber, Low-Loss Niobium-Tin Compound for Superconducting Alternating-Current Power Transmission Applications, *Science*, vol. 183, pp. 952–954, 1974.
78. C. P. Bean, Magnetization of High-Field Superconductors, *Rev. Modern Phys.*, vol. 36, pp. 31–39, 1964.
79. W. M. Rohsenow, Boiling, in *Handbook of Heat Transfer Fundamentals*, 2nd ed., (W. M. Rohsenow, J. P. Hartnett, and E. N. Ganic, eds.), chap. 12, McGraw-Hill, New York, 1985.
80. R. Cole, Boiling Nucleation, in *Advances in Heat Transfer*, Vol. 10 (J. P. Hartnett and T. F. Irvine, eds.), Academic Press, New York, pp. 86–164, 1974.
81. J. W. Westwater, Nucleate Pool Boiling, in *Lectures on Advanced Heat Transfer* (B. T. Chao, ed.), University of Illinois Press, pp. 217–232, 1969.
82. Y. Y. Hsu and R. W. Graham, *Transport Processes in Boiling and Two-Phase Systems*, Hemisphere, Washington, 1976.
83. W. M. Rohsenow, A Method of Correlating Heat Transfer Data for Surface Boiling of Liquids, *Trans. ASME*, vol. 74, pp. 969–976, 1952.
84. W. T. Brown Jr., Study of Flow Surface Boiling, Ph.D. thesis, Massachusetts Institute of Technology, Cambridge, Mass., 1967.

85. J. J. Lorenz, B. B. Mikic, and W. M. Rohsenow, The Effect of Surface Conditions on Boiling Characteristics, *Proc. 5th Int. Heat Transfer Conf.*, vol. 9, pp. 35–39, Tokyo, 1974.
86. B. B. Mikic and W. M. Rohsenow, A New Correlation of Pool-Boiling Data Including the Effect of Heating Surface Characteristics, *J. Heat Transfer*, vol. 91, pp. 245–250, 1969.
87. T. Miyachi and S. Yagi, Nucleate Boiling Heat Transfer on Horizontal Flat Surfaces, *Soc. Chem. Eng. Jpn.*, vol. 25, pp. 18–30, 1961.
88. J. P. Crutchfield, J. D. Farmer, N. H. Packard, and R. S. Shaw, Chaos, *Sci. Am.*, vol. 255, pp. 36–57, 1986.
89. H. Froehling, J. P. Crutchfield, J. D. Farmer, N. H. Packard, and R. S. Shaw, On Determining the Dimension of Chaotic Flows, *Physica 3D*, pp. 605–617, 1981.
90. H. G. Schuster, *Deterministic Chaos: An Introduction*, Physik-Verlag, Weinheim, West Germany, 1984.
91. F. A. Williams, *Combustion Theory*, 2nd Ed., Benjamin Cummings, Menlo Park, Calif., 1985.
92. J. Mantzaras, P. G. Felton, and F. V. Bracco, Fractals and Turbulent Premixed Engine Flames, *Combustion and Flame*, vol. 77, pp. 295–310, 1989.
93. B. B. Mandelbrot, On the Geometry of Homogeneous Turbulence with Stress on the Fractal Dimension of the Iso-Surfaces of Scalars, *J. Fluid Mech.*, vol. 72, pp. 401–416, 1975.
94. B. B. Mandelbrot, Intermittent Turbulence and Fractal Dimensions: Kurtosis and the Spectral Exponent 5/3 + B, in *Turbulence and Navier-Stokes Equations, Lecture Notes in Mathematics* (R. Teman, ed.), vol. 565, pp. 121–145, Springer, New York, 1976.
95. A. J. Chorin, Scaling Laws in the Vortex Lattice Model of Turbulence, *Commun. Math. Phys.*, vol. 114, pp. 167–176, 1988.
96. A. J. Chorin, Spectrum, Dimension and Polymer Analogies in Fluid Turbulence, *Phys. Rev. Lett.*, vol. 60, pp. 1947–1949, 1988.
97. S. Lovejoy, Area-Perimeter Relation for Rain and Cloud Areas, *Science*, vol. 216, pp. 185–187, 1982.
98. H. G. E. Hentschel and I. Procaccia, Relative Diffusion in Turbulent Media: The Fractal Dimension of Clouds, *Phys. Rev. A*, vol. 29, pp. 1461–1470, 1984.
99. F. C. Gouldin, S. M. Hilton, and T. Lamb, Experimental Evaluation of the Fractal Geometry of Flamelets, *22nd Symp. (Int.) on Combustion*, pp. 541–550, 1988.
100. M. Murayama and T. Takeno, Fractal-like Character of Flamelets in Turbulent Premixed Combustion, *22nd Symp. (Int.) on Combustion*, pp. 551–559, 1988.
101. N. Peters, Laminar Flamelet Concepts in Turbulent Combustion, *21st Symp. (Int.) on Combustion*, pp. 1231–1250, 1986.
102. D. C. Siegla and G. W. Smith, *Particulate Carbon Formation During Combustion*, Plenum, New York, 1983.
103. S. C. Lee and C. L. Tien, Optical Constants of Soot in Hydrocarbon Flames, *18th Symp. (Int.) on Combustion*, pp. 1159–1166, 1981.
104. B. L. Drolen and C. L. Tien, Absorption and Scattering of Agglomerated Soot Particulate, *J. Quant. Spectroscopy Radiative Transfer*, vol. 37, pp. 433–448, 1987.
105. S. Kumar and C. L. Tien, Dependent Scattering and Absorption of Radiation by Small Particles, *Proc. 24th ASME/AIChE National Heat Transfer Conf.*, Pittsburgh, 1987.
106. R. Richter, L. M. Sander, and Z. Cheng, Computer Simulations of Soot Aggregation, *J. Colloid Interface Sci.*, vol. 100, pp. 203–209, 1984.
107. T. A. Witten and L. M. Sander, Diffusion-Limited Aggregation, A Kinetic Critical Phenomenon, *Phys. Rev. Lett.*, vol. 47, pp. 1400–1403, 1981.
108. P. Meakin, Diffusion-Controlled Deposition on Fibres and Surfaces, *Phys. Rev. A*, vol. 27, pp. 2616–2623, 1983.
109. L. Niemeyer, L. Pietronero, and H. G. Wiesmann, Fractal Dimension of Dielectric Breakdown, *Phys. Rev. Lett.*, vol. 52, pp. 1033–1036, 1984.
110. L. Pietronero, A. Erzan, and C. Evertsz, Theory of Laplacian Fractals: Diffusion-Limited Aggregation and Dielectric Breakdown Model, *Physica A*, vol. 151, pp. 207–245, 1988.
111. P. Meakin, Formation of Fractal Clusters and Networks by Irreversible Diffusion-Limited Aggregation, *Phys. Rev. Lett.*, vol. 51, pp. 1119–1122, 1983.

112. M. Kolb, R. Botet, and R. Jullien, Scaling of Kinetically Growing Clusters, *Phys. Rev. Lett.*, vol. 51, pp. 1123–1126, 1983.
113. R. Jullien, Fractal Aggregates, *Comments on Condensed Matter Physics*, vol. 13, pp. 177–205, 1987.
114. M. von Smoluchowski, Drei Vortrage uber Diffusion, Brownsche Molekularbewegung und Koagulation von Kolloidteilchen, *Phys. Z.*, vol. 17, pp. 557–599, 1916.
115. M. V. Berry and I. C. Percival, Optics of Fractal Clusters such as Smoke, *Optica Acta*, vol. 33, pp. 577–591, 1986.
116. D. W. Schaefer, J. E. Martin, P. Wiltzius, and D. Cannell, Fractal Geometry of Colloidal Aggregates, *Phys. Rev. Lett.*, vol. 52, pp. 2371–2374, 1984.
117. D. A. Weitz and M. Oliveria, Fractal Structures formed by Kinetic Aggregation of Aqueous Gold Colloids, *Phys. Rev. Lett.*, vol. 52, pp. 1433–1436, 1984.
118. M. Y. Lin, H. M. Lindsay, D. A. Weitz, R. C. Ball, R. Klein, and P. Meakin, Universality of Fractal Aggregates as Probed by Light Scattering, *Proc. Roy. Soc. London A*, vol. 423, pp. 71–87, 1989.
119. W. L. Grosshandler and A. T. Modak, Radiation from Nonhomogeneous Combustion Products, *18th Symp. (Int.) on Combustion*, pp. 601–609, 1981.
120. M. E. Kounalakis, J. P. Gore, and G. M. Faeth, Turbulence/Radiation Interactions in Nonpremixed Hydrogen/Air Flames, *22nd Symp. (Int.) on Combustion*, pp. 1281–1290, 1988.
121. W. L. Grosshandler, Radiative Heat Transfer in Nonhomogeneous Gases: A Simplified Approach, *Int. J. Heat Mass Transfer*, vol. 23, pp. 1147–1459, 1980.
122. V. I. Tatarski, *Wave Propagation in a Turbulent Medium*, McGraw-Hill, New York, 1961.
123. P. Cheng, Heat Transfer in Geothermal Systems, *Adv. Heat Transfer*, vol. 14, pp. 1–105, 1978.
124. K. S. Udell, Heat Transfer in Porous Media Heated from Above with Evaporation, Condensation and Capillary Effects, *J. Heat Transfer*, vol. 105, pp. 485–492, 1983.
125. A. E. Scheidegger, *The Physics of Flow Through Porous Media*, Toronto University Press, Toronto, 1957.
126. D. Wilkinson and J. F. Willemsen, Invasion Percolation: A New Form of Percolation Theory, *J. Phys. A*, vol. 16, pp. 3365–3376, 1983.
127. J. C. Slattery, *Momentum, Energy and Mass Transfer in Continua*, Kreiger, New York, 1978.
128. S. Whitaker, Diffusion and Dispersion in Porous Media, *AIChE J.*, vol. 13, pp. 420–427, 1967.
129. P. Cheng and C. T. Hsu, Applications of Van Driest's Mixing Length Theory to Transverse Thermal Dispersion in Forced Convection Flow Through a Packed Bed, *Int. Commun. Heat Mass Transfer*, vol. 13, pp. 613–625, 1986.
130. S. M. Kuo and C. L. Tien, Transverse Dispersion in Packed Sphere Beds, *Proc. 25th Natl. Heat Transfer Conf.*, vol. 1, pp. 629–634, 1988.
131. M. L. Hunt and C. L. Tien, Diffusion and Dispersion for Boundary Layer Flow in a Porous Medium, *Proc. 25th Natl. Heat Transfer Conf.*, vol. 1, pp. 635–642, 1988.
132. C. Sammis, G. King, and R. Biegel, The Kinematics of Gouge Deformation, *Pure Appl. Geophys.*, vol. 125, pp. 777–812, 1987.
133. A. H. Thompson, A. J. Katz, and C. E. Krohn, The Microgeometry and Transport Properties of Sedimentary Rock, *Adv. Physics*, vol. 36, pp. 625–694, 1987.
134. L. W. Gelhar, A. Mantoglou, C. Welty, and K. R. Rehfeldt, A Review of Field Scale Physical Solute Transport Processes in a Saturated and Unsaturated Porous Media, Rep. EA-4190, Electric Power Research Institute, Palo Alto, Calif., 1985.
135. B. Ross, Dispersion in Fractal Fracture Networks, *Water Resources Res.*, vol. 22, pp. 823–827, 1986.
136. S. W. Wheatcraft and S. C. Tyler, An Explanation of Scale-Dependent Dispersivity in Heterogeneous Aquifers Using Concepts of Fractal Geometry, *Water Resources Res.*, vol. 24, pp. 566–578, 1988.
137. E. Guyon, L. Oger, and T. J. Plona, Transport Properties in Sintered Porous Media Composed of Two Particle Sizes, *J. Phys. D*, vol. 20, pp. 1637–1644, 1987.
138. R. Lenormand, Flow Through Porous Media: Limits of Fractal Patterns, *Proc. Roy. Soc. London A*, vol. 423, pp. 159–168, 1989.
139. P. Z. Wong, The Statistical Physics of Sedimentary Rock, *Physics Today*, pp. 24–32, December, 1988.

140. K. J. Maloy, J. Feder, and G. Jossang, Viscous Fingering Fractals in Porous Media, *Phys. Rev. Lett.*, vol. 55, pp. 2688–2691, 1985.
141. L. Paterson, Diffusion-Limited Aggregation and Two-Fluid Displacements in Porous Media, *Phys. Rev. Lett.*, vol. 52, pp. 1621–1624, 1984.
142. K. J. Maloy, F. Boger, J. Feder, and G. Jossang, Dynamics and Structure of Viscous Fingers in Porous Media, in *Time-Dependent Effects in Disordered Materials* (R. Pynn and T. Reste, eds.), Plenum, New York, pp. 111–138, 1987.
143. A. Majumdar and C. L. Tien, Effects of Surface Tension on Film Condensation in Porous Media, *J. Heat Transfer*, vol. 112, pp. 751–757, 1990.
144. T. M. Shaw, Drying as an Immiscible Displacement Process with Fluid Counterflow, *Phys. Rev. Lett.*, vol. 59, pp. 1671–1674, 1987.
145. W. J. Hamilton, *Textbook of Human Anatomy*, St. Martin's Press, New York, 1956.
146. R. G. Kessel and R. H. Kardon, *Tissues and Organs: A Text Atlas of Scanning Electron Microscopy*, W. H. Freeman, San Francisco, 1979.
147. R. C. Eberhart, Thermal Models of Single Organs, in *Heat Transfer in Medicine and Biology* (A. Shitzer and R. C. Eberhart, eds.), Plenum, New York, vol. 1, chap. 12, 1985.
148. M. M. Chen, The Tissue Energy Balance Equation, in *Heat Transfer in Medicine and Biology* (A. Shitzer and R. C. Eberhart, eds.), vol. 1, chap. 7, 1985.
149. S. Weinbaum and L. M. Jiji, A New Simplified Bioheat Equation for the Effect of Blood Flow on Local Average Tissue Temperature, *J. Biomech. Eng.*, vol. 107, pp. 131–139, 1985.
150. A. S. Poppel, Network Models of Perpheral Circulation, in *Handbook of Bioengineering* (R. Shalak and S. Chien, eds.), McGraw-Hill, New York, 1987.
151. Y. H. Liu and E. L. Rittman, Branching Pattern of Pulmonary Arterial Tree in Anesthetized Dogs, *J. Biomech. Eng.*, vol. 108, pp. 289–293, 1986.
152. M. M. Chen and K. R. Holmes, Microvasculature Contributions in Heat Transfer, in *Thermal Characteristics of Tumors: Applications in Detection and Treatment* (R. K. Jain and P. M. Gullino, eds.), Annals of New York Academy of Science, vol. 335, pp. 137–150, 1980.
153. S. Weinbaum, L. M. Jiji, and D. E. Lemons, Theory and Experiment for the Effect of Vascular Microstructure on Surface Tissue Heat Transfer—Part I: Anatomical Foundation and Model Conceptualization, *J. Biomech. Eng.*, vol. 106, pp. 321–330, 1984.
154. L. M. Jiji, S. Weinbaum, and D. E. Lemons, Theory and Experiment for the Effect of Vascular Microstructure on Surface Tissue Heat Transfer—Part II: Model Formulation and Solution, *J. Biomech. Eng.*, vol. 106, pp. 331–341, 1984.
155. J. W. Baish, A Fractal-Based Theory of Bioheat Transfer, *AIChE Symp. Ser.*, vol. 85, pp. 406–411, 1989.
156. E. N. Lightfoot, *Transport Phenomena and Living Systems*, John Wiley, New York, 1979.
157. B. Rubinsky and E. G. Carvalho, Transient Mass Transfer Processes During the Perfusion of a Biological Organ with a Cryophylactic Agent Solution, *Cryobiology*, vol. 19, pp. 70–82, 1982.
158. J. B. Bassingthwaite, A Concurrent Flow Model for Extraction During Transcapillary Passage, *Circulation Res.*, vol. 35, pp. 483–503, 1979.

CHAPTER
THREE

THERMAL SHOCK PHENOMENA UNDER HIGH RATE RESPONSE IN SOLIDS

Da Yu Tzou

NOMENCLATURE

a_i	coefficients in the general boundary condition at the crack surface
A, B	arbitrary constant in the eigenfunction $H(\theta)$
c	$v/2\alpha$, a parameter used in the thermal wave equation, 1/m
C	speed of heat propagation in the solid, m/s
C_p	heat capacity, kJ/kg °C
C_s	shear wave speed $\sqrt{\mu/\rho}$, m/s
E	Young's modulus, Pa
f	transformation variable in the hypospace r
f_g	the Green's function for f
F	a coefficient defined as $[1 - h^2(1 - M^2)]/h$
g_i	mathematical coefficients defined in Eq. (57)
h	scaling factor in the stretching coordinates
H	angular variations of the near-tip temperature
k	isotropic thermal conductivity, W/m °C
k_x, k_y	principal values of the orthotropic thermal conductivity, W/m °C
K_i	modified Bessel function of the second kind of order i

M	thermal Mach number defined as v/C
M_s	Mach number relative to the shear wave speed, v/C_s
n	integer in the expression of eigenvalues
q, q_i	heat flux vector and its components in the ith direction, W/m^2
Q_i	angular variations of the near-tip heat flux components
r	radial distance measured from the moving heat source or the crack tip
r_ζ	transformation function to the hypospace, m
R	radius of the circle for observation, m
S	general heat source term, W/m^3
t	physical time, s
T	temperature, deg
T_g	the Green's function for T, deg
u_i	displacement vector, m
v	speed of the moving heat source or propagating crack tip, m/s
α	thermal diffusivity, m^2/s
β	a function of Poisson's ratio, $(1 + \nu)/(1 - \nu)$
γ	transformation function on the independent variable θ
Γ	time function in the near-tip temperature distribution
δ	Dirac delta function; Kronecker delta function
ϵ	coefficient of thermal expansion, (mm/mm)/°C
ζ	transformation function on the independent variable θ
ζ_i	stretching coordinates, m
η	dummy variable of integrations
θ	polar angle of the moving coordinate system, deg
κ	a function of Poisson's ratio, $(1 - 2\nu)/[2(1 - \nu)]$
λ	eigenvalues; exponent of r for the angular variation of near-tip temperature
μ	shear modulus, Pa
ν	Poisson's ratio
ξ_i	$i = 1, 2$. Moving coordinates centered at the crack tip, m
ρ	mass density, kg/m^3
σ	stresses, Pa
τ	relaxation time, s
Φ	thermoelastic displacement potential, transformation function on the dependent variable H, N
Φ_0	thermoelastic potential solely contributed by the apparent heat source, N
ω	angular velocity of the moving crack, rad/s
∇	gradient operator, 1/m
∇^2	Laplacian operator

Subscripts and Superscripts

$(\ )_{,i}$	$\partial/\partial\xi_i$, $i = 1, 2$
$(\ )_D$	the quantity in the thermal diffusion model
X_i	the component of X in the ith direction
X_{ij}	tensorial quantity of X
$\mathbf{X}$	vector quantity of X

1 INTRODUCTION

The thermal shock wave studied in this chapter results from the accumulation of thermal energy in a preferential direction around either a *rapidly moving heat source* or a *propagating crack tip* in the solid medium. It is a direct consequence of the thermal inertia effect in the wave theory for heat conduction. For this reason, this chapter starts with a brief review of the hyperbolic theory of heat conduction with emphasis on the research developed in the past for engineering applications. Discussions of the physical significance of the thermal waves follow in Sec. 3 to emphasize the physical mechanisms implemented in the wave theory which are excluded in Fourier's law of heat conduction. The conditions for the dominance of the wave nature over the diffusion behavior in the process of heat transport are examined to reveal the engineering environment under which considerations of thermal waves are necessary. The thermal shock phenomena are then introduced in Sec. 4. Starting with two-dimensional formulations in terms of both temperature and heat flux vector, the formation of thermal shock waves is investigated for problems involving a rapidly moving heat source and a fast propagating crack tip in the solid. As a combination, they are called *moving origins* in this chapter. The thermal field is characterized by the thermal Mach number, M, which is defined as the ratio of the speed of the moving origin, v, to that of the heat propagation, C, in the solid. Mathematically, $M = v/C$. In transition from the subsonic ($M < 1$) to the transonic ($M = 1$) to the supersonic ($M > 1$) ranges, the evolution of the heat-affected zone versus the thermally undisturbed zone in the physical domain, the discontinuous variations of the physical quantities across the thermal shock surfaces, and the swinging phenomenon for the variations of the temperature level in transition of the thermal Mach number are the salient features of the theory which cannot be depicted by the classical diffusion model. The direct influence of the thermal shock formation is examined in Sec. 5 by considering the resulting thermal stresses around the moving heat source. Mathematically, the thermal inertia effect is shown to present as the *highest-order differential* in the field equation governing the Neuber-Papkovich thermoelastic displacement potential. The resulting thermal stress field is therefore affected in an intrinsic fashion. Especially in the presence of the thermal shock waves at the transonic and in the supersonic ranges, with $M \geq 1$, the thermal cracks emanating from the heat source display a different damage pattern than those predicted by the diffusion model, which not only has practical values in defense-oriented industries but also provides a measurable phenomenon for the experimental confirmation of thermal shock formation. With the knowledge thus developed for two-dimensional problems, a general formulation for the thermal shock formation in a three-dimensional solid is provided in Sec. 6. The purpose is to study the effects of the additional dimensionality on the temperature level in the heat-affected zone and the geometric configuration of the thermal shock surface. Due to the three-dimensional structure of the thermal shock waves, a special phenomenon in the supersonic range that the temperature level *increases* with the distance away from the heat source in the direction perpendicular to the plane containing the moving heat source is discussed. Finally, the three-dimensional formulation for the thermoelastic stresses is provided. The intention is to develop fundamental understandings of the interactions between the thermal and the stress waves interwoven in the heat-affected zone.

2 BRIEF REVIEW OF RESEARCH ON THERMAL WAVE PROPAGATION

The merit of the thermal wave model lies in its unique way of describing the thermal disturbance propagating in the solid with a finite speed. It is an improved model intended to remove the assumption of infinite speed of heat propagation in the classical diffusion theory. The model was originally proposed by Maxwell [1] via kinetic considerations of molecules. The earliest experimental proof for the thermal disturbances propagating as waves at finite speed was made by Peshkov [2]. At a temperature of 1.4 K, the speed of heat propagation in liquid helium was determined to be 19 m/s. Based on a heuristic basis that the onset of heat to flow must occur in a finite period of time, Morse and Feshbach [3], Vernotte [4–6], and Cattaneo [7] further proposed a modified model which introduced the concept of thermal relaxation time. By describing the diffusion process as a discontinuous random migration, Fourier's law of heat conduction appears to be a low-order approximation in this model [8–10].

The physical contents of the thermal wave model have been investigated from various physical points of view. It includes, for example, a collision model established on the basis of statistical mechanics [11–14], identification of the analogy between the random walk and the diffusion process [10], a concept based on the scatter center in an otherwise continuous medium [15], modification of thermodynamics with fading memory [16] and irreversibility [17], considerations based on the special theory of relativity [18, 19], and interpretations in terms of the kinetic theory of molecules [20, 21] and the quantum mechanical form of the Boltzmann transport equation [22]. Some theoretical calculations were also made for the most crucial parameter—the relaxation time in the thermal wave phenomenon [23–26]. The concept of transport of a phonon density disturbance was adopted, and the Debye-type model in these works leads directly without any recourse to the two-fluid model proposed by Landau [27]. Second-sound propagation in dielectric solids was also discussed by Prohofsky and Krumhansl [28].

In terms of the temperature representation, the consideration given to the thermal disturbance propagating with a finite speed in the solid leads essentially to a hyperbolic-type energy equation [29–32]. Along with the physical approaches, the mathematical structure and the physical interpretations of the thermal wave model have also been explored extensively. Based on the thermal wave equation with temperature representation, the temperature and heat flux distributions in a semi-infinite medium due to a step change at the boundary temperature were studied by Baumeister and Hamill [33, 34]. The transient behavior of two contacting materials at short times is an important subject in thermal wave propagation. Fauske [35] investigated the correlation between the instantaneous interface temperature of two suddenly contacting liquids to avoid the rapid development of spontaneous nucleation in the cooler liquid. Kazimi and Erdman [36], in 1975, studied the time interval in which non-Fourier effects are important in assessing the interface temperature between uranium dioxide and sodium. Maurer and Thompson [37] considered another important situation involving high heat flux imposed on the surface of the medium. In another work by Chen [38], the temperature rate was further emphasized in the history of heat propagation. Under various initial and boundary conditions, the nature of thermal waves in short time response was further studied by other researchers [39–41].

Luikov et al. [42] further expanded the theoretical structures of the hyperbolic theory of heat conduction by incorporating the time rate change of thermal wave speed in the unsteady state of heat transfer. In the resulting energy equation with temperature representation, it involves a modified coefficient of $C_{,t}/C^3$, with C being the finite wave speed, in front of the diffusion term. Along with considerations of relaxation time in the thermal wave model, some other efforts were also devoted to modifications of Fourier's law. The purpose, of course, is to remove the discrepancy of assuming infinite speed of heat propagation in diffusion theory. To be mentioned are the work done by Zel'dovich and Raizer [43], in which a power-type dependence of heat capacity and thermal conductivity of the medium on temperature was adopted, and that by Letcher [44, 45] for boundary-layer theory in heat conduction. The latter essentially led to work done by Kao [46], which incorporated the finite speed of heat propagation in the analysis of temperature in thin surface layers. The method of characteristics was applied to the hyperbolic theory of heat conduction by Wiggert [47]. The stability criterion for the numerical analysis in his work explicitly involves the relaxation time. The integrodifferential relation for the internal energy and heat flux was considered by Amos and Chen [48], and the method of Laplace transform was applied to solve the integrodifferential equation. On the numerical site, because the propagation of thermal waves in a solid medium involves a discontinuity at the thermal wavefront, special considerations on the finite-difference simulations were given by Taitel [49] and by Carey and Tsai [50]. Some recent work by Özisik and Vick predicted the growth and decay of a thermal pulse in both a semi-infinite [51] and a finite [52] medium. In their work the temperature pulse propagating at a finite speed has finite height and width, which is qualitatively similar to that observed experimentally by Bertman and Sandiford [53]. They also found that the hyperbolic heat conduction equation with body heat generation includes an additional term involving the time derivative of the energy generation. Later, in 1985, Frankel et al. [54] further demonstrated that the energy equation with flux representation in the thermal wave model is actually more convenient to use than that with the temperature representation for the problems involving heat flux-specified boundary conditions.

The finite speed of thermal wave propagation has also attracted some attention in the fluid motion and the thermoelastic response of the solids. Nettleton [55] considered that the thermal energy in a liquid is carried by very high-frequency waves and showed that the pressure and the internal energy depend on the temperature in a nonlinear fashion. Boley [56], Boley and Tolins [57], Brazel and Nolan [58], and Kao [59] studied the significance of non-Fourier effects in assessing the thermal stresses generated by a suddenly changed surface temperature. A coupled formulation of thermoelasticity incorporating the finite wave speed of heat propagation was proposed by Lord and Shulman [60]. In addition to the diffusion and the wave terms, the energy equation also includes additional terms involving the first and the second time derivatives of the dilatation (ϵ_{kk}). The stress jump in a one-dimensional solid induced by the finite wave speed of heat propagation was investigated by Achenbach [61]. Two wavefronts were shown to exist in the physical domain. Also, in the absence of the coupling interactions, he showed that the speed of the fast wavefront is greater than the larger of C_L and C, with C_L being the speed of the normal stress wave, and the slow wavefront is less than that of the smaller of C_L and C. Basically, the theoretical structure considered by Popov [62] is similar to that of Lord and Shulman [60], and other work done by Wilms [63] provided an approximate method of solution to the problem considered by Boley and Tolins [57].

From the literature summarized above, we may perceive that significant research effort has been devoted to the wave characteristics of heat propagation in the engineering environment with low temperature, short time response, high heat flux, high rate of temperature change, thin surface layers, and material interfaces. These conditions reflect the important situations considered in thermal resistance design for high-performance structures.

3 PHYSICAL SIGNIFICANCE OF THERMAL WAVES

The physical basis of thermal wave mechanics can be illustrated from the viewpoints of the kinetic theory of molecules conducting the thermal energy and irreversible thermodynamics with fading memory. A summary of these approaches is helpful in illuminating the essential mechanisms implemented in the thermal wave model.

3.1 Causality Restriction on Finite Wave Speed

A typical example employing the kinetic theory of molecules is the work done by Kelly [18], who reformulated the covariant form of the Boltzmann equation and imposed the causality restriction—namely, that the speed of any object in the universe cannot exceed the speed of light in vacuum—on the process of heat transport. By expanding the relativistic distribution function f and the elastic collision factor S in terms of the spherical harmonics in the momentum space $\mathbf{p}$, a generalized equation governing the averaged number of thermally activated particles $n(r, t)$ over the entire momentum space was obtained:

$$\frac{\partial^2 n}{\partial t^2} + \langle \nu \rangle \frac{\partial n}{\partial t} - \langle \nu S \rangle \, n = \frac{1}{3} \langle v^2 \rangle \, \nabla^2 n \tag{1}$$

where ν is the relativistic counterpart of the momentum transfer collision frequency, v is the group velocity of the thermally activated particles, and the brackets indicate kinetic averages,

$$\langle \nu, v^2, \nu S \rangle = \frac{1}{n} \int [\nu, v^2, \nu S] \, f_0 \, d^3\mathbf{p} \tag{2}$$

with f_0 being the dominant term of f in the expansion by spherical harmonics. Equation (1) has indeed the proposed form of the thermal wave equation in the hyperbolic theory of heat conduction. In addition to the wave term appearing as a direct consequence of the causality condition, it contains also an extra term "$\langle \nu S \rangle n$" in the equation, which is out of the expectation of the proposal made originally by Morse and Feshbach

[3]. In the situation, however, where the collision frequency ν among the particles conducting the thermal energy is a constant in the momentum space, ν can be taken out of the brackets and, due to the normalization condition of $n^{-1} \int f_0 \, d^3\mathbf{p} = 1$,

$$\langle \nu \rangle = \nu \left\{ \frac{1}{n} \int f_0 \, d^3\mathbf{p} \right\} = \nu \tag{3}$$

The averaged value of $\langle \nu S \rangle$, the coefficient in front of n in Eq. (1), in this case is essentially $\nu \langle S \rangle$, which is equal to 0 because the number of particles is not altered in the elastic collision. Equation (1), under the assumptions that the collision frequency ν is a constant and the averaged number of thermally activated particles in the momentum space is proportional to the temperature, then reduces to the thermal wave equation in the hyperbolic theory of heat conduction and the thermal diffusion coefficient,

$$\alpha = \frac{\langle v^2 \rangle}{3\nu} \tag{4}$$

and the speed of heat propagation in the solid,

$$C = \frac{\langle v \rangle}{\sqrt{3}} \tag{5}$$

are obtained automatically as a consequence. Most important, from a strictly theoretical point of view, a solid medium with a finite value of the thermal diffusivity (α), that is, $\langle v^2 \rangle$ is finite and $\nu \neq 0$ referring to Eq. (4), should not have an infinite value for the speed of heat propagation C. Two special features are also observed from Eqs. (4) and (5). First, Eq. (4) for the thermal diffusivity has exactly the same form as that obtained by the nonrelativistic Boltzmann equation formulated in the classical diffusion theory [12]. As a consequence, the diffusion behavior for heat transport is incorporated in the thermal wave model. Second, the speed of heat propagation in the solid is proportional to the algebraic mean of the group velocity of the thermally activated particles in the three-dimensional momentum space, which not only possesses a finite value under the constraint of the causality restriction, but also depends on the substructure of the solid medium transferring the thermal energy. Combining these two features, therefore, the hyperbolic theory of heat conduction is indeed a model describing the interactions between the *diffusion behavior* and the *wave nature* of a thermal disturbance propagating at a finite speed in a solid medium [64–77]. For problems involving a rapidly propagating crack tip, for example, an asymptotic expansion was made by Tzou [66, 67] to examine the relative contributions from the diffusion and the wave behavior. It has been found that in the early stage of heat conduction with the response time being short, the wave nature is highly pronounced in the tip region and the heat transport process is completely dominated by wave propagation rather than diffusion. For irreversible physical processes occurred in a shorter time interval than that required for the diffusion behavior to be retrieved, such as crack initiation around a rapidly moving heat source [68–72] or crack kinking and curving [73] in a brittle material, the thermal wave model accounting for the finite speed of heat propagation seems to be more appropriate to use in describing the interactions between the thermal and the mechanical fields.

3.2 Path Dependency of Heat Flux Vector Versus Temperature Gradient

Another approach, adopted by Gurtin and Pipkin [16], is the typical example deriving the constitutive relation for the thermal wave model from the first and the second laws of irreversible thermodynamics. Starting with a general formulation for nonlinear materials with memory which involves consideration of the admissible state functionals in the local Clausius-Duhem inequality, the heat flux as well as the entropy were determined in an energy-bearing medium. The nonlinear theory thus developed provides a linear description of heat conduction by preserving up to the first-order term of the temperature gradient in the series expansion of the memorial functional. The linearized version of the heat flux vector $\mathbf{q}$ derived in their work,

$$\mathbf{q}(t) = -\int_0^\infty a(s)\,\nabla T(t-s)\,ds \tag{6}$$

with $a(s)$ being called the heat-flux relaxation function depending on the material types, is indeed the constitutive relation employed in the hyperbolic theory of heat conduction. The physical essence behind Eq. (6) can be better illustrated by considering the differential form of the constitutive relation in the thermal wave model [64, 65, 67]. With details discussed later, in Sec. 4, the heat flux vector in the thermal wave model can be arranged to relate to the temperature gradient by

$$\mathbf{q}(t) = -\left(\frac{kC^2}{\alpha}\right)\exp\left(-\frac{C^2 t}{\alpha}\right)\int_0^t \exp\left(\frac{C^2\eta}{\alpha}\right)\nabla T(r,\eta)\,d\eta \tag{7}$$

From the mean-value theorem, then, an algebraic expression,

$$\mathbf{q}(t) = -\left[\left(\frac{kC^2}{\alpha\tau}\right)\exp\left(\frac{C^2(\tau-t)}{\alpha}\right)\right]\nabla T(\mathbf{r},\tau) \qquad \text{for } \tau \in [0, t] \tag{8}$$

is immediately obtained in a form parallel to Fourier's law of heat conduction. This equation clearly indicates the path dependency of the heat flux vector in relation to the temperature gradient established between two adjacent material points. First of all, it shows a *phase lag* between the heat flux vector $\mathbf{q}(t)$ at time t and the temperature gradient $\nabla T(\mathbf{r}, \tau)$ established at a previous time instant $\tau < t$. This is why some researchers call the hyperbolic theory of heat conduction the one with *thermal inertia*. Second, due to the effect of thermal inertia, the "apparent" thermal conductivity denoted by the quantity enclosed in brackets in Eq. (8) not only changes with time but also depends on the entire history of the temperature gradient from 0 to t—as reflected in the value of τ determined from the mean-value theorem. This observation leads to the history dependency of the heat flux vector, and Gurtin and Pipkin called this behavior the one with *memory*.

In brief, the thermal inertia is a unique feature in the hyperbolic theory of heat conduction. The phase lag between the heat flux vector and the temperature gradient is a direct consequence of this physical phenomenon which results from a description of the thermal disturbance propagating with a finite speed in a solid.

4 THERMAL SHOCK PHENOMENA

Due to the similarity of heat and mass transfer in the medium, the thermal shock formation around a moving origin could be illustrated qualitatively by considering a water jet impinging on the surface of a big chunk of sponge as shown in Fig. 1. For a stationary water jet, as shown in Fig. 1*a*, a distinct water diffusion pattern emanating from the point of impingement can be observed. For the water jet moving with a velocity v as shown in Fig. 1*b*, on the other hand, two distinct "water lines" are expected which separate the dry area (moisture-undisturbed zone) from the region with water concentration (moisture-affected zone). Obviously, the water concentration has a discontinuous variation across the water lines, and the location of the water lines thus formed depends on the moving speed of the water jet. Because the wave speed of mass transfer in the sponge is relatively slow, the shock phenomenon across the water line is easier to explain according to common sense. For the corresponding problem in heat transfer involving an intensified heat source impinging on the surface of a structure, because the wave speed of heat propagation is much faster than that of the mass transfer in the sponge, the experiments would become more involved and some advanced techniques may be needed. The analytical model proposed in this section aims to describe thermal shock phenomena in the process of heat transport on a quantitative basis. The moving origins under consideration include both a moving heat source and a propagating crack tip in a solid medium. At this stage of development, only the linear theory is considered and the thermal properties of the solid medium are assumed to be isotropic and homogeneous.

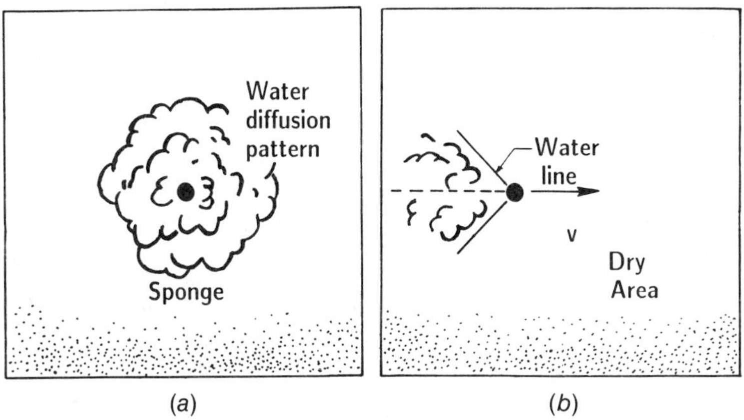

Figure 1 A chunk of sponge impinged by (*a*) a stationary water jet and (*b*) a moving water jet with a velocity v.

4.1 Mathematical Formulations in Terms of Temperature and Heat Flux Vector

The energy and constitutive equations in the thermal wave model can be expressed as

$$- \nabla \cdot \mathbf{q} + S = \rho C_p T_{,t}$$
$$\left(\frac{\alpha}{C^2}\right) \mathbf{q}_{,t} + \mathbf{q} = - k \nabla T \tag{9}$$

with ρ being the density of the medium, k the thermal conductivity, C_p the heat capacity, S the body heat source, α the thermal diffusivity, and C the finite speed of heat propagation in the solid. In the case that the wave speed C approaches infinity, the assumption made in the diffusion theory, the second of Eq. (9) reduces to Fourier's law of heat conduction and the diffusion behavior is retrieved. The quantity α/C^2 has the dimension of time and it is defined as the *relaxation time* in the thermal wave model. It is traditionally used as the physical parameter for the importance of considering thermal wave propagation. For a solid medium with large relaxation time, i.e., either the thermal diffusivity α is large or the speed of heat propagation C is small, the heat transport process in the solid is dominated by the wave propagation rather than diffusion. The relationship between the heat flux **q** and the temperature gradient ∇T can be obtained by integrating the second of Eq. (9). The result is shown by Eq. (7).

Eliminating either temperature T or heat flux **q** from Eq. (9) results in a field equation with either flux (**q**) or temperature (T) representations:

$$\nabla[\nabla \cdot \mathbf{q}] - \nabla S = \left(\frac{1}{\alpha}\right) \left[\left(\frac{\alpha}{C^2}\right) \mathbf{q}_{,tt} + \mathbf{q}_{,t}\right] \tag{10}$$

$$\alpha \nabla^2 T + \left(\frac{1}{\rho C_p}\right) \left[S + \left(\frac{\alpha}{C^2}\right) S_{,t}\right] = \left(\frac{\alpha}{C^2}\right) T_{,tt} + T_{,t} \tag{11}$$

For problems involving heat flux-specified boundary conditions, it has been found [54] that the **q** representation is more convenient to use than the T representation. This is due to the complicated integral relationship between the heat flux vector and the temperature gradient in the thermal wave model.

From a mathematical point of view, the wave characteristics in the process of heat transport dominate over the diffusion behavior if

$$\left(\frac{\alpha}{C^2}\right) T_{,tt} \gg T_{,t} \qquad \text{or} \qquad T_{,t} \gg \exp\left[\left(\frac{C^2}{\alpha}\right) t\right] \tag{12}$$

The same criterion applies to the heat flux vector **q** because the mathematical structure of the time derivatives in the **q** and T representations is identical. Equation (12) states that the wave nature in the process of heat transport can be important if (1) the critical frequency C^2/α (the reciprocal of the relaxation time) is small, (2) the response time t is short, or (3) the time rate change of temperature (heat flux) $T_{,t}$ is high. The first condition involves the intrinsic properties of the solid medium, while the second and the third conditions also depend on the combined effect of the geometric configuration and the thermal loading imposed on the system. In the neighborhood of a localized moving energy source, for example, the time rate of change of temperature is extremely high and the wave nature could be pronounced even if the relaxation time of the solid medium is small. The criterion established by Eq. (12) is therefore more general than the traditional one, which only takes the intrinsic properties of the solid medium into consideration.

In the T representation, Eq. (11), the presence of an *apparent* heat source involving the first-order time derivative of the real heat source applied to the solid medium is another salient feature of the thermal wave model. It implies that under high rate response in the solid, the established temperature level depends not only on the intensity but also on the time rate of change of the heat source. Again, this is a result attributed to the finite wave speed C. In the case that C approaches infinity, Eqs. (10) and (11) reduce to those in the classical diffusion theory employing Fourier's law.

The mathematical structures of the $\mathbf{q}$ and T representations shown by Eqs. (10) and (11) will be analyzed in detail by considering the thermal field around a moving heat source and a fast-propagating crack tip in a solid medium. For the first problem, involving a moving heat source, the high rate response results from the short time duration of contact between the localized moving energy source and the material continuum. For the second problem, involving a propagating crack tip, on the other hand, the high rate response results from the abrupt change of geometric curvature at the sharp crack tip. The material coordinate system convecting with the moving origin will be used throughout the analysis, because it is more appropriate to use in describing the high rate response in the neighborhood of the moving origin.

4.2 Thermal Shock Formation Around a Rapidly Moving Heat Source

Thermal shock phenomena will be first examined by considering a point heat source moving with a constant speed v in the x_1 direction in a two-dimensional, infinite medium as shown in Fig. 2. Due to the finite wave speed, the heat-affected zone is bounded by a circular thermal wavefront propagating in the physical domain with an expanding radius Ct. The contact or response time of a material continuum beneath the heat source obviously depends on the speed of the moving heat source. When the speed increases, the response time of the material continuum becomes short and the thermal wave characteristics are expected to be pronounced. Because this problem is considered popularly in the classical diffusion theory, it serves as a good example to illustrate the importance of considering thermal wave propagation with short response time.

The energy equation with T representation will be used to find the temperature field around the moving heat source. For the coordinate system shown in Fig. 2, the point heat source can be represented by

$$S(x_1, x_2, t) = Q\,\delta(x_1 - vt)\,\delta(x_2) \tag{13}$$

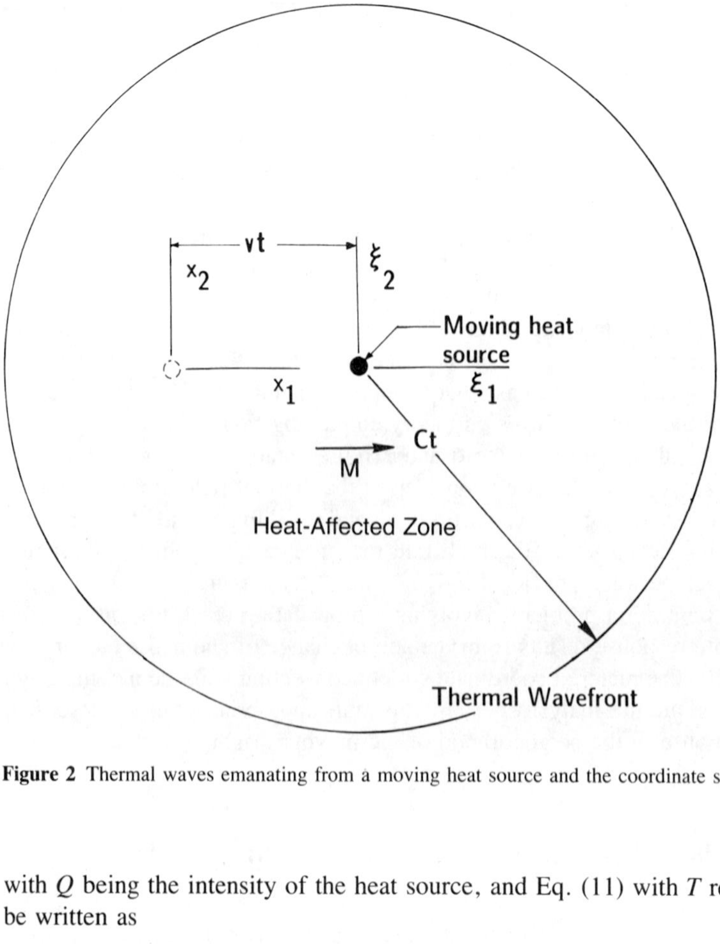

Figure 2 Thermal waves emanating from a moving heat source and the coordinate systems.

with Q being the intensity of the heat source, and Eq. (11) with T representation can be written as

$$\alpha \nabla^2 T - \left(\frac{\alpha}{C^2}\right) T,_{tt} - T,_t$$
$$= -\left(\frac{Q}{\rho C_p}\right)\left[\delta(x_1 - vt) + \left(\frac{\alpha}{C^2}\right)\delta,_t(x_1 - vt)\right]\delta(x_2) \quad (14)$$

To an observer moving with the heat source, Galiléi transformation,

$$x_1 = \xi_1 + vt \qquad \text{and} \qquad x_2 = \xi_2 \quad (15)$$

can be applied, and the energy equation (14) can be further expressed in terms of the material coordinates ξ_i as

$$\alpha[(1 - M^2)T,_{11} + T,_{22}] + 2c\alpha T,_1$$
$$= -\left(\frac{Q}{\rho C_p}\right)[\delta(\xi_1) - (M^2/2c)\,\delta,_1(\xi_1)]\,\delta(\xi_2) \quad (16)$$

where the quantity $(\ )_{,i}$ denotes partial differentiation with respect to ξ_i for $i = 1, 2$, and c a parameter defined as $v/2\alpha$. A thermal Mach number M, defined as v/C, is introduced in Eq. (16) to characterize the thermal field. It is the ratio of the speed of the moving heat source relative to the heat propagation speed in the solid. Before a general method is attempted to solve Eq. (16), we notice that the characteristic of the solution for Eq. (16) is governed by $\Delta = -4(1 - M^2)$. In the *subsonic* range with $M < 1$, the quantity Δ is smaller than zero and Eq. (16) is *elliptic*. In transition to the *transonic* stage with $M = 1$ and the *supersonic* range with $M > 1$, the quantity Δ is equal to and greater than zero, and the mathematical type of Eq. (16) is *parabolic* and *hyperbolic*, respectively. Due to such an intrinsic transition of the mathematical characteristics of the energy equation, the thermal field around the moving heat source needs to be obtained individually in the respective ranges of the thermal Mach number. In all cases, the temperature and its gradient are assumed to vanish at distances far away from the heat source.

4.2.1 Subsonic Range with $M < 1$. The speed of the moving heat source v is slower than the heat propagation speed C in the subsonic range with $M < 1$, and the energy equation (16) is elliptic. A closed-form solution for the temperature distribution governed by Eq. (16) in this case can be obtained by applying successive transformations for the dependent variable from $T(\xi_i)$ to $f(\xi_i)$,

$$T(\xi_i) = \exp\left[-\frac{c\xi_1}{(1 - M^2)}\right] f(\xi_i) \tag{17}$$

and for the independent variables from ξ_i to r:

$$r = \sqrt{\left(\frac{\xi_1^2}{1 - M^2}\right) + \xi_2^2} \tag{18}$$

The resulting energy equation in the one-dimensional *hypospace* r then takes the form of

$$f_{,rr} - \frac{1}{r} f_{,r} - \left(\frac{c^2}{1 - M^2}\right) f$$
$$= \exp\left(\frac{c\xi_1}{1 - M^2}\right)\left[-\left(\frac{Q}{\rho\alpha C_p}\right)\delta(\xi_1) + \left(\frac{QM^2}{2\rho c\alpha C_p}\right)\delta_{,1}(\xi_1)\right]\delta(\xi_2) \tag{19}$$

which facilitates the use of the method of Green's functions to solve the function $f(r)$. By noticing that Eq. (19) is governed by the modified Bessel operator, whose Green's function f_g is [74]

$$f_g(r|r_\eta) = -K_0\left[\frac{c|r - r_\eta|}{(1 - M^2)^{1/2}}\right] \tag{20}$$

with K_0 being the modified Bessel function of the second kind and $r_\eta = [\eta_1^2/(1 - M^2) + \eta_2^2]^{1/2}$, and the integral properties for the Dirac delta functions [75],

$$\int G[r(\xi_1, \xi_2)|r_\eta(\eta_1, \eta_2)]\ \delta(\eta_1)\ \delta(\eta_2)\ d\eta_1\ d\eta_2 = G[r(\xi_1, \xi_2)|r_\eta(0, 0)]$$

$$\int G[r(\xi_1, \xi_2)|r_\eta(\eta_1, \eta_2)]\ \delta_{,1}(\eta_1)\ \delta(\eta_2)\ d\eta_1\ d\eta_2 = -\left\{\frac{\partial}{\partial\eta_1} G[r(\xi_1, \xi_2)|r_\eta(\eta_1, 0)]\right\}_{\eta_1\to 0} \tag{21}$$

the solution of Eq. (19) for $f(r)$ is

$$f[r(\xi_i)] = \left(\frac{Q}{\rho\alpha C_p}\right)\left\{\frac{2 - M^2}{2(1 - M^2)} K_0\left[\frac{cr}{(1 - M^2)^{1/2}}\right] - \frac{M^2}{2(1 - M^2)} K_1\left[\frac{cr}{(1 - M^2)^{1/2}}\right]\right\} \tag{22}$$

By the inverse transformations of Eqs. (17) and (18), then, the temperature distribution in the subsonic range is obtained:

$$\frac{T(\xi_i)}{Q/\rho\alpha C_p} = \exp\left(-\frac{c\xi_1}{1 - M^2}\right)\left\{\frac{2 - M^2}{2(1 - M^2)} K_0\left[\frac{cr}{(1 - M^2)^{1/2}}\right] - \frac{M^2}{2(1 - M^2)} K_1\left[\frac{cr}{(1 - M^2)^{1/2}}\right]\right\} \qquad M < 1 \tag{23}$$

The thermal Mach number M appears as the *explicit* variable in characterizing the temperature field. In the limiting case that the wave speed C approaches infinity and hence the value of M approaches zero, the temperature distribution represented by Eq. (23) reduces to that employing Fourier's law of heat conduction.

In the subsonic range with $M < 1$, the temperature is bounded everywhere except for the location right beneath the point heat source at $r = 0$. For the thermal waves propagating in aluminum, as an example,

$$\rho = 2.72 \times 10^3 \text{ kg/m}^3 \qquad k = 220 \text{ W/m K} \qquad C_p = 0.895 \text{ kJ/kg K} \tag{24}$$

Figure 3 shows the temperature distributions along the circumference of a circle centered at the heat source. The radius of the circle is taken to be 10^{-4} m, which is approximately the lower bound of continuum mechanics beyond which the microstructural effects can be neglected in the formulation made on the macroscopic level. As expected, the near-field temperature level decreases as the thermal Mach number, and hence the speed of the moving heat source, increases. For the heat source moving with a high speed—$M = 0.9$, for example—the material continua in the neighborhood of the heat source do not have sufficient time to respond to the appropriate temperature level before the heat source moves away and a significant deviation from classical diffusion theory results. In all cases with different values of M, the temperature level reaches maximum at the *trailing edge* of the moving heat source, i.e., at $\theta = 180°$. By examining these results, it seems that the wave effects are not as pronounced in the

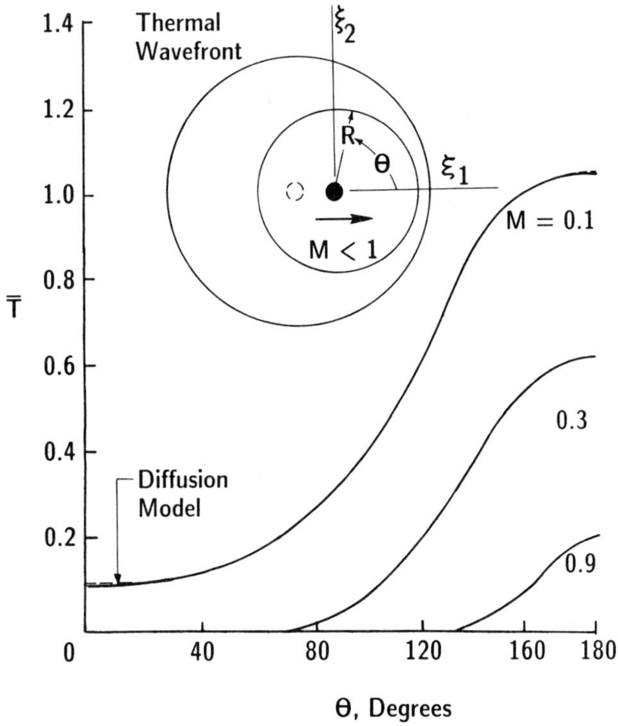

Figure 3 Temperature distribution along the circumference of the continuum circle centered at the heat source. Subsonic range with $M > 1$. $\bar{T} = T/(Q/\rho\alpha C_p)$.

low range of the thermal Mach number, say, $M < 0.1$. Fourier's law of heat conduction may yield satisfactory results when the speed of the heat source is low or the heat propagation speed in the solid medium is high.

4.2.2 Supersonic Range with $M > 1$. The supersonic case with $M > 1$ will be discussed first because it provides a conjugate situation of the subsonic case. When the speed of the moving heat source is faster than the heat propagation speed in the solid, the thermal Mach number is greater than 1 and the energy equation (16), which is hyperbolic in this case, can be rearranged to the following form:

$$\alpha[(M^2 - 1)T_{,11} - T_{,22}] - 2c\alpha T_{,1}$$
$$= \left(\frac{Q}{\rho C_p}\right)\left[\delta(\xi_1)\,\delta(\xi_2) - \left(\frac{M^2}{2c}\right)\delta_{,1}(\xi_1)\,\delta(\xi_2)\right] \quad (25)$$

By the hypovariable transformations, similarly, the corresponding pair of Eqs. (17) and (18) for the dependent and independent variables in the hypospace r is

$$T(\xi_i) = \exp\left[\frac{c\xi_1}{(M^2 - 1)}\right] f(\xi_i) \qquad r = \left[\frac{\xi_1^2}{(M^2 - 1)} - \xi_2^2\right]^{1/2} \qquad \text{for } M > 1 \quad (26)$$

The corresponding equation governing the function of $f(\xi_i)$ becomes

$$f_{,rr} + \frac{1}{r} f_{,r} - \left(\frac{c^2}{M^2 - 1}\right) f = \left(\frac{Q}{\rho C_p \alpha}\right) \exp\left(-\frac{c\xi_1}{M^2 - 1}\right) \delta(\xi_1)\,\delta(\xi_2) - \left(\frac{QM^2}{2c\rho C_p \alpha}\right) \exp\left(-\frac{c\xi_1}{M^2 - 1}\right) \delta_{,1}(\xi_1)\,\delta(\xi_2) \quad (27)$$

and the Green's function in this case is

$$f_g(r|r_\eta) = K_0\left[\frac{c|r - r_\eta|}{(M^2 - 1)^{1/2}}\right] \quad (28)$$

By the same procedure as that described in the previous case, then, the temperature field in the supersonic range can be obtained as:

$$\frac{T(\xi_i)}{Q/\rho\alpha C_p} = -\exp\left(\frac{c\xi_1}{M^2 - 1}\right)\left\{\frac{2 - M^2}{2(M^2 - 1)} K_0\left[\frac{cr}{(M^2 - 1)^{1/2}}\right] - \frac{M^2}{2(M^2 - 1)} K_1\left[\frac{cr}{(M^2 - 1)^{1/2}}\right]\right\} \quad M > 1 \quad (29)$$

Unlike the subsonic case, we first notice that the radial distance r defined in Eq. (26) restricts the applicable range of the temperature expression, namely, Eq. (29), to the domain

$$\frac{\xi_1^2}{M^2 - 1} > \xi_2^2 \quad (30)$$

Before an expression for the temperature distribution suitable for the rest of the physical domain is sought, we first examine the condition depicted by Eq. (30). The physical interpretation of Eq. (30) can be nicely made if we represent the condition in terms of the polar coordinate (R, θ) centered at the heat source, as shown in Fig. 4. The equivalent expression of Eq. (30) is

$$|\tan\theta| < \left(\frac{1}{M^2 - 1}\right)^{1/2} \quad (31)$$

or, in a more familiar form,

$$0 < \theta < \sin^{-1}\left(\frac{1}{M}\right) \quad \text{or} \quad \pi - \sin^{-1}\left(\frac{1}{M}\right) < \theta < \pi \quad (32)$$

where the ranges for θ in the third and fourth quadrants are omitted due to the symmetry of the problem. In Eq. (32), the first domain is the *heat-affected zone* for an energy source moving to the left, while the second is that for an energy source moving to the right. The second case is assumed for the present study under the Galiléi transformation defined in Eq. (15). It shows that the heat-affected zone is bounded by two straight lines defined by $\theta = \pm \sin^{-1}(1/M)$ measured from the trailing edge of the heat source. This is an identical phenomenon to the aerodynamic shock formation in a supersonic flow field [76], and it is informative for us to conclude that the temperature

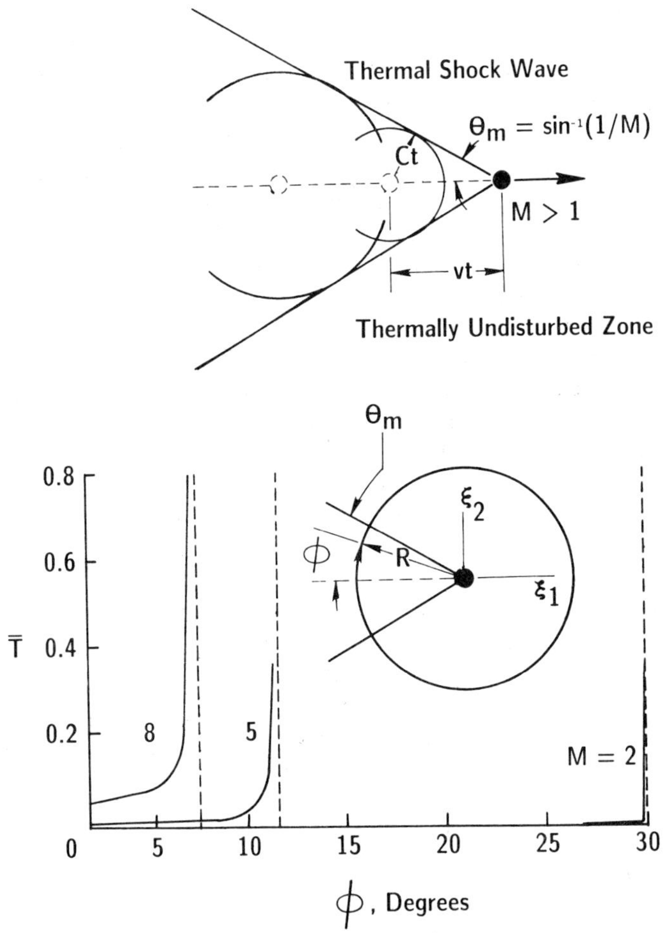

Figure 4 Temperature distribution along the circumference of the continuum circle centered at the heat source. Supersonic range with $M > 1$. $\bar{T} = T/(Q/\rho\alpha C_p)$.

field outside the heat-affected zone stays undisturbed as the heat source moves with supersonic speed. It is in this sense that the physical domain is called the *thermally undisturbed zone*, where the temperature stays at the reference value and the material continua do not sense the existence of the heat source. As the angle θ approaches the values of $\pm \sin^{-1}(1/M)$, $\xi_1^2/(M^2 - 1)$ approaches ξ_2^2 and r approaches 0 according to Eq. (26). Because the Bessel functions K_0 and K_1 approach infinity with different rates when their arguments approach zero, Eq. (29) depicts a temperature singularity as the *thermal shock surface* located at $\theta = \pm\sin^{-1}(1/M)$ is approached. At $\theta = \pm\sin^{-1}(1/M)$, a large value of temperature reflects significant amount of thermal energy cumulated thereby and it is the preferential direction of thermal energy accumulation around a heat source moving in the supersonic range. In summary, the temperature distribution in the heat-affected zone is represented by Eq. (29), while it stays at the reference value (assumed to be zero without loss in generality) in the thermally undisturbed zone. The temperature approaches infinity at the thermal shock surface, and a discontinous variation of the temperature field across the shock surface is clear. The physical domain of the heat-affected zone is 2 $\sin^{-1}(1/M)$ behind the moving heat source, which becomes slender as the thermal Mach number increases.

Figure 4 displays the temperature distributions represented by Eq. (29). The angle ϕ is measured from the trailing edge of the moving heat source. When the thermal Mach number increases from 2 to 5 to 8, distinct thermal shock waves are formed at $\phi = 30°$, 11.54°, and 7.18° and the heat-affected zone reduces from 60° to 23.07° to 14.36°. The temperature approaches infinity at the thermal shock surface and reaches a *minimum* at the trailing edge of the heat source. We also notice that the temperature level in the heat-affected zone around the moving heat source *increases* with the thermal Mach number—a phenomenon contrary to the corresponding response in the subsonic range. The physical interpretation lies in the thermal shock formation in the supersonic range. When the thermal Mach number increases, the physical domain of the heat-affected zone reduces and the material continua thereby are closer to the temperature singularity existing at the shock surfaces. In other words, the temperature singularity at the thermal shock surface has a stronger influence on the temperature level in the heat-affected zone when the thermal Mach number increases. In combination with the corresponding response in the subsonic range, this is a *swinging* phenomenon for the variation of the temperature level in transition of the thermal Mach number from the subsonic to the supersonic ranges. Disregarding the singularities in the vicinity of the thermal shock waves, it depicts that in the subsonic range with $M < 1$, the temperature level around the moving heat source decreases as the thermal Mach number increases. After entering the supersonic range with $M > 1$, on the other hand, the temperature level in the heat-affected zone increases with the thermal Mach number. According to the physical parameters considered in this example, the thermal Mach number at which the temperature level around the heat source reaches a minimum is approximately 2.

4.2.3 Transonic Stage with $M = 1$. When the heat source is moving at the same speed as that of the heat propagation in the solid, the value of M equals 1 and the energy equation degenerates into a parabolic type:

$$T,_{22} + 2cT,_1 = -\left(\frac{Q}{\rho\alpha C_p}\right)\left[\delta(\xi_1)\,\delta(\xi_2) - \left(\frac{M^2}{2c}\right)\delta,_1(\xi_1)\,\delta(\xi_2)\right] \tag{33}$$

With the derivative on the variable ξ_1 being in the position of the time derivative, Eq. (33) is essentially a diffusion equation whose Green's function is readily available [77] as

$$T_g(\xi_i|\eta_i) = \left[-\frac{1}{(8\pi c)^{1/2}}\right]\exp\left[-\frac{c(\xi_2 - \eta_2)^2/2|\xi_1 - \eta_1|}{(|\xi_1 - \eta_1|)^{1/2}}\right] \tag{34}$$

and the temperature can thus be integrated directly as

$$\frac{T(\xi_i)}{Q/\rho\alpha C_p} = \left[\frac{1}{(8\pi c)^{1/2}}\right]\exp\left(\frac{c\xi_2^2}{2\xi_1}\right)\left[\frac{1}{(-\xi_1)^{1/2}} + \frac{c\xi_2^2 + \xi_1}{4c(-\xi_1)^{5/2}}\right] \tag{35}$$

Again, the applicable range of Eq. (35) is restricted to the domain of

$$\xi_1 < 0 \qquad \text{or} \qquad \frac{\pi}{2} < \theta < \pi \tag{36}$$

and the temperature has a singularity at $\xi_1 = 0$. The physical domain represented by Eq. (36) is essentially the limiting case of Eq. (32) with M approaching 1 for a heat source moving to the right. The concept of the heat-affected zone is therefore independently confirmed by the two different approaches adopted in the transonic and the supersonic ranges. In the transonic stage with $M = 1$, a normal shock wave exists at $\theta = 90°$ ($\xi_1 = 0$), where the temperature approaches infinity. In the supersonic range, the thermal shock wave separates the heat-affected zone from $\pi/2$ to π from the thermally undisturbed zone from 0 to $\pi/2$. The temperature distribution in the heat-affected zone is represented by Eq. (35), while the temperature in the thermally undisturbed zone stays at the reference level. This situation is displayed graphically in Fig. 5. Note that the physical domain of the heat-affected zone is comparatively wide in this case (180°), and the influence of the temperature singularity existing at the shock surface is not as strong as that in the supersonic range. The temperature in the heat-affected zone, as a continuation of the corresponding phenomenon in the subsonic range, reaches a *maximum* at the trailing edge of the moving heat source.

4.2.4 Heat Flux Vector. Based on the temperature field obtained in the subsonic, transonic, and supersonic ranges, the heat flux vector in the various ranges of the thermal Mach number can be found immediately. At the transonic and in the supersonic ranges with $M \geq 1$, because the temperature field possesses a discontinuous change across the thermal shock surfaces, the corresponding variation for the heat flux vector is certainly of interest.

Consistent with the description of the temperature on the material coordinate system, the constitutive equation, namely, the second of Eq. (9), is first represented in terms of the ξ_i coordinates:

$$\left(\frac{M^2}{2c}\right)\mathbf{q}_{,1} - \mathbf{q} = k\,\nabla T \tag{37}$$

To an observer moving with the heat source, this is the equation governing the heat flux vector, and the integral representations of the solutions are

$$\{q_1, q_2\} = k\int_{\infty}^{0} e^{-\zeta}\left[T_{,1}\left(\xi_1 + \frac{M^2}{2c}\zeta, \xi_2\right), T_{,2}\left(\xi_1 + \frac{M^2}{2c}\zeta, \xi_2\right)\right]d\zeta \tag{38}$$

where q_1 and q_2 are the components of the heat flux vector in the ξ_1 and ξ_2 directions and $\nabla T \equiv \{T_{,1}, T_{,2}\}$. Due to the intrinsic transition of the energy equation from the subsonic to the supersonic ranges, the temperature gradients should be determined by the temperature distributions in the respective range of the thermal Mach number. They are Eq. (23) for the subsonic range with $M < 1$, Eq. (35) for the transonic stage with $M = 1$, and Eq. (29) for the supersonic range with $M > 1$. Without going into the details, the results are summarized as follows:

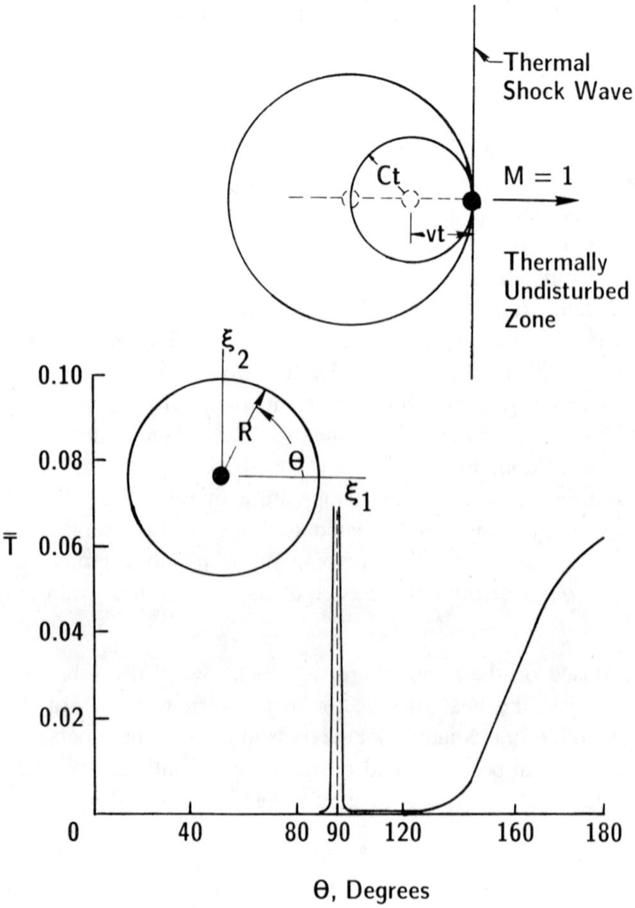

Figure 5 Temperature distribution along the circumference of the continuum circle centered at the heat source. Transonic stage with $M = 1$. $\bar{T} = T/(Q/\rho\alpha C_p)$.

Subsonic range with $M < 1$:

$$\frac{T,_1}{Q/\rho\alpha C_p} = \exp\left(\frac{c\xi_1}{M^2-1}\right)\left\{-\left[\frac{M^2 c\xi_1}{2r(M^2-1)^{5/2}} + \frac{(2-M^2)c}{2(M^2-1)^2}\right]K_0\left[\frac{cr}{(M^2-1)^{1/2}}\right] + \left[\frac{M^2 c}{2(M^2-1)^2} + \frac{(2-M^2)c\xi_1}{2r(M^2-1)^{5/2}} - \frac{M^2\xi_1}{2r^2(M^2-1)^2}\right]K_1\left[\frac{cr}{(M^2-1)^{1/2}}\right]\right\}$$

$$\frac{T,_2}{Q/\rho\alpha C_p} = \exp\left(\frac{c\xi_1}{M^2-1}\right)\left\{\left[\frac{M^2 c\xi_2}{2r(M^2-1)^{3/2}}\right]K_0\left[\frac{cr}{(M^2-1)^{1/2}}\right] + \left[\frac{M^2\xi_2}{2r^2(M^2-1)} - \frac{(2-M^2)c\xi_2}{2r(M^2-1)^{3/2}}\right]K_1\left[\frac{cr}{(M^2-1)^{1/2}}\right]\right\} \quad (39)$$

Transonic stage with $M = 1$:

$$\frac{T_{,1}}{Q/\rho\alpha C_p} = \left[\frac{1}{(8\pi c)^{1/2}}\right] \exp\left(\frac{c\xi_2^2}{2\xi_1}\right) \left[\frac{1}{8c(-\xi_1)^{7/2}}\right] \left\{c(4\xi_1^2 + 5\xi_2^2) + 3\xi_1 + \left(\frac{c\xi_2^2}{\xi_1}\right) [c(4\xi_1^2 + \xi_2^2) + \xi_1]\right\} \tag{40}$$

$$\frac{T_{,2}}{Q/\rho\alpha C_p} = \left[-\frac{1}{(8\pi c)^{1/2}}\right] \exp\left(\frac{c\xi_2^2}{2\xi_1}\right) \left[\frac{\xi_2}{4(-\xi_1)^{7/2}}\right] [c(4\xi_1^2 + \xi_2^2) + 3\xi_1]$$

Supersonic range with $M > 1$:

$$\frac{T_{,1}}{Q/\rho\alpha C_p} = \exp\left(\frac{c\xi_1}{M^2 - 1}\right) \left\{-\left[\frac{M^2 c\xi_1}{2r(M^2-1)^{5/2}} + \frac{(2-M^2)c}{2(M^2-1)^2}\right] K_0\left[\frac{cr}{(M^2-1)^{1/2}}\right] + \left[\frac{M^2 c}{2(M^2-1)^2} + \frac{(2-M^2)c\xi_1}{2r(M^2-1)^{5/2}} - \frac{M^2\xi_1}{2r^2(M^2-1)^2}\right] K_1\left[\frac{cr}{(M^2-1)^{1/2}}\right]\right\}$$

$$\frac{T_{,2}}{Q/\rho\alpha C_p} = \exp\left(\frac{c\xi_1}{M^2 - 1}\right) \left\{\left[\frac{M^2 c\xi_2}{2r(M^2 - 1)^{3/2}}\right] K_0\left[\frac{cr}{(M^2 - 1)^{1/2}}\right] + \left[\frac{M^2\xi_2}{2r^2(M^2 - 1)} - \frac{(2 - M^2)c\xi_2}{2r(M^2 - 1)^{3/2}}\right] K_1\left[\frac{cr}{(M^2 - 1)^{1/2}}\right]\right\} \tag{41}$$

With these expressions for the temperature gradients, Eq. (38) can be integrated accordingly for the heat flux components in the respective range of the thermal Mach number. Although analytical solutions for q_1 and q_2 are difficult to obtain due to the complexity of the integrands, the integral involved in Eq. (38) obviously exists, since all the functions vanish at a large value of ζ.

The thermal properties specified in Eq. (24) are used again here to illustrate the numerical results for the heat flux vector. Along the same circle centered at the heat source, the distributions of the heat flux components q_R and q_θ in various ranges of the thermal Mach number are shown in Figs. 6 through 9. They are related to the components of q_1 and q_2 by

$$q_R = q_1 \cos\theta + q_2 \sin\theta \qquad \text{and} \qquad q_\theta = -q_1 \sin\theta + q_2 \cos\theta \tag{42}$$

The lower bound ∞ appearing in the improper integral of Eq. (38) is approached by an integer parameter B such that the difference between the results of two successive integrations, with B and $B - 1$ as the lower bounds, respectively, is less than 1% of the current value. For all the values of M ranging from 0.1 to 5 considered in the present study, the values for B that achieve such accuracy are less than 10 by employing Simpson's rule. All the physical trends existing in the temperature field are preserved in the distributions of q_R and q_θ. Due to the presence of the thermal shock waves at the transonic stage, q_θ switches to positive in the vicinity of the shock wave singularity, as shown by Fig. 7. In the supersonic range with $M > 1$, as shown by Fig. 8, q_θ stays

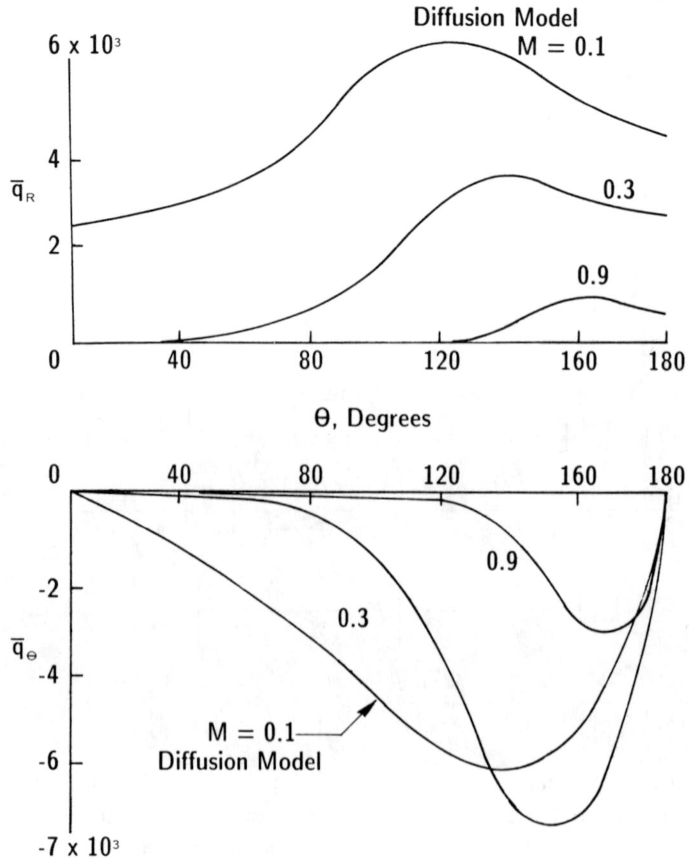

Figure 6 Distributions of the heat flux components q_R and q_θ in the subsonic range. $\bar{q}_R = q_R/Q$ and $\bar{q}_\theta = q_\theta/Q$.

positive in the entire domain of the heat-affected zone due to the strong influence of the singularity at the thermal shock surface. Also, the swinging phenomenon can be clearly observed from Figs. 8 and 9. Disregarding the singularities in the vicinity of the thermal shock waves, the value of M beyond which the heat flux increases with the thermal Mach number is approximately 3.

Unlike the ordinary situation where multidimensional problems only add additional functional dependence on the additional space variables, the considerations given to the multidimensional structures of thermal waves essentially lead to several important and distinctive features which cannot be depicted by one-dimensional studies or the classical diffusion theory. In addition to the direct influences on the thermal field, the thermal shock formation also affects the thermoelastic stress pattern around the moving heat source. This is the major topic of Secs. 5 and 6. Due to the formation of thermal shock waves in the transonic and supersonic ranges with $M \geq 1$, we will see that the present model predicts a wider domain of the damage zone spreading in the neigh-

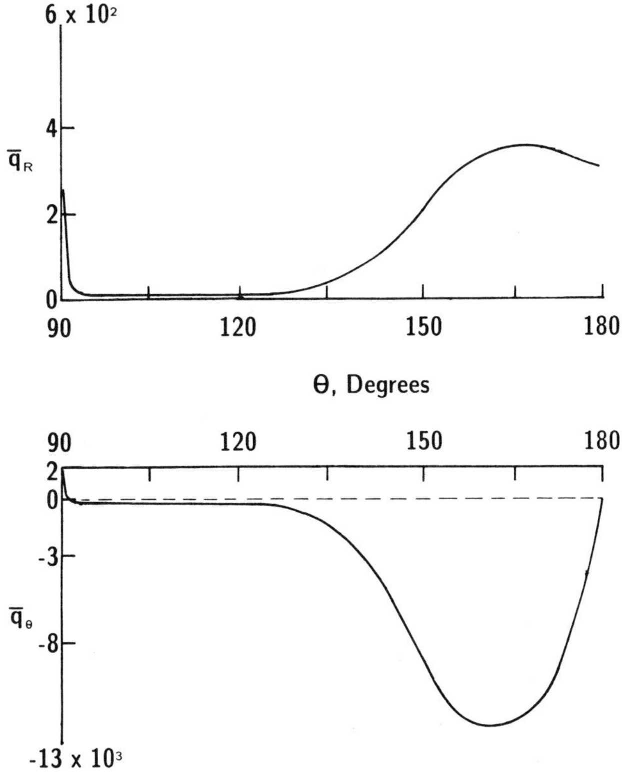

Figure 7 Distributions of the heat flux components q_R and q_θ at the transonic stage. $\bar{q}_R = q_R/Q$ and $\bar{q}_\theta = q_\theta/Q$.

borhood of the trailing edge of the heat source. This is quite a different pattern from that predicted by classical diffusion theory.

4.3 Thermal Shock Formation Around a Rapidly Propagating Crack Tip

A propagating crack tip presents another type of moving origin in a solid medium. Similar to the previous case involving a moving heat source, we are to examine the physical phenomena of thermal shock formation around the crack tip as it propagates at a speed faster than the heat propagation speed in the solid. As global instability of a macrocrack is developed under external excitation, the terminating velocity of the crack can be estimated as $0.38\sqrt{E/\rho}\ (1 - a_c/a)$, according to the energy-balance principles in linear elastic fracture mechanics [78, 79]. Here $\sqrt{E/\rho}$ is the longitudinal wave speed in the solid medium, a is the instantaneous length of the crack, and a_c is the critical crack length characterizing the brittle behavior of the crack growth. For metals, the terminating velocity is on the order of 10^3 m/s, and consideration of the

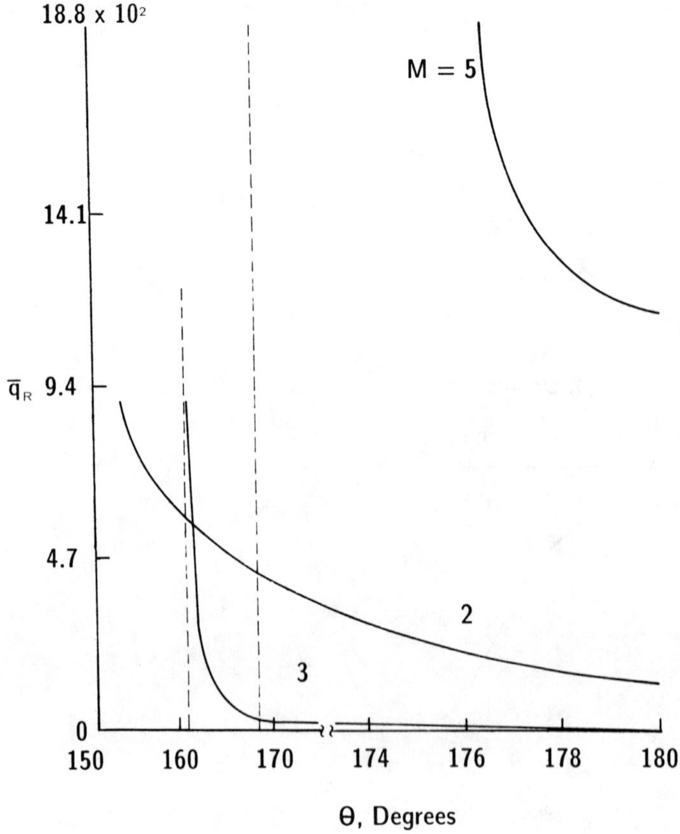

Figure 8 Distributions of the heat flux components q_R in the supersonic range. $\bar{q}_R = q_R/Q$.

thermal shock formation around the crack tip is of great importance for nonisothermal crack arresting in engineering structures. In the sequel, both the temperature and flux formulations will be studied to examine the thermal shock phenomena.

Consider in general a macrocrack propagating along a smooth but arbitrary trajectory in an infinite medium with finite speed of heat propagation, as shown in Fig. 10. The linear and angular velocities v and ω of the moving crack are assumed to be constant. Assuming that the energy release at the propagating crack tip due to creation of new crack surfaces is negligible, the heat source terms in Eq. (11) vanish and the energy equation with T representation reduces to the following form:

$$\alpha \nabla^2 T = \left(\frac{\alpha}{C^2}\right) T_{,tt} + T_{,t} \qquad (43)$$

Similar to the previous case with a moving heat source, the quasi-stationary description of the temperature field can be made by introducing the material coordinate system ξ_i

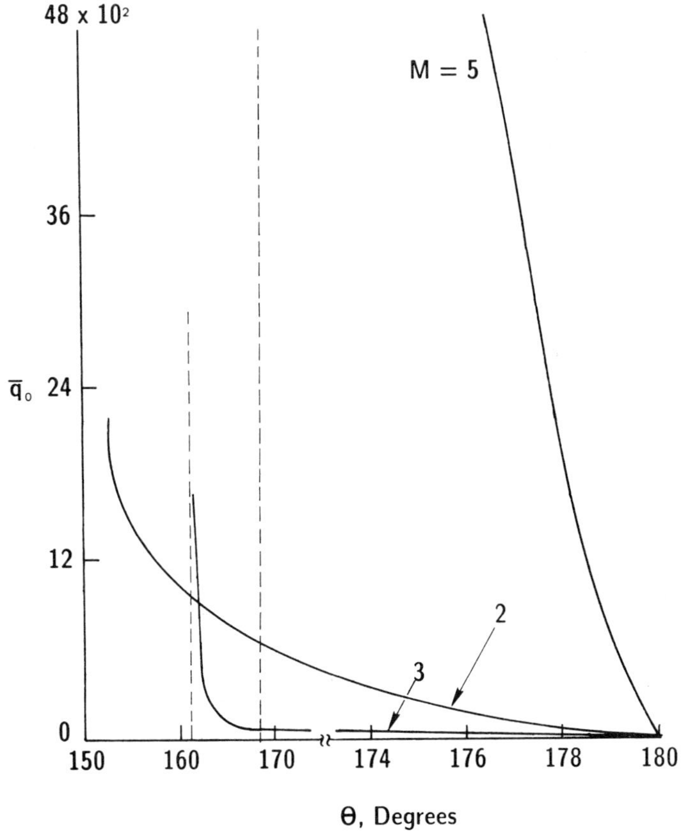

Figure 9 Distributions of the heat flux components q_θ in the supersonic range. $\bar{q}_\theta = q_\theta/Q$.

centered at the moving crack tip. While the Laplacian operator keeps an invariant form in the ξ_i coordinates, the terms containing derivatives with respect to time in Eq. (43) are transformed into the material derivatives,

$$T_{,t} \rightarrow T_{,t} - vT_{,1} - \omega(\xi_1 T_{,2} - \xi_2 T_{,1}) \tag{44}$$

$$\begin{aligned} T_{,tt} \rightarrow T_{,tt} &- 2vT_{,t1} + v^2 T_{,11} - \omega\xi_1[T_{,t2} - (v - \omega\xi_2)T_{,12} + \omega T_{,1} \\ &- \omega\xi_1 T_{,22}] + \omega\xi_2[T_{,t1} - (v - \omega\xi_2)T_{,11} - vT_{,2} - v\xi_1 T_{,12}] \\ &+ \omega[\xi_2 T_{,t1} - \xi_1 T_{,t2}] + v\omega[\xi_1 T_{,12} + T_{,2} - \xi_2 T_{,11}] \end{aligned} \tag{45}$$

and the equation governing the temperature field observed from the material coordinate system is obtained by substituting Eqs. (44) and (45) into Eq. (43). The subscripts in these equations denote differentiation with respect to the ξ_i coordinates.

The resulting equation is to be solved subject to the boundary conditions specified at the crack surface and the crack line instantaneously tangent to the crack trajectory

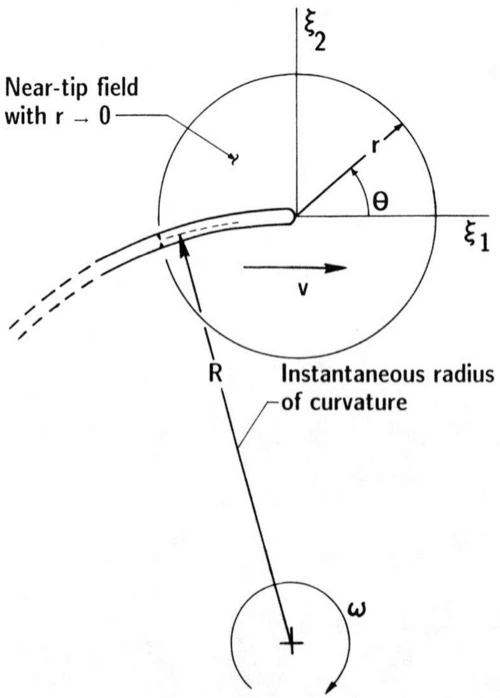

Figure 10 Geometric configuration of a moving crack and the material coordinate system convecting with the crack tip.

along the ξ_1 axis, as shown in Fig. 10. For an observer standing at the crack tip, we may impose a symmetrical condition

$$T_{,2} = 0 \qquad \text{at } \xi_2 = 0 \text{ and } \xi_1 > 0 \tag{46}$$

This concept was also adopted by Williams [80] and by Achenbach and Bažant [81] in investigating the elastodynamic near-tip stress and displacement fields. At the surface of the crack, on the other hand, a general form for the energy balance across the crack surface is

$$a_1T + a_2q_2 = a_3 \qquad \text{at } \xi_2 = 0 \text{ and } \xi_1 < 0 \tag{47}$$

In the initial stage of the development with temperature formulation, however, we assume the coefficients a_2 and a_3 to be zero and Eq. (47) becomes

$$T = 0 \qquad \text{at } \xi_2 = 0 \text{ and } \xi_1 < 0 \tag{48}$$

It is a simplified form assuming that both the heat flux due to irreversible energy dissipating into the surrounding media and the energy release due to creation of new crack surfaces are negligible in the history of crack propagation. Equations (46) and (48) compose a mixed-type boundary condition at $\xi_2 = 0$ which must be satisfied by the temperature field.

The energy equation (43) with $T_{,t}$ and $T_{,tt}$ being replaced by the material derivatives shown by Eqs. (44) and (45) takes the following form:

$$\begin{aligned}\alpha \nabla^2 T = {} & \alpha M^2 T_{,11} + \left[\left(\frac{\alpha}{C^2}\right)T_{,tt} - \left(\frac{M^2}{c}\right)T_{,1t}\right] + (T_{,t} - vT_{,1}) \\ & - \omega(\xi_1 T_{,2} - \xi_2 T_{,1}) + \left(\frac{M^2}{2cR}\right)\{\xi_2[T_{,t1} - (v - \omega\xi_2)T_{,11} \\ & - \omega T_{,2} - \omega\xi_1 T_{,12}] - \xi_1[T_{,t2} - (v - \omega\xi_2)T_{,12} + \omega T_{,1} \\ & - \omega\xi_1 T_{,22}] - (\xi_1 T_{,t2} - \xi_2 T_{,t1}) + v(T_{,2} + \xi_1 T_{,12} - \xi_2 T_{,11})\}\end{aligned} \tag{49}$$

where a thermal Mach number M, defined as v/C in parallel to the previous study, is introduced and c is a parameter defined as $v/2\alpha$. For a stationary crack with $v = \omega = 0$ or a solid medium with the speed of heat propagation approaching infinity, i.e., $C \rightarrow \infty$, M approaches zero and Eq. (49) reduces to that employing Fourier's law of heat conduction.

In trying for the fundamental solution of Eq. (49), a particular product form for T,

$$T(r, \theta, t) = r^\lambda \, \Gamma(t) \, H(\theta) \tag{50}$$

is sought. The polar coordinates (r, θ) centered at the crack tip are especially useful in describing the state of the affairs in the vicinity of the crack tip. To be noticed, however, is that such a product form is valid only as the r dependency of the temperature field is identical along various directions approaching the crack tip [82]. The exponent λ, though considered as yet unknown, should be an isotropic constant in this sense. As usual, the derivatives in the polar coordinates r and θ are related to those in the ξ_i coordinates by

$$\begin{aligned} T_{,1} &= \cos\theta \, T_{,r} - \left(\frac{\sin\theta}{r}\right) T_{,\theta} \\ T_{,2} &= \sin\theta \, T_{,r} + \left(\frac{\cos\theta}{r}\right) T_{,\theta} \end{aligned} \tag{51}$$

etc. Substituting Eqs. (50) into (49) and using Eq. (51) for the differentiations, the r dependency of the terms involved in the resulting equation can be summarized as follows:

$$\begin{aligned} & \nabla^2 T, \; T_{,11} \sim r^{\lambda-2} \\ & T_{,1}, \; T_{,2}, \; T_{,t1}, \; \xi_2 T_{,11}, \; \xi_1 T_{,12}, \; \xi_2 T_{,11} \sim r^{\lambda-1} \\ & T_{,t}, \; \xi_1 T_{,1}, \; \xi_2 T_{,2}, \; \xi_1 T_{,2}, \; \xi_2 T_{,1}, \; T_{,tt}, \; \xi_2 T_{,t1}, \; \xi_2^2 T_{,11}, \; \xi_1\xi_2 T_{,12}, \; \xi_1 T_{,t2}, \\ & \xi_2 T_{,t1}, \; \xi_1^2 T_{,22} \sim r^\lambda \end{aligned} \tag{52}$$

Multiplying the entire equation by $r^{2-\lambda}$, then, we can see that the terms proportional to $r^{\lambda-1}$ and r^{λ} in Eq. (52) are proportional to r and r^2, respectively. As r approaches zero, i.e., in the immediate vicinity of the crack tip, these terms vanish and the asymptotic form of Eq. (49) reduces to

$$\nabla^2 T = M^2 T_{,11} \tag{53a}$$

or

$$(1 - M^2 \sin^2 \theta)\, H_{,\theta\theta} - [M^2(1 - \lambda) \sin 2\theta]\, H_{,\theta} + \lambda\{\lambda + M^2[(2 - \lambda)\cos^2 \theta - 1]\}\, H = 0 \tag{53b}$$

upon substitution of Eq. (50). Similar to the case involving a moving heat source, Eq. (53*a*) transits from an elliptic to a parabolic to a hyperbolic type as the thermal Mach number transits from the subsonic ($M < 1$) to the transonic ($M = 1$) to the supersonic ($M > 1$) ranges. Some intrinsic variations of the near-tip behavior of temperature are thus expected in transition of the thermal Mach number. Again, the thermal Mach number M appears as the explicit variable characterizing the angular distribution of the near-tip temperature. Equation (53*b*) is valid for any function of $\Gamma(t)$. The time independence of this equation indicates that the near-tip characteristics of the H function [and hence the near-tip temperature according to Eq. (50)] are the same for a crack either propagating in a transient or steady state. Also, we notice that the angular velocity ω of the moving crack with smooth trajectory is a higher-order effect as far as the near-tip temperature is concerned. These are the common features of the near-tip displacement field [81]. Equation (53*b*), governing the angular distribution of the near-tip temperature, is a second-order ordinary differential equation with variable coefficients. It has to be solved subject to the boundary conditions

$$H_{,\theta} = 0 \text{ at } \theta = 0 \qquad \text{and} \qquad H = 0 \text{ at } \theta = \pi \tag{54}$$

resulting from Eqs. (46) and (48).

Similar to the previous case with a moving heat source, analytical solutions for $H(\theta)$ can be found by successive transformations. We first introduce a transformation for the dependent variable from H to Φ:

$$H(\theta) = (1 - M^2 \sin^2 \theta)^{\lambda/2}\Phi(\theta) \tag{55}$$

The Φ representation of Eq. (53*b*) then becomes:

$$(g_1 f)\Phi_{,\theta\theta} - (f M^2 \sin 2\theta)\Phi_{,\theta} + (g_1 f_{,\theta\theta} + g_2 f_{,\theta} + g_3 f)\Phi = 0 \tag{56}$$

with

$$\begin{aligned} g_1 &= 1 - M^2 \sin^2 \theta & g_2 &= - M^2(1 - \lambda) \sin 2\theta \\ g_3 &= \lambda\{\lambda + M^2[(2 - \lambda)\cos^2 \theta - 1]\} & f &= g_1^{\lambda/2} \end{aligned} \tag{57}$$

In order to further simplify Eq. (56), we now introduce another transformation for the independent variable from θ to γ. Let us consider in general a function $\gamma \equiv \gamma(\theta)$. By applying the chain rule, Eq. (56) can be expressed in γ-space as

$$\Phi_{,\gamma\gamma} + \left[\frac{(g_1 f)\gamma_{,\theta\theta} - (fM^2 \sin 2\theta)\gamma_{,\theta}}{g_1 f \gamma_{,\theta}{}^2}\right]\Phi_{,\gamma} + \left(\frac{g_1 f_{,\theta\theta} + g_2 f_{,\theta} + g_3 f}{g_1 f \gamma_{,\theta}{}^2}\right)\Phi = 0 \quad (58)$$

Obviously, if we choose the coefficient of the first-order derivative to be 0, i.e.,

$$(g_1 f)\gamma_{,\theta\theta} - (fM^2 \sin 2\theta)\gamma_{,\theta} = 0 \quad (59)$$

the resulting equation reveals the highlight of possessing a closed-form solution. Equation (59) governs the transformation γ for the independent variable θ. According to Eq. (57), it strongly depends on the thermal Mach number of the propagating crack tip.

4.3.1 Subsonic Range with $M < 1$. For a crack propagating in the subsonic range, the crack velocity is slower than the heat propagation speed in the solid. The function $\gamma(\theta)$ satisfying Eq. (59) can be integrated to give

$$\gamma(\theta) = \tan^{-1}[(1 - M^2)^{1/2} \tan \theta]$$

and consequently, (60)

$$\gamma_{,\theta} = \frac{(1 - M^2)^{1/2}}{1 - M^2 \sin^2 \theta}$$

This is an expression identical to that observed by Achenbach and Bažant [81] for the out-of-plane displacement field based on the Lorentz transformation. Substitution of Eq. (60) into (58) then yields

$$\Phi_{,\gamma\gamma} + \lambda^2 \Phi = 0 \quad (61)$$

Equation (61) reduces the present problem to an eigenvalue problem for λ. The near-tip temperature distribution, based on the solution of Eq. (61) and the assistance of Eqs. (50), (55), and (60), consequently appears as the eigenfunction of the problem. The simple form of Eq. (61) results from the special coefficients appearing in Eq. (58), which is not generally the case for any second-order ordinary differential equation with variable coefficients.

Determination of the eigenvalues λ lies in the boundary conditions, Eq. (54). In the transformed space $\Phi(\gamma)$, it appears in the following form:

$$\Phi_{,\gamma} = 0 \text{ at } \gamma = 0 \qquad \text{and} \qquad \Phi = 0 \text{ at } \gamma = \pm\pi \quad (62)$$

where the principal values of Eq. (60) for γ are adopted. Equations (61) and (62) compose the desired eigenvalue problem. The eigenfunction Φ can be integrated immediately as

$$\Phi = A \cos \lambda\gamma \quad (63)$$

with the eigenvalues being

$$\lambda = \frac{2n + 1}{2} \qquad \text{for } n = 0, 1, 2, \ldots \tag{64}$$

As r approaches zero in the near-tip region, we notice that the terms of r^{λ}, with λ shown in Eq. (64) with $n \neq 0$, will approach zero at a rate faster than that of $r^{1/2}$ corresponding to $n = 0$ and the near-tip solution is dominant by the fundamental solution with $\lambda = \frac{1}{2}$. The near-tip solution for the temperature, therefore, will be investigated by taking the lowest value of n being zero. Substituting Eq. (63) into Eq. (55) and taking the inverse transformation from γ to θ according to Eq. (60), the final form of the solution for $H(\theta)$ is

$$H(\theta) = \frac{A[(1 - M^2 \sin^2 \theta)^{1/2} + \cos \theta]^{1/2}}{\sqrt{2}} \qquad \text{for } M < 1 \tag{65}$$

The coefficient A is an arbitrary constant in the eigenvalue system, and the expression enclosed in brackets governs the angular variation of the temperature field in the core region. In the subsonic range with $M < 1$, clearly, the angular distribution of the near-field temperature represented by Eq. (65) is well defined everywhere and there is no singularity existing in the physical domain. The corresponding expression in the thermal diffusion model, $H_D(\theta)$, can be retrieved by substituting 0 for M in Eq. (65):

$$H_D(\theta) = \frac{A[1 + \cos \theta]^{1/2}}{\sqrt{2}} \tag{66}$$

With the expressions shown by Eqs. (65) and (66), it is noted from Eq. (50) that both the thermal wave and the thermal diffusion models display a $\sqrt{r}$ temperature field in the vicinity of the crack tip. A $1/\sqrt{r}$-type of singularity for the temperature gradient at the crack tip thus exists in both models.

For a crack moving to the right, graphical representations for $H(\theta)$ and $H_D(\theta)$ with A being unity are shown in Fig. 11. As expected, the thermal diffusion model ($M = 0$) employing Fourier's law of heat conduction overestimates the near-tip temperature level. The deviation between the two models becomes significant as the thermal Mach number increases. For all the cases with $M = 0$ (the diffusion model), 0.1, 0.5, and 0.9, the temperature reaches its *maximum* at the leading edge of the moving crack at $\theta = 0°$. Similar to the subsonic response in the previous case involving a moving heat source, the H function, and hence the temperature level, decreases as the thermal Mach number increases. Also, at the trailing edge of the moving crack tip at $\theta = \pi$, the slope of the H function ($H_{,\theta}$) has a value of $-(1 - M^2)^{1/2}/2$. As the thermal Mach number increases from 0 to 1 (the transonic stage), this value varies from $-\frac{1}{2}$ to zero.

4.3.2 Supersonic Range with $M > 1$. For a crack tip moving to the right at a speed faster than that of the heat propagation in the solid, the transformation shown in Eq.

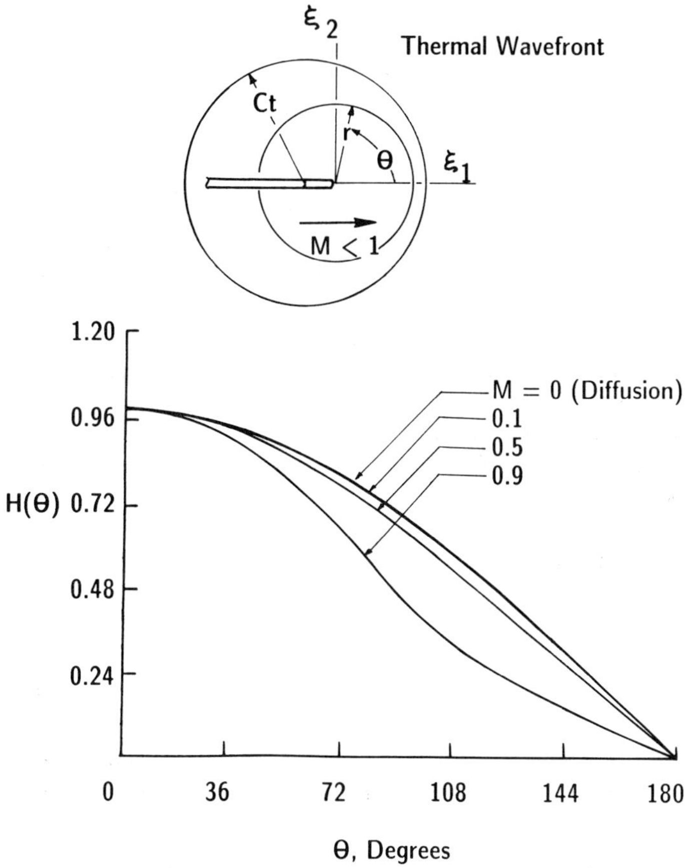

Figure 11 Angular variations of the near-tip temperature in the subsonic range with $M < 1$.

(55) can still be used but the temperature distribution under investigation must be confined to the domain of

$$1 - M^2 \sin^2 \theta > 0 \qquad \text{or} \qquad 0 \leq \theta \leq \sin^{-1}\left(\frac{1}{M}\right) \tag{67}$$

with θ measured from the trailing edge of the crack tip. Moreover, in the supersonic range with $M > 1$, the transformation $\gamma(\theta)$ satisfying Eq. (59) alternatively takes the form

$$\gamma(\theta) = \frac{1}{2} \ln\left[\frac{(M^2 - 1)^{1/2} \tan \theta + 1}{(M^2 - 1)^{1/2} \tan \theta - 1}\right] \qquad \text{and} \qquad \gamma,_\theta = \frac{(M^2 - 1)^{1/2}}{1 - M^2 \sin^2 \theta} \tag{68}$$

Substituting Eq. (68) into (58) then renders an equation governing the Φ function for the supersonic range:

$$\Phi,_{\gamma\gamma} - \lambda^2 \Phi = 0 \qquad \text{for } M > 1 \tag{69}$$

This is the corresponding equation to Eq. (61) for the subsonic range with $M < 1$. Obviously, the eigenvalue system formulated in the subsonic range with $M < 1$ is dramatically changed in the supersonic range with $M > 1$.

Although a different approach is adopted for the present problem, we first notice that the physical domain represented by Eq. (67) for $M > 1$ is the same as that for the *heat-affected zone* induced by a moving heat source [refer to Eq. (32)]. By the same argument, therefore, it is conceivable that *thermal shock waves* exist at $\theta = \pm \sin^{-1}(1/M)$ which separate the *heat-affected zone* from the *thermally undisturbed zone*. The thermal shock angle $\sin^{-1}(1/M)$ inclined to the trailing edge of the moving crack is thus the same as that induced by a moving heat source in the supersonic range. The temperature distribution in the heat-affected zone ranging from $\theta = -\sin^{-1}(1/M)$ to $\sin^{-1}(1/M)$ is determined by the Φ function satisfying Eq. (69) and the use of Eqs. (50), (55), and (68). The thermally undisturbed zone, on the other hand, occupies the rest of the physical domain, where the temperature stays at a reference value.

The temperature field in the heat-affected zone relies on the solution of Eq. (69). Defining now a parameter ζ,

$$\zeta(\theta) = \frac{1 + (M^2 - 1)^{1/2} \tan \theta}{1 - (M^2 - 1)^{1/2} \tan \theta} \tag{70}$$

such that, from Eq. (68),

$$\gamma = \pm \frac{i\pi}{2} + \frac{1}{2} \ln(\zeta) \tag{71}$$

with $i = \sqrt{-1}$ and ζ being greater than zero for θ in the heat-affected zone defined by Eq. (67). The solution of Eq. (69) in terms of the variable ζ can then be obtained as

$$\begin{aligned} \Phi[\zeta(\theta)] &= \cos\left(\frac{\lambda\pi}{2}\right)\left\{A \exp\left[\frac{\lambda \ln(\zeta)}{2}\right] + B \exp\left[-\frac{\lambda \ln(\zeta)}{2}\right]\right\} \\ &\quad + i \sin\left(\frac{\lambda\pi}{2}\right)\left\{A \exp\left[\frac{\lambda \ln(\zeta)}{2}\right] - B \exp\left[-\frac{\lambda \ln(\zeta)}{2}\right]\right\} \end{aligned} \tag{72}$$

Since the temperature function must be real, Eq. (72) requires that

$$\sin\left(\frac{\lambda\pi}{2}\right) = 0 \quad \text{or} \quad \lambda = 2n \text{ with } n = 0, 1, 2, 3, \ldots \tag{73}$$

For a nontrivial solution, therefore, the lowest eigenvalue λ in the supersonic range is 2. This is a result intrinsically different from the near-tip behavior of the temperature field in the subsonic range. As discussed previously, the r dependency of the near-tip temperature is $\sqrt{r}$ in the subsonic range with $M < 1$, and the temperature gradient has a $1/\sqrt{r}$-type singularity at the crack tip. In the supersonic range with $M > 1$, alternatively, such a dependency transits to r^2 [refer to Eq. (50)] and, consequently, the temperature gradient behaves as r which is bounded at the crack tip. This is a unique feature induced by the thermal shock formation in the near-tip region which cannot be depicted by the thermal diffusion model. Notice also that only one boundary condition at $\theta = 0$ (the trailing edge of the moving crack), or $\zeta = 1$ and $\gamma = \pm i(\pi/2)$ according to Eq. (71), remains in the present case with $M > 1$:

$$\Phi = 0 \qquad \text{at } \gamma = \pm i\left(\frac{\pi}{2}\right) \text{ and } \zeta = 1 \tag{74}$$

Another boundary at $\theta = \pi$ (the leading edge of the moving crack) stays in the thermally undisturbed zone, and the symmetrical condition $\Phi_{,\gamma} = \Phi_{,\theta} = 0$ is automatically satisfied for a uniform temperature field at the reference level. Substituting Eq. (72) into Eq. (74) renders a result of $A = -B$. The Φ function in the supersonic range is thus obtained as

$$\Phi[\zeta(\theta)] = B\left(\zeta - \frac{1}{\zeta}\right) \tag{75}$$

with B being an arbitrary constant. With the lowest value of λ being 2, combination of Eq. (55) with Eq. (75) then gives the angular distribution of the temperature field in the heat-affected zone,

$$H(\theta) = B(1 - M^2 \sin^2 \theta)\left(\zeta - \frac{1}{\zeta}\right) \tag{76}$$

which, upon substitution of Eq. (70), can be expressed in terms of the variable θ:

$$H(\theta) = 2B(M^2 - 1)^{1/2} \sin(2\theta) \qquad \text{for } M > 1 \tag{77}$$

As the thermal shock wave is approached from the site in the heat-affected zone, namely, $\theta \to \sin^{-1}(1/M)$, $\sin(2\theta)$ approaches $2(M^2 - 1)^{1/2}/M^2$ and we have

$$H(\theta) \to \frac{4B(M^2 - 1)}{M^2} \qquad \text{as } \theta \to \sin^{-1}\left(\frac{1}{M}\right) \tag{78}$$

Equation (78) indicates that the temperature remains *bounded* at the thermal shock wave located at $\theta = \sin^{-1}(1/M)$ with $M > 1$. It has a limit value of $4B$ as M approaches infinity. We thus conclude that the temperature is *discontinuous* but *finite* across the thermal shock surface. As the thermal Mach number increases, the H function represented by Eq. (77) [and hence the temperature level according to Eq. (50)] increases. Together with the corresponding response in the subsonic range with $M < 1$ that the temperature decreases as the thermal Mach number increases, this is the *swinging* phenomenon for the temperature field in transition of the thermal Mach number from the subsonic to the supersonic ranges. These phenomena are shown in Fig. 12 with M being 2, 5, and 10 in the supersonic range. Similar to the previous case with $M < 1$, the results are normalized with respect to the arbitrary constant B. The thermal shock waves are located, respectively, at $\theta = 30°$, 11.54°, and 5.74°, and the temperature in the heat-affected zone increases monotonically from the crack surface to the thermal shock waves.

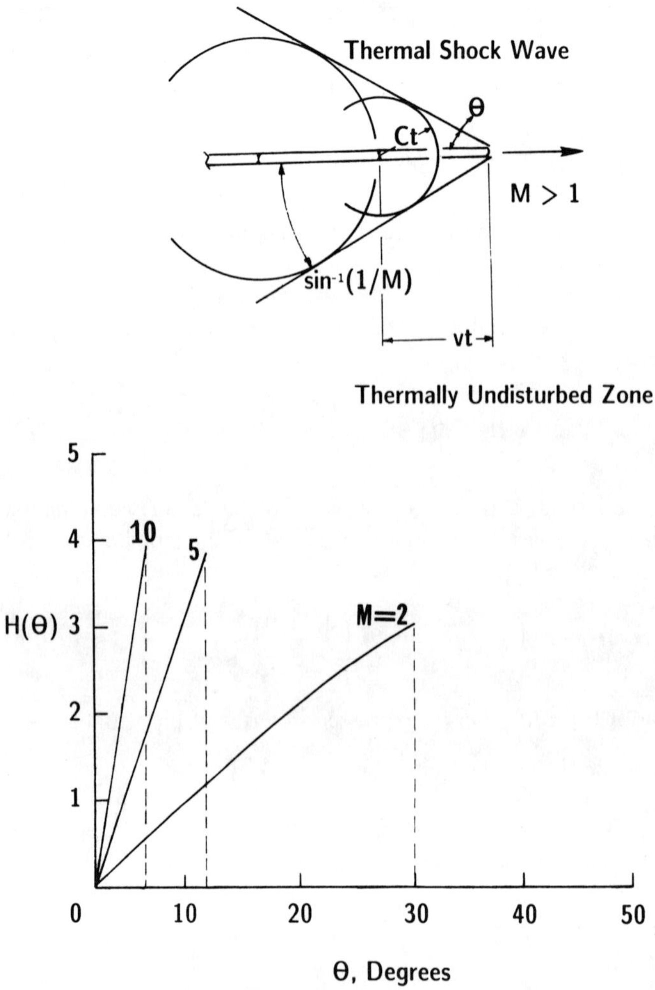

Figure 12 Angular variations of the near-tip temperature in the supersonic range with $M > 1$.

4.3.3 Transonic Stage with $M = 1$. As the speed of the moving crack is the same as that of the heat propagation in the solid, Eq. (57) reduces to the form

$$g_1 = \cos^2\theta \qquad g_2 = (\lambda - 1)\sin 2\theta$$

$$g_3 = \lambda[\lambda \sin^2\theta + \cos(2\theta)] \qquad f = \cos^\lambda \theta \tag{79}$$

and the equation governing the Φ function becomes

$$\Phi_{,\gamma\gamma} = 0 \tag{80}$$

The solution of Eq. (80) is simply

$$\Phi(\gamma) = A\gamma + B \tag{81}$$

and the transformation function γ, from Eq. (59), is

$$\gamma(\theta) = \tan\theta \tag{82}$$

With this relationship, the boundary condition at the crack surface, the second of Eq. (54), becomes

$$\Phi = 0 \qquad \text{at } \gamma = 0 \tag{83}$$

which yields a result of $B = 0$ from Eq. (81). Substituting Eq. (82) into Eq. (81) and the result into Eq. (55), the H function at the transonic stage with $M = 1$ is obtained:

$$H(\theta) = A\cos^{\lambda}\theta\tan\theta \qquad \text{for } M = 1 \tag{84}$$

Determination of the value of λ in this case relies on the argument for the *boundness* of the temperature field across the shock surface. Referring to Eq. (78) for the supersonic range with $M > 1$, the temperature is bounded when the thermal shock wave is approached from the site in the heat-affected zone, which should also be the case if the transonic stage with $M = 1$ is approached from the supersonic site. Based on this observation, the *lowest* value of λ in Eq. (84) is 1, and the resulting H function is

$$H(\theta) = \text{A}\sin\theta \qquad \text{for } M = 1 \tag{85}$$

where the coefficient A is an arbitrary constant as appeared in the eigenvalue system previously. With $A = 1$, the fundamental mode of the angular distribution of the near-tip temperature is displayed in Fig. 13. Note that, according to Eq. (67), a *normal* shock actually starts to form at $\theta = 90°$ at the transonic stage with $M = 1$, and the heat-affected zone with half-symmetry consequently ranges from 0° to 90° measured from the crack surface. The boundary at $\theta = 0$ is therefore in the thermally undisturbed zone, and the first boundary condition in Eq. (54) is automatically satisfied for a uniform temperature kept at the reference level. Most important, we notice that the r dependency of the near-field temperature represented by Eq. (50) transits from $r^{1/2}$ to r as the thermal Mach number transits from the subsonic range to the transonic stage. As a result, the temperature gradient at the transonic stage is *independent of* r in the near-tip region.

In summary, the similarities in the thermal shock waves induced by a rapidly moving heat source and a fast-propagating crack tip in solids include the thermal shock angle, the physical domain of the heat-affected zone, and the swinging phenomenon for the temperature variation in transition of the thermal Mach number from the subsonic to the supersonic ranges. The difference, however, lies in the discontinuous variation of the near-field temperature across the shock surfaces. For problems involving a moving heat source, the temperature jumps from *infinity* in the heat-affected zone to the reference level in the thermally undisturbed zone. For problems involving a moving crack, on the other hand, the temperature jump across the shock surface is *finite* and depends on the thermal Mach number, as shown by Eq. (78). Besides, in transition of the thermal Mach number from the subsonic to the transonic to the supersonic ranges, the r dependency of the near-tip temperature changes from $r^{1/2}$ to r to r^2. As the radial distance r measured from the crack tip approaches zero, the well-known

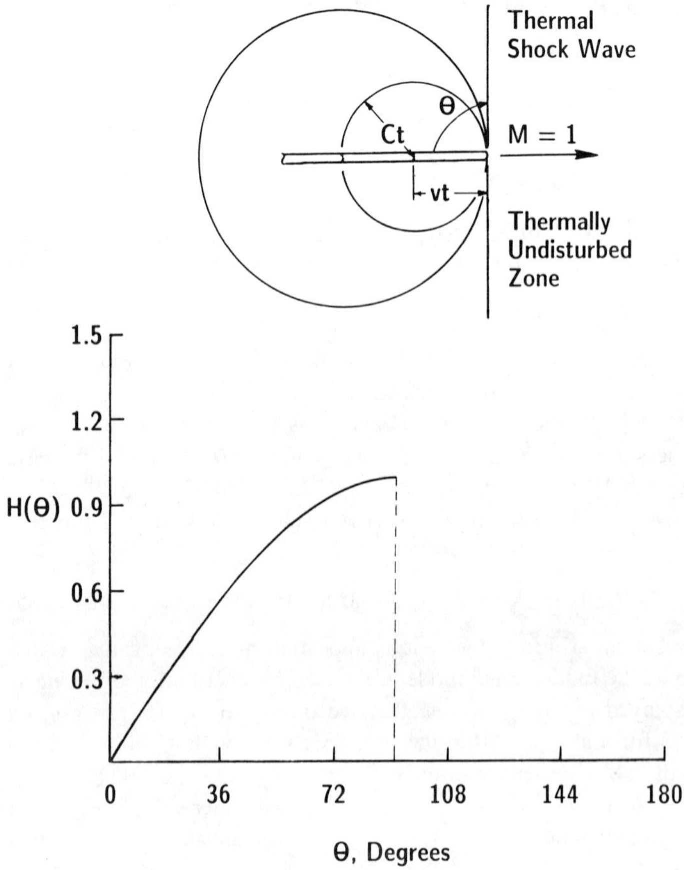

Figure 13 Angular variations of the near-tip temperature at the transonic stage with $M = 1$.

$1/\sqrt{r}$ singularity for the temperature gradient exists *only* in the subsonic range with $M < 1$. These special features result from the thermal shock formation in the transonic and supersonic ranges.

4.3.4 Similarity Between Flux and Temperature Formulations. From an analytical point of view, the thermal shock phenomenon around a rapidly propagating crack has been investigated by employing the temperature formulation. In order to examine this important physical phenomenon further, the flux formation will be used in this section for the propagating crack subject to a *flux-specified* condition at the crack surface. The geometric and dynamic configurations of the propagating crack are the same as those shown in Fig. 10.

The energy equation with **q** representation has been derived in Eq. (10). In the absence of heat generation from the crack tip during propagation, it takes the form of

$$\alpha\nabla[\nabla \cdot \mathbf{q}] = \left[\left(\frac{\alpha}{C^2}\right)\mathbf{q},_{tt} + \mathbf{q},_t\right] \tag{86}$$

Except for the spatial operators on the left-hand side of the equation, the time derivatives involved are the same as those in Eq. (43) with T representation. In terms of the material coordinates ξ_i convecting with the crack tip, therefore, Eq. (86) can be expressed as

$$\begin{aligned}\alpha\nabla[\nabla\cdot\mathbf{q}] &= \alpha M^2\mathbf{q},_{11} + \left[\left(\frac{\alpha}{C^2}\right)\mathbf{q},_{tt} - \left(\frac{M^2}{c}\right)\mathbf{q},_{1t}\right] + (\mathbf{q},_t - v\mathbf{q},_1)\\ &- \omega(\xi_1\mathbf{q},_2 - \xi_2\mathbf{q},_1) + \left(\frac{M^2}{2cR}\right)\{\xi_2[\mathbf{q},_{t1} - (v - \omega\xi_2)\mathbf{q},_{11}\\ &- \omega\mathbf{q},_2 - \omega\xi_1\mathbf{q},_{12}] - \xi_1[\mathbf{q},_{t2} - (v - \omega\xi_2)\mathbf{q},_{12} + \omega\mathbf{q},_1\\ &- \omega\xi_1\mathbf{q},_{22}] - (\xi_1\mathbf{q},_{t2} - \xi_2\mathbf{q},_{t1}) + v(\mathbf{q},_2 + \xi_1\mathbf{q},_{12} - \xi_2\mathbf{q},_{11})\}\end{aligned} \tag{87}$$

where, as before, M is the thermal Mach number measuring the speed of crack propagation relative to the heat propagation speed in the solid. Because the right-hand side of this equation is identical to that in Eq. (49) with the heat flux $\mathbf{q}$ replacing the temperature T, a similar product form for the heat flux can be assumed:

$$\mathbf{q}(r, \theta, t) = r^{\lambda}\,\Gamma(t)\,\mathbf{Q}(\theta) \tag{88}$$

with $\mathbf{Q}(\theta)$ governing the angular distributions of the heat flux vector. Asymptotic analysis of the terms involved in Eq. (87) in the near-tip region then yields

$$\begin{aligned}&\nabla(\nabla\cdot\mathbf{q}),\ \mathbf{q},_{11} && \sim r^{\lambda-2}\\ &\mathbf{q},_1,\ \mathbf{q},_2,\ \mathbf{q},_{t1},\ \xi_2\mathbf{q},_{11},\ \xi_1\mathbf{q},_{12},\ \xi_2\mathbf{q},_{11} && \sim r^{\lambda-1}\\ &\mathbf{q},_t,\ \xi_1\mathbf{q},_1,\ \xi_2\mathbf{q},_2,\ \xi_1\mathbf{q},_2,\ \xi_2\mathbf{q},_1,\ \mathbf{q},_{tt},\ \xi_2\mathbf{q},_{t1},\ \xi_2^2\mathbf{q},_{11},\\ &\xi_1\xi_2\mathbf{q},_{12},\ \xi_1\mathbf{q},_{t2},\ \xi_2\mathbf{q},_{t1},\ \text{and}\ \xi_1^2\mathbf{q},_{22} && \sim r^{\lambda}\end{aligned} \tag{89}$$

When the crack tip is closely approached, i.e., r approaches zero, the terms proportional to $r^{\lambda-1}$ and r^{λ} in Eq. (89) vanish and the dominant terms are $\nabla(\nabla\cdot\mathbf{q})$ and $\mathbf{q},_{11}$. The asymptotic form of Eq. (87) in the near-tip region is then

$$\nabla(\nabla\cdot\mathbf{q}) = M^2\mathbf{q},_{11} \qquad \text{with } r \rightarrow 0 \tag{90}$$

or, in terms of its components,

$$(1 - M^2)q_{1,11} + q_{2,12} = 0 \tag{91}$$

$$q_{1,12} + q_{2,22} - M^2 q_{2,11} = 0 \tag{92}$$

presenting a set of coupled differential equations for the components q_1 and q_2 of $\mathbf{q}$. In a similar manner to obtaining the T or $\mathbf{q}$ representation for the energy equation, a single equation governing q_2 can be obtained by eliminating q_1 from Eqs. (91) and (92). Combining the differential of Eq. (91) with respect to ξ_2 and that of Eq. (92) with respect to ξ_1, we obtain

$$q_{2,122} + (1 - M^2)q_{2,111} = 0 \tag{93}$$

A particular form for this equation is

$$q_{2,22} + (1 - M^2)q_{2,11} = 0 \qquad \text{or} \qquad \nabla^2 q_2 = M^2 q_{2,11} \tag{94}$$

which is sufficient for our purpose with regard to the eigenstate of the near-tip heat flux. Equation (94) is to be solved subject to the flux-specified conditions along the crack line at $\xi_2 = 0$:

$$\begin{aligned} q_2 &= 0 \qquad \text{at } \xi_2 = 0 \text{ and } \xi_1 < 0 \\ q_{2,2} &= 0 \qquad \text{at } \xi_2 = 0 \text{ and } \xi_1 > 0 \end{aligned} \tag{95}$$

The first condition assumes an insulated crack surface during propagation, while the second results from the symmetry of the problem. In terms of the polar coordinate system centered at the crack tip and the heat flux component Q_2, the boundary conditions can be expressed as

$$Q_{2,\theta} = 0 \text{ at } \theta = 0 \qquad \text{and} \qquad Q_2 = 0 \text{ at } \theta = \pi \tag{96}$$

One should notice that Eqs. (94) and (96) governing the heat flux component q_2 have *exactly* the same form as Eqs. (53*a*) and (54) governing the near-tip temperature. The rest of the procedure in the previous section, therefore, can be extended to the present problem with q_2 replacing T and Q_2 replacing H in all the corresponding derivations. Substituting Eq. (88) for q_2 into Eq. (94), for example, the equation governing the angular distribution $Q_2(\theta)$ is

$$(1 - M^2 \sin^2 \theta)Q_{2,\theta\theta} - [M^2(1 - \lambda) \sin 2\theta]Q_{2,\theta} + \lambda\{\lambda + M^2[(2 - \lambda) \cos^2 \theta - 1]\}Q_2 = 0 \tag{97}$$

which replaces the H function governing the angular distribution of the near-tip temperature in Eq. (53*b*) by the Q_2 function governing that of the heat flux component q_2.

Due to the analogy between the temperature and flux formulations, Figs. 11 through 13 can also be used to illustrate the angular variations of the heat flux component Q_2 in the near-tip region. The H functions in these figures are replaced by the Q_2 functions, and the thermal shock formation in the transonic and supersonic ranges with $M \geq 1$ is quite clear from Figs. 12 and 13. The analogy between the temperature (T) and the heat flux vector ($\mathbf{q}$) in the near-tip region, including the transition of the r dependency from $\sqrt{r}$ to r to r^2 from the subsonic, transonic, to the supersonic ranges of the thermal Mach number, is another salient feature of the thermal wave theory for solids under short time response. Even considering the transient variation, these features cannot be depicted by classical diffusion theory.

Another component Q_1 of the heat flux vector can be obtained by the Q_2 function in the respective ranges of the thermal Mach number. Substituting Eq. (88) into Eqs. (91) and (92) and eliminating the terms containing $Q_{2,\theta}$ from the resulting equations give a first-order differential equation governing $Q_1(\theta)$:

$$(1 - M^2 \sin^2 \theta)Q_{1,\theta} + \left(\frac{\lambda M^2}{2} \sin 2\theta\right)Q_1 = \lambda Q_2 \tag{98}$$

The angular distribution of the heat flux component $Q_1(\theta)$ can then be integrated directly as

$$Q_1(\theta) = \lambda(1 - M^2 \sin^2 \theta)^{\lambda/2} \int^{\theta} \frac{Q_2(\eta)}{(1 - M^2 \sin^2 \eta)^{[(\lambda+2)/2]}} d\eta \tag{99}$$

Note that in addition to the Q_2 function, the eigenvalue λ also has different values in transition of the thermal Mach number from the subsonic to the transonic to the supersonic ranges.

5 EFFECT OF THERMAL SHOCK ON FRACTURE DAMAGE AROUND A RAPIDLY MOVING HEAT SOURCE

When the speed of the moving heat source exceeds that of the heat propagation in the solid, the thermal shock phenomenon essentially depicts a preferential direction along which intensified thermal energy tends to cumulate. As a result of the thermal energy accumulation, the intensification of the *deformation energy*, and hence the resulting thermal stresses, induced by the thermal shock waves are an immediate topic of major concern. This section will first provide a stationary formulation for the thermoelastic stress analysis incorporating the effect of finite wave speed. The emphasis is then placed on the resulting fracture pattern emanating from the moving heat source due to the large temperature gradient existing in the vicinity of the thermal shock waves. Both the strain energy density [83] and the maximum circumferential stress [84] criteria are used for comparison. The near-field fracture pattern around a rapidly moving heat source provides not only a visible means for the determination of the thermal wave speed in the solid medium but also a potential proof for the existence of the thermal shock waves.

5.1 Thermoelastic Formulation

The representation of the Neuber-Papkovich thermoelastic displacement potential [85] will be modified to account for the finite wave speed. The case of plane strain will be adopted in the analysis, which describes a more brittle response for crack initiation than the case of plane stress.

On the material coordinate system ξ_i convecting with the same speed as the heat source, as a first-order approximation, the thermoelastic potential function Φ relates to the temperature by the quasi-stationary equation

$$\nabla^2\Phi = \epsilon\beta T(\xi_i), \text{ for } i = 1, 2 \qquad \text{and} \qquad \nabla^2 \equiv \frac{\partial^2}{\partial \xi_1^2} + \frac{\partial^2}{\partial \xi_2^2} \tag{100}$$

where ϵ is the thermal expansion coefficient of the medium and β is a function of Poisson's ratio ν, defined as $(1 + \nu)/(1 - \nu)$. The temperature distributions $T(\xi_i)$ are described by an observer standing on the moving heat source and have been obtained analytically in the respective ranges of the thermal Mach number in Sec. 4.2. In summary, they are represented by Eq. (23) for the subsonic range with $M < 1$, Eq. (29) for the supersonic range with $M > 1$, and Eq. (35) for the transonic stage with $M = 1$. Instead of solving Eq. (100) by Poisson's integral [86], we may combine Eq. (100) with the energy equation with T representation, Eq. (16), to yield an ordinary differential equation for Φ. Eliminating the temperature T in Eq. (100) by taking the appropriate differentiations with respect to ξ_1 and ξ_2 according to the main operator of Eq. (16), the following ordinary differential equation governing the potential function Φ results:

$$M^2\Phi,_{11} - 2c\Phi,_1 = \epsilon\beta\left(T + \frac{1}{\alpha}\Phi_0\right) \tag{101}$$

where the function Φ_0 satisfies

$$\nabla^2\Phi_0 = \left(\frac{Q}{\rho C_p}\right)\left[\delta(\xi_1) - \left(\frac{M^2}{2c}\right)\delta,_1(\xi_1)\right]\delta(\xi_2) \tag{102}$$

is the elastic potential solely contributed by the *apparent* heat source. The effect of the finite wave speed is more clearly seen in this formulation. In addition to that in the temperature distributions, the thermal Mach number M also appears explicitly in the field equation governing the thermoelastic displacement potential. In the case that either v approaches zero or C approaches infinity, M approaches zero and Eq. (101) reduces to the Neuber-Papkovich representation employing the thermal diffusion model [87].

The modified term in Eq. (101) is led by M^2, which appears as the *highest*-order differential in the field equation. In comparison with the corresponding equation with $M = 0$ for the thermal diffusion model, it intrinsically alters the characteristic of the solution for Φ in transition of the thermal Mach number from the subsonic to the transonic to the supersonic ranges. Once the solution for Φ_0 is obtained by solving Eq. (102), with the assistance of the temperature expressions obtained in the respective range of the thermal Mach number, Eq. (101) can be solved for $\Phi,_1(\xi_i)$. Through the relations

$$u_i(\xi_i) = \Phi,_i \tag{103}$$

$$\sigma_{ij}(\xi_i) = 2\mu(\Phi,_{ij} - \delta_{ij}\nabla^2\Phi) \qquad i = 1, 2 \text{ for in-plane components} \tag{104}$$

and

$$u_3 = 0 \qquad \sigma_{33}(\xi_i) = -2\mu\nabla^2\Phi \qquad \text{under plane strain conditions} \tag{105}$$

then the thermoelastic displacement field u_i and the thermal stress field σ_{ij} can be calculated accordingly.

The solution for $\Phi,_1(\xi_i)$ is straightforward. By arranging the left-hand side of Eq. (101) into an exact differential, we obtain

$$\Phi_{,1}(\xi_i) = \left(\frac{\epsilon\beta}{M^2}\right) \int_\infty^{\xi_1} \exp\left[\frac{2c(\xi_1 - \eta)}{M^2}\right] \left[T(\eta, \xi_2) + \frac{1}{\alpha}\Phi_0(\eta, \xi_2)\right] d\eta \quad (106)$$

where the physical condition that $\Phi_{,1}$ keeps bounded at a large distance from the heat source has been imposed. From Eqs. (104) and (105), the stress components can be obtained by direct differentiations on $\Phi_{,1}(\xi_i)$. The results are

$$\begin{aligned}
\sigma_{11}(\xi_i) &= \left(\frac{2\mu\epsilon\beta}{M^2}\right)\left[Z(\xi_i) + (1 - M^2)\,T(\xi_i) + \frac{1}{\alpha}\Phi_0(\xi_i)\right] \\
\sigma_{22}(\xi_i) &= \left(-\frac{2\mu\epsilon\beta}{M^2}\right)\left[Z(\xi_i) + T(\xi_i) + \frac{1}{\alpha}\Phi_0(\xi_i)\right] \\
\sigma_{12}(\xi_i) &= \left(\frac{2\mu\epsilon\beta}{M^2}\right)\left[\left(\frac{M^2}{2c}\right) Z_{,2}(\xi_i)\right] \\
\sigma_{33}(\xi_i) &= -2\mu\epsilon\beta T(\xi_i)
\end{aligned} \quad (107)$$

with

$$\begin{aligned}
Z(\xi_i) &= \int_\infty^0 e^{-\zeta}\left[T\left(\xi_1 + \frac{M^2}{2c}\zeta, \xi_2\right) + \frac{1}{\alpha}\Phi_0\left(\xi_1 + \frac{M^2}{2c}\zeta, \xi_2\right)\right] d\zeta \\
Z_2(\xi_i) &= \int_\infty^0 e^{-\zeta}\left[T_{,2}\left(\xi_1 + \frac{M^2}{2c}\zeta, \xi_2\right) + \frac{1}{\alpha}\Phi_{0,2}\left(\xi_1 + \frac{M^2}{2c}\zeta, \xi_2\right)\right] d\zeta
\end{aligned} \quad (108)$$

The Φ_0 function appearing in these equations can be found directly from the Green's function for the two-dimensional Laplacian operator [74]. The result is

$$\Phi_0(\xi_i) = \frac{Q}{2\pi\rho C_p}\left(\ln R - \frac{M^2}{2c}\frac{\xi_1}{R^2}\right) \qquad \text{with } R = \sqrt{\xi_1^2 + \xi_2^2} \quad (109)$$

and its derivatives:

$$\Phi_{0,1}(\xi_i) = \frac{Q}{4\pi\rho C_p}\left[\frac{2\xi_1}{R^2} - \frac{M^2(\xi_2^2 - \xi_1^2)}{cR^4}\right] \quad (110)$$

$$\Phi_{0,2}(\xi_i) = \frac{Q}{4\pi\rho C_p}\left(\frac{2\xi_2}{R^2} + \frac{2M^2\xi_1\xi_2}{cR^4}\right) \quad (111)$$

Along with the temperature distributions represented by Eqs. (23), (29), and (35) in the respective range of the thermal Mach number, the equations summarized above provide sufficient information to integrate the thermal stresses. Although analytical forms for σ_{ij} can hardly be obtained due to complexity of the integrand involved in Eq. (108), the integrations of $Z(\xi_i)$ and $Z_{,2}(\xi_i)$ obviously exist since all the functions involved in the integration approach zero at large value of ζ. To be noticed is that the thermal stresses employing the diffusion model cannot be retrieved by setting $M = 0$ in Eqs. (107) and (108). Instead, one should substitute zero for M in Eq. (101) and take the spatial differentiations directly on the resulting expression for $\Phi_{,1}(\xi_i)$ in calculating the stresses.

5.2 Orientation of Crack Initiation and Crack Trajectories Emanating from the Heat Source

The thermal stresses induced by the temperature field incorporating the finite wave speed can be found in general from Eqs. (107) through (111). With emphasis on the thermal shock formation in the transonic and supersonic ranges, we shall present the distributions of the circumferential stress σ_θ and the strain energy density function dW/dV along the same continuum circle as that considered in Sec. 4.2 for the near-field temperature. For nonisothermal linear elastic deformations, they can be expressed, respectively, as

$$\sigma_\theta = \sigma_{11}\sin^2(\theta) + \sigma_{22}\cos^2(\theta) - \sigma_{12}\sin(2\theta) \tag{112}$$

$$\frac{dW}{dV} = \left(\frac{1}{2E}\right)(\sigma_{11}^2 + \sigma_{22}^2 + \sigma_{33}^2) - \left(\frac{\nu}{E}\right)(\sigma_{11}\sigma_{22} + \sigma_{11}\sigma_{33} + \sigma_{22}\sigma_{33}) + \left(\frac{1+\nu}{E}\right)\sigma_{12}^2 + \mu(1+\nu)(\epsilon T)^2 \tag{113}$$

which combine the effects from all the stress components.

The circumferential stress and the strain energy density function are the crucial quantities for predicting the orientations of crack initiation by the *minimum* strain energy density [83] and the *maximum* circumferential stress [84] criteria. In estimating the crack trajectories emanating from the heat source according to the circumferential stress criterion, the distributions of σ_θ are first obtained along the circumferences of several circles centered at the moving heat source with gradually increasing radii. Every radius of the circle reflects the physical scale on which observation on crack initiation is made. The location possessing the *maximum* value of the circumferential stress on every circle is then identified and connected to yield a spatial curve for the crack trajectory. This is also the procedure when the strain energy density criterion is employed, but the location possessing the maximum circumferential stress is replaced by that possessing the *minimum* strain energy density. Such a computational philosophy has been used widely in linear elastic fracture mechanics [88, 89, for example] and is similar to that adopted by Liebowitz et al. [90]. Because the change of crack geometry is not accounted for in this approach, the trajectories obtained in this fashion correspond to those emanating *instantaneously* from the heat source under high rate response. In the present analysis, various radii are taken in the respective range of the thermal Mach number such that the characteristics of the resulting fracture pattern are most clearly represented. In addition to the thermal properties described in Eq. (24), the following elastic properties of the solid medium are assumed in the numerical computations:

$$\alpha = 8 \times 10^{-5}\ \text{m}^2/\text{s} \qquad \nu = 0.3 \qquad E = 1.0 \times 10^5\ \text{MPa}$$
$$\epsilon = 1.0 \times 10^{-6}\ \text{m/m/C} \tag{114}$$

Because the temperature has different expressions in transition of the thermal Mach number, the thermoelastic stress analysis must be made *individually* in the subsonic, transonic, and the supersonic ranges.

5.2.1 Subsonic Range with $M < 1$. When the speed of the moving heat source is slower than that of the heat propagation in the solid, the temperature T and its gradient $T_{,2}$ around the heat source are represented by Eqs. (23) and (39). Along with the potential function Φ_0 and its derivative $\Phi_{0,2}$ shown, respectively, by Eqs. (109) and (111), the thermal stress components in the subsonic range can be calculated according to Eqs. (107) and (108). Based on Eqs. (112) and (113), then, the circumferential stress and the strain energy density function are obtained.

In the subsonic range, the crack trajectories emanating from the heat source are examined at $M = 0.1$ and 0.7. At $M = 0.1$, Fig. 14 shows the distributions of σ_θ and dW/dV along the circumferences of circles centered at the heat source with radii

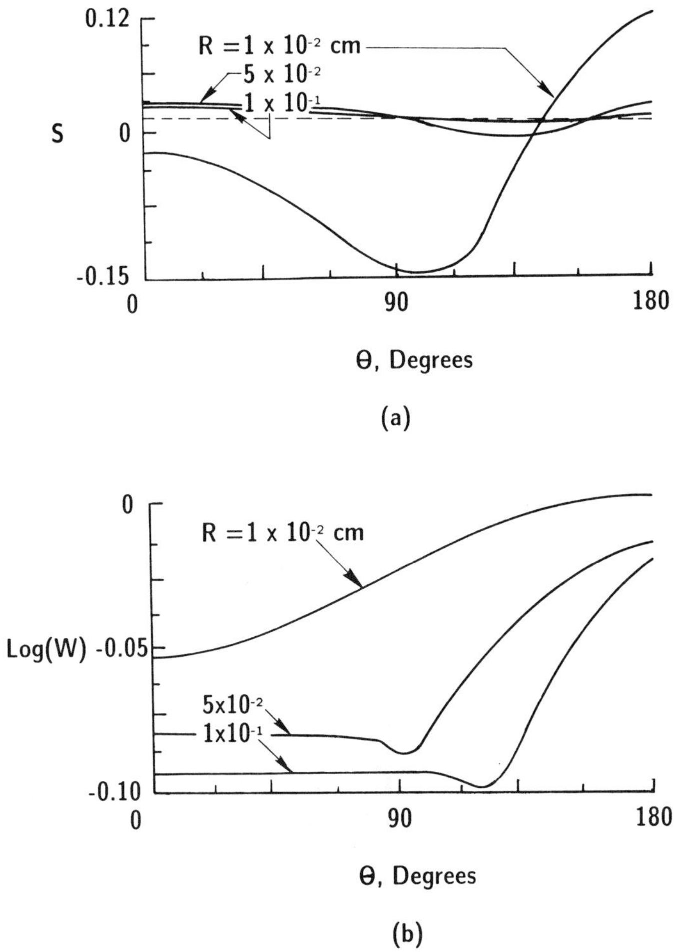

Figure 14 The distributions of (*a*) the circumferential stress σ_θ and (*b*) the strain energy density function dW/dV around a heat source moving in the subsonic range with $M = 0.1$. $S = \sigma_\theta/(2\mu\epsilon\beta Q/\rho\alpha C_p)$ and $W = (dW/dV)/[(2\mu\epsilon\beta Q/\rho\alpha C_p)^2/E]$.

1×10^{-2}, 5×10^{-2}, and 1×10^{-1} cm. Due to the symmetry of the problem, only the distributions in the upper half-plane with $0 \leq \theta \leq 180°$ are presented. From Fig. 14*a*, we notice that, as expected, the magnitude of the compressive and tensile stresses decreases when the distance away from the heat source increases. This trend is also shown in Fig. 14*b* for the strain energy density distributions where the log scale for dW/dV is adopted to show the locations possessing the minimum values more clearly. The location possessing a maximum value of σ_θ is at $\theta = 180°$ ($S = 0.1181$), 0 (0.0127), and 0 (0.0064) as the radius of the circle increases from 1×10^{-2} to 5×10^{-2} to 1×10^{-1} cm. According to the maximum circumferential stress criterion, then, a spatial locus is obtained by connecting these points on the circles as shown in Fig. 15. Because the polar coordinates (R, θ) are convecting with the heat source in

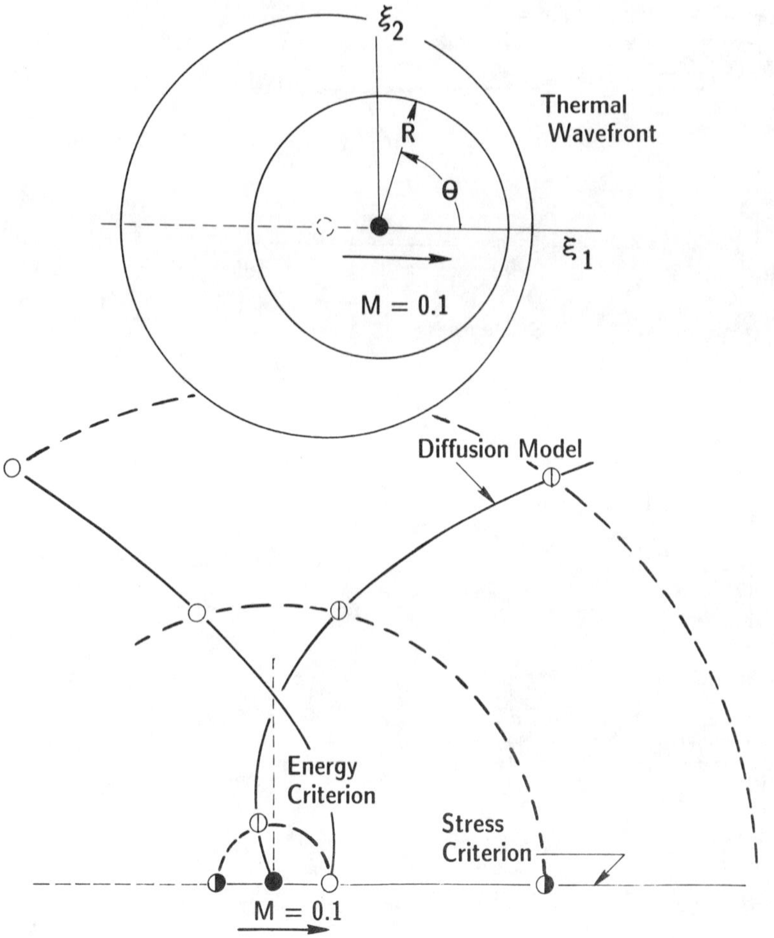

Figure 15 The crack trajectory emanating from a heat source moving in the subsonic range with $M = 0.1$.

nature, the crack orientation at $\theta = 180°$ for $R = 1 \times 10^{-2}$ cm essentially depicts a fracture path coincident with the trace of the moving heat source and the stress criterion basically predicts a crack trajectory *ahead of* the heat source. The energy criterion, on the other hand, describes a more capricious cracking pattern. As shown in Fig. 14*b*, the location possessing a minimum value of dW/dV changes from 0 ($W = 0.4326 \times 10^{-2}$) to 111° ($0.1311 \times 10^{-3}$) to 131° ($0.3374 \times 10^{-4}$) as the radius of the circle increases from 1×10^{-2} to 5×10^{-2} to 1×10^{-1} cm. The strain energy density criterion, therefore, depicts a crack trajectory initiated at the leading edge of the heat source and turning sharply toward the trailing edge as the value of R increases as shown in Fig. 15. The stress and the energy criteria predict a similar fracture pattern on a smaller scale, with R being approximately less than 1×10^{-2} cm, while on a larger scale, with R being greater than this threshold value, the deviation for the fracture patterns predicted by the two criteria becomes significant.

The classical diffusion model assumes an infinite speed for heat propagation and the existence of a thermal disturbance, as a direct consequence of such an assumption, can be detected anywhere in the solid right after its presence. According to the energy density criterion, the crack trajectory predicted by the diffusion model (assuming that $v = 1$ cm/s) is also displayed in Fig. 15 for comparison. It depicts a profile passing $\theta = 82°$, 68°, and 64° as the radius R increases from 1×10^{-2} to 5×10^{-2} to 1×10^{-1} cm. Any value of R smaller than 1×10^{-2} cm predicts a crack initiation angle of 90°. For a value of R larger than 1 cm, as another extreme, the crack initiation angle turns toward the trailing edge of the moving heat source at 140°. In view of the crack trajectories displayed in Fig. 15, we can see that the effect of finite wave speed is important, especially in the neighborhood of the heat source where large temperature gradient exists. Neglecting such an effect may lead to significant deviations in predicting the crack trajectory emanated under high rate response.

When the thermal Mach number increases to 0.7, the distributions of σ_θ and dW/dV and the resulting crack trajectories predicted by the stress and energy criteria are displayed in Figs. 16 and 17. Because the stress and energy levels decrease as the thermal Mach number increases, the radii of the circles in the present case are selected to be smaller than those in the previous case. They are 5×10^{-4}, 1×10^{-3}, and 5×10^{-3} cm. On these scales of observation, the locations possessing maximum values of σ_θ are all at the leading edge of the moving heat source, $\theta = 0°$. The maximum values are $S_{max} = 0.6371 \times 10^{-2}$, 0.6437×10^{-1}, and 0.1819×10^{-1} as the radius increases from 5×10^{-4} to 1×10^{-3} to 5×10^{-3} cm. Again, the stress criterion predicts a fracture path ahead of the moving heat source as shown in Fig. 17, but the crack orientation angle at $\theta = 180°$ predicted in the previous case disappears. The location possessing a minimum value of dW/dV, as shown in Fig. 16*b*, sweeps from 0 to 45° to 109° toward the trailing edge of the moving heat source as the radius increases. Their values are, respectively, $W_{min} = 0.2802 \times 10^{-1}$, 0.7175×10^{-2}, and 0.2756×10^{-3}. By comparing the fracture patterns shown in Figs. 15 and 17, the crack trajectories predicted by the energy criterion at $M = 0.1$ and 0.7 are qualitatively similar, but shift more toward the leading edge of the moving heat source as the thermal Mach number increases.

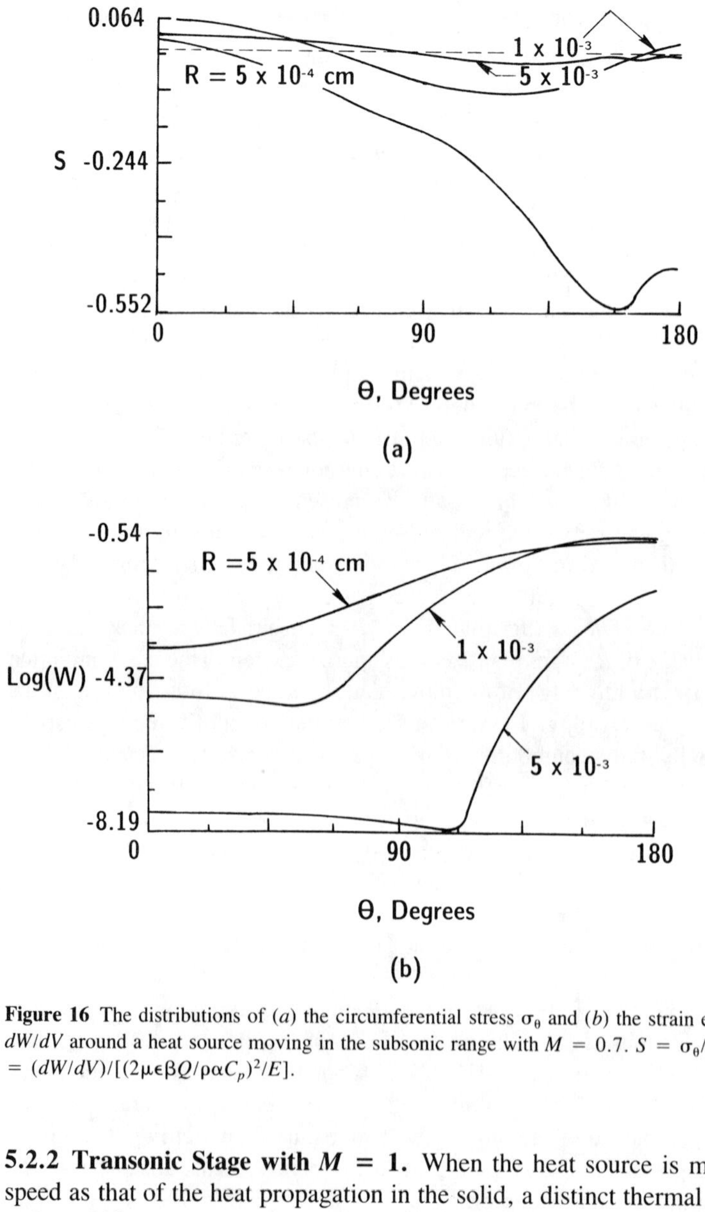

Figure 16 The distributions of (*a*) the circumferential stress σ_θ and (*b*) the strain energy density function dW/dV around a heat source moving in the subsonic range with $M = 0.7$. $S = \sigma_\theta/(2\mu\epsilon\beta Q/\rho\alpha C_p)$ and $W = (dW/dV)/[(2\mu\epsilon\beta Q/\rho\alpha C_p)^2/E]$.

5.2.2 Transonic Stage with $M = 1$. When the heat source is moving at the same speed as that of the heat propagation in the solid, a distinct thermal shock wave forms at $\theta = 90°$. As a result, intensified deformation energy and thermal stresses present in the neighborhood of 90°. For a heat source moving to the right, the physical domain of the heat-affected zone is from 90° to 180° measured from the leading edge of the heat source. The distributions of σ_θ and dW/dV are displayed in Fig. 18 for the observation circles with radii of 1×10^{-2}, 2×10^{-2}, and 3×10^{-2} cm. The location possessing the maximum value of σ_θ, to the contrary of the subsonic case, shifts to the *trailing edge* of the moving heat source, and the crack trajectory predicted in this

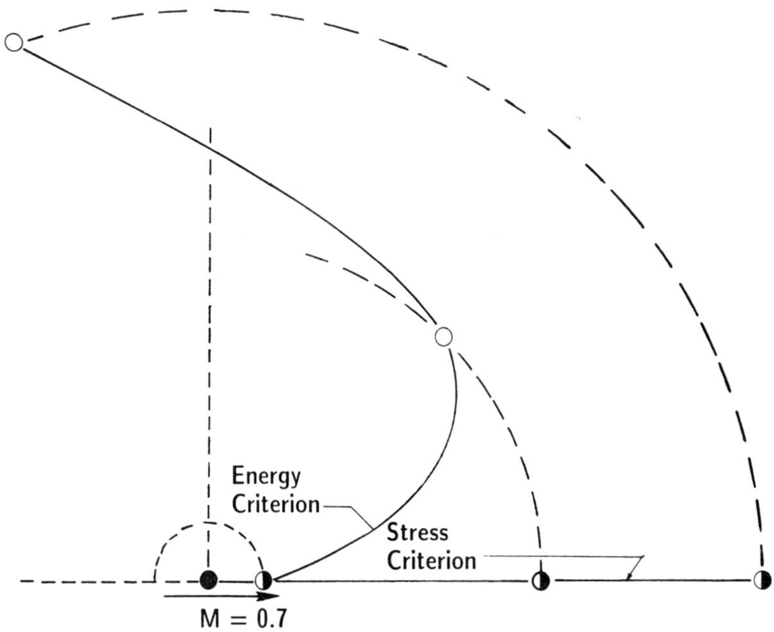

Figure 17 The crack trajectory emanating from a heat source moving in the subsonic range with $M = 0.7$.

case coincides with the trace of the moving heat source as shown in Fig. 19. Room for doubt, however, should be allowed for this fracture pattern, because another peak possessing a relative maximum exits on the same circle which has almost the same value as that at $\theta = 180°$. For $R = 1 \times 10^{-2}$ cm, the S values at $\theta = 130°$ and 180° are, respectively, 0.3676×10^{-2} and 0.5334×10^{-2}, which still provide acceptable resolutions for distinguishing the *absolute* maximum of S. For $R = 2 \times 10^{-2}$ cm, however, the difference between the maximum values existing at 180° ($S_{max} = 0.2787 \times 10^{-2}$) and 143° ($S_{max} = 0.2418 \times 10^{-2}$) becomes small, which implies a possibility of dual crack formation from both 143° and 180°. This situation becomes more pronounced as the radius R further increases to 3×10^{-2} cm. The maximum value of S at 180° is 0.1897×10^{-2} in this case, while another existing at 149° is 0.1761×10^{-2}. When the energy intensity of the heat source impinged on the surface of the structure gradually increases, the maximum values at the two locations may arrive at the threshold value of fracture stress simultaneously, and the second profile shown in Fig. 19 connecting the locations at $\theta = 143°$ for $R = 2 \times 10^{-2}$ cm and 149° for $R = 3 \times 10^{-2}$ cm displays another possible crack trajectory emanating from the heat source. Similar ambiguity exists in the fracture pattern predicted by the energy criterion. A plateau exists in the distributions of dW/dV for all the values of R under consideration. For $R = 1 \times 10^{-2}$ cm, the value of W reaches a minimum, being 0.5248×10^{-4} at $\theta = 125°$, but this value remains unchanged in the intermediate domain from 95° to 125°. When the radius increases from 2×10^{-2} to 3×10^{-2} cm, the minimum values of W reduce to 0.1318×10^{-4} and 0.5856×10^{-5},

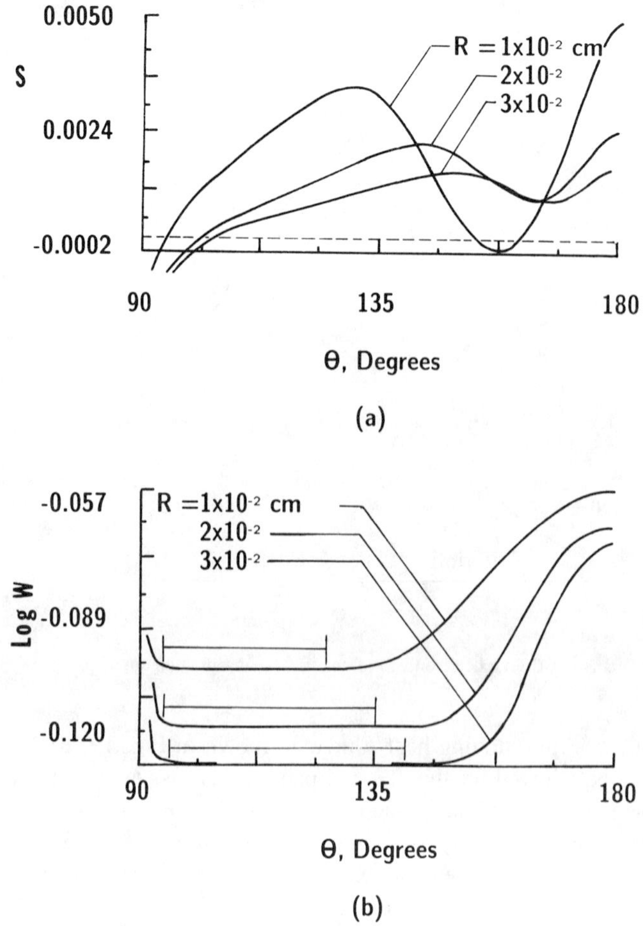

Figure 18 The distributions of (*a*) the circumferential stress σ_θ and (*b*) the strain energy density function dW/dV around a heat source moving at the transonic stage with $M = 1$. $S = \sigma_\theta/(2\mu\epsilon\beta Q/\rho\alpha C_p)$ and $W = (dW/dV)/[(2\mu\epsilon\beta Q/\rho\alpha C_p)^2/E]$.

respectively, while the angular domains of the plateau are broadened from $\theta \in [96°, 135°]$ to $[98°, 141°]$. This situation is illustrated by the fracture surfaces in Fig. 19 at various values of R. The material continua along the shaded surfaces are subject to large amounts of volume expansion, and the surfaces describe the potential area of material damage by cracking.

At the transonic stage with $M = 1$, it seems that both the stress and energy criteria do not predict a distinct crack trajectory emanating from the heat source. Before obtaining direct confirmation from experiment, it is worthwhile to pursue this phenomenon further by comparing the other fracture criteria. Besides, the large value of the compressive stress in the neighborhood of the thermal shock waves (refer to Fig. 18*a*), may lead to a damage pattern by shear failure. Along with the microcracking

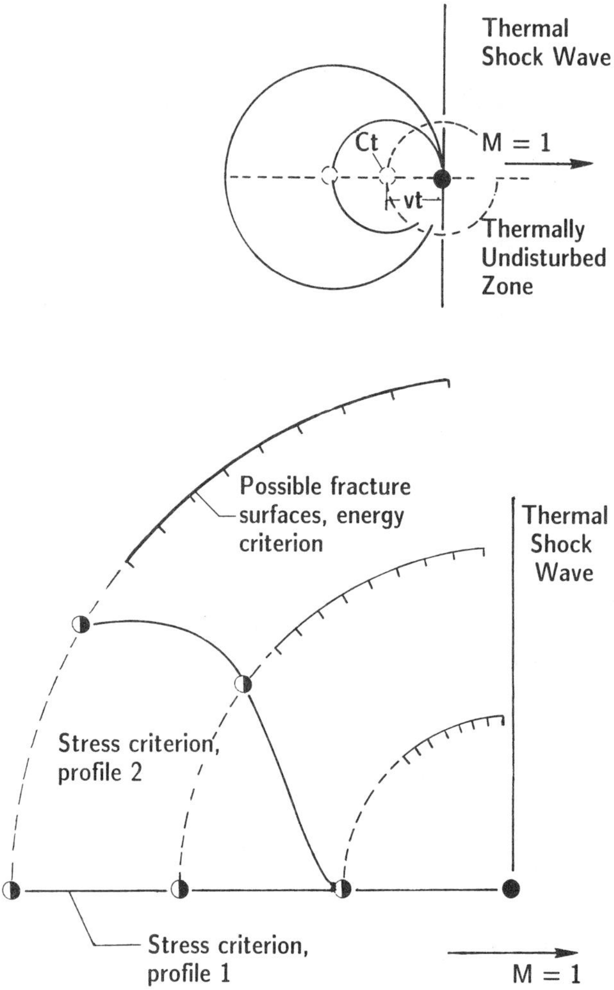

Figure 19 The fracture damage pattern in the neighborhood of a heat source moving at the transonic range with $M = 1$.

damage induced by excessive volumetric strain established in the shock area [91], this will be a central topic to be discussed in the near future.

5.2.3 Supersonic Range with $M > 1$. When the speed of the moving heat source exceeds that of the heat propagation in the solid, distinct thermal shocks are formed at $\theta = \sin^{-1}(1/M)$ measured from the trailing edge. Figure 20 shows the distributions of σ_θ and dW/dV at $M = 5$. The thermal shock angle in this case is $\theta = 168.46°$, and the radii of the circles for observation are taken to be 5×10^{-3}, 8×10^{-3}, and 1×10^{-2} cm. Due to the strong influence of the thermal shock waves, we notice that

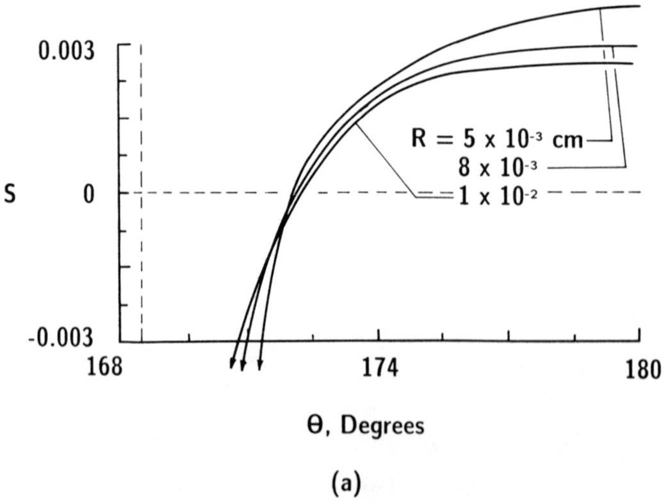

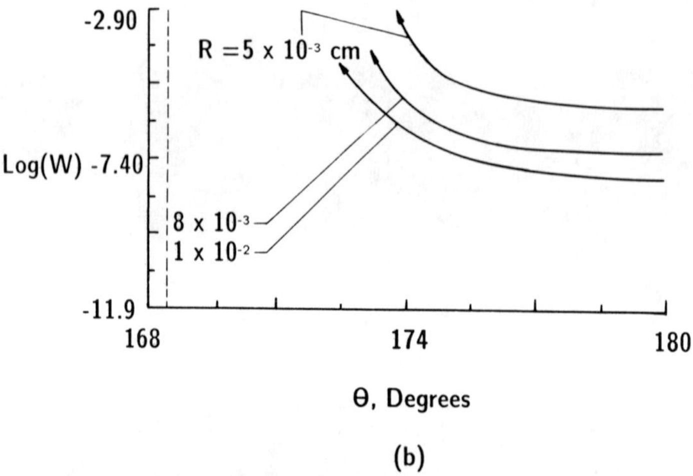

Figure 20 The distributions of (*a*) the circumferential stress σ_θ and (*b*) the strain energy density function dW/dV around a heat source moving in the supersonic range with $M = 5$. $S = \sigma_\theta/(2\mu\epsilon\beta Q/\rho\alpha C_p)$ and $W = (dW/dV)/[(2\mu\epsilon\beta Q/\rho\alpha C_p)^2/E]$.

the physical domain, subject to a compressive state of circumferential stress (from 168.46° to 172°, penetrates approximately one-third of the way into the heat-affected zone (from 168.46° to 180°). For the three values of R under considerations, the locations possessing the maximum values of S, 0.3748×10^{-2} for $R = 5 \times 10^{-3}$ cm, 0.3015×10^{-2} for $R = 8 \times 10^{-3}$ cm, and 0.2884×10^{-2} for $R = 1 \times 10^{-2}$ cm, are all at the trailing edge of the heat source at $\theta = 180°$ and consequently the crack trajectory predicted is coincident with the rear end of the moving heat source as depicted in Fig. 21. The crack trajectory predicted by the energy criterion displays

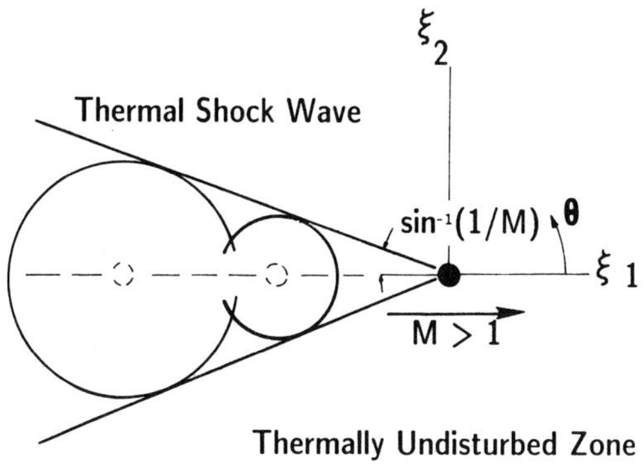

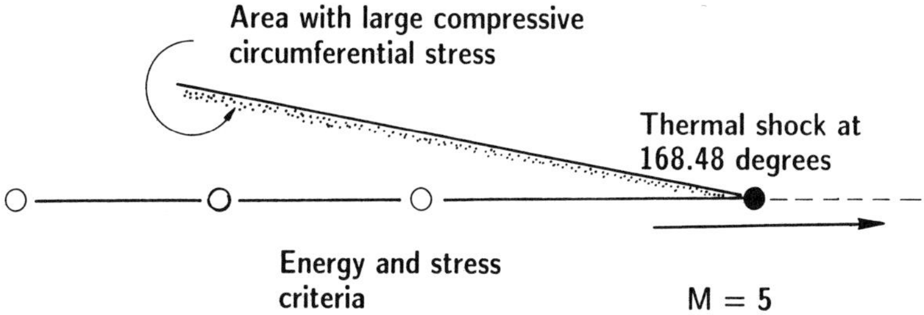

Figure 21 The material damage pattern in the heat affected zone behind a heat source moving in the supersonic range with $M = 5$.

an identical pattern in this case. As shown in Fig. 20*b*, when the radius of the circle increases from 5×10^{-3} to 8×10^{-3} to 1×10^{-2} cm, the minimum value of W at $\theta = 180°$ decreases from 0.3698×10^{-2} to 0.6052×10^{-3} to 0.4987×10^{-3}, respectively. Because the crack trajectory predicted in this case coincides with the path traveled through by the moving heat source, the damage pattern of *shear slippage* due to large values of the compressive circumferential stress adjacent to the thermal shock waves becomes relatively important. Determination of the shear failure pattern includes the orientation of the shear plane in the material continuum as well as the maximum shear stress existing on it. A thorough study of this subject will appear soon in the open literature and also emphasizes the evolution of the shear failure pattern in transition of the thermal Mach number [72].

Because the crack trajectories emanating from the heat source are an irreversible physical process which depend strongly on the fracture path developed in the early stage, the damage pattern in the vicinity of a rapidly moving heat source seems to be a viable measure for confirming the existence of the thermal shock waves. Especially for a heat source moving at the transonic and in the supersonic ranges, the fracture/damage patterns predicted by the thermal wave model deviate significantly from those predicted by diffusion theory. In the presence of thermal shocks, the wave theory in general predicts a *wider* damage zone in the neighborhood of the trailing edge of the moving heat source.

In closing, it should also be noticed that the thermal stresses induced by the apparent heat source [refer to Eqs. (102), (109), and (111) for the potential function Φ_0] do not vanish in the thermally undisturbed zone although there is no temperature rise thereby. Mathematically, this can be seen clearly from Eq. (108), where a zero temperature level still renders a nonzero value of the Z function, and hence the resulting thermal stresses according to Eq. (107), due to the contributions from the potential function Φ_0. Physically, this reflects the fact that the stress wave in the solid propagates at a *faster* speed than the thermal wave such that the thermal stresses induced by the apparent heat source present in the thermally undisturbed zone although the thermal disturbance is absent. At the transonic and in the supersonic ranges with M equal to 1, 2, 3, and 5, Fig. 22 displays the circumferential stress contributed solely by the apparent heat source in the thermally undisturbed zone. The angle θ is measured from the leading edge of the moving heat source in this figure. Unlike the total stress level contributed by both the temperature and the apparent heat source, the magnitude of this portion of the stress decreases monotonically as the thermal Mach number increases. Besides, although the evolution of the near-field stresses and energy with respect to the thermal Mach number is not discussed in this section, the *swinging* phenomenon corresponding to that in the thermal field exists as well. In the subsonic range with $M < 1$, both the circumferential stress and the strain energy density around the moving heat source decrease as the thermal Mach number increases. Entering the

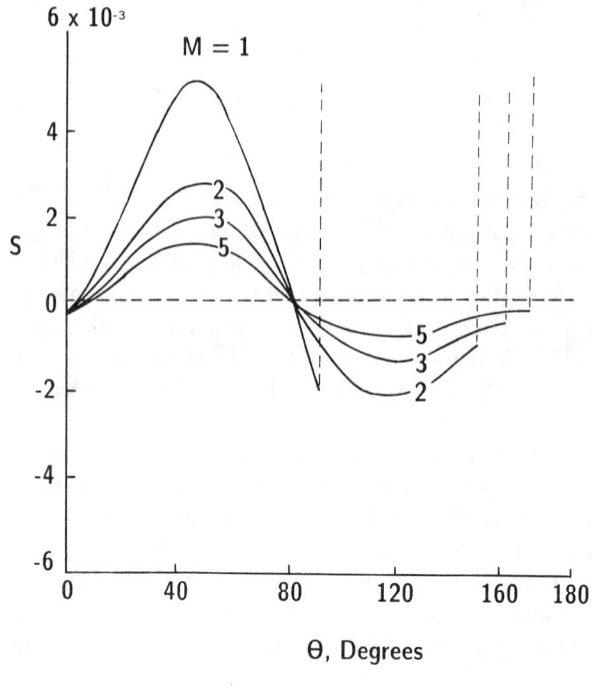

Figure 22 The circumferential stress solely contributed by the function Φ_0 in the thermally undisturbed zone. $S = \sigma_\theta/(2\mu\epsilon\beta Q/\rho\alpha C_p)$.

supersonic range with $M > 1$, on the other hand, both the circumferential stress and the strain energy density increase with the thermal Mach number. More detailed discussions of this phenomenon on a quantitative basis can be found in two papers by Tzou [68, 69].

6 THREE-DIMENSIONAL THEORY OF THERMAL SHOCK WAVES

Under high rate response, the description for the thermal shock formation in a two-dimensional solid has been made in Secs. 4 and 5. Through both temperature and flux formulations, the thermal shock waves induced by a rapidly moving heat source and a fast-propagating crack tip have been studied and compared in great detail. Due to the presence of several unique features in the thermal shock phenomenon, such as the thermal energy cumulated in a preferential direction around the rapidly moving origin, the discontinuous variations of the physical quantities across the thermal shock surfaces, and the swinging phenomenon, etc., the present section further develops the theory into a three-dimensional space and examines these phenomena in a more general sense. Similar to the approach adopted for the two-dimensional theory, a three-dimensional description of the shock formation in the thermal field will be made first. Based on the temperature distributions obtained in the various ranges of the thermal Mach number, then, a more general thermoelastic formulation involving the shear wave speed will follow, which will depict the interactions between the *thermal* and the *shear* waves interwoven in the heat-affected zone.

6.1 Thermal Shock Formation in a Three-Dimensional Solid

Let us consider the thermal field around a moving heat source in a three-dimensional space x_i as shown in Fig. 23. Without loss of generality, the point heat source is assumed to move along the x_1 axis so that, mathematically,

$$S(x_1, x_2, x_3, t) = Q\,\delta(x_1 - vt)\,\delta(x_2)\,\delta(x_3) \tag{115}$$

The physical meanings of the mathematical quantities used in this section, unless stated otherwise, are the same as in Sec. 4. In a three-dimensional solid, the energy equation with T representation, Eq. (11), can be written as

$$\alpha\nabla^2 T - \left(\frac{\alpha}{C^2}\right) T_{,tt} - T_{,t} = -\left(\frac{Q}{\rho C_p}\right)\left[\delta(x_1 - vt) + \left(\frac{\alpha}{C^2}\right)\delta_{,t}(x_1 - vt)\right]\delta(x_2)\,\delta(x_3) \tag{116}$$

with ∇^2 being the Laplacian operator in the three-dimensional space. The quasi-stationary form of the energy equation, similar to the two-dimensional analysis, can be obtained by applying the Galiléi transformation,

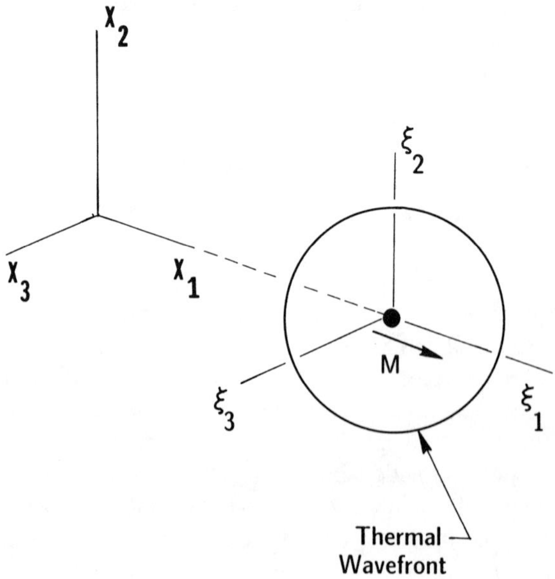

Figure 23 Spherical thermal wavefront emanating from a moving heat source and the coordinate system.

$$x_1 = \xi_1 + vt \qquad x_2 = \xi_2 \qquad x_3 = \xi_3 \tag{117}$$

and Eq. (116) becomes

$$\alpha[(1 - M^2)T_{,11} + T_{,22} + T_{,33}] + 2c\alpha T_{,1} = -\left(\frac{Q}{\rho C_p}\right)\left[\delta(\xi_1) - \left(\frac{M^2}{2c}\right)\delta_{,1}(\xi_1)\right]\delta(\xi_2)\,\delta(\xi_3) \tag{118}$$

This equation parallels that obtained in the two-dimensional formulation, Eq. (16). Again, the thermal Mach number M dominates the characteristics of the temperature field, since it appears in one of the *highest*-order differentials in the energy equation.

Due to the lack of a priori knowledge about the transitional behavior of the temperature as the thermal Mach number varies, the method of hypovariable transformations will be employed again to yield closed-form solutions. As already shown in Sec. 4, the method is capable of transforming Eq. (118) into a *one-dimensional* equation in the hypospace r, which significantly reduces the mathematical effort. Also, this is more convenient for the subsequent thermal stress analysis, because the temperature appears as the nonhomogeneous term in the field equation governing the displacement potential.

Because Eq. (118) varies intrinsically as a function of the thermal Mach number, the temperature field should be investigated individually for $M < 1$, $M = 1$, and $M > 1$. In all the cases under consideration, the physical quantities are assumed to vanish at infinity.

6.1.1 Subsonic Range with $M < 1$. For a heat source moving in the subsonic range with $M < 1$, we first consider a transformation on the dependent variable $T(\xi_i)$:

$$T(\xi_i) = \exp\left(-\frac{c\xi_1}{1 - M^2}\right) f(\xi_i) \qquad \text{for } M < 1 \text{ and } i = 1, 2, 3 \tag{119}$$

By substituting Eq. (119) into (118), the equation governing $f(\xi_i)$ can be obtained as

$$(1 - M^2)f,_{11} + f,_{22} + f,_{33} - \left(\frac{c^2}{1 - M^2}\right) f$$
$$= -\left(\frac{Q}{\rho C_p \alpha}\right) \exp\left(\frac{c\xi_1}{1 - M^2}\right) \left[\delta(\xi_1) - \left(\frac{M^2}{2c}\right) \delta,_1(\xi_1)\right] \delta(\xi_2)\, \delta(\xi_3) \tag{120}$$

By applying another transformation on the independent variables from ξ_i to r,

$$r = \sqrt{\left(\frac{\xi_1^2}{1 - M^2}\right) + \xi_2^2 + \xi_3^2} \qquad \text{for } M < 1 \tag{121}$$

similar to Eq. (18) for the two-dimensional analysis, Eq. (120) can be reduced to a one-dimensional equation in the hypospace r:

$$f,_{rr} + \frac{2}{r} f,_r - \left(\frac{c^2}{1 - M^2}\right) f = \exp\left(\frac{c\xi_1}{1 - M^2}\right) \left[-\left(\frac{Q}{\rho C_p \alpha}\right) \delta(\xi_1)\right.$$
$$\left. + \left(\frac{QM^2}{2c\rho C_p \alpha}\right) \delta,_1(\xi_1)\right] \delta(\xi_2)\, \delta(\xi_3) \qquad M < 1 \tag{122}$$

where the independent variables ξ_i in the nonhomogeneous terms of the equation are temporarily retained for later use. In comparing Eq. (122) with that obtained previously for the two-dimensional problems, Eq. (19), we notice that except for the presence of $\delta(\xi_3)$ due to the additional dimensionality, the only difference of the resulting equation in the r-space lies in the coefficient of $2/r$. In the two-dimensional case, the corresponding coefficient is $1/r$ and the resulting equation was governed by the modified Bessel operator. The solution of Eq. (19) consists of two parts. One is from the real heat source applied to the solid, while another is from the *apparent* heat source due to the effect of finite wave speed. The solutions for both parts can be found if the Green's function $f_g(r|r_\eta)$ satisfying the equation

$$f_{g,rr} + \frac{2}{r} f_{g,r} - \left(\frac{c^2}{1 - M^2}\right) f_g = \delta(r - r_\eta) \qquad \text{for } M < 1 \tag{123}$$

is obtained. Eq. (123) is governed by the generalized Bessel operator [27]. Its homogeneous solutions can be found as

$$f_H \sim r^{-1/2} I_{1/2}\left[\frac{cr}{(1 - M^2)^{1/2}}\right] \qquad \text{and} \qquad r^{-1/2} I_{-1/2}\left[\frac{cr}{(1 - M^2)^{1/2}}\right] \tag{124}$$

or, equivalently [92],

$$f_{H1} \sim \frac{\sinh[cr/(1 - M^2)^{1/2}]}{r} \quad \text{and} \quad f_{H2} \sim \frac{\cosh[cr/(1 - M^2)^{1/2}}{r} \tag{125}$$

By observing that the linear combination of the two solutions shown in Eq. (125) also provides a homogeneous solution for Eq. (123), we may construct our Green's function by the difference of the two homogeneous solutions:

$$f_g(r|r_\eta) = \frac{A\{\cosh[c|r - r_\eta|/(1 - M^2)^{1/2}] - \sinh[c|r - r_\eta|/(1 - M^2)^{1/2}]\}}{c|r - r_\eta|} \tag{126}$$

where $r_\eta = [\eta_1^2/(1 - M^2) + \eta_2^2 + \eta_3^2]^{1/2}$ denotes the position of the heat source in the three-dimensional solid. For the point heat source coincident with the origin of the moving coordinates ξ_i, r_η equal to zero and it is straightforward to show that

$$\lim_{r\to\infty} f_g(r|0) = 0 \quad \text{and} \quad \lim_{r\to 0} f_g(r|0) \to \infty \tag{127}$$

which is the general characteristic for the Green's function. The constant A in Eq. (126) is determined from Abel's relationship [74] for the Wronskian between the functions $f_g(r|0)$ and either f_{H1} or f_{H2}:

$$W[f_g(r|0), f_{H1}] = W[f_g(r|0), f_{H2}] = \frac{A}{r^2} \tag{128}$$

which gives $A = c/(1 - M^2)^{1/2}$.

By using Eq. (126) as the Green's function for Eq. (122) and the integral properties of the Dirac delta function and its derivative:

$$\int G[r(\xi_1, \xi_2, \xi_3)|r_\eta(\eta_1, \eta_2, \eta_3)]\, \delta(\eta_1)\, \delta(\eta_2)\, \delta(\eta_3)\, d\eta_1\, d\eta_2\, d\eta_3 = G[r(\xi_1, \xi_2, \xi_3)|r_\eta(0, 0, 0)]$$

$$\int G[r(\xi_1, \xi_2, \xi_3)|r_\eta(\eta_1, \eta_2, \eta_3)]\, \delta_{,1}(\eta_1)\, \delta(\eta_2)\, \delta(\eta_3)\, d\eta_1\, d\eta_2\, d\eta_3 = -\left\{\frac{\partial}{\partial\eta_1} G[r(\xi_1, \xi_2, \xi_3)|r_\eta(\eta_1, 0, 0)]\right\}_{\eta_1\to 0} \tag{129}$$

the solution of $f(r)$ is then obtained:

$$f[r(\xi_i)] = \left(\frac{Q}{\rho\alpha C_p}\right)\{\cosh[cr/(1 - M^2)^{1/2}] - \sinh[cr/(1 - M^2)^{1/2}]\}\frac{[2cr(1 - M^2)^{1/2} + M^2]}{2c^2r^2} \tag{130}$$

and the temperature $T(\xi_i)$ in the physical space ξ_i is obtained from Eq. (119):

$$\frac{T(\xi_i)}{(Q/\rho\alpha C_p)} = \exp[-c\xi_1/(1-M^2)]\,\{\cosh[cr/(1-M^2)^{1/2}]$$

$$-\frac{\sinh[cr/(1-M^2)^{1/2}]\}[2cr(1-M^2)^{1/2}+M^2]}{2c^2r^2} \quad \text{for } M<1 \qquad (131)$$

In the case that M approaches 0, i.e., the thermal wave speed C approaches infinity, Eq. (131) reduces to the solution employing the diffusion theory.

In order to compare the two-dimensional temperature field obtained previously, the distributions on the ξ_1–ξ_2 plane with $\xi_3 = 0$ will be displayed. Figure 24 shows the near-field temperature for the subsonic range with $M < 1$ represented by Eq. (131). The curve with $M = 0.1$ is very close to that obtained by classical diffusion theory. First, we observe that the temperature level around the moving heat source decreases as the value of M increases. In all the cases with different values of M, we notice that the temperature level increases with a higher rate as the trailing edge of the moving heat source at $\theta = 180°$ is approached. The thermally undisturbed zone is in fact

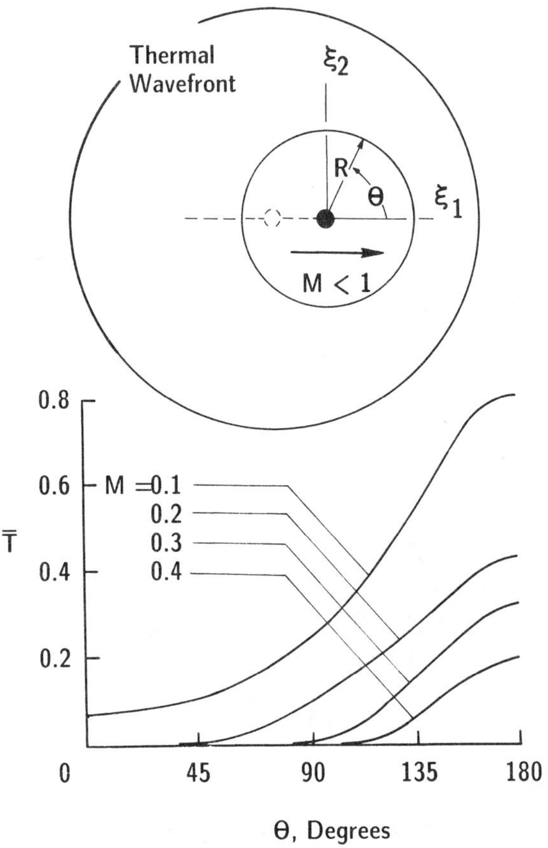

Figure 24 Temperature distribution along the circumference of the circle centered at the heat source on the (ξ_1, ξ_2) plane with $\xi_3 = 0$. The subsonic range with $M < 1$ and $\bar{T} = T/(Q/\rho\alpha C_p)$.

gradually formed as the thermal Mach number of the heat source approaches the transonic stage. This is reflected by the presence of a domain of θ (from $M = 0.2$ and thereafter) within which only a minor temperature rise is observed. This domain increases approximately from [0°, 45°] to [0°, 90°] to [0°, 115°] as the value of M increases from 0.2 to 0.3 to 0.4. Physically, these trends are identical to those in the two-dimensional responses. In comparison with the distributions shown in Fig. 3, however, the two-dimensional analysis seems to *overestimate* the temperature levels. The effect of the ξ_3 coordinate, the distance away from the heat source in the ξ_3 direction, is shown in Fig. 25 for the thermal Mach number $M = 0.4$. As expected, the temperature level decreases as the value of ξ_3 increases. Also, due to the finite wave speed of the thermal disturbance propagating in the solid, the temperature level decreases dramatically as the distance from the heat source (ξ_3) increases from 0 to 0.1 mm. Under the present conditions, the presence of the heat source cannot be detected at a distance of 0.16 mm away from the heat source because the temperature level there is almost zero.

6.1.2 Supersonic Range with $M > 1$. As $M > 1$, Eq. (118) can be rearranged to the following form:

$$\alpha[(M^2 - 1)T_{,11} - T_{,22} - T_{,33}] - 2c\alpha T_{,1} = \left(\frac{Q}{\rho C_p}\right)\{\delta(\xi_1) - \left(\frac{M^2}{2c}\right)\delta_{,1}(\xi_1)\}\,\delta(\xi_2)\,\delta(\xi_3) \quad (132)$$

The transformations on the dependent (from T to f) and independent (from ξ_i to r) variables in this case are

$$T(\xi_i) = \exp\left(\frac{c\xi_1}{M^2 - 1}\right) f(\xi_i) \quad \text{and}$$

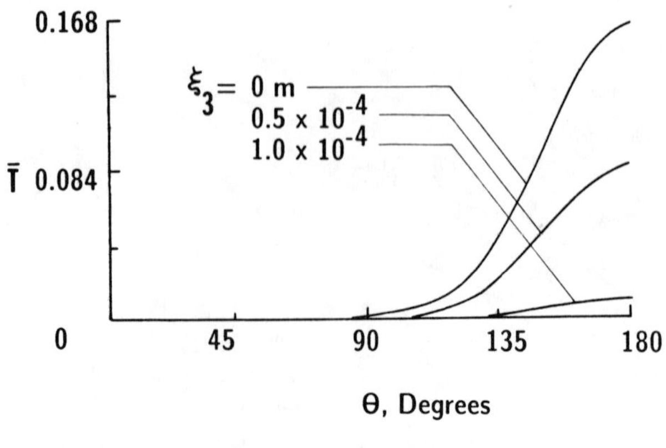

Figure 25 The effect of ξ_3 values on the temperature distribution along the circumference of the circle centered at the heat source. The subsonic range with $M = 0.4$ and $\bar{T} = T/(Q/\rho\alpha C_p)$.

$$r = \left[\frac{\xi_1^2}{(M^2 - 1)} - \xi_2^2 - \xi_3^2 \right]^{1/2} \qquad \text{for } M > 1 \tag{133}$$

The corresponding equation governing the function of $f(\xi_i)$ becomes

$$f_{,rr} + \frac{2}{r} f_{,r} - \left(\frac{c^2}{M^2 - 1} \right) f$$
$$= \left(\frac{Q}{\rho C_p \alpha} \right) \exp\left(- \frac{c\xi_1}{M^2 - 1} \right) \left[\delta(\xi_1) - \left(\frac{M^2}{2c} \right) \delta_{,1}(\xi_1) \right] \delta(\xi_2)\,\delta(\xi_3) \tag{134}$$

and the Green's function is

$$f_g(r|r_\eta) = \frac{A\{\cosh[c|r - r_\eta|/(M^2 - 1)^{1/2}] - \sinh[c|r - r_\eta|/(M^2 - 1)^{1/2}]\}}{c|r - r_\eta|} \tag{135}$$

By employing the same procedure as that used in the previous case, the temperature distribution can be obtained as

$$\frac{T(\xi_i)}{Q/\rho\alpha C_p} = \exp\left(\frac{c\xi_1}{M^2 - 1} \right)$$
$$\times \frac{\{\cosh[cr/(M^2 - 1)^{1/2}] - \sinh[cr/(M^2 - 1)^{1/2}]\}[2cr(2M^2 - 1) + M^2(M^2 - 1)^{1/2}]}{[2c^2r^2(M^2 - 1)^{1/2}]}$$
$$\text{for } M > 1 \tag{136}$$

The applicable range of Eq. (136), as expected, is restricted under the constraint of $r > 0$. From the second of Eq. (133), this implies that

$$\frac{\xi_1^2}{(M^2 - 1)} \geq \xi_2^2 + \xi_3^2 \tag{137}$$

In a three-dimensional solid, the equality represents a *quasi-hyperboloidal* surface with its vertex coincident with the moving heat source. On the ξ_1–ξ_2 plane with $\xi_3 = 0$, in making contact with the two-dimensional result, the representation of Eq. (137) in terms of the polar coordinates (R, θ) centered at the origin with the angle θ measured from the trailing edge of the heat source is

$$\pi - \sin^{-1}\left(\frac{1}{M} \right) < \theta < \pi \qquad \text{for a heat source moving to the right} \tag{138}$$

which has the same form as Eq. (32). According to the same argument made in the two-dimensional problem, therefore, Eq. (136) is the temperature distribution in the *heat-affected zone* defined by Eq. (137). The temperature stays at the reference value in the rest of the physical domain, which is defined as the *thermally undisturbed zone*. At any other plane with ξ_3 being nonzero, Eq. (137) represents a hyperbola with the lengths of the semimajor and semiminor axes being $\xi_3(M^2 - 1)^{1/2}$ and ξ_3, respectively. The asymptotes are $\xi_1/(M^2 - 1)^{1/2} = \pm\xi_2$, which coincide with the locations of the thermal shock waves on the ξ_1–ξ_2 plane with $\xi_3 = 0$, as shown by Eq. (138).

The geometric configuration of the quasi-hyperboloidal surface is further illustrated in Fig. 26. Another possible solution of Eq. (137) is simply the image of the surface with respect to the ξ_2–ξ_3 plane which corresponds to the conjugate situation for a heat source moving to the left. The major axis of the shock surface coincides with the direction of the moving heat source and its length increases with the thermal Mach number. The projection of the quasi-hyperboloidal surface onto the ξ_2–ξ_3 plane is a circle, while it presents a wedge with the vertex moving with the heat source on the ξ_1–ξ_2 or ξ_1–ξ_3 plane. Apparently, the projections of the thermal shock surface possess the same characteristics as the *conical* shock waves induced by the high-speed aerodynamic flow field, while the quasi-hyperboloidal surface is a unique feature pertinent to the thermal shock waves.

On the ξ_1–ξ_2 plane with $\xi_3 = 0$, the temperature distribution represented by Eq. (136) in the supersonic range is displayed in Fig. 27. Distinct shock waves are formed at $\theta = 150°$, 161°, 166°, and 169° as the thermal Mach number M increases from 2 to 5. The physical domain of the heat-affected zone consequently decreases from 60° to 22°. By comparing the temperature distributions at various values of M, we also notice that the temperature level in the heat-affected zone *increases* with the thermal Mach number. The swinging phenomenon for the temperature variation in transition of the thermal Mach number from the subsonic to the supersonic ranges thus exits in both two- and three-dimensional formulations. For $M = 4$, the effect of ξ_3 in the supersonic range is shown in Fig. 28. Surprisingly, the temperature level in the heat-affected zone *increases* with the out-of-plane distance (ξ_3) away from the heat source. The physical interpretation for this unusual response again lies in the influence of the thermal shock waves, as illustrated in Fig. 29. When the value of ξ_3 increases, the lengths of both the major and minor axes of the hyperbola increase. By noticing that the hyperbolic surface is indeed the thermal shock surface where the temperature approaches infinity, this observation reveals that the material continuum at a fixed location on the circle is *closer* to the thermal shock wave with a larger value of ξ_3.

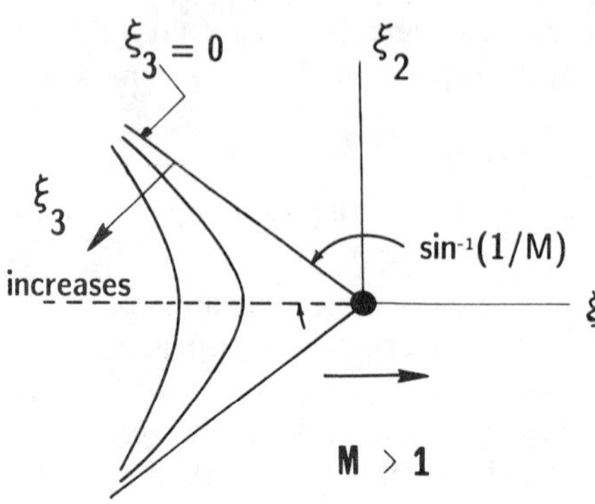

Figure 26 Quasi-hyperboloidal shock surfaces in a three-dimensional solid with $M \geq 1$.

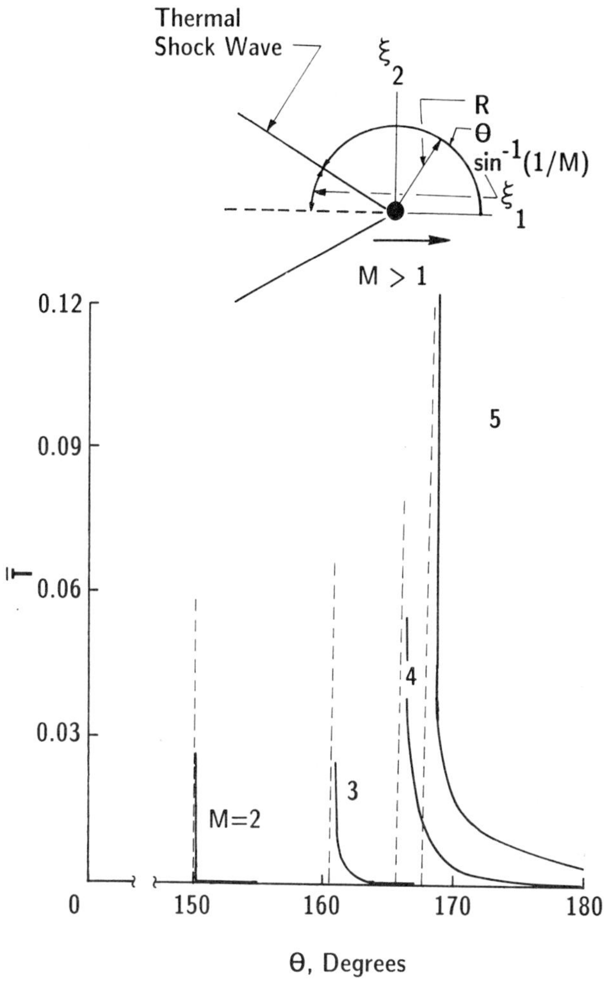

Figure 27 Formation of the oblique thermal shock waves on the (ξ_1, ξ_2) plane with $\xi_3 = 0$ and the near-field temperature distributions. The supersonic range with $M > 1$.

As a consequence, the local temperature level in the heat-affected zone is higher and the physical domain of the heat-affected zone in terms of the angle θ is narrower. For $M = 4$, for example, Fig. 28 shows that the physical domain of the heat-affected zone reduces from [166°, 180°] to [167°, 180°] to [171°, 180°] as the value of ξ_3 increases from 0 to 0.01 to 0.02 mm. To be noticed is that such a phenomenon exists only when the circle for observation with a radius R (refer to Fig. 29) intersects with the hyperbolic shock surface with a certain value of ξ_3. In the case that ξ_3 increases to a *critical* value of $R/(M^2 - 1)^{1/2}$, which is approximately 0.26×10^{-4} m for $M = 4$ and $R = 10^{-4}$ m, the heat-affected zone shrinks to a point at $\theta = 180°$ where the temperature approaches infinity. For a value of ξ_3 larger than the critical value,

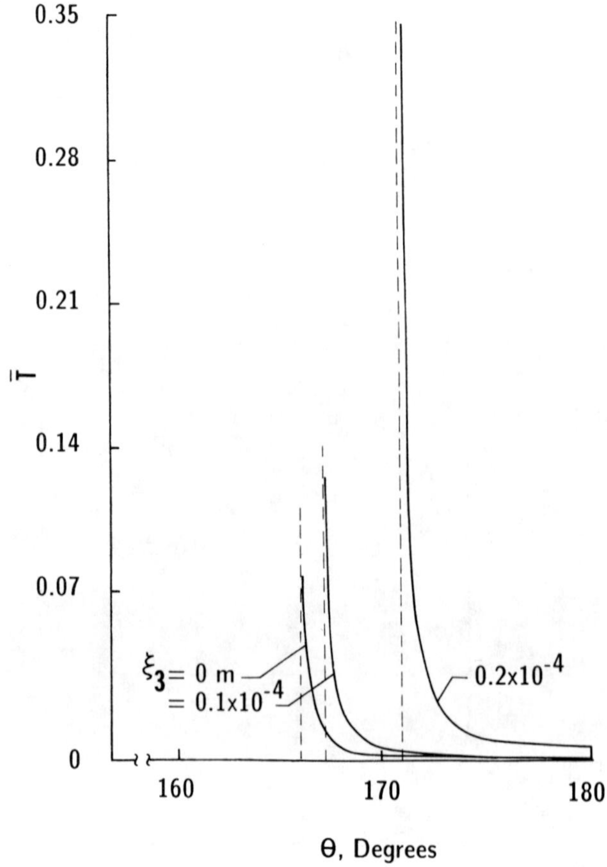

Figure 28 Increase of the near-field temperature with the distance away from the heat source (the ξ_3 values) in the heat-affected zone. The supersonic range with $M = 4$.

under the same radius R, the thermal shock wave departs from the observation circle and the entire circle is located in the thermally undisturbed zone where the temperature stays at the reference value. This is a unique phenomenon pertinent to the three-dimensional theory of thermal shock waves.

6.1.3 Transonic Stage with $M = 1$. As the heat source moves at the same speed as that of the heat propagation in the solid, a normal shock wave exists at $\xi_1 = 0$ according to Eq. (137) or $\theta = \pi/2$ according to Eq. (138) on the ξ_1–ξ_2 plane. For a heat source moving to the right, the heat-affected zone is confined to the domain of $\xi_1 < 0$. Equation (118) in this case degenerates into an elliptic type:

$$T,_{22} + T,_{33} + 2cT,_1 = -\left(\frac{Q}{\rho\alpha C_p}\right)\left[\delta(\xi_1) - \left(\frac{1}{2c}\right)\delta,_1(\xi_1)\right]\delta(\xi_2)\,\delta(\xi_3) \quad (139)$$

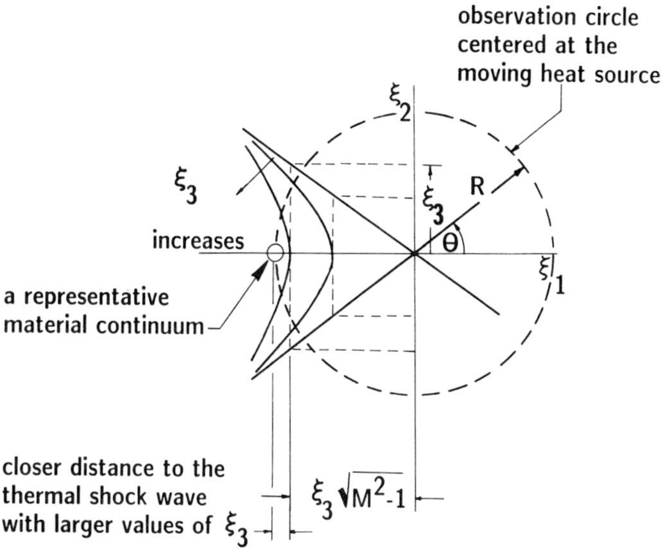

Figure 29 The stronger influence of the temperature singularity existing at the thermal shock wave with a larger value of ξ_3 on the temperature level in a material continuum on the circle.

The Green's function for Eq. (139) is simply that for the *two-dimensional* diffusion equation with ξ_1 replacing the position of the time variable. The result is [75]

$$T_g(\xi_i|\eta_i) = \left(\frac{1}{4\pi c}\right) \exp\left\{\frac{c[(\xi_2 - \eta_2)^2 + (\xi_3 - \eta_3)^2]/2|\xi_1 - \eta_1|}{|\xi_1 - \eta_1|}\right\} \tag{140}$$

The temperature distribution in the heat-affected zone with $\xi_1 < 0$ can then be integrated in the same fashion as that in the previous cases, which yields

$$\frac{T}{Q/\rho\alpha C_p} = -\left(\frac{1}{4\pi c}\right) \exp\left[\frac{c(\xi_2^2 + \xi_3^2)}{2\xi_1}\right] \left\{1/\xi_1 + \frac{[c(\xi_2^2 + \xi_3^2) + 2\xi_1]}{4c\xi_1^3}\right\} \quad \text{for } M = 1 \tag{141}$$

On the ξ_1–ξ_2 plane with $\xi_3 = 0$, the temperature distribution represented by Eq. (141) is shown in Fig. 30. A normal shock is formed at $\theta = 90°$ and moving with the heat source. The thermal field in the domain of $0° < \theta < 90°$ stays undisturbed as the heat source moves ahead. We observe that except for the singularity existing at the thermal shock wave located at $\theta = 90°$, the temperature level in the heat-affected zone from 90° to 180° decreases continuously from those in the subsonic range. Again, the effect of ξ_3 coordinate is shown in Fig. 31. As the value of ξ_3 increases from 0 to 0.025 to 0.05 mm, the temperature level decreases with the distance away from the heat source—a continuation of the response from the subsonic range.

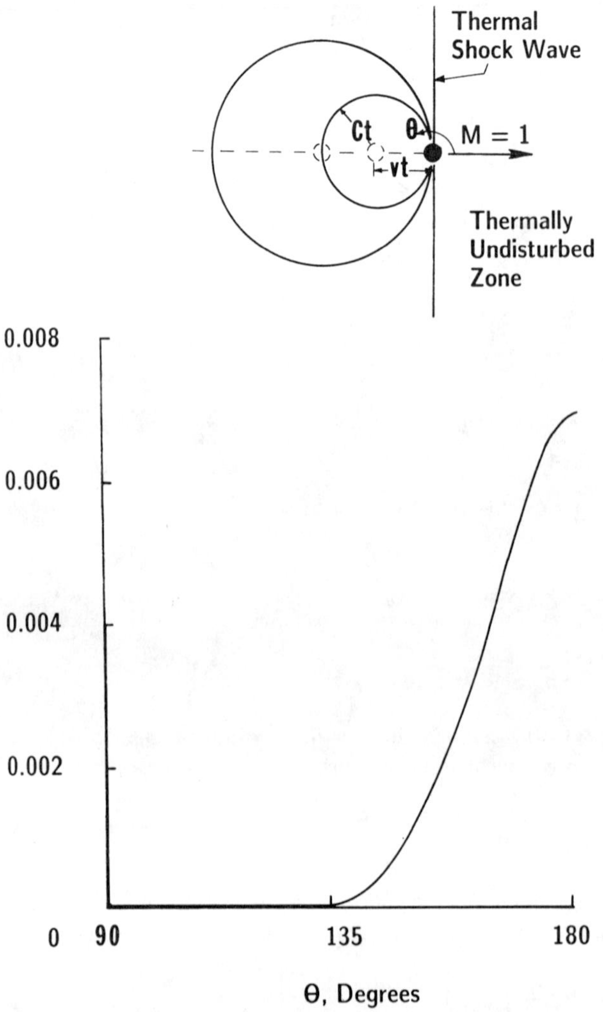

Figure 30 The normal shock formation on the (ξ_1, ξ_2) plane with $\xi_3 = 0$ and the near-field temperature distribution. The transonic stage with $M = 1$.

At this point, we may conclude that the thermal shock formation around a rapidly moving heat source in a solid is a direct consequence for material continua under high rate response. The temperature rate equation reflecting the importance of considering the thermal wave nature under this condition has been shown in Eq. (12). Under a quasi-stationary description, Eq. (12) can be expressed in terms of the temperature gradient in the ξ_1 direction:

$$T_{,1} >> \exp\left(\frac{2c\xi_1}{M^2}\right) \tag{142}$$

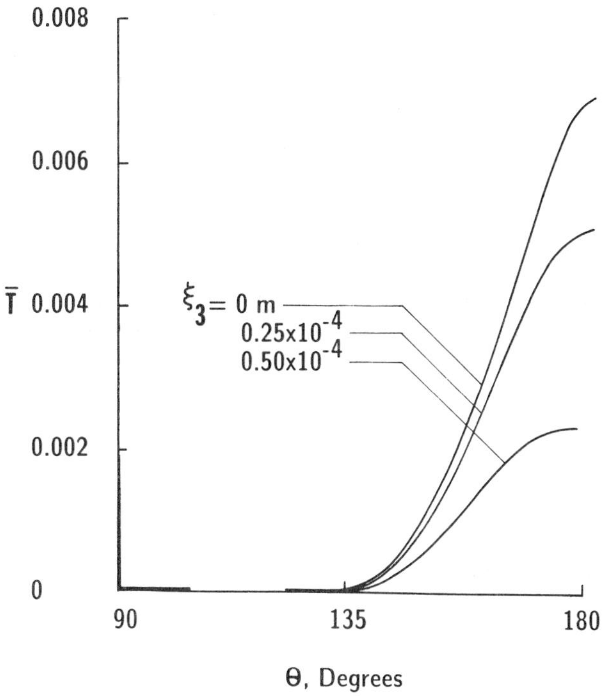

Figure 31 The effect of ξ_3 values on the temperature distribution along the circumference of the circle centered at the heat source. The transonic stage with $M = 1$.

In the neighborhood of the moving heat source, we have shown that the temperature gradient $T_{,1}$ approaches infinity and the condition stated in Eq. (142) is satisfied. For a heat source moving with a high speed at the transonic and in the supersonic range, the wave behavior in the process of heat transport becomes even more pronounced due to the large temperature gradient existing in the neighborhood of the thermal shock waves.

6.2 Effect of Shear Wave Speed on Thermoelastic Displacement Potential

The quasi-stationary formulation for the thermoelastic stress analysis made in Sec. 5.1 is actually a first-order description neglecting the thermal stress waves propagating in the solid. In developing the three-dimensional theory for the thermal stresses induced by the thermal shock waves, we will provide a more general formulation which accounts for the interactions between the thermal stress and shock waves.

For a three-dimensional solid under transient response, the Neuber-Papkovich thermoelastic displacement potential is governed by a wave equation [93]:

$$\nabla^2\Phi - \left(\frac{\kappa}{C_s^2}\right)\Phi_{,tt} = \epsilon\beta T \qquad (143)$$

where κ is a function of Poisson's ratio defined as $(1 - 2\nu)/[2(1 - \nu)]$, C_s is the shear wave speed defined as $\sqrt{\mu/\rho}$ with μ and ρ being respectively the shear modulus and mass density of the solid, and β relates to Young's modulus by $E/(1 - \nu)$. To an observer moving with the heat source, the quasi-stationary form of Eq. (143) can be obtained by applying the Galiléi transformation, Eq. (117),

$$\nabla^2\Phi - \kappa M_s^2 \Phi_{,11} = \epsilon\beta T \tag{144}$$

with $\nabla^2 \equiv \partial^2/\partial\xi_1^2 + \partial^2/\partial\xi_2^2 + \partial^2/\partial\xi_3^2$ and M_s denoting the Mach number of the moving heat source relative to the *shear wave speed*. Mathematically, $M_s = v/C_s$. It also relates to the thermal Mach number M by $M_s = M(C/C_s)$. For the present formulation, a similar form equivalent to Eq. (100) can be achieved by coordinate stretching. Introducing ζ coordinates with

$$\zeta_1 = \frac{\xi_1}{(1 - \kappa M_s^2)^{1/2}} \qquad \zeta_2 = \xi_2 \qquad \zeta_3 = \xi_3 \tag{145}$$

Eq. (144) becomes

$$\nabla_\zeta^2\Phi = \epsilon\beta T \qquad \text{with } \nabla_\zeta^2 \equiv \frac{\partial^2}{\partial\zeta_1^2} + \frac{\partial^2}{\partial\zeta_2^2} + \frac{\partial^2}{\partial\zeta_3^2} \tag{146}$$

Before combining Eq. (146) with the energy equation to obtain an ordinary differential equation governing the Φ function, the energy equation (118) has to be transformed into the same ζ-space. By applying the chain rule for the differentiations with respect to ξ_i involved in Eq. (118) according to Eq. (145), the representation of the energy equation in the ζ-space can be obtained as

$$\nabla_\zeta^2 T + [h^2(1 - M^2) - 1]T_{,11} + 2chT_{,1}$$
$$= -\left(\frac{Q}{\rho\alpha C_p}\right)\left[\delta\left(\frac{\zeta_1}{h}\right) - \left(\frac{hM^2}{2c}\right)\delta_{,1}\left(\frac{\zeta_1}{h}\right)\right]\delta(\zeta_2)\,\delta(\zeta_3) \tag{147}$$

with h being a scaling factor defined as $1/(1 - \kappa M_s^2)^{1/2}$ and all the differentiations taken with respect to the ζ_i coordinates. Mathematically, for example, $T_{,1} \equiv \partial T/\partial\zeta_1$. Note also that the condition that κM_s^2 being smaller than 1 is temporarily imposed in the following discussions. Relating the terms involving $T_{,1}$ and $T_{,11}$ in Eq. (147) to the Φ function by taking the appropriate differentiations with respect to ζ_1 on Eq. (146), then, the differential equation corresponding to Eq. (101) is obtained:

$$\left[\frac{1 - h^2(1 - M^2)}{h}\right]\Phi_{,11} - 2c\Phi_{,1} = \frac{\epsilon\beta}{h}\left[T(\zeta_i) + \frac{1}{\alpha}\Phi_0(\zeta_i)\right] \tag{148}$$

where the Φ_0 function satisfying

$$\nabla_\zeta^2\Phi_0 = \left(\frac{Q}{\rho C_p}\right)\left[\delta\left(\frac{\zeta_1}{h}\right) - \left(\frac{hM^2}{2c}\right)\delta_{,1}\left(\frac{\zeta_1}{h}\right)\delta(\zeta_2)\,\delta(\zeta_3)\right. \tag{149}$$

is the elastic potential solely contributed by the apparent heat source. Due to the similar forms of Eqs. (148) and (149) to Eqs. (101) and (102) in the two-dimensional formulation, the rest of the mathematical procedure used in Sec. 5.1 can be extended from here on. Based on the Φ function thus obtained in the ζ-space,

$$\Phi_{,1}(\xi_i) = \left(\frac{\epsilon\beta}{hF}\right) \int_{\infty}^{\xi_1} \exp\left[\frac{2c(\zeta_1 - \eta)}{F}\right] \left[T(\eta, \zeta_2) + \frac{1}{\alpha} \Phi_0(\eta, \zeta_2)\right] d\eta \quad (150)$$

with $F = [1 - h^2(1 - M^2)]/h$, the corresponding expression in the moving coordinates ξ_i can be obtained according to Eq. (145). The thermoelastic displacement field can then be determined according to

$$2\mu u_i = \Phi_{,i} \quad (151)$$

and from the general Hooke's law, the thermal stress components are

$$\sigma_{ij} = \Phi_{,ij} + \delta_{ij}\left\{\left(\frac{\nu}{1 - 2\nu}\right) \nabla^2\Phi - \left[\frac{2(1 + \nu)}{1 - 2\nu}\right] \mu\epsilon T\right\} \quad \text{for } i, j = 1, 2, 3 \quad (152)$$

In Eqs. (151) and (152), to reiterate, the differentiations are taken with respect to the ξ_i coordinates. The expressions for temperature obtained in the ξ_i coordinates, namely, Eq. (131) for the subsonic range with $M < 1$, Eq. (136) for the supersonic range with $M > 1$, and Eq. (141) for the transonic stage with $M = 1$, can thus be used directly.

The thermoelastic stress analysis depends on the solution of Eq. (148) for the Φ function. For the case with $\eta M_s^2 < 1$, the scaling factor h is well defined and the solution for $\Phi_{,1}$ is shown by Eq. (150). In addition to the thermal Mach number M, the Mach number of the moving heat source relative to the shear wave speed M_s also appears in the coefficient leading the *highest*-order differential in the field equation (148). Intrinsic variation in the solution for Φ is thus expected. In the case that M_s approaches zero, i.e., the speed of the thermal wave (C) is much smaller than that of the shear wave (C_s) propagation in the solid, the scaling factor h approaches 1. The ζ_i coordinates in this case coincide with the ξ_i coordinates [refer to Eq. (145)] and the first coefficient in brackets in Eq. (148) reduces to M^2. Equation (148) in this case reduces to the form of Eq. (101), which was derived previously based on a stationary formulation. Physically, this situation is especially true in a low-temperature environment where the thermal wave speed C is approximately two orders of magnitude smaller than the shear wave speed C_s.

The present formulation is limited to the case with κM_s^2 being smaller than 1 such that the scaling factor h is well defined. For aluminum with $\nu = \frac{1}{3}$ and $C_s = 3200$ m/s, this condition implies a speed of the moving heat source slower than 6400 m/s, which fulfills most conditions in current industrial applications. For the rigorosity of the mathematical formulation, however, the field equation governing the thermoelastic displacement potential in the case of $\kappa M_s^2 > 1$ can be obtained by slightly modifying the formulation from the stage of stretching the coordinates from ξ_i to ζ_i.

For $\kappa M_s^2 > 1$, Eq. (144) in the ξ_i coordinates can first be rearranged into the following form:

$$\kappa M_s^2 \Phi_{,11} - \nabla^2 \Phi = -\epsilon \beta T \tag{153}$$

By defining alternatively the stretching coordinates ζ_i as

$$\zeta_1 = \frac{\xi_1}{(\kappa M_s^2 - 1)^{1/2}} \qquad \zeta_2 = \xi_2 \qquad \zeta_3 = \xi_3 \tag{154}$$

and applying a transformation for the independent variables ζ_i to the hypospace r_ζ,

$$r_\zeta = \sqrt{\zeta_1^2 - \zeta_2^2 - \zeta_3^2} \tag{155}$$

the potential equation (153) is reduced to an ordinary differential equation equivalent to Eq. (148)

$$\frac{d^2\Phi}{dr_\zeta^2} + \left(\frac{1}{2}\right)\frac{d\Phi}{dr_\zeta} = -\epsilon\beta T(\zeta_i) \tag{156}$$

which can be solved by the same technique. To be noticed is that the solution of Eq. (156) for Φ, and hence the thermoelastic stress and displacement fields determined from Eqs. (151) and (152), is limited to the physical domain of

$$\zeta_1^2 > \zeta_2^2 + \zeta_3^2 \tag{157}$$

or, in terms of the ξ_i coordinates, by Eq. (154),

$$\frac{\xi_1^2}{(\kappa M_s^2 - 1)} > \xi_2^2 + \xi_3^2 \tag{158}$$

For a heat source moving with an extremely high speed, i.e., $v > C_s/\sqrt{\kappa}$ obtained from the condition of $\kappa M_s^2 > 1$, Eq. (158) depicts the physical domain penetrated by the thermal stress wave and it is the *mechanically affected zone* corresponding to Eq. (137) for the heat-affected zone penetrated by the thermal wave. The thermoelastic potential Φ in this zone is governed by Eq. (156). The physical domain outside that specified by Eq. (158) is the *mechanically undisturbed zone*, where the thermoelastic displacements and stresses stay at the reference values. For metals with Poisson's ratio ν smaller than 0.5, we can also show in general that the physical domain represented by Eq. (158) includes that represented by Eq. (137) and that the mechanically affected zone is usually larger than the heat-affected zone. It implies that the shear stress wave propagates in the solid with a speed faster than that of the thermal wave. In the common region belonging to both the mechanically and the heat-affected zones, the thermal stress field will depend on both Mach numbers M and M_s, because the temperature distribution appears in the nonhomogeneous terms in the potential equation (156). The capricious interaction between the thermal and the shear stress waves in this case is a complicated but interesting phenomenon to be studied in the near future. In reality, however, the speed of the moving heat source seldom falls in this range, so the research value of this problem remains only on an academic level.

7 CONCLUSIONS

From a theoretical point of view, the thermal shock phenomena resulting from the accumulation of thermal energy around a rapidly moving origin in a solid medium have been investigated rigorously. The moving origins under consideration include a moving heat source and a propagating crack tip. In each case, the shock phenomena are examined by both temperature and heat flux formulations. In summary, the special features in the thermal shock theory developed so far include the following.

1. *The thermal Mach number*. For both a moving heat source and a propagating crack tip, the thermal Mach number (M) is the most important physical parameter characterizing the thermal field around the moving origin. Mathematically, the thermal Mach number appears in the coefficient of the *highest-order differential* in the energy equation. In transition from the subsonic ($M < 1$) to the transonic ($M = 1$) to the supersonic ($M > 1$) ranges, therefore, intrinsic variations of the temperature field result.

2. *Thermal shock waves*. For a heat source moving at the transonic and in the supersonic range with $M \geq 1$, large amounts of thermal energy tend to accumulate on a *quasi-hyperboloidal* surface represented by $\xi_1^2/(M^2 - 1) = \xi_2^2 + \xi_3^2$, with ξ being the material coordinates convecting with the heat source, and *thermal shock waves* are formed thereby. On the ξ_1–ξ_2 plane with ξ_3 being zero, the thermal shock surface degenerates into a wedge shape and the thermal shock angle can be expressed in terms of the polar coordinates centered at the heat source. Mathematically, this is $\sin^{-1}(1/M)$ measured from the *trailing edge* of the moving heat source. This is also the result for a rapidly propagating crack tip in a two-dimensional medium. The thermal shock waves separate the heat-affected zone from the thermally undisturbed zone in the physical domain. The temperature level in the thermally undisturbed zone stays at the reference value, and the material continua therefore do not sense the existence of the moving origin.

3. *Discontinuities across the shock surfaces*. The thermodynamic quantities possess discontinuous variations across the thermal shock surfaces. For the thermal field induced by a rapidly moving heat source, both the temperature and the heat flux vector approach infinity as the thermal shock wave is approached from a site in the heat-affected zone. They remain at the reference values, on the other hand, as the thermal shock wave is approached from a site in the thermally undisturbed zone. For the problem involving a rapidly propagating crack tip, the thermodynamic jump across the shock surface is *finite*. The near-tip temperature and the heat flux component perpendicular to the crack surface, for example, jump from a value of $4(M^2 - 1)/M^2$ to the reference value across the thermal shock waves. There is a limiting value of 4 as the thermal Mach number approaches infinity.

4. *The swinging phenomenon*. In transition of the thermal Mach number from the subsonic to the supersonic ranges, a *swinging* phenomenon exists for the variations of the physical quantities in the thermomechanical field. In the subsonic range with $M < 1$, the near-field temperature around the moving origin, for example, *decreases* as the thermal Mach number increases. But in the supersonic range with $M > 1$, the

temperature level in the heat-affected zone *increases* with the thermal Mach number. The physical quantities possessing such a special phenomenon include the heat flux vector, the circumferential stress, and the strain energy density function. Physically, the swinging phenomenon results from the thermal shock formation in the solid when the speed of the moving origin exceeds that of the heat propagation.

5. *Transition of the r dependency in the near-tip region.* For a crack tip propagating in a two-dimensional solid, the *r dependency* of the near-tip temperature changes from $r^{1/2}$ to r to r^2 in transition of the thermal Mach number from the subsonic to the transonic to the supersonic ranges. The well-known $1/\sqrt{r}$ singularity of the temperature gradient at the crack tip with $r = 0$ exists *only* in the subsonic range with $M < 1$. In the supersonic range with $M > 1$, moreover, because the temperature gradient $T_{,r}$ *increases with r* in the near-tip region and must decay to zero as r approaches infinity due to the physical boundary conditions, the temperature gradient must reach a *maximum* somewhere ahead of the crack tip. This distinct phenomenon should be investigated by employing the full-field formulation rather than the asymptotic formulations given by Eqs. (53*a*) and (90). Also, the r dependency of the resulting thermal stresses in the near-tip region is an immediate research topic worthy of being pursued further. A detailed understanding of this behavior contributes directly to arresting a crack propagating in a solid in a nonisothermal environment.

6. *Interactions between thermal and stress waves.* In incorporating the thermal wave speed into the consideration of thermoelastic deformation around a moving heat source, the modified term accounting for such an effect appears as the *highest*-order differential in the resulting field equation governing the thermoelastic displacement potential. The coefficient leading the modified term involves both Mach numbers M (relative to the thermal wave speed) and M_s (relative to the shear wave speed). In the case that M_s approaches zero, the coefficient reduces to M^2 and the thermal stress and the fracture patterns thus developed under high rate response vary intrinsically from those predicted by classical diffusion theory. Qualitatively, the thermal shock theory predicts a *wider* damage zone spreading in the neighborhood of the trailing edge of the heat source.

Basically, the thermal shock phenomenon is pertinent to the high rate response where the wave nature in the process of heat transport dominates over the diffusion behavior. When the speed of the moving heat source, for instance, exceeds that of the heat propagation in the solid, the material continua in the neighborhood do not have sufficient time to respond to the appropriate temperature level before moving away, and the thermal energy is thus cumulated around the heat source. The quasi-stationary formulation provided in this analysis, as shown by Eq. (142), further relates the high rate change of temperature to the large temperature gradient. For the problem involving a moving heat source, the large temperature gradient results from the intensified energy deposited onto the localized area of thermal irradiation. For the problem involving a propagating crack tip, on the other hand, the large temperature gradient is induced by the abrupt change of geometric curvature at the crack tip. Both problems are thus illustrative in emphasizing the importance of considering the thermal wave characteristics since the criterion of Eq. (142) is satisfied.

Although the experimental evidence for the thermal shock formation in solids is still unavailable at this point, the present work has studied such a possible phenomenon from various points of view on an analytical basis. The purpose is to identify the potential physical measures, such as the temperature swing in the supersonic range or the permanent damage pattern emanating from a rapidly moving heat source, for the experimental confirmation of the thermal shock waves. Obviously, direct experimental evidence is very much desired at this stage of development.

For a solid under high rate response, there are still many fundamental aspects to be further understood in the thermal shock theory. The following topics are some typical examples.

1. *The nonlinear theory of thermal shock formation.* To be noticed is that the constitutive equation, the second of Eq. (9), is a linear approximation for the solid with fading memory. The leading coefficient α/C^2 is indeed the relaxation time τ defined as an intrinsic thermal property in the hyperbolic theory of heat conduction. For a more general form of the heat flux vector depending on the time lag between the observation and the relaxation time, i.e.,

$$\mathbf{q}(t - \tau) = -k\,\nabla T(t) \tag{159}$$

the Taylor's series expansion of the heat flux vector with respect to $\tau = 0$ yields

$$\mathbf{q}(t - \tau) = \mathbf{q}(t) + \mathbf{q}_{,t}(t)\tau + \mathbf{q}_{,tt}(t)\frac{\tau^2}{2} + \cdots = -k\,\nabla T(t) \tag{160}$$

By truncating the higher-order terms of τ in Eq. (160), then, the resulting equation has the same form as that in Eq. (9) and the heat flux vector and the temperature gradient are related to each other at *the same* instant of time. The thermal shock phenomenon discussed in this chapter, therefore, presents a first-order description neglecting the contributions from the higher-order terms in Eq. (160). On a quantitative basis, how the higher-order terms influence the thermal shock phenomenon is a research topic to be pursued further in the near future.

2. *Reflection and refraction patterns of the thermal shock waves.* With regard to the engineering applications of the thermal shock theory in a more rigorous sense, the interaction between the thermal shock wave and a *free surface* or an *interface* between dissimilar materials is another important topic to be fully understood. The scattering pattern of the thermal shock waves by a free surface or a material interface reflects the amount of the thermal energy dissipating into the material continua in the neighborhood, which dominates the physical processes of subsurface crack formation and interfacial delamination in a nonisothermal environment with a large temperature gradient. These failure modes are of major concern in the early-time response of high-performance structures. For the problem involving a material interface, due to the presence of dissimilar materials, the subsonic, transonic, and supersonic ranges for the velocity of the moving head source must be defined with respect to the thermal wave speed in each material. A heat source moving in the supersonic range to one material involved in the system, for example, may therefore be subsonic, transonic, or supersonic in another, so a total of nine possible cases result.

3. *Anisotropic effect on the thermal shock formation*. For composite materials with anisotropic thermal conductivity, understanding of the physical mechanism of thermal shock formation needs to be further developed. For an orthotropic material with k_x and k_y being the principal values of the thermal conductivity along the x and y axes of material orthotropy, for example, the thermal shock angle has been derived as $\theta = \pm \tan^{-1}[(k_y/k_x)/(M^2 - 1)]^{1/2}$ measured from the trailing edge of the moving heat source. For the isotropic case with $k_x = k_y$, this expression reduces to $\theta = \pm \sin^{-1}(1/M)$, which is the result obtained previously in Eq. (32). For a heat source moving in the supersonic range with $M > 1$, clearly, the physical domain of the heat-affected zone *increases* with the ratio of k_y/k_x between the principal values. This phenomenon would become much more complicated should the orthotropy of the thermal wave speed be further incorporated into consideration.

With regard to the thermal stress induced by the thermal shock waves, the non-isothermal *plastic* deformation developed under a large temperature gradient needs to be taken into account in estimating the damage pattern around a rapidly moving heat source. Especially in the neighborhood of the thermal shock waves, the variation of the material effective stress and effective strain curves [94] must be updated in an incremental manner according to the local temperature level. Due to the nonlinearity of the problem, an isoparametric finite-element model incorporating with the incremental theory of plasticity and the J_2–I_1 flow rule is currently under development. The mean stress (I_1) effect on the effective stress J_2 for the onset of plastic deformation is implemented in the model to account for the high rate response of the material continua under thermal excitations.

REFERENCES

1. J. C. Maxwell, On the Dynamic Theory of Gases, *Phil. Trans. Roy. Soc. London*, vol. 157, pp. 49–88, 1867.
2. V. Peshkov, "Second Sound" in Helium II, *J. Phys., USSR*, vol. VIII, p. 381, 1944.
3. P. M. Morse and H. Feshbach, *Methods of Theoretical Physics*, McGraw-Hill, New York, vol. 1, p. 865, 1953.
4. P. Vernotte, Les Paradoxes de la Théorie Continue de l'équation de la chaleur, *Compt. Rend.*, vol. 246, pp. 3154–3155, 1958.
5. P. Vernotte, The Ture Heat Equation, *Compt. Rend.*, vol. 247, p. 2103, 1958.
6. P. Vernotte, Some Possible Complications in the Phenomena of Thermal Conduction, *Compt. Rend.*, vol. 252, pp. 2190–2191, 1961.
7. C. Cattaneo, A Form of Heat Conduction Equation Which Eliminates the Paradox of Instantaneous Propagation, *Compt. Rend.*, vol. 247, pp. 431–433, 1958.
8. G. I. Taylor, Diffusion by Continuous Movements, *Proc. London Math. Soc.*, vol. 20, pp. 196–201, 1922.
9. S. Goldstein, On Diffusion by Discontinuous Movements, and on the Telegraph Equation, *Quart. J. Mech. Appl. Math.*, vol. IV(2), pp. 129–156, 1951.
10. H. D. Weymann, Finite Speed of Propagation in Heat Conduction, Diffusion, and Viscus Shear Motion, *Am. J. Phys.*, vol. 36, pp. 488–496, 1967.
11. M. Chester, Second Sound in Solids, *Phys. Rev.*, vol. 131, pp. 2013–2015, 1963.
12. H. Grad, in *Handbuch der Physik* (ed. S. Flügge), Springer-Verlag, Berlin, vol. 12, p. 271, 1958.
13. J. M. Ziman, *Electrons and Phonons*, Clarendon Press, Oxford, 1960.

14. P. G. Klemens, in *Solid State Physics* (ed. F. Seitz and Dd. Turnbull), Academic Press, New York, vol. 7, p. 4, 1958.
15. K. H. Mueller, *Z. Angew. Math. Phys.*, vol. 14, p. 748, 1963.
16. M. E. Gurtin and A. C. Pipkin, A General Theory of Heat Conduction with Finite Wave Speeds, *Arch. Rat. Mech. Anal.*, vol. 31, pp. 113–126, 1968.
17. S. Kaliski, Wave Equation of Heat Conduction, *Bull. Acad. Pol. Sci. Ser. Sci. Tech.*, vol. 13, pp. 211–219, 1965.
18. D. C. Kelly, Diffusion: A Relativistic Appraisal, *Am. J. Phys.*, vol. 36, pp. 585–591, 1968.
19. N. G. Van Kampen, A Model for Relativistic Heat Transfer, *Physica*, vol. 46, pp. 315–332, 1970.
20. V. A. Bubnov, Wave Concepts in the Theory of Heat, *Int. J. Heat Mass Transfer*, vol. 19, pp. 175–184, 1976.
21. B. M. Berkovsky and V. G. Bashtovoi, The Finite Velocity of Heat Propagation from the Viewpoint of Kinetic Theory, *Int. J. Heat Mass Transfer*, vol. 20, pp. 621–626, 1977.
22. M. J. Maurer, Relaxation Model for Heat Conduction in Metals, *J. Appl. Phys.*, vol. 40, pp. 5123–5130, 1969.
23. F. London, *Superfluids*, John Wiley, New York, vol. II, p. 101, 1954.
24. J. C. Ward and J. Wilks, The Velocity of Second Sound in Liquid Helium Near Absolute Zero, *Phil. Mag.*, vol. 42, pp. 314–316, 1951.
25. J. C. Ward and J. Wilks, Second Sound and the Thermo-mechanical Effect at Very Low Temperature, *Phil. Mag.*, vol. 43, pp. 48–50, 1952.
26. R. B. Dingle, Derivation of the Velocity of the Second Sound from Maxwell's Equation of Transfer, *Proc. Phys. Soc. London*, vol. A65, pp. 374–376, 1952.
27. L. Landau, *J. Phys. USSR*, vol. 5, pp. 71–74, 1941.
28. E. W. Prohofsky and J. A. Krumhansl, Second-Sound Propagation in Dielectric Solids, *Phys. Rev.*, vol. 133(A), pp. 1403–1410, 1964.
29. A. V. Luikov, *Heat Conduction Theory*, Izd. Vyssh. Shkola, Moscow, chap. 3, 1967.
30. A. V. Luikov, *Heat and Mass Transfer Handbook*, Izd. Energiya, Moscow, 1972.
31. A. V. Luikov and B. M. Berkovsky, *Convection and Thermal Waves*, Izd. Energiya, Moscow, 1974.
32. A. V. Luikov, *Analytical Heat Diffusion Theory*, Academic Press, New York, pp. 9–12 and pp. 20–21, 1968.
33. K. J. Baumeister and T. D. Hamill, Hyperbolic Heat-Conduction Equation—A Solution for the Semi-Infinite Body Problem, *J. Heat Transfer*, vol. 91, pp. 543–548, 1969.
34. K. J. Baumeister and T. D. Hamill, On the Hyperbolic Heat Conduction Equations, *J. Heat Transfer*, vol. 93, pp. 126–127, 1971.
35. H. K. Fauske, On the Mechanism of Uranium Dioxide Sodium Explosive Interactions, *Nucl. Sci. Eng.*, vol. 51, pp. 95–101, 1973.
36. M. S. Kazimi and C. A. Erdman, On the Interface Temperature of Two Suddenly Contacting Materials, *J. Heat Transfer*, vol. 97, pp. 615–617, 1975.
37. M. J. Maurer and H. A. Thompson, Non-Fourier Effects at High Heat Flux, *J. Heat Transfer*, vol. 95, pp. 284–286, 1973.
38. P. J. Chen, On the Growth and Decay of Temperature Rate Waves of Arbitrary Form, *ZAMM*, vol. 20, pp. 448–453, 1969.
39. E. Lorenzini and M. Spiga, *Warme, Stoffubertrag*, vol. 16, p. 113, 1982.
40. R. Domanski, *6t Int. Heat Transfer Conf., National Resource Council of Canada*, Toronto, Ontario, p. 275, 1978.
41. S. H. Chan, J. D. Low, and W. K. Mueller, Hyperbolic Heat Conduction in Catalytic Supported Crystallites, *AIChE J.*, vol. 17, pp. 1499–1501, 1971.
42. A. V. Luikov, V. A. Bubnov, and I. A. Soloviev, On Wave Solutions of the Heat-Conduction Equation, *Int. J. Heat Mass Transfer*, vol. 19, pp. 245–248, 1976.
43. Y. B. Zel'dovich and Y. P. Raizer, Shock Waves, Heavy Explosions, Physical Processes in Gas Flows, *Mechanics in USSR for 50 Years*, Izd. Nauka, Moscow, vol. 2, 1970.
44. J. S. Letcher, On Boundary Layers in the Theory of Heat Conduction in Solids, *SIAM Rev.*, vol. II, pp. 20–29, 1969.

45. J. S. Letcher, An Improved Theory for Heat Conduction in Thin Layers, *J. Heat Transfer*, vol. 91, pp. 585–587, 1969.
46. T. Kao, Non-Fourier Heat Conduction in Thin Surface Layers, *J. Heat Transfer*, vol. 99, pp. 343–345, 1977.
47. D. C. Wiggert, Analysis of Early-Time Transient Heat Conduction by Method of Characteristics, *J. Heat Transfer*, vol. 99, pp. 35–40, 1977.
48. D. E. Amos and P. J. Chen, Transient Heat Conduction with Finite Wave Speeds, *J. Heat Transfer*, vol. 37, pp. 1145–1146, 1970.
49. Y. Taitel, On the Parabolic, Hyperbolic, and Discrete Formulation of the Heat Conduction Equation, *Int. J. Heat Mass Transfer*, vol. 15, pp. 369–371, 1972.
50. G. F. Carey and M. Tsai, *Numer. Heat Transfer*, vol. 5, pp. 309–314, 1982.
51. B. Vick and M. N. Özisik, Growth and Decay of a Thermal Pulse Predicted by the Hyperbolic Heat Conduction Equation, *J. Heat Transfer*, vol. 105, pp. 902–907, 1983.
52. M. N. Özisik and B. Vick, Propagation and Relection of Thermal Waves in a Finite Medium, *Int. J. Heat Mass Transfer*, vol. 27, pp. 1845–1854, 1984.
53. B. Bertman and D. J. Sandiford, Second Sound in Solid Helium, *Sci. Am.*, vol. 222, pp. 92–101, 1970.
54. J. I. Frankel, B. Vick, and M. N. Özisik, Flux Formulation of Hyperbolic Heat Conduction, *J. Appl. Phys.*, vol. 58, pp. 3340–3345, 1985.
55. R. E. Nettleton, Relaxation Theory of Thermal Conduction in Liquids, *Phys. Fluid*, vol. 3, pp. 216–225, 1960.
56. B. A. Boley, The Analysis of Problems of Heat Conduction and Melting, *High Temperature Structures and Materials*, Pergamon Press, New York, 1964.
57. B. A. Boley and I. S. Tolins, Transient Coupled Thermoelastic Boundary Value Problems in the Halh-Space, *J. Appl. Mech.*, vol. 29, pp. 637–646, 1962.
58. J. P. Brazel and E. J. Nolan, Non-Fourier Effects in the Transmission of Heat, *Symp. 6th Conf. on Thermal Conductivity*, Dayton, Ohio, p. 237, 1966.
59. Tsai-Tse Kao, On the Thermally Induced Non-Fourier Stress Waves in a Semi-Infinite Medium, *AIAA J.*, vol. 14, pp. 818–820, 1976.
60. H. W. Lord and Y. Shulman, A Generalized Dynamic Theory of Thermoelasticity, *J. Mech. Phys. Solids*, vol. 15, pp. 299–309, 1967.
61. J. D. Achenbach, The Influence of Heat Conduction on Propagating Stress Jumps, *J. Mech. Phys. Solids*, vol. 16, pp. 273–282, 1968.
62. E. B. Popov, Dynamic Coupled Problem of Thermoelasticity for a Half-Space Taking into Account of the Finiteness of the Heat Propagation Velocity, *J. Appl. Math. Mech. (PMM)*, vol. 31, pp. 349–356, 1967.
63. E. V. Wilms, On Coupling Effects in Transient Thermoelastic Problems, *J. Appl. Mech.*, vol. 31, pp. 719–722, 1964.
64. D. Y. Tzou, On the Thermal Shock Waves Induced by a Moving Heat Source, *J. Heat Transfer*, vol. 111, pp. 232–238, 1989.
65. D. Y. Tzou, Shock Wave Formation Around a Moving Heat Source in a Solid with Finite Speed of Heat Propagation, *Int. J. Heat Mass Transfer*, vol. 32, pp. 1979–1987, 1989.
66. D. Y. Tzou, Thermal Shock Waves Induced by a Moving Crack, *J. Heat Transfer*, vol. 112, pp. 21–27, 1990.
67. D. Y. Tzou, Thermal Shock Waves Induced by a Moving Crack—A Heat Flux Formulation, *Int. J. Heat Mass Transfer*, vol. 33, pp. 877–885, 1990.
68. D. Y. Tzou, Thermoelastic Fracture Induced by the Thermal Shock Waves Around a Moving Heat Source, *Heat Transfer in Manufacturing and Materials Processing*, 1989 ASME National Heat Transfer Conference, HTD vol. 113, pp. 11–17, 1989.
69. D. Y. Tzou, The Effects of Thermal Shock Waves on the Orientations of Crack Initiations Around a Moving Heat Source, *J. Eng. Fracture Mech.*, vol. 34, pp. 1109–1118, 1989.
70. D. Y. Tzou, Thermal Shock Formation in a Three Dimensional Solid Due to a Rapidly Moving Heat Source, *J. Heat Transfer*, vol. 113, pp. 242–244, 1991.
71. D. Y. Tzou, Three Dimensional Structures of the Thermal Shock Waves Around a Rapidly Moving Heat Source, *Int. J. Eng. Sci.*, vol. 28, pp. 1003–1017, 1990.

72. D. Y. Tzou, Three Dimensional Formation of Thermal Shock Waves: The Effects of Thermoelastic Stresses. *Int. J. Eng. Sci.*, submitted for publication.
73. B. Cotterell and J. R. Rice, Slightly Curved or Kinked Cracks, *Int. J. Fracture*, vol. 16, pp. 155–161, 1980.
74. G. Arfken, *Mathematical Methods for Physicists*, Academic Press, New York, pp. 763 and 760, 1970.
75. P. M. Morse and H. Feshbach, *Methods of Theoretical Physics*, McGraw-Hill, New York, vol. 1, pp. 865 and 837–839, 1953.
76. A. Shapiro, *The Dynamics and Thermodynamics of Compressible Fluid Flow*, Ronald Press, New York, vol. 1, p. 304, 1954.
77. H. S. Carslaw and J. C. Jaeger, *Conduction of Heat in Solids*, Oxford University Press, 1959.
78. J. P. Berry, Some Kinetic Considerations of the Griffith Criterion for Fracture, *J. Mech. Phys. Solids*, vol. 8, pp. 194–216, 1960.
79. J. I. Bluhm, Fracture Arrest, in *Fracture* (ed. H. Liebowitz), Academic Press, New York, vol. V, pp. 1–63, 1969.
80. M. L. Williams, Stress Singularities Resulting from Various Boundary Conditions in Angular Corners in Extension, *J. Appl. Mech.*, vol. 19, pp. 526–536, 1952.
81. J. D. Achenbach and Z. P. Bažant, Elastodynamic Near-Tip Stress and Displacement Fields for Rapidly Propagating Cracks in Orthotropic Materials, *J. Appl. Mech.*, vol. 42, pp. 183–189, 1975.
82. D. Y. Tzou, Qualitative Assessment on the Three Dimensional Configuration of the Material Boundary Layer and the Core Region, *Int. J. Eng. Sci.*, vol. 26, pp. 635–647, 1988.
83. G. C. Sih, A Special Criterion for Crack Propagation, in *Methods of Analysis and Solutions of Crack Problems, Mechanics of Fracture* (ed. G. C. Sih), Noordhoff, the Neatherlands, Leyden, vol. 1, pp. xxi–xlv, 1973.
84. F. Erdogan and G. C. Sih, On the Crack Extension in Plates Under Plane Loading and Transverse Shear, *J. Basic Eng.*, vol. 85, pp. 519–527, 1963.
85. J. N. Goodier, Integration of Thermoelastic Equations, *Phil. Mag.*, vol. 23, pp. 1017–1024, 1937.
86. O. D. Kellogg, *Foundations of Potential Theory*, Dover, New York, pp. 150–156, 1953.
87. M. Matczyński and G. C. Sih, Prediction of Damage Sites Ahead of a Moving Heat Source, in *Defects, Fracture, and Fatigue* (ed. G. C. Sih and J. W. Provan), Martinus Nijhoff, the Hague, pp. 211–222, 1983.
88. D. Y. Tzou and G. C. Sih, Discussion on Criteria for Brittle Fracture in Biaxial Tension, *J. Eng. Fracture Mech.*, vol. 21, pp. 977–981, 1985.
89. D. Y. Tzou and G. C. Sih, Crack Growth Prediction of Subsurface Crack in Yielded Material, *J. Appl. Mech.*, vol. 90, pp. 237–240, 1985.
90. H. Liebowitz, J. D. Lee, and N. Subramonian, Criteria for Predicting Crack Extension Angle and Path in Plane Crack Problem, in *Analytical and Experimental Fracture Mechanics* (ed. G. C. Sih), Sijthoff and Noordhoff, the Netherlands, Alphen aan den Rijn, p. 81, 1980.
91. D. Y. Tzou, Accumulative Material Damage Induced by Microcracking Under Extreme Loading Conditions, Technical Report No. SURP-05-9859, Sandia National Laboratories, Albuquerque, N.M., 1989.
92. F. B. Hildebrand, *Advanced Calculus for Applications*, Prentice-Hall, Englewood Cliffs, N.J., p. 156, 1974.
93. H. Parkus, Thermal Stresses, in *Handbook of Engineering Mechanics* (ed. W. Flügge), McGraw-Hill, New York, p. 43–2, 1963.
94. G. C. Sih and D. Y. Tzou, Arrest Characteristics of an Impacted Crack Dissipating Plastic Energy, *J. Theor. Appl. Fracture Mech.*, vol. 6, pp. 29–37, 1986.

CHAPTER
FOUR

TRANSPORT FROM CONTINUOUSLY MOVING MATERIALS UNDERGOING THERMAL PROCESSING

Yogesh Jaluria

ABSTRACT

This chapter presents a detailed consideration of the transport processes associated with a continuously moving material undergoing thermal processing. This problem is of interest in processes such as extrusion, fiber drawing, hot rolling, and continuous casting. Three main approaches are outlined. The first is directed at the heat transfer in the material, assuming convective coefficients at the surface. The second solves for the flow generated by an isothermal or uniform-heat-flux moving surface, focusing on the fluid flow and not on the transport in the material. The third approach is a combination of the first two and considers the conjugate problem that couples the flow and transport in the fluid with those in the material. Therefore, this is the most realistic formulation with respect to practical materials processing systems. The flow may be treated as a boundary layer, or the full, elliptic equations may be solved. Both formulations are discussed, along with the relevant numerical procedures.

A detailed review of the work done on this problem, using the various mathematical formulations outlined above, is presented. The results obtained from several studies are outlined, and the physical mechanisms that govern the transport processes are discussed. Additional effects, such as thermal buoyancy, transients, and additional forced flow in the ambient medium, are considered. The experimental work done on this problem is also reviewed. The main focus of this chapter is on the physical considerations that are important in this and other similar thermal transport processes. Also of particular interest is a critical evaluation of the present status of information on this topic and the areas where a detailed and planned future effort is needed. Several important, new, and interesting considerations in the thermal transport from a continuously moving material are brought out, with respect to a basic understanding of the underlying physical processes as well as to practical applications in the thermal processing of materials.

NOMENCLATURE

A	cross-sectional area of moving material
Bi, $\hat{\mathrm{B}}$i	Biot number, defined in Eqs. (5*c*) and (9*d*)
Bi_L	Biot number at the leading end of a moving rod
C	specific heat of the moving solid material; specific heat at constant pressure of the fluid
$\hat{C}_f$	friction coefficient $[= \hat{\tau}_s/(\frac{1}{2}\rho_f U_s^2)]$
D	diameter of cylindrical rod or half-thickness of plate
$\tilde{D}$	thickness of a moving plate
F	dimensionless stream function, defined in Eq. (11)
Gr	Grashof number, defined in Eq. (28)
Gr_x	local Grashof number $[= g\beta(T - T_\infty)x^3/\nu^2]$
h	local convective heat transfer coefficient
k	thermal conductivity
ℓ	characteristic length dimension, defined in Eq. (16)
L	dimensionless length, defined in Eq. (5c)
L'	physical length of rod or plate
L_T	total dimensionless distance between slot and spool, Fig. 2
Nu	local Nusselt number $[= h(x)x/k_f]$
P	perimeter of moving material
Pe, $\hat{\mathrm{P}}$e, $\tilde{\mathrm{P}}$e	Peclet number, defined in Eqs. (3), (9*d*), and (18*b*)
Pe_g	grid Peclet number $(= U_s\, \Delta x/\alpha_s)$
Pr	Prandtl number $(= \nu/\alpha_f)$
q	local surface heat flux from the moving material to the fluid
r	radial coordinate distance
R	property parameter $\{= [(k\rho C_f/(k\rho C)_s]^{1/2}$
$\hat{R}$	dimensionless radial coordinate distance, defined in Eq. (3)
Re, $\tilde{\mathrm{R}}$e	Reynolds number, defined in Eqs. (23) and (16)
Re_x	local Reynolds number $(= U_s x/\nu)$
T	local temperature
T_s	surface temperature
u	velocity component in the direction of material motion
U	dimensionless u velocity component, defined in Eqs. (16) and (23)
U_s	solid surface speed
U_∞	forced-flow velocity in ambient medium
v	velocity component normal to the direction of material motion
V, $\tilde{V}$	dimensionless v velocity component, defined in Eqs. (23) and (16)
$\overline{V}$	velocity vector in the flow
x	coordinate distance from the slot in the direction of material motion
X, $\hat{X}$, $\tilde{X}$	dimensionless x coordinate distance, defined in Eqs. (3), (9*d*), and (16)
X_b	upstream location, $X = -X_b$, where $T = T_0$ condition is applied
ΔX, $\Delta\tilde{X}$	grid sizes in direction of material motion
y	coordinate distance normal to the moving surface
Y, $\tilde{Y}$	dimensionless y coordinate distance, defined in Eqs. (23) and (16)

Y_s	coordinate distance within solid from the midplane of a plate
ΔY, $\Delta\tilde{Y}$	grid sizes in direction normal to plate motion
α	thermal diffusivity
β	coefficient of thermal expansion of the fluid
γ	characteristic length, defined in Eq. (9*d*)
δ	parameter, defined in Eq. (9*d*)
η	similarity variable, defined in Eq. (11)
$\hat{\eta}$	similarity variable, defined in Fig. 29
θ	dimensionless temperature, defined in Eq. (3)
θ_L	dimensionless temperature at the leading edge of the moving material
$\hat{\theta}$	dimensionless temperature, $\hat{\theta} = T/T_\infty$, used when radiative heat transfer is included
ν	kinematic viscosity of the fluid
ξ	dimensionless distance in the direction of material motion $[= x(\rho Ck)_f/(U_s D^2 \rho_s^2 C_s^2)$
ρ	density
$\hat{\rho}$	parameter, defined in Eq. (9*d*)
σ	parameter, defined in Eq. (9*d*)
τ'	physical time
τ, $\bar{\tau}$	dimensionless time, defined in Eqs. (23) and (3)
$\hat{\tau}_s$	surface shear stress
$\Delta\tau$	time step
ψ'	physical stream function, defined in Eqs. (11) and (23)
ψ	dimensionless stream function, defined in Eq. (23)
ω'	physical vorticity, defined in Eq. (23)
ω	dimensionless vorticity, defined in Eq. (23)

Subscripts

c	midplane or axis
f	fluid
0	slot, $x = 0$
s	surface of moving material
∞	ambient

1 INTRODUCTION

In a wide variety of manufacturing processes, heat and mass transfer considerations are of crucial importance in the design, optimization, and control of the process and of the relevant system. There has been growing interest and research activity in this area, which is generally referred to as thermal processing of materials and which includes processes such as casting, extrusion, bonding, heat treatment, crystal growing, glass fiber drawing, and plastic forming [1–3]. The increased interest in thermal processing is a consequence of the advent of new materials and processes, as well as of the global competition with respect to product quality and cost.

A problem of considerable practical and fundamental interest in the area of thermal processing of materials is that of heat transfer from a continuously moving material which loses energy by convection and radiation at the surface to the ambient medium. The thermal transport from such a moving material is important in processes such as hot rolling, wire and fiber drawing, continuous casting, plastic extrusion, and metal forming [4–6]. The speed at which the material moves can be as low as a few centimeters per hour, as in the case of the Czochralski crystal growth process [7], or as high as a few meters per second, as in the case of hot rolling or drawing in cable manufacture [8]. For short periods of time, following the start of the process, the thermal field is not fully established and the transient aspects of the process are of interest. However, as the time elapsed increases, the process generally approaches a steady-state circumstance. For instance, in an extrusion process, as the free end of the extruded material moves farther away from the die, a steady-state circumstance is eventually attained for constant boundary conditions. Figure 1 shows sketches of a few processes in which heat transfer from a continuously moving material is of interest.

Due to the importance of this heat transfer problem in many manufacturing processes, several studies have considered the geometry, boundary conditions, material characteristics, and other aspects that apply to specific processes. For instance, the formation and cooling of glass and metal fibers have been studied by Glicksman [9, 10], Matovich and Pearson [11], and Gifford [12]. Metal forming has been considered by Tseng [13, 14] and solidification of a moving sheet by Chopra and Niessen [15] and Kuiken [16]. The cooling of flat sheets and cylinders, as related to hot rolling and metal forming, has been studied extensively [17–20]. Several of these studies, along with others in relevant areas, will be considered again later in this chapter.

There are three main approaches that have been adopted in the study of transport from a continuously moving material. The first is directed at the conduction heat transfer in the solid, modified by the convective transport within the material due to its motion, without solving for the flow and the convection in the adjacent fluid. Convective heat transfer coefficients, based on empirical correlations derived from experimental results in similar configurations, are employed to impose the thermal boundary conditions at the surface. The second approach focuses on the flow induced in the ambient fluid due to material movement and determines the resulting thermal field and the associated heat transfer rate. These results can then be employed to compute the local convective heat transfer coefficient for, say, an isothermal metal sheet as a function of distance from the inlet point, which represents a die or a set of rollers. The third approach considers the conjugate transport resulting from the convection-conduction transport within the moving material and the convective heat transfer in the fluid, the two mechanisms being coupled through the boundary conditions at the surface of the material. This approach, therefore, yields the flow and the thermal field in the fluid, as well as the resulting thermal field in the moving material. Different characteristic quantities arise in each of these approaches, leading to different nondimensional variables and parameters. However, these different variables do result in the most efficient representation and solution of particular aspects of the overall problem.

This chapter presents a detailed review of work done on the thermal transport

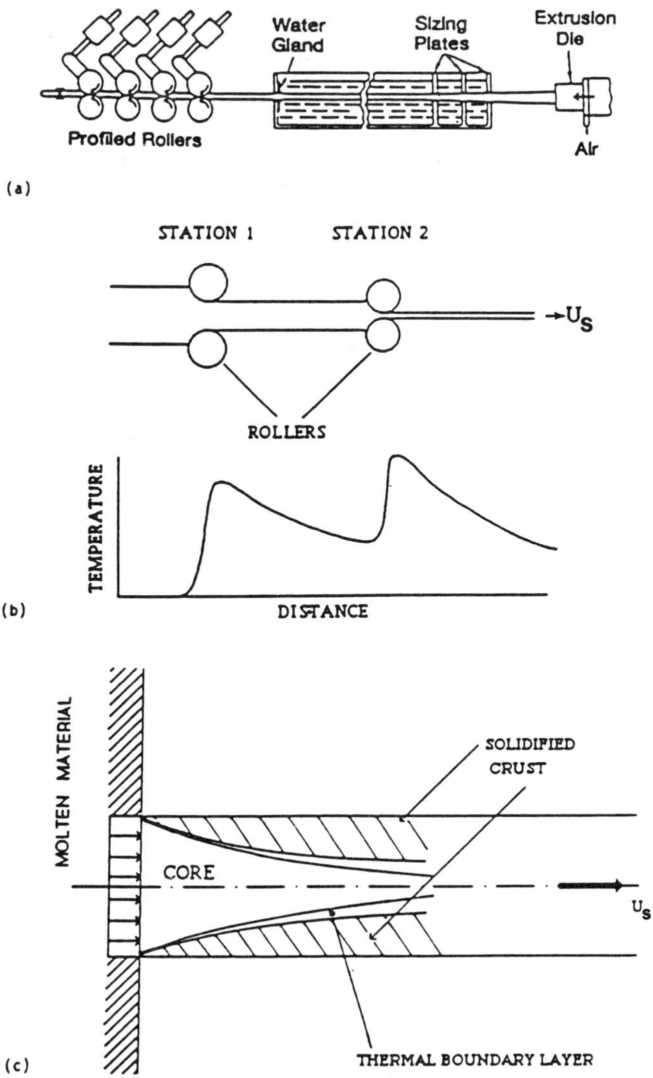

Figure 1 Examples of a few manufacturing systems for the thermal processing of materials that involve thermal transport in a moving material: (*a*) wire drawing; (*b*) hot rolling; (*c*) continuous casting.

associated with a continuously moving material. The analysis, numerical simulation, and results from all three approaches are presented. However, the focus is on the conjugate problem, since it is the most realistic and versatile one among these approaches. Though the numerical simulation of this heat transfer circumstance is of particular interest in this chapter, analytical and experimental results, wherever available, are also presented for validation of the numerical model. The present status of available information on heat transfer from a continuously moving material is discussed. Also presented in detail are the important unanswered questions, the nature of future work needed, additional aspects that need to be considered, new areas and directions that arise from the work done so far, underlying physical mechanisms in this transport process, and innovations that may stem from an effort on this problem. Though this chapter is directed at heat transfer from a moving material to the ambient, the approach and the basic considerations can easily be extended to the circumstance where the material is being heated, instead of being cooled, as it moves downstream.

This chapter addresses three main considerations with respect to the problem under study. These refer to the fundamental questions associated with transport from a moving material, the mathematical and numerical modeling of the underlying physical processes, and the results for a wide variety of conditions that are of interest in many applied areas mentioned earlier. The three main approaches, outlined above, are discussed in detail, indicating the important basic and applied results obtained, as well as the physical circumstances for which a particular model may be applied. Though simple models may be employed to obtain the basic trends, more complex heat transfer models are generally needed to provide accurate, quantitative information for the design and optimization of the relevant processes and systems. All these aspects are considered in this chapter, and the relevance of the various models, with respect to the fundamental and applied concerns, is outlined.

2 HEAT TRANSFER WITHIN THE MOVING MATERIAL

The temperature distribution within the moving material is of particular importance in materials processing, since it relates to the thermal stresses that arise in the material, to the material microstructure that governs the product quality, to the distance needed for achieving a desired temperature level, to the energy transferred by the moving material, and to several other relevant aspects, depending on the process under consideration [7, 8, 20–22]. Consequently, several studies have investigated the heat transfer within the moving material, employing a chosen convective heat transfer coefficient at the surface and, in some cases, the additional heat loss due to thermal radiation. The boundary conditions, material properties, geometry, temperature level, etc., are all determined by the chosen thermal process, and the resulting governing equation is solved for the temperature T analytically or numerically.

2.1 Analysis

Let us consider a cylindrical rod of diameter D moving at a constant speed U_s, as shown in Fig. 2. Then, the time-dependent temperature $T(r, x, \tau')$ in the material is governed, for constant material properties, by the following equation:

$$\frac{\partial T}{\partial \tau'} + U_s \frac{\partial T}{\partial x} = \alpha_s \left(\frac{\partial^2 T}{\partial r^2} + \frac{1}{r}\frac{\partial T}{\partial r} + \frac{\partial^2 T}{\partial x^2} \right) \tag{1}$$

where r is the radial coordinate distance from the axis, x is the axial coordinate distance from the point of emergence of the solid material, τ' is the physical time, and α_s is

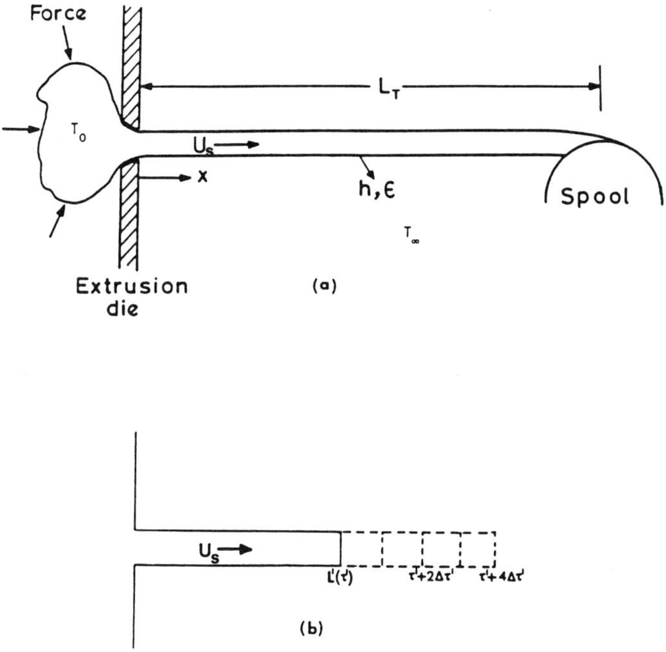

Figure 2 (*a*) Sketch of the wire extrusion process. (*b*) The transient process, showing the edge of the moving material at different time intervals.

the thermal diffusivity of the material. This equation may be nondimensionalized with D and U_s as characteristic quantities to yield

$$\frac{\partial \theta}{\partial \bar{\tau}} + \text{Pe}\,\frac{\partial \theta}{\partial X} = \frac{\partial^2 \theta}{\partial \hat{R}^2} + \frac{1}{\hat{R}}\frac{\partial \theta}{\partial \hat{R}} + \frac{\partial^2 \theta}{\partial X^2} \tag{2}$$

where Pe is the Peclet number, T_∞ is the ambient temperature, T_0 is the inlet temperature at $x = 0$, and

$$\bar{\tau} = \frac{\alpha_s \tau'}{D^2} \qquad \theta = \frac{T - T_\infty}{T_0 - T_\infty} \qquad X = \frac{x}{D} \qquad \hat{R} = \frac{r}{D} \qquad \text{Pe} = \frac{U_s D}{\alpha_s} \tag{3}$$

Similarly, for a plate which is wide in the transverse or z direction, the two-dimensional, transient heat conduction equation is

$$\frac{\partial \theta}{\partial \bar{\tau}} + \text{Pe}\,\frac{\partial \theta}{\partial X} = \frac{\partial^2 \theta}{\partial X^2} + \frac{\partial^2 \theta}{\partial Y^2} \tag{4}$$

where Pe is now based on the plate thickness $\tilde{D}$ and $Y = y/\tilde{D}$, $X = x/\tilde{D}$.

The corresponding initial and boundary conditions may also be written for these equations. For Eq. (2), these are

$$\bar{\tau} = 0: \qquad L(\bar{\tau}) = 0 \tag{5a}$$

$$\bar{\tau} > 0: \quad \text{at } X = 0, \quad \theta = 1.0 \text{ for } 0 \le \hat{R} \le 0.5$$

$$\text{at } X = L(\bar{\tau}), \quad \frac{\partial \theta}{\partial X} = -\text{Bi}_L\, \theta \text{ for } 0 \le \hat{R} \le 0.5 \tag{5b}$$

$$\text{at } \hat{R} = 0, \quad \frac{\partial \theta}{\partial \hat{R}} = 0 \text{ for } 0 \le X \le L(\bar{\tau})$$

$$\text{at } \hat{R} = 0.5, \quad \frac{\partial \theta}{\partial \hat{R}} = -\text{Bi}\, \theta \text{ for } 0 \le X \le L(\bar{\tau})$$

where

$$L = \frac{L'}{D} \quad \text{and} \quad \text{Bi} = \frac{hD}{k_s} \tag{5c}$$

Here L' is the physical length of the rod at a given time τ', that is, $L' = U_s\tau'$, the Biot number $\text{Bi} = hD/k_s$, and Bi_L is for convection at the end of the rod. The solid thermal conductivity is denoted by k_s and the convective heat transfer coefficient at the surface by h. As L increases, it is expected that the steady-state circumstance would be attained eventually, so that as $L \to \infty$, $\partial\theta/\partial X \to 0$, implying that the end heat loss is negligible for a very long plate or rod. Also, the temperature attains the ambient temperature, so that $\theta \to 0$ as $L \to \infty$.

The heat transfer problem outlined above may also be considered in terms of a simpler one-dimensional model, with temperature $\theta(X, \bar{\tau})$ taken as uniform over a given cross section. This model will apply for small values of Bi, for example, for thin plates of high thermal conductivity material. The governing equation becomes

$$\frac{\partial \theta}{\partial \bar{\tau}} + \text{Pe}\,\frac{\partial \theta}{\partial X} = \frac{\partial^2 \theta}{\partial X^2} - \frac{PD}{A}\,\text{Bi}\,\theta \tag{6}$$

where P is the perimeter and A is the cross-sectional area of the moving material. The initial condition is the same as Eq. (5*a*), and the boundary conditions become

$$\bar{\tau} > 0: \quad \text{at } X = 0,\ \theta = 1.0 \quad \text{and} \quad \text{at } X = L(\bar{\tau}),\ \frac{\partial \theta}{\partial X} = -\text{Bi}_L\,\theta \tag{7}$$

Therefore, the Peclet number Pe and the Biot number Bi arise as the main governing parameters. If the end convective heat transfer coefficient is different from h, Bi_L is an additional parameter. The geometry is reflected in the boundary conditions or in the value of PD/A in the one-dimensional model. If radiative loss at the surface is also included, the Biot number may be defined in terms of the overall heat transfer coefficient for small temperature differences or a corresponding radiation parameter arises and must be included in the analysis [23].

The temperature θ at the point of emergence of the material, $X = 0$, is taken as 1.0 in the above formulation. However, a more realistic boundary condition would be one that takes into account the upstream thermal conditions in, say, a furnace, oven, or extruder. Also the temperature θ may be imposed as 1.0 far upstream, that is, $X \to -\infty$, indicating the temperature at which the material is maintained before moving toward the point of emergence. This is an important consideration, since the temperature at $X = 0$ is a consequence of material motion and the downstream heat transfer mechanisms. This aspect is again considered in detail with respect to the conjugate problem. However, for the present, the situation shown in Fig. 3 is considered. Regions 1 and 3 have no surface heat transfer, i.e., the surfaces are adiabatic, and convective heat transfer exists in region 2, which is of length L'. The temperatures T_1, T_2, and T_3 in the three regions are governed by the following equations for the one-dimensional model [24]:

$$\frac{d^2T_1}{dx^2} - \frac{U_s}{\alpha_s}\frac{dT_1}{dx} = 0 \tag{8a}$$

$$\frac{d^2T_2}{dx^2} - \frac{U_s}{\alpha_s}\frac{dT_2}{dx} - \frac{hP}{k_sA}(T_2 - T_\infty) = 0 \tag{8b}$$

$$\frac{d^2T_3}{dx^2} - \frac{U_s}{\alpha_s}\frac{dT_3}{dx} = 0 \tag{8c}$$

The temperature T is assumed to approach T_0 as $x \to -\infty$ and to remain finite as $x \to \infty$. Continuity of temperature and heat flux at $x = 0$ and $x = L'$ provide the remaining boundary conditions.

2.2 Numerical Simulation and Results

For the transient problem, the length L is a function of time $\bar{\tau}$. This is simulated numerically by taking the dimensionless length L as unchanged over a small time

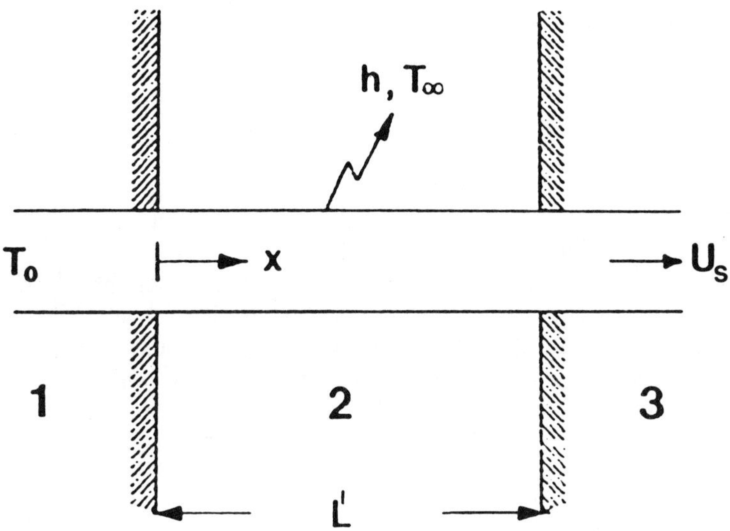

Figure 3 Sketch of a continuously moving material whose surface is insulated, except over a region of length L', where convective heat loss occurs, and which has a specified temperature T_0 far upstream.

increment $\Delta\bar{\tau}$ and then taking the increased value for the next time step, as shown in Fig. 2. Thus, the physical length L' is varied in steps as $L' = U_s\,(i\,\Delta\tau')$, where i varies from 0 at the initial condition, through 1, 2, 3, . . . , yielding $\tau' = i\,\Delta\tau'$ and $L' = U_s\,\Delta\tau', 2U_s\,\Delta\tau', 3U_s\,\Delta\tau'$, . . . , as time elapses. This procedure has also been employed in earlier studies of moving-boundary problems [25]. The initial temperature of the element adjacent to $X = 0$ is taken as the boundary condition $\theta(0) = 1.0$. The dimensionless time step $\Delta\bar{\tau}$ must be varied to ensure a negligible dependence of the numerical results on the value chosen. As computation proceeds, the length of the rod or plate increases and the solution is obtained until the steady-state conditions are attained at large values of L, or of $\bar{\tau}$. The numerical solution may be obtained by employing the alternating direction implicit (ADI) finite-difference method of Peaceman and Rachford [26] for two-dimensional problems and the Crank-Nicolson method for the one-dimensional model. Both these methods are second-order accurate in time and space and yield tridiagonal algebraic systems, which can be easily solved by the Thomas algorithm (TDMA). Therefore, efficient algorithms are obtained [27]. Central differences may be employed for the spatial derivatives at small values of the Peclet number Pe. Due to numerical instability at large values of the grid Peclet number $\mathrm{Pe}_g = U_s\,\Delta x/\alpha_s$, the grid may need to be refined or the various other schemes, such as the exponential and the power-law differencing schemes, may be employed. For further details on this aspect, see [27, 28].

Some of the typical results obtained from the numerical simulation outlined above are shown in Figs. 4 and 5. Figure 4*a* shows the temperature variation with the downstream distance X for the one-dimensional steady-state circumstance at different values of the dimensionless distance L_T between the inlet and the location beyond which the adiabatic boundary condition is imposed. Here, $L_T = L_T'/D$, L_T' being the physical distance. The solution is obtained by dropping the transient term in Eq. (6) and imposing the condition $d\theta/dX = 0$ at $X = L_T$. Clearly, the temperature θ_L at the end depends on the distance L_T. As L_T increases to large values, being 7.0 for the conditions employed here, the end temperature attains the ambient value, that is, $\theta_L = 0$. Any further increase in L_T would, therefore, have no effect on the temperature distribution. This figure is important in choosing the distance L_T for a given physical process. Similar trends are obtained for the two-dimensional problem [23].

Figure 4*b* shows the calculated time-dependent temperature distributions for a cylindrical rod from the one-dimensional model, indicating the approach to steady-state conditions at large time $\bar{\tau}$. The length $L(\bar{\tau})$ increases with time, ultimately becoming large enough for the end temperature to attain the ambient value T_∞. This represents the attainment of the steady-state circumstance for given values of Pe, Bi, and Bi_L. Similar behavior is observed for a flat plate and for two-dimensional and axisymmetric cases. Figure 5 shows the calculated isotherms for some of these cases. Here, the diameter D of the cylindrical rod and thickness $\tilde{D}$ of the flat plate are employed as characteristic lengths. It is interesting to note that, as expected, the one-dimensional model is appropriate for small values of Bi. Also, the time-dependent isotherms vary gradually as time elapses until the steady-state situation is attained.

Equations (8*a*)–(8*c*) may be solved analytically to yield the temperature distribution. The solution is obtained as [24]

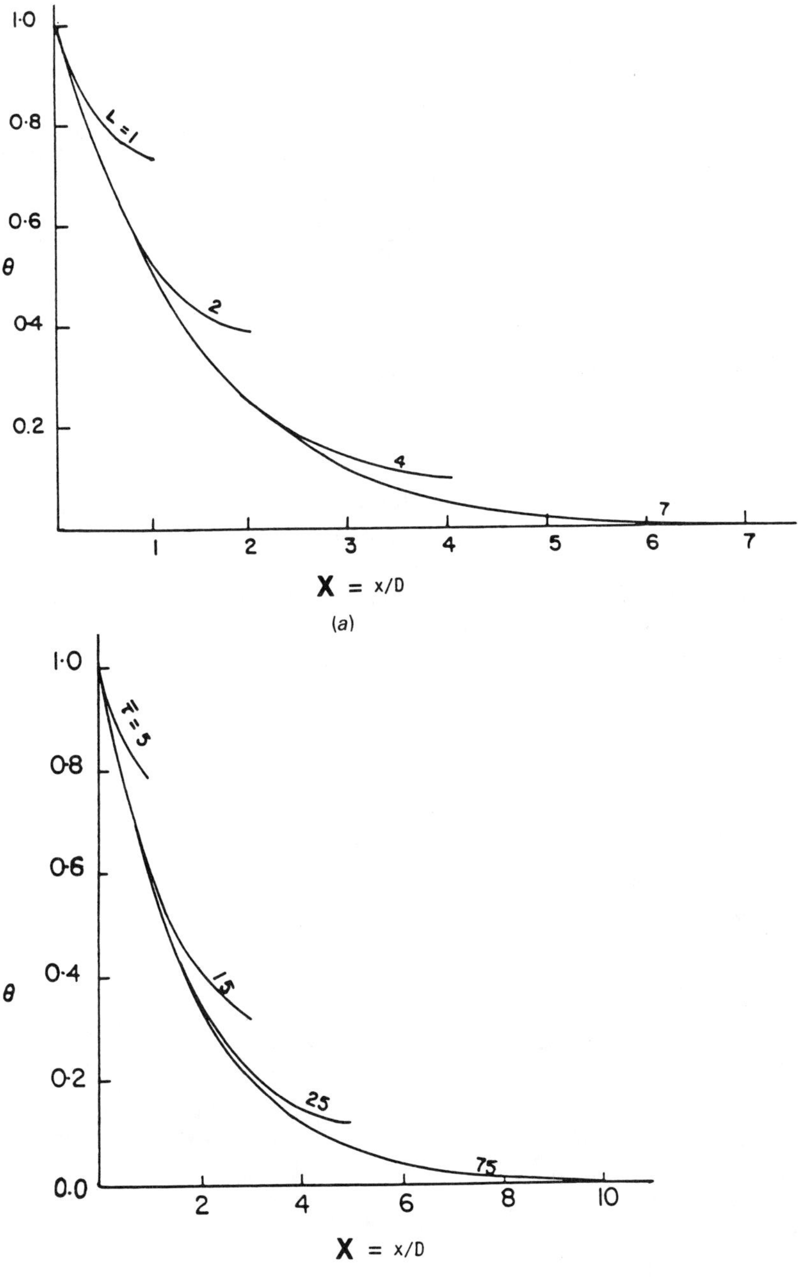

Figure 4 (*a*) Steady-state temperature distributions in a continuously moving cylindrical rod at different values of L, the dimensionless distance between the die and the spool, where an adiabatic condition is applied, as shown in Fig. 2. Here, Pe = 0.4 and Bi = 0.2. (*b*) Transient temperature distributions in a continuously moving cylindrical rod at various dimensionless time intervals $\bar{\tau}$ after the start of the process, assuming one-dimensional conduction, for Pe = 0.2 and Bi = 0.1.

$$\frac{T_1 - T_0}{T_\infty - T_0} = \left[1 + \frac{2\,\hat{\mathrm{Pe}}(\sigma e^{\sigma/2\delta} - \hat{\rho} e^{\hat{\rho}/2\delta})}{\hat{\rho}^2 e^{\hat{\rho}/2\delta} - \sigma^2 e^{\sigma/2\delta}}\right] e^{\hat{\mathrm{Pe}}\,\hat{X}} \tag{9a}$$

$$\frac{T_2 - T_0}{T_\infty - T_0} = 1 + \frac{2\,\hat{\mathrm{Pe}}(\sigma e^{\sigma/2\delta} e^{\hat{\rho}\hat{X}/2} - \hat{\rho} e^{\hat{\rho}/2\delta} e^{\sigma\hat{X}/2})}{\hat{\rho}^2 e^{\hat{\rho}/2\delta} - \sigma^2 e^{\sigma/2\delta}} \tag{9b}$$

$$\frac{T_3 - T_0}{T_\infty - T_0} = 1 + \frac{2\,\hat{\mathrm{Pe}}(\sigma - \hat{\rho}) e^{\hat{\mathrm{Pe}}/\delta}}{\hat{\rho}^2 e^{\hat{\rho}/2\delta} - \sigma^2 e^{\sigma/2\delta}} \tag{9c}$$

where

$$\hat{X} = \frac{x}{\gamma} \qquad \delta = \frac{\gamma}{L'} \qquad \gamma = \frac{A}{P} \qquad \hat{\mathrm{Pe}} = \frac{U_s \gamma}{\alpha_s} \qquad \hat{\mathrm{Bi}} = \frac{4h\gamma}{k_s}$$

$$\hat{\rho} = \hat{\mathrm{Pe}} + (\hat{\mathrm{Pe}}^2 + \hat{\mathrm{Bi}})^{1/2} \qquad \sigma = \hat{\mathrm{Pe}} - (\hat{\mathrm{Pe}}^2 + \hat{\mathrm{Bi}})^{1/2} \tag{9d}$$

Here, γ is taken as the characteristic length. This analytical result is given largely to show that some situations, particularly the one-dimensional steady-state case, are amenable to an analytical solution. Such solutions can thus be employed for validation of the numerical model and for providing physical insight into the underlying transport processes. The length scale is chosen as γ, since these results may be applied to different cross sections, besides those for the plate and the rod; see Eq. (6).

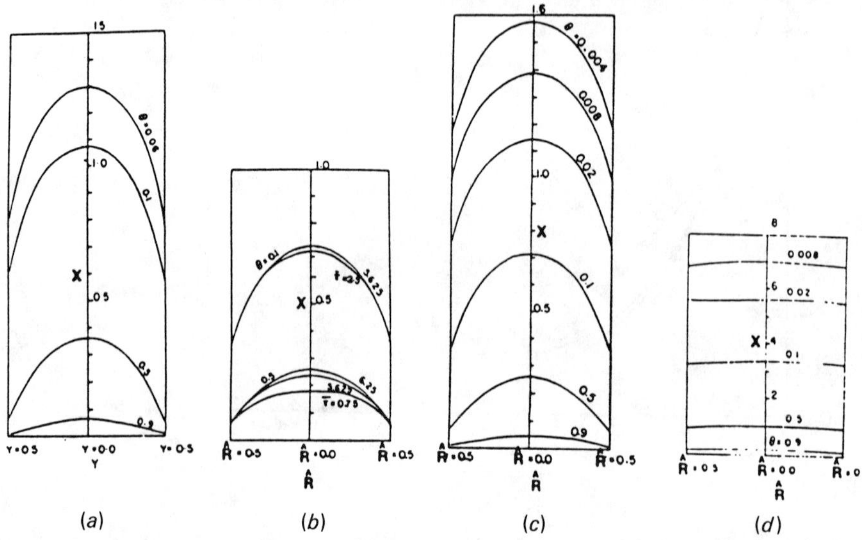

(a) (b) (c) (d)

Figure 5 (*a*) Isotherms in a long, moving, two-dimensional strip at Bi = 10.0 and Pe = 0.4. (*b*) Time-dependent isotherms for the axisymmetric problem at Bi = 10.0 and Pe = 0.4. (*c*) Isotherms in a long, moving, cylindrical rod at Bi = 10.0 and Pe = 0.4. (*d*) Isotherms in a long, moving, cylindrical rod at Bi = 0.2 and Pe = 0.4.

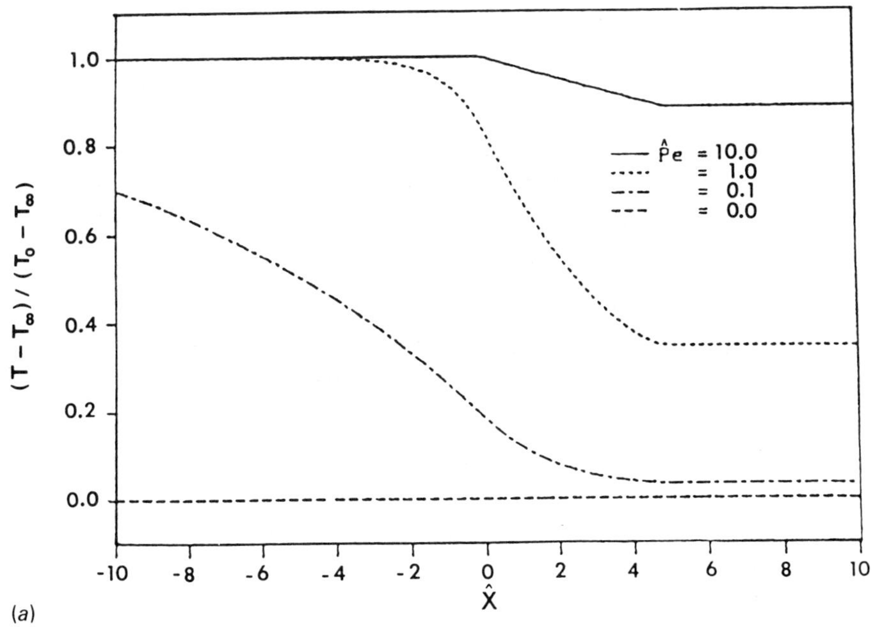

(a)

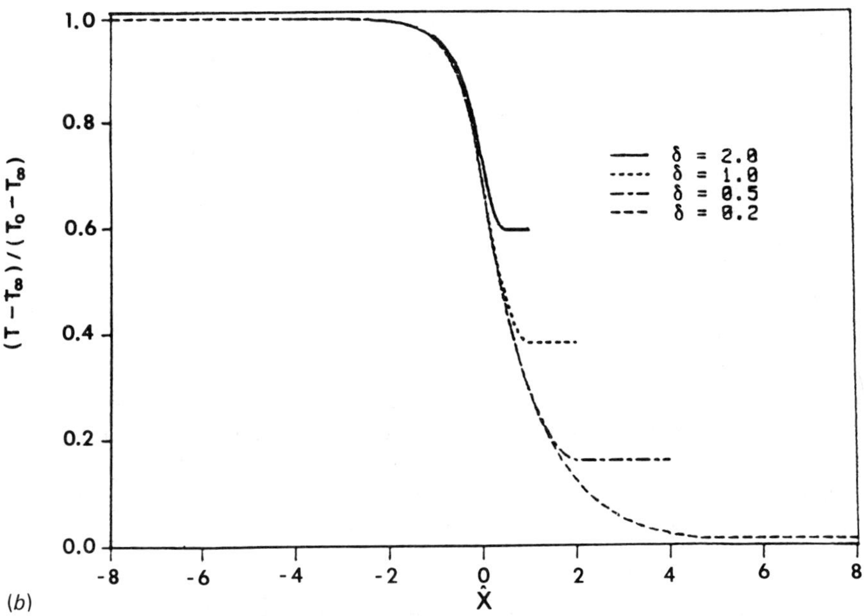

(b)

Figure 6 (*a*) Effect of varying the Peclet number $\hat{\text{Pe}}$ on the temperature distribution for the problem shown in Fig. 3, at $\hat{\text{Bi}} = 1.0$ and $\delta = 0.2$. (*b*) Effect of varying the parameter δ on the temperature distribution, at $\hat{\text{Pe}} = 2.0$ and $\hat{\text{Bi}} = 10.0$.

Thus, $\hat{\mathrm{Pe}}$ and $\hat{\mathrm{Bi}}$ emerge as the main governing parameters that characterize the operating conditions. The geometry is represented by the parameter δ. Figure 6 shows these results graphically for typical values of these parameters. Several interesting physical trends are seen. The downstream temperature decay becomes more gradual as $\hat{\mathrm{Pe}}$ is increased, at constant $\hat{\mathrm{Bi}}$ and δ; see Fig. 6*a*. This may be explained in terms of increasing speed U_s as $\hat{\mathrm{Pe}}$ increases, resulting in shorter time for heat loss up to a given distance $\hat{X}$. Therefore, the temperature level at a given location $\hat{X}$ increases as $\hat{\mathrm{Pe}}$ is increased. The penetration of the diffusion effects into the negative $\hat{X}$ direction, i.e., upstream, is lower for larger $\hat{\mathrm{Pe}}$, indicative of decreasing conduction effects, as compared to convection, as $\hat{\mathrm{Pe}}$ increases. A larger $\hat{\mathrm{Bi}}$ implies more rapid cooling downstream, and the upstream penetration of thermal effects was also found to increase with increasing $\hat{\mathrm{Bi}}$. Figure 6*b* shows the effect of varying δ on the temperature distribution. A decrease in δ implies an increase in L', for a given cross section. This gives rise to a lower temperature in region 3, as seen in the figure, since a larger distance is available for heat loss. For very small δ, being around 0.2 in this figure, the temperature at the end of region 2, $x = L'$, is close to the ambient temperature and a further decrease in δ will not affect the temperature distribution. This physically expected trend was also observed earlier in the results shown in Fig. 4*a*.

The above discussion brings out some of the salient physical features of this problem. These include the approach to steady state at large time $\bar{\tau}$, the upstream penetration of thermal effects and the resulting temperature at $X = 0$, increase in the rate of temperature decay with X as the Biot number is increased or the Peclet number is decreased, and the distance needed for the temperature level to attain the ambient value. Clearly, this is a very brief outline of the work done on heat transfer within a moving material. The various other references mentioned here may be consulted for further details on the results obtained for a variety of applications in the area of the thermal processing of materials. Several aspects need further detailed investigation. These include transient and three-dimensional effects, inclusion of material property variations, the dependence of the material microstructure and properties on the thermal field that arises, different thermal conditions at the surface, other geometries and configurations, proper imposition of boundary conditions in the region upstream of and at the point of emergence, and inclusion of other modes of heat transfer. A few of these aspects have received some attention in the literature as discussed later in this chapter. But many of these considerations must be included in a realistic numerical simulation to provide inputs for improving existing processes and designing new ones. In particular, the heat transfer coefficient at the surface is not known, but arises from the fluid flow adjacent to the surface. Therefore, the modeling of the transport phenomena in the fluid is crucial to a realistic simulation of the problem under consideration. This aspect is considered in the next section.

3 CONVECTIVE TRANSPORT FROM A MOVING SURFACE

Several studies have considered the flow induced by a moving surface in a quiescent ambient medium and have determined the resulting heat transfer for isothermal sur-

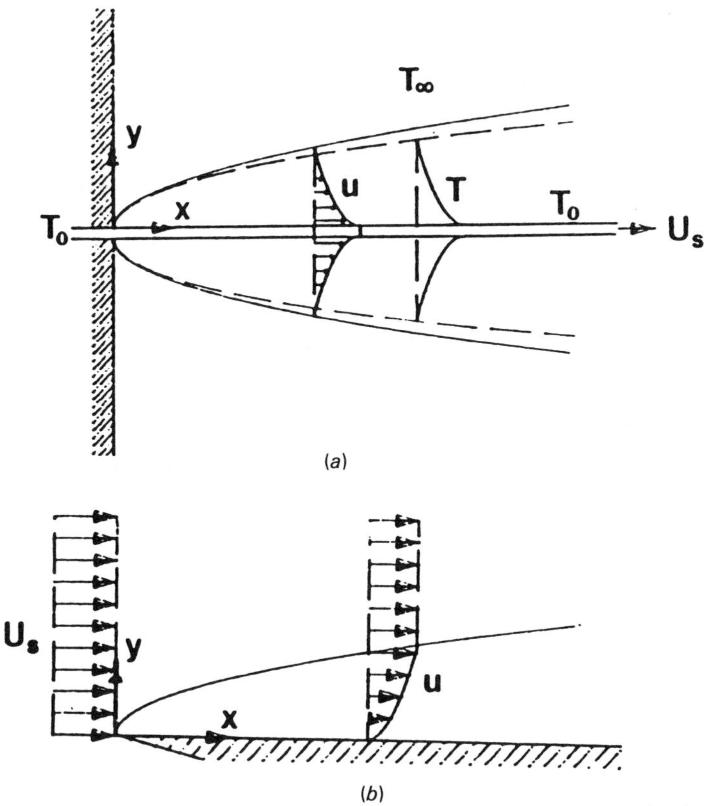

Figure 7 (*a*) Sketch of the velocity and temperature profiles due to a moving isothermal surface at temperature T_0. (*b*) Sketch of the velocity profiles for a uniform parallel flow over a stationary surface.

faces. Thus, the transport within the solid material is not considered and the effort is directed at the fluid flow that arises. Most of these studies have considered steady two-dimensional, or axisymmetric, boundary-layer flows. Figure 7*a* shows the velocity and temperature boundary layers generated by a continuous, isothermal, moving flat surface. This problem was studied by Sakiadis [29, 30], who employed a similarity transformation to obtain ordinary differential equations from the governing partial differential equations for a semi-infinite surface. The boundary layer is seen to grow in the direction of motion of the surface, and the velocity decreases from U_s at the surface to zero in the ambient medium. This velocity variation is quite different from that obtained for a semi-infinite stationary flat surface in a uniform flow parallel to the surface; see Fig. 7*b*.

Tsou et al. [31] showed experimentally that a boundary-layer flow of the form shown in Fig. 7*a* does arise. A very good agreement between the measured velocity profile, for laminar flow, and the theoretical results was obtained. They employed a rotating drum system with a skimmer plate in order to simulate the boundary-layer flow on a continuously moving surface. The radius of the rotating cylinder was kept large so that the effects of curvature could be neglected. A fair agreement was also obtained for turbulent flow. Griffin and Thorne [32] observed the growth of the thermal boundary layer on a continuous, flat, moving belt using a shadowgraph. Their results were in fair agreement with the theoretical results of Erickson et al. [33]. Griffin and Thorne have also discussed the effect of the wall, above and below the slot through which the belt emerges, on the heat transfer coefficient near the slot.

Local similarity and nonsimilarity methods have been employed for moving horizontal and inclined isothermal surfaces, respectively, by Chen and Strobel [34] and by Moutsoglou and Chen [35]. The flow and temperature fields were determined, from which the heat transfer rate was obtained. Buoyancy effects were included to consider a mixed-convection circumstance. Among other investigators, Koldenhof [36] employed integral techniques to solve the laminar boundary-layer equations for flow generated by moving flat plates and cylinders.

Assuming a boundary-layer flow, the governing equations for constant fluid properties and for negligible viscous dissipation and pressure work effects are obtained for laminar, steady, two-dimensional flow due to a moving isothermal surface at temperature T_s (Fig. 7*a*) as

$$\frac{\partial u}{\partial x} + \frac{\partial v}{\partial y} = 0 \tag{10a}$$

$$u\frac{\partial u}{\partial x} + v\frac{\partial u}{\partial y} = \nu\frac{\partial^2 u}{\partial y2} \tag{10b}$$

$$u\frac{\partial T}{\partial x} + v\frac{\partial T}{\partial y} = \alpha_f\frac{\partial^2 T}{\partial y2} \tag{10c}$$

where u and v are the velocity components in the x and y directions, respectively, ν is the kinematic viscosity, and α_f is the fluid thermal diffusivity. These equations may be reduced to ordinary differential equations by using the following similarity transformation [24, 30]:

$$\psi' = \sqrt{2\nu U_s x}\,F(\eta) \qquad \eta = \frac{y}{x}\sqrt{\frac{U_s x}{2\nu}} \qquad \theta = \frac{T - T_\infty}{T_s - T_\infty} \tag{11}$$

where ψ' is the physical stream function, η is the similarity variable, and $F(\eta)$ is the dimensionless stream function. Thus, $u = \partial\psi'/\partial y = U_s F'(\eta)$ and $v = -\partial\psi'/\partial x$. The resulting governing equations are

$$F''' + FF'' = 0 \tag{12}$$

$$\theta'' + \Pr F\theta' = 0 \tag{13}$$

where the primes in these equations indicate differentiation with respect to η. Here, Pr is the Prandtl number, taken as 0.7 and 7.0 for air and water at normal temperatures, respectively.

These equations are the same as those for forced-convection heat transfer from a stationary, semi-infinite flat surface to a uniform flow over it, as shown in Fig. 7*b*. However, the boundary conditions are different, since $u \rightarrow 0$ as $y \rightarrow \infty$ in the moving-surface case, whereas for a stationary plate $u \to U_\infty$, the free-stream velocity. The physical boundary conditions $u = U_s$, $v = 0$, $T = T_s$ at the plate surface, and $u \to 0$, $T \to T_\infty$ far away from the plate surface, give

$$F(0) = 0 \qquad F'(0) = 1 \qquad \theta(0) = 1 \qquad F'(\infty) = 0 \qquad \theta(\infty) = 0 \tag{14}$$

where the quantity within each set of parentheses indicates the η value at which the given boundary condition is applied.

Therefore, governing equations are ordinary differential equations, with Eq. (12) being a third-order nonlinear equation which is independent of the temperature field. The temperature θ is obtained from the energy equation, Eq. (13), which needs the solution, in terms of $F(\eta)$, from Eq. (12). Therefore, Eq. (12) is solved first, breaking it into three first-order equations and applying a shooting method based on standard techniques such as the Runge-Kutta or predictor-corrector method [27, 37]. Since this is a boundary-value problem, with boundary conditions at $\eta = 0$ and ∞, a correction scheme, such as the Newton-Raphson method, is applied for satisfying the boundary conditions. Finite-difference methods may also be used. After the numerical solution to Eq. (12) has been obtained, Eq. (13) is solved to obtain $\theta(\eta)$.

Figure 8 shows the results obtained at Pr = 0.7 for this circumference, in terms of the calculated distributions of $F'(\eta)$, which represents the dimensionless velocity component u/U_s in the x direction, and $\theta(\eta)$. The corresponding distributions for the flow over a stationary plate are also shown for comparison. Table 1 presents the

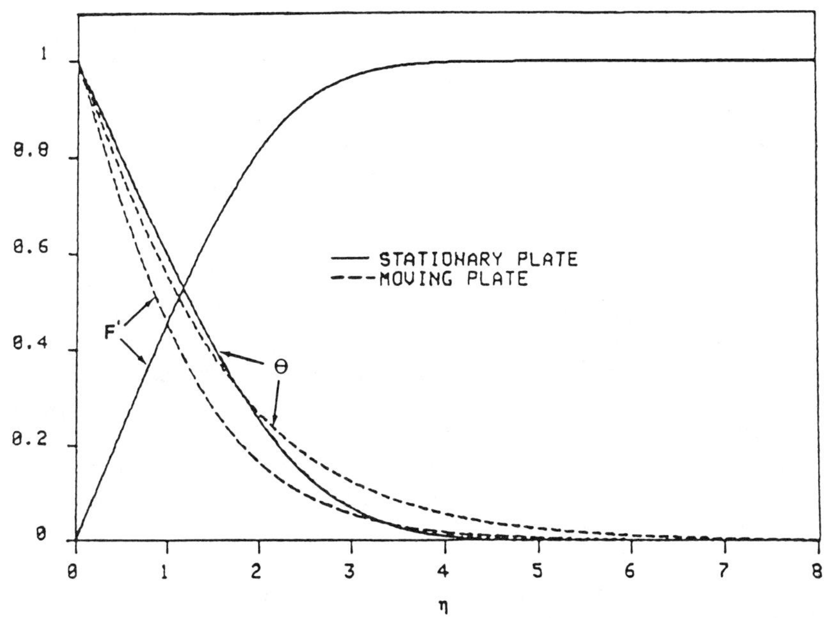

Figure 8 Variation of F' and θ with the similarity variable η for the moving plate in a quiescent medium, as well as for a stationary plate with a uniform forced flow over it.

Table 1 Comparison of the similarity solution for flow induced by a continuously moving plate with that for flow over a semi-infinite, stationary plate

	Flow over a semi-infinite, stationary plate	Flow induced by a continuously moving plate
	Pr = 0.7	
$F''(0)$	0.4692	−0.6280
$\theta'(0)$	−0.4122	−0.4939
η_{eT}	3.98	5.89
η_{ev}	3.47	4.48
	Pr = 7.0	
$F''(0)$	0.4692	−0.6280
$\theta'(0)$	−0.9068	−1.9457
η_{eT}	1.74	1.14
η_{ev}	3.47	4.48

important variables obtained from this computation, for Pr = 0.7 and 7.0. These include $F''(0)$, which determines the surface shear, $\theta'(0)$, which determines the surface heat flux, and η_{eT} and η_{eV}, which yield the location of the edge of the thermal and velocity boundary layers, respectively. Thus, the shear stress at the plate surface is higher for the moving plate. Similarly, the local heat transfer coefficient $h(x)$, which is proportional to $[-\theta'(0)]$, is higher for the moving surface. These effects are due to entrainment that brings the colder ambient fluid toward the moving surface. For a stationary plate in a uniform flow, the normal velocity v is positive and the ambient fluid is not drawn toward the surface. This effect increases the temperature gradient at the surface and, hence, the heat transfer from a moving surface. This implies that for a given fluid and velocity U_s, a moving plate loses energy at a faster rate than a semi-infinite stationary plate with uniform flow at the same velocity over it.

These results are important in the determination of the heat transfer rate from a moving, isothermal, flat surface. The temperature distribution in the solid material is not considered. However, at large Peclet numbers and small Biot numbers, the down-stream temperature decay is small, as seen earlier, and these results may be used to obtain the heat transfer rate. Similarly, moving surfaces which are maintained iso-thermal due to the thermal process under consideration, for instance, phase change at the surface, can be considered in terms of these results. For further details on other studies on this problem, the various references mentioned here may be consulted. Further work is needed on moving finite surfaces, with a leading edge, buoyancy effects, applicability of the boundary-layer approximations, and effect of the inclusion of an additional forced flow on the transport processes. However, the conjugate prob-lem, which is considered next, is of much greater practical interest than the transport phenomena discussed in this and the preceding sections, with respect to the thermal processing of materials. This is particularly true of processes such as extrusion and fiber drawing of low-thermal-conductivity materials, such as polymers and glass, since the quality of the product is a strong function of the temperature distribution in the solid material.

4 CONJUGATE TRANSPORT IN A STATIONARY AMBIENT MEDIUM

We now come to the conjugate problem in which the heat transfer within the moving material is coupled with the convective flow and heat transfer in the fluid. This implies that the flow generated by the moving material is computed, along with the thermal field, in order to obtain the heat transfer rate from the material. Therefore, the heat transfer coefficient at the surface of the material is not assumed, as done for the results presented in Sec. 2, but is obtained from a solution of the governing equations for the fluid flow. In a sense, this combines the two approaches outlined in Secs. 2 and 3, and presents a more realistic model for heat transfer from a moving material undergoing thermal processing.

Very few studies have considered the conjugate problem for a continuous, moving material. Chida and Katto [17] considered the flow induced by a continuously moving plate of finite thickness. Thus, the effect of conduction within the plate was included in their calculations. However, the study was largely directed at large Peclet number Pe. Kuiken [38, 39] considered the thermal transport from a low-heat-resistance cylinder, or sheet, moving through a fluid. The temperature of the material was taken as uniform at a given downstream location. More recently, extensive numerical work has

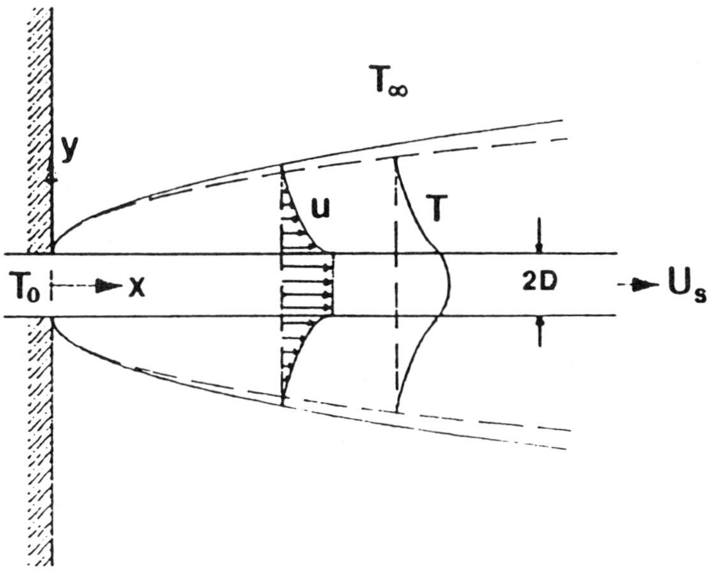

Figure 9 Schematic of the velocity and temperature profiles in the conjugate problem for a continuously moving flat plate.

been done on this conjugate problem by Karwe and Jaluria [24, 40, 41]. The basic formulation and the numerical results from this work are presented in detail here.

4.1 Boundary-Layer (Parabolic) Formulation

If boundary-layer assumptions are made for the flow generated by the moving material, the governing equations are Eqs. (10*a*)–(10*c*). In addition, the heat transfer within a moving plate is governed by the equation

$$U_s \frac{\partial T}{\partial x} = \alpha_s \frac{\partial^2 T}{\partial y_s^2} \tag{15}$$

The circumstance under consideration is shown in Fig. 9, along with the coordinate system. Here, y is measured outward from the plate surface and y_s is the coordinate distance within the solid material measured from the middle plane of the plate. The diffusion in the direction of motion is neglected in the governing equations in order to obtain a parabolic boundary-layer formulation which can be solved by marching in the x direction, as outlined below.

The governing equations may be nondimensionalized by using the following transformations:

$$\tilde{X} = \frac{x}{\ell} \qquad \tilde{Y} = \frac{y}{\ell}\sqrt{\tilde{\mathrm{Re}}} \qquad \tilde{\mathrm{Re}} = \frac{U_s \ell}{\nu} \qquad U = \frac{u}{U_s}$$

$$\tilde{V} = \frac{v}{U_s}\sqrt{\tilde{\mathrm{Re}}} \qquad \theta = \frac{T - T_\infty}{T_0 - T_\infty} \qquad Y_s = \frac{y_s}{D} \tag{16}$$

where ℓ is an arbitrary length scale in the x direction for the semi-infinite surface under consideration and $\tilde{\mathrm{Re}}$ is the Reynolds number. The governing equations for steady state are as follows.

For the fluid:

$$\frac{\partial U}{\partial \tilde{X}} + \frac{\partial \tilde{V}}{\partial \tilde{Y}} = 0 \tag{17a}$$

$$U \frac{\partial U}{\partial \tilde{X}} + \tilde{V} \frac{\partial U}{\partial \tilde{Y}} = \frac{\partial^2 U}{\partial \tilde{Y}^2} \tag{17b}$$

$$U \frac{\partial \theta}{\partial \tilde{X}} + \tilde{V} \frac{\partial \theta}{\partial \tilde{Y}} = \frac{1}{\mathrm{Pr}} \frac{\partial^2 \theta}{\partial \tilde{Y}^2} \tag{17c}$$

Within the plate:

$$\frac{\partial \theta}{\partial \tilde{X}} = \frac{1}{\tilde{\mathrm{Pe}}} \frac{\partial^2 \theta}{\partial Y_s^2} \tag{18a}$$

where

$$\tilde{\text{Pe}} = \frac{U_s D^2}{\alpha_s \ell} \tag{18b}$$

Here, ℓ is chosen as the characteristic dimension in keeping with the usual formulation for boundary-layer flow over a semi-infinite surface. Clearly, the equations are obtained in a very simplified form with this nondimensionalization.

The boundary conditions for Eqs. (17) and (18) are obtained from the no-slip conditions at the surface of the plate, the quiescent ambient medium far from the flow, and the continuity of heat flux at the surface. These are

$$\begin{aligned}
&\text{At } \tilde{X} = 0, && \tilde{Y} > 0: U = \tilde{V} = \theta = 0 \\
& && 0 \le Y_s \le 1: \theta = 1 \\
&\text{At } \tilde{X} > 0, && \tilde{Y} = 0 \text{ or } Y_s = 1: U = 1,\ \tilde{V} = 0,\ \frac{\partial \theta}{\partial Y_s} = \frac{\partial \theta}{\partial \tilde{Y}} R \sqrt{\frac{\tilde{\text{Pe}}}{\text{Pr}}} \\
& && \tilde{Y} \to \infty: U \to 0,\ \theta \to 0 \\
& && Y_s = 0: \frac{\partial \theta}{\partial Y_s} = 0
\end{aligned} \tag{19}$$

where $R = [(k\rho C)_f/(k\rho C)_s]^{1/2}$ is termed the property parameter. The last condition arises from symmetry about the $\tilde{X}$ axis. No boundary conditions are needed downstream, since the governing equations are parabolic in $\tilde{X}$.

The equations for the parabolic formulation, Eqs. (17) and (18), along with the boundary conditions given by Eq. (19), are easily solved by marching in the $\tilde{X}$ direction, using a finite-difference technique such as the Crank-Nicholson method, which has a truncation error of order $(\Delta\tilde{X})^2$ and $(\Delta\tilde{Y})^2$, where $\Delta\tilde{X}$ and $\Delta\tilde{Y}$ are the grid sizes in the two directions. A grid employing $\Delta\tilde{X} = 0.05$ and $\Delta\tilde{Y} = 0.2$ was found to yield accurate results, when the finite-difference results were compared with the similarity solution for an isothermal plate [24]. Also, a further refinement of the grid did not significantly affect the numerical results.

Some of the typical results obtained from this formulation are shown in Figs. 10–12, for various values of the governing parameters $\tilde{\text{Pe}}$, Pr, and R. It is seen from Fig. 10*a* that, for very small values of R, which would arise for, say, a highly conducting plate material, the centerline temperature of the plate drops very gradually and the downstream penetration of thermal effects is large, as expected. Such results can therefore be used to determine the distance needed for the material to cool to a desired level. Figure 10*b* shows the computed temperature profiles at $\tilde{X} = 2.4$ within the plate and the fluid. For the parametric values employed, the temperature distribution within the material is fairly uniform. A larger temperature variation arises for a material of smaller thermal conductivity. The temperature gradient at the surface is larger for smaller values of R, indicating higher heat transfer coefficients for a given fluid if the

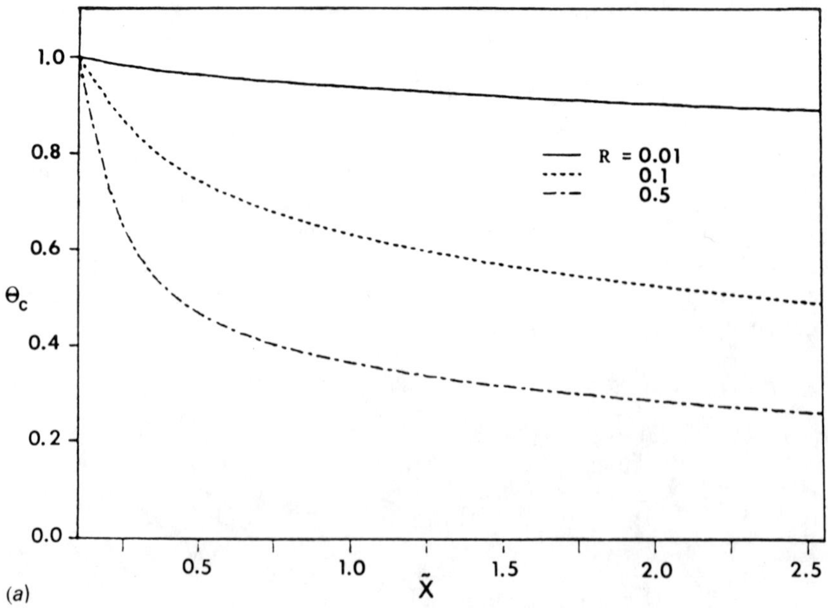

(a)

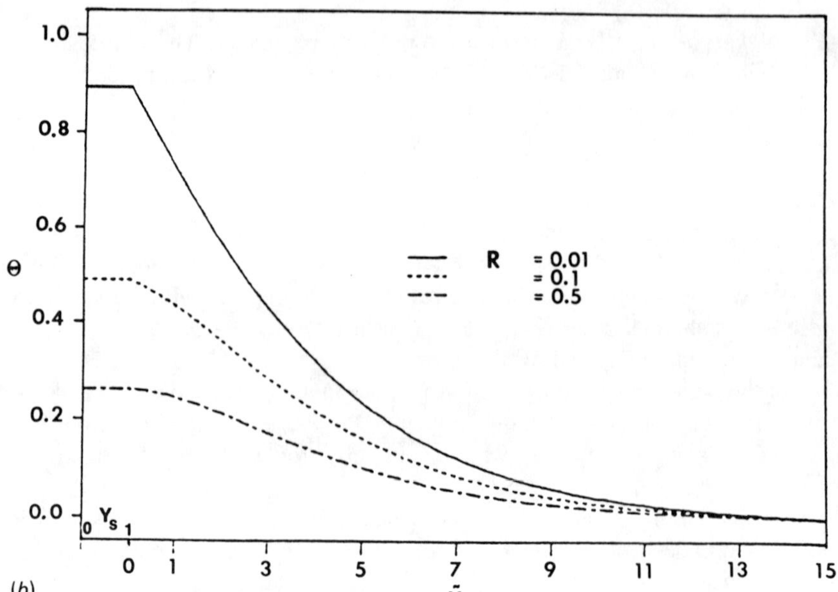

(b)

Figure 10 (*a*) Downstream variation of the centerline temperature θ_c for the conjugate boundary-layer formulation at Pr = 0.7 and $\tilde{\mathrm{Pe}}$ = 0.01. (*b*) Temperature variation along the *y* direction within the plate and the fluid at $\tilde{X}$ = 2.4.

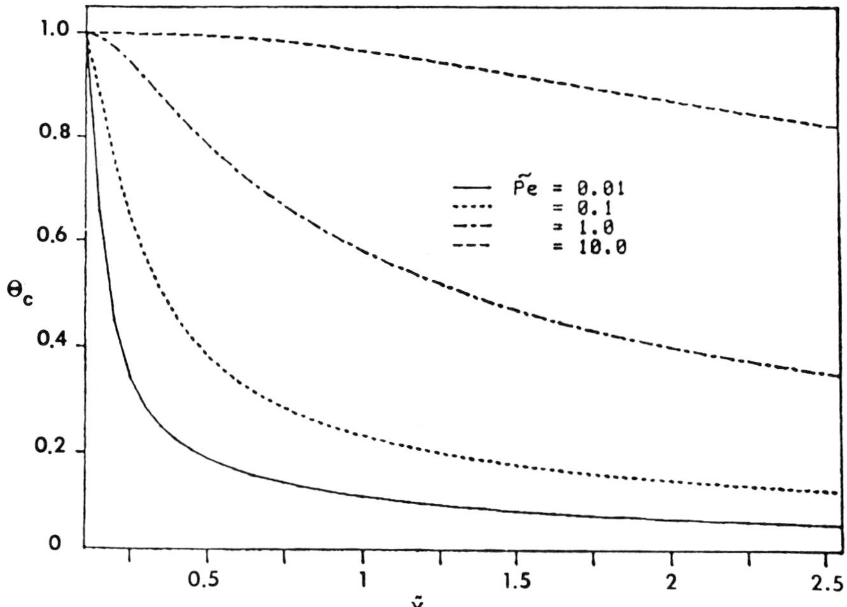

(a)

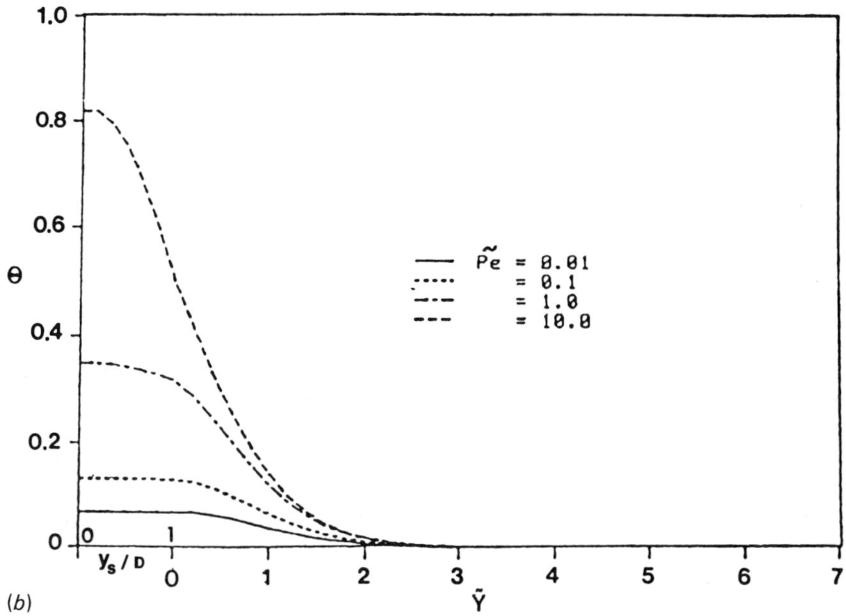

(b)

Figure 11 (*a*) Downstream variation of the centerline temperature θ_c for a heated glass plate moving in water, assuming a two-dimensional boundary-layer flow and conjugate transport; Pr = 7.0 and R = 1.092. (*b*) Temperature distributions in the y direction at $\tilde{X}$ = 2.5.

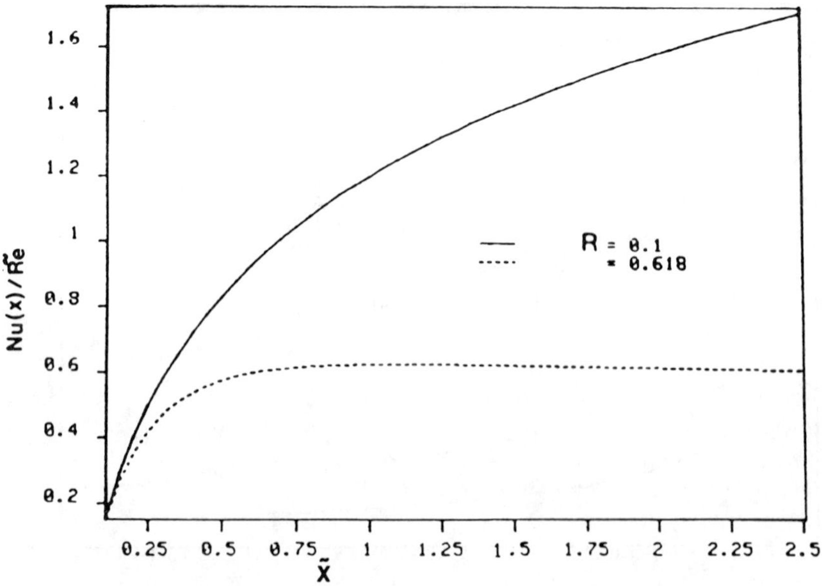

Figure 12 Downstream variation of the Nusselt number Nu(x) at $\tilde{\text{Pe}} = 0.1$ in the conjugate boundary-layer-flow case.

quantity $(k\rho C)_s$ of the moving material is increased. Again, this trend reflects the higher temperature levels downstream for smaller R.

The results for a typical case of a moving glass plate, with water as the fluid, are shown in Fig. 11. As seen earlier in Sec. 2, the downstream temperature decay decreases as the Peclet number $\tilde{\text{Pe}}$ increases. Also, the temperature distribution within the material is more uniform, at a given downstream location, for smaller $\tilde{\text{Pe}}$, indicating the result of a longer exposure of the plate to the ambient at the corresponding smaller velocity U_s. Thus, to achieve a desired cooling rate or temperature uniformity within the material, the speed of the moving material U_s may be varied. Basically, for a plate of given material, thickness, and inlet temperature T_0, the only variables are the fluid and the speed U_s. Temperature uniformity across the material is particularly important in the thermal processing of glass, ceramics, and other such brittle materials. The rate of temperature decay is again important in determining the distance that must be provided for the material to reach a given level, which is usually the temperature by which the process is complete and further cooling is not needed.

Figure 12 shows the downstream variation of the local Nusselt number $\text{Nu}(x) = h(x)x/k_f$, where $h(x) = q/(T_s - T_\infty)$. Here, q is the surface heat transfer flux, which is obtained from the calculated temperature distributions as $q = -k_f(\partial T/\partial y)_0$, where the temperature gradient is computed at the surface, $y = 0$. It is seen that Nu(x) becomes nearly constant for $\tilde{X} > 0.6$ for $R = 0.618$, at the $\tilde{\text{Pe}}$ and Pr values considered. However, Nu(x) is larger at $R = 0.1$ and does not become constant downstream over the $\tilde{X}$ range considered. A decrease in R here implies an increase in $(k\rho C)_s$, which increases the downstream penetration of the thermal field and, thus, the temperature level. This is reflected in a higher Nu(x), which continues to vary downstream.

The parabolic conjugate formulation presented here yields the temperature variation in the moving material as well as the flow and thermal fields in the fluid. Thus, it focuses on the main physical variables in the problem, namely, the fluid and the speed U_s for a plate of given material and thickness. The heat transfer coefficient at the surface is not assumed but is obtained from the solution. Therefore, a more realistic simulation than those presented in Secs. 2 and 3 of the thermal transport from a moving material undergoing thermal processing is obtained. The formulation can easily be extended to axisymmetric cases, such as transport from a moving cylindrical rod. However, this parabolic formulation neglects the axial diffusion effects, which are seen to be important near the inlet, i.e., for small x, due to the large temperature gradients that arise; see Figs. 10 and 11. Also, the thermal conditions in the region upstream of the point of emergence and the upstream penetration of thermal effects cannot be considered by this approach. Similarly, the fluid flow in the region close to $x = 0$ is not simulated accurately. The regions near dies or rollers are often designed differently, and the flow and heat transfer in these regions are important for the thermal process to which the material is subjected. All these considerations make it imperative to solve the full equations, particularly in the region close to the inlet.

4.2 Solution of the Full Equations (Elliptic Formulation)

The need for the solution of the full governing equations for the fluid flow as well as for the heat transfer within the moving material, particularly in regions close to the slot, $x = 0$, from which the material emerges, has been clearly indicated from the limitations of the boundary-layer, or parabolic, formulation. However, this approach is a complicated one and, as such, only a few specific cases have been considered in the literature [40, 41], leaving the problem open for a wide range of future work.

The full governing equations, including the transient term, are as follows.

For the fluid:

$$\nabla \cdot \overline{V} = 0 \tag{20a}$$

$$\frac{\partial \overline{V}}{\partial \tau'} + \overline{V} \cdot \nabla \overline{V} = -\frac{1}{\rho_f} \nabla p + \nu \nabla^2 \overline{V} \tag{20b}$$

$$\frac{\partial T}{\partial \tau'} + \overline{V} \cdot \nabla T = \alpha_f \nabla^2 T \tag{20c}$$

For the solid material:

$$\frac{\partial T}{\partial \tau'} + U_s \frac{\partial T}{\partial x} = \alpha_s \nabla_s^2 T \tag{21}$$

where $\overline{V}$ is the velocity vector in the flow, p is the local pressure, and ∇, ∇_s are the gradient operators in the fluid and in the solid. Figure 13 shows the coordinate system,

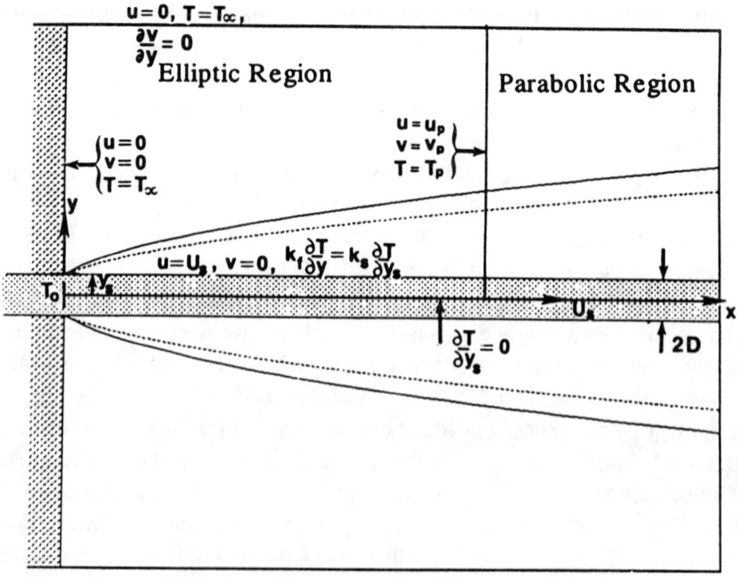

Figure 13 The computational domain for the solution of the full equations, indicating the elliptic and parabolic regions.

along with the various boundary conditions. Far downstream, the axial diffusion terms are expected to become small and the parabolic formulation is employed. Very close to the slot, the full equations are solved. The two formulations are matched at the interface, as discussed in detail here.

The boundary conditions are similar to those employed for the parabolic formulation, except that $T = T_0$ may be imposed at an upstream location, $x = -x_b$. Also, the value of x_b may be increased until the downstream numerical results are largely unaffected by a further increase. Within the region from $x = -x_b$ to $x = 0$, the surface of moving material may be taken as adiabatic. Thus,

$$\begin{aligned} &\text{For } -x_b \le x \le 0, && y_s = D: \frac{\partial T}{\partial y_s} = 0 \\ &\text{For } x = -x_b, && 0 \le y_s \le D: T = T_0 \end{aligned} \tag{22}$$

Similarly, other upstream thermal conditions may be employed for different processes. The other boundary conditions are given in Fig. 13. However, it must be noted that the normal velocity component v is not taken as zero at the outer edge of the computational domain in order to allow for entrainment, and $\partial v/\partial y = 0$ is used instead to provide the appropriate conditions there [42].

For a numerical solution of the full equations, these are recast in terms of the vorticity ω' and the stream function ψ', by taking the curl of the momentum equation. The nondimensionalization used is given by

$$X = \frac{x}{D} \qquad Y = \frac{y}{D} \qquad Y_s = \frac{y_s}{D} \qquad U = \frac{u}{U_s} \qquad V = \frac{v}{U_s}$$

$$\theta = \frac{T - T_\infty}{T_0 - T_\infty} \qquad \psi = \frac{\psi'}{U_s D} \qquad \omega = \frac{\omega' D}{U_s} \qquad \tau = \frac{\tau' U_s}{D} \tag{23}$$

$$\text{Re} = \frac{U_s D}{\nu} \qquad \text{Pe} = \frac{U_s D}{\alpha_s}$$

where

$$u = \frac{\partial \psi'}{\partial y} \qquad v = -\frac{\partial \psi'}{\partial x} \qquad \omega' = \frac{\partial v}{\partial x} - \frac{\partial u}{\partial y}$$

The dimensionless equations, in the vorticity stream-function formulation, are obtained for laminar flow of the fluid as

$$\frac{\partial \omega}{\partial \tau} + \nabla \cdot (\overline{V}\omega) = \frac{1}{\text{Re}} \nabla^2 \omega \tag{24a}$$

$$\nabla^2 \psi = -\omega \tag{24b}$$

$$\frac{\partial \theta}{\partial \tau} + \nabla \cdot (\overline{V}\theta) = \frac{1}{\text{Re Pr}} \nabla^2 \theta \tag{24c}$$

The energy equation for the plate is

$$\frac{\partial \theta}{\partial \tau} + \frac{\partial \theta}{\partial X} = \frac{1}{\text{Pe}} \nabla_s^2 \theta \tag{25}$$

where the conservative formulation has been used for the convection terms to obtain $\nabla \cdot (\overline{V}\omega)$ and $\nabla \cdot (\overline{V}\theta)$, instead of $\overline{V} \cdot \nabla \omega$ and $\overline{V} \cdot \nabla \theta$, respectively, by employing the continuity equation. This formulation gives local conservation and, thus, generally leads to more accurate results [27]. Similarly, the boundary conditions are obtained in terms of ω and ψ.

For the solution of the full equations, the transient vorticity equation, Eq. (24*a*), is solved along with the stream-function equation, Eq. (24*b*). With inputs from this solution, the energy equation, Eq. (24*c*), is solved, along with Eq. (25), which is coupled with the temperature field in the flow through the boundary conditions. Several numerical methods are available for solving these equations. For the transient vorticity transport equation, the ADI method is among the most efficient finite-difference methods [26]. The stream-function equation is a Poisson equation and may be solved by iterative methods, such as the successive overrelaxation (SOR) method, or direct solvers such as the cyclic reduction method [27]. The solutions from the elliptic and parabolic regions are matched at the interface, whose location is varied numerically to ensure that the results are essentially independent of the chosen position. The initial conditions are assumed to be zero flow and uniform ambient temperature in the fluid. Time marching is continued until steady-state conditions, as indicated by convergence criteria of order 10^{-3} on the temperature and the velocity gradients, with respect to

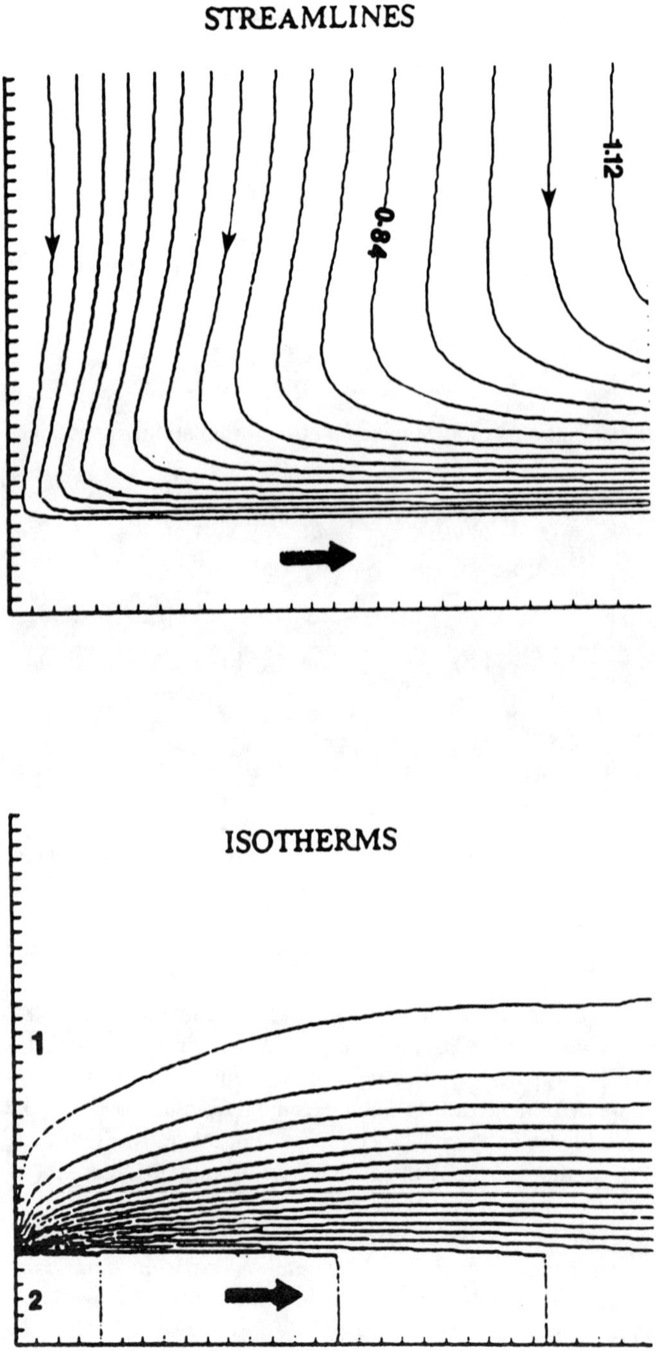

Figure 14 Resulting streamlines and isotherms for a heated aluminum plate moving in air, as obtained from the solution of the full equations at Pe = 0.05, Pr = 0.7, $k_f/k_s = 1.27 \times 10^{-4}$, and Re = 20.0.

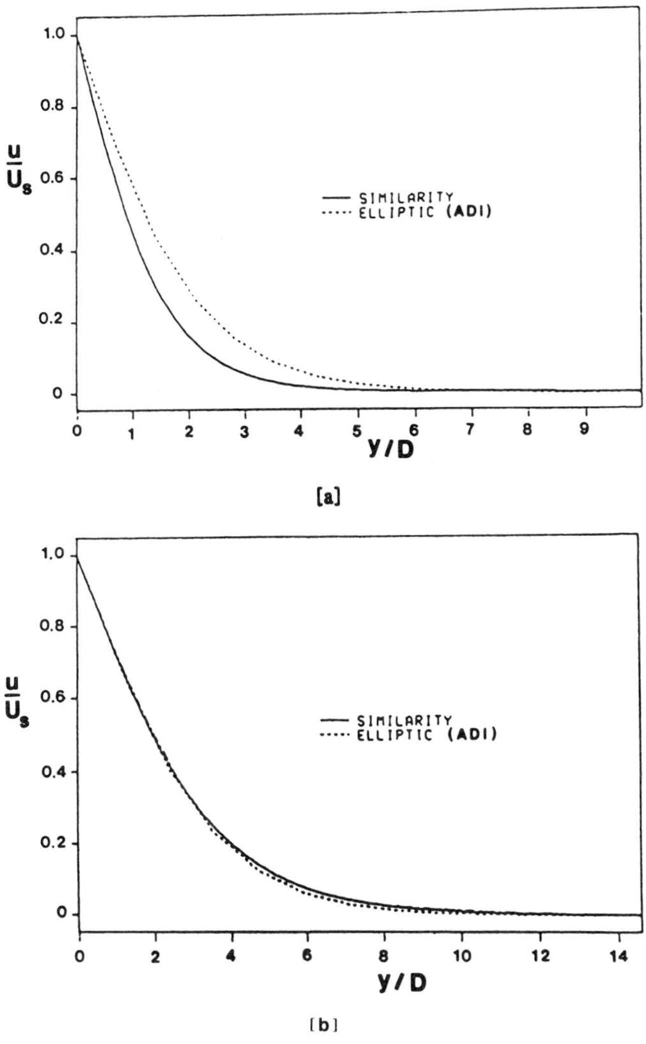

Figure 15 A comparison between the results from the boundary-layer similarity formulation and those from the elliptic formulation, in terms of the variation of u/U_s along y at (*a*) $x/D = 2.5$ and (*b*) $x/D = 12.5$.

time, are attained. The various numerical parameters, such as grid size, convergence criteria, etc., are varied till the numerical results are found to be essentially independent of the values selected. Several results were obtained in [40, 41] using this scheme. Some of the typical physical trends are discussed here.

Figure 14 shows the streamlines and the isotherms obtained for a heated aluminum plate being cooled in air at typical values of the governing parameters. The cold ambient fluid is drawn toward the moving surface, increasing the heat transfer rate from the surface beyond that for a stationary plate in a parallel fluid stream, as discussed earlier. The temperature distribution within the plate is almost uniform due to the high thermal conductivity of aluminum. The boundary-layer thickness is seen to grow in the direction of motion. The isotherms indicate that large temperature gradients exist near the slot, implying large thermal stresses and structural changes in this region. It is also seen that the ambient fluid is drawn toward the slot and, then, it flows along the moving surface. This effect is a consequence of the large pressure gradients directed toward the slot due to the flow. This results in a small reverse flow close to the die. Far downstream, this effect dies down and the flow approaches the characteristics of a boundary-layer flow. Similar trends have been obtained from the solution of the full governing equations in other flows [42]. These results clearly indicate the importance of an accurate simulation of the transport processes near the slot for practical materials processing systems.

A comparison between the results from the boundary-layer similarity formulation for a moving isothermal surface and those from a solution of the full equations is shown in Fig. 15, in terms of the calculated variation of U with Y at two X locations. It is again seen that the elliptic, or nonboundary-layer, effects are important near the slot, as expected. These effects decay downstream and the difference between the results from the two formulations becomes small at large X. This implies that the boundary-layer formulation of the preceding subsection is quite satisfactory for calculating the transport from the moving surface far downstream. However, if interest lies in the region close to the die or slot, the elliptic formulation is needed.

Another important consideration, from a fundamental as well as a practical standpoint, is the imposition of the thermal conditions at the slot or die. As mentioned earlier, the thermal field penetrates upstream due to diffusion and the temperature at the slot, $X = 0$, is a consequence of upstream conditions and downstream heat transfer.

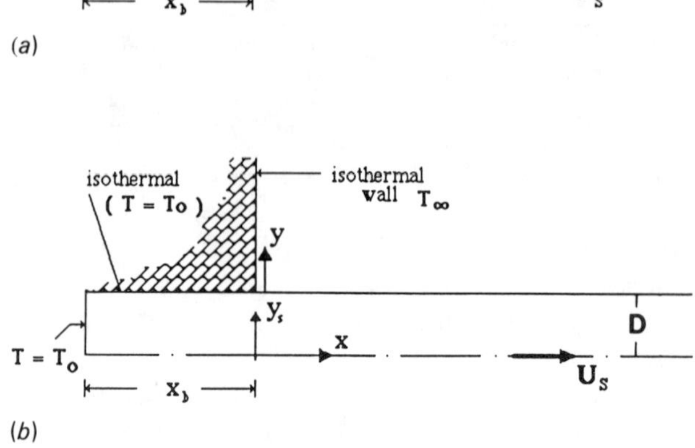

Figure 16 Two circumstances in the simulation of heat transfer conditions upstream of the point of emergence of the plate: (*a*) adiabatic surface boundary condition; (*b*) isothermal surface boundary condition.

Due to the associated transport processes, the temperature distribution over the plate cross section is not uniform and the imposition of a given, uniform temperature at $X = 0$ is not realistic with respect to practical processes. In order to simulate the upstream conditions, two possible circumstances are shown in Fig. 16. The temperature condition $\theta = 1.0$ is applied at $X = -X_b$, where the value of X_b is increased until the numerical results become essentially independent of a further increase. This simulates the uniform temperature in a furnace, oven, or extruder. The two cases shown simulate adiabatic and uniform temperature conditions at the plate surface in the upstream region. Obviously, the actual conditions in the upstream region for a given process may also be simulated for a more accurate representation. However, the two limiting cases shown in Fig. 16 allow one to study the basic trends by computing the drop in the temperature level at the slot due to downstream heat transfer processes. Also, the total thermal energy entering the material at the slot is due only to convection in the moving material, at temperature T_0, far upstream. This again represents the physical conditions in several practical processes closely.

Figures 17*a* and 17*b* show the effect of the upstream thermal conditions on the numerical results. The effect of varying the location $X = -X_b$, where the temperature boundary condition is applied, is seen in Fig. 17*a*. Here, the nondimensional temperature is taken as $\hat{\theta}$, where $\hat{\theta} = T/T_\infty$, in order to include radiative loss from the surface. The condition at $X = -X_b$ is taken as $\hat{\theta} = T/T_\infty = 2.0$ [41]. As X_b increases to large values, the results become independent of a further increase, giving the behavior for $X_b \to \infty$, which is representative of the situation where the material emerges from a long furnace, oven, or extruder. In many practical situations, X_b is finite, representing, say, a heated region for the thermal processing of the material. Figure 17*b* shows the effect of the boundary conditions in the upstream region $-X_b \leq X \leq 0$ on the results, taking $X_b = 10.0$. The adiabatic and the isothermal conditions at the surface, as shown in Fig. 16, are considered. The centerline temperature is higher for the isothermal condition, implying that the back-conduction effects are smaller in this case. Various other boundary conditions arise in the upstream region, depending on the thermal process under consideration. For instance, a uniform heat flux condition may be applied at the plate surface for radiative heating in an electric furnace. These upstream conditions will influence the length of the cooling trough, the distance between the die and the spool, cooling rates needed, etc. Clearly, this is an area in which a detailed future effort is needed to provide both a better understanding of the underlying processes near the point of emergence of the material and quantitative inputs for design.

5 OTHER EFFECTS

The formulation and the numerical solution of the full equations for the conjugate problem outlined in the preceding section can be employed for simulating a wide variety of thermal processes that involve a continuously moving material. In this section, the results for a few additional effects which are important in several practical situations and also from a fundamental consideration of the transport process, are presented.

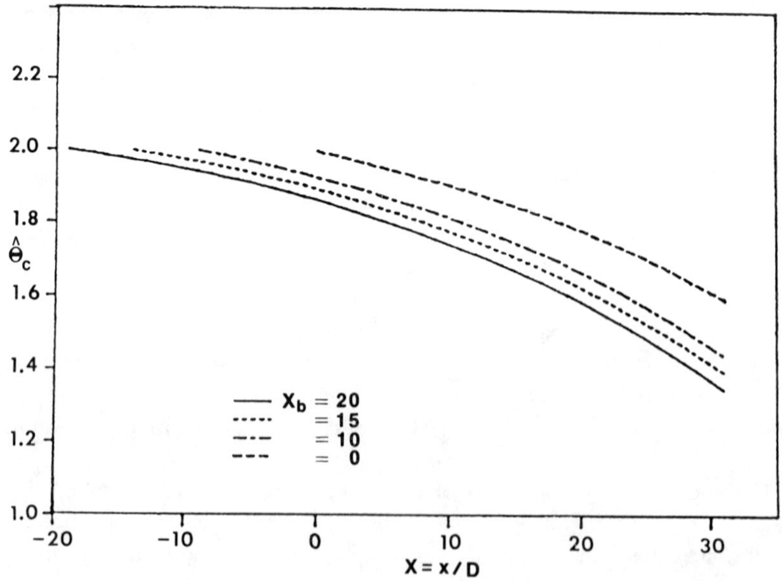

(*a*)

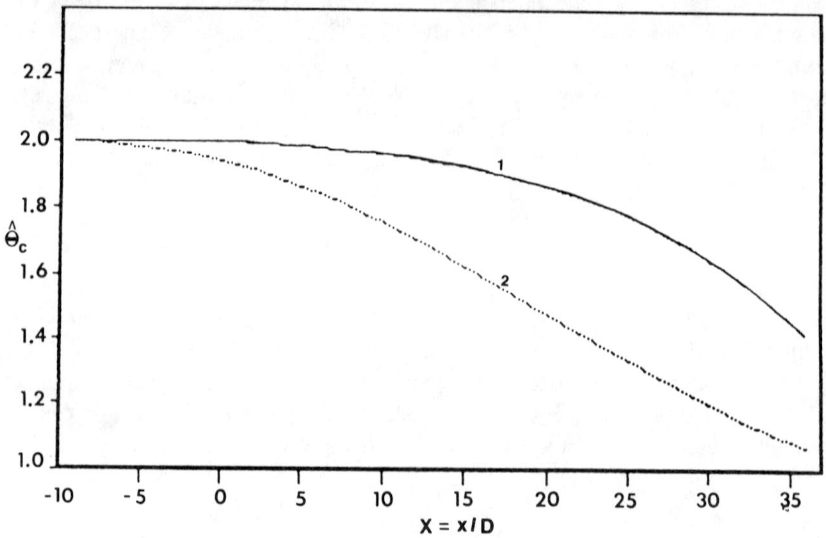

(*b*)

Figure 17 (*a*) Centerline temperature variation for the case when the temperature $\hat{\theta}_0$ is specified at different locations upstream, $X = -X_b$. The plate surface is assumed to be adiabatic in the region $-X_b \leq X \leq 0$. Here, $Pr = 0.7$, $Pe = 0.05$, $Re = 20.0$, $Gr = 0.0$, $k_f/k_s = 1.27 \times 10^{-4}$, and $\hat{\theta}_0 = T_0/T_\infty = 2.0$. (*b*) Effect of adiabatic and isothermal surface boundary conditions in the region $-X_b \leq X \leq 0$ on the downstream axial temperature variation: (1) $\hat{\theta} = \hat{\theta}_0$ at $Y_s = 1$, $-X_b \leq X \leq 0$; (2) $\partial\hat{\theta}/\partial Y_s = 0$ at $Y_s = 1$, $-X_b \leq X \leq 0$.

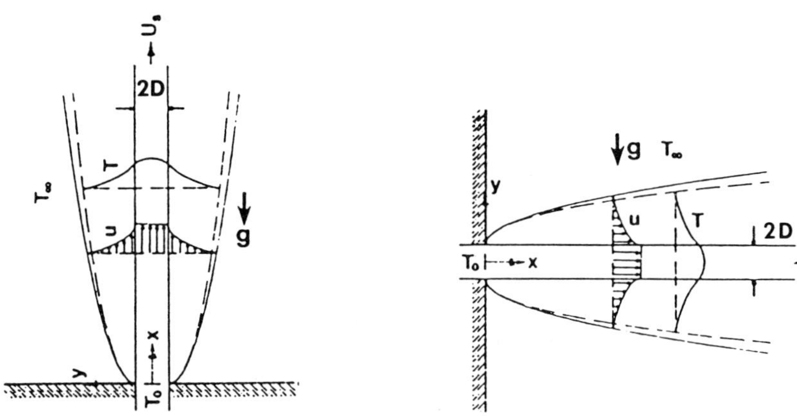

Figure 18 Flow geometry near the die showing the plate, the slot and the walls: (*a*) plate moving vertically upward; (*b*) plate moving horizontally.

5.1 Thermal Buoyancy

The inclusion of thermal buoyancy in the calculations of the fluid flow induced by a moving, heated surface is important in many applications, particularly at low speeds U_s and high temperature levels T_0 at the inlet. Figure 18 shows the sketches of the flows for vertically and horizontally moving plates, along with the coordinate system. If buoyancy effects are important in the flow, the orientation will influence the flow and, consequently, the heat transfer.

The inclusion of buoyancy gives rise to an additional term in the momentum equation, if Boussinesq approximations, which neglect the effect of density change in the continuity equation, are applied [43, 44]. Thus, an additional term $-\bar{g}\beta(T - T_\infty)$ arises on the right-hand side of Eq. (20*b*), where $\bar{g}$ is the gravitational vector and β is the coefficient of thermal expansion of the fluid. Therefore, for the vertical case of Fig. 18*a*, $\bar{g}$ is directed in the negative x direction and an additional term $g\beta(T - T_\infty)$, where g is the magnitude of gravitational acceleration, arises in the x component of the momentum equation.

In nondimensionalized terms, the inclusion of thermal buoyancy effects leads to Eq. (24*a*), becoming

$$\frac{\partial \omega}{\partial \tau} + \nabla \cdot (\bar{V}\omega) = \frac{1}{\text{Re}} \nabla^2 \omega - \frac{\text{Gr}}{\text{Re}^2} \frac{\partial \theta}{\partial Y} \tag{26}$$

and

$$\frac{\partial \omega}{\partial \tau} + \nabla \cdot (\bar{V}\omega) = \frac{1}{\text{Re}} \nabla^2 \omega + \frac{\text{Gr}}{\text{Re}^2} \frac{\partial \theta}{\partial X} \tag{27}$$

for the plate moving vertically upward and for the flow over the upper surface when it moves horizontally, respectively. The sign of the additional term in Eq. (26) is reversed for a plate moving downward. Similarly, the flow adjacent to the lower surface of a horizontally moving plate is governed by Eq. (27) with the sign of the buoyancy term reversed. Here, Gr is the Grashof number, defined as

$$\mathrm{Gr} = \frac{g\beta(T_0 - T_\infty)D^3}{\nu^2} \tag{28}$$

For a heated, moving surface, buoyancy effects were considered in [34, 35], as mentioned earlier. The conjugate problem, with buoyancy effects, was studied in [40, 41, 45]. The results from these latter studies are presented here.

The inclusion of buoyancy couples the momentum and energy equations for the fluid flow, through the buoyancy term. This implies that all the governing equations must be solved together, since conjugate conditions at the material surface couple the energy equations for the fluid and the solid. The problem thus becomes more complicated than if the buoyancy effects are neglected, since the flow is independent of the thermal field in that case and is the same for different thermal conditions in the process. The flow was calculated first and the results employed in solving the coupled energy equations. With thermal buoyancy effects included, all the equations must be solved simultaneously. However, the numerical techniques that may be applied are the same as before. At each time step, the vorticity transport, the stream function, and the energy equations are solved to obtain the flow and thermal fields at the next time step. However, if the thermal conditions are varied, the flow will be affected and has to be solved again for the new conditions. Some of the important physical characteristics of this circumstance are discussed here.

Figure 19 shows the computed results for a plate moving vertically upward, at different values of the mixed-convection parameter $\mathrm{Gr/Re}^2$. In this case, the buoyancy force aids the flow induced by the plate motion. This results in a flow velocity that exceeds the plate speed U_s, being much larger than U_s at large values of $\mathrm{Gr/Re}^2$. Also shown here are the results for the case when buoyancy effects are neglected, $\mathrm{Gr/Re}^2 = 0$. A higher velocity level leads to a higher surface heat transfer, which is reflected in lower temperature levels downstream, as seen in Fig. 19*b*. For horizontal movement of the plate, the buoyancy force is normal to the plate velocity. The thermal buoyancy affects the pressure distribution, which in turn affects the flow [43]. Therefore, the effect of thermal buoyancy is smaller for the horizontal orientation, as compared to that for the vertical orientation, as confirmed by the numerical results shown in Fig. 20. Here, only the flow over the upper surface is considered for the horizontal orientation, assuming symmetry at the x axis. This condition applies if the bottom surface is insulated. If transport occurs on both top and bottom surfaces, the two regions on either side of the plate must be considered in conjunction. For the cylindrical rod circumstance, the inclusion of thermal buoyancy makes the flow asymmetric and the three-dimensional problem must be solved.

For the aiding circumstance of Fig. 19, the variation of the local Nusselt number $\mathrm{Nu}(x) = hx/k_f$ is shown in Fig. 21 for a 5-mm-thick aluminum plate moving in water at two speeds. As expected, the local Nusselt number is smaller, at a given downstream location, for the smaller speed U_s. Far downstream, the surface temperature approaches the ambient temperature T_∞ and the heat transfer rate also approaches zero. Conse-

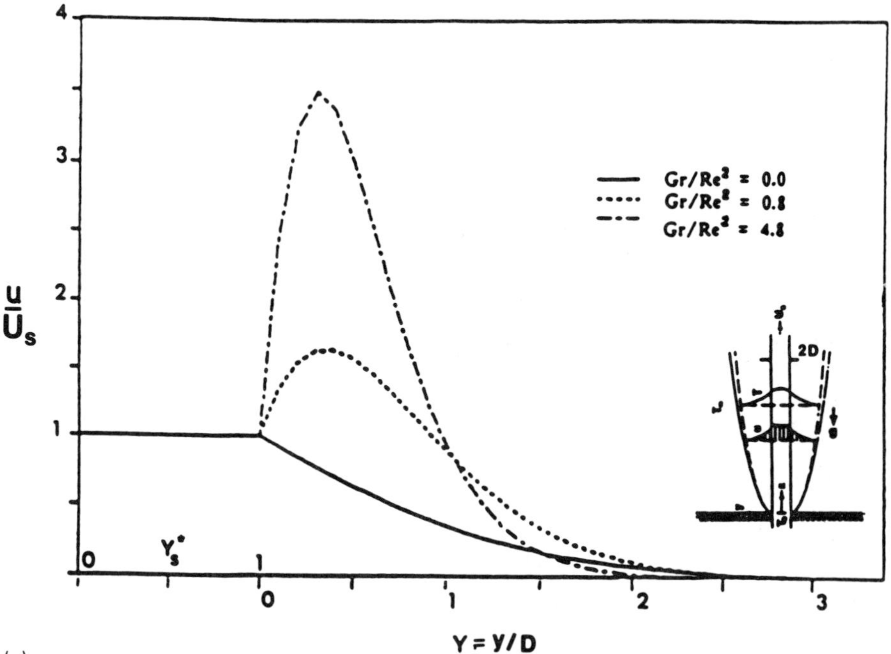

(*a*)

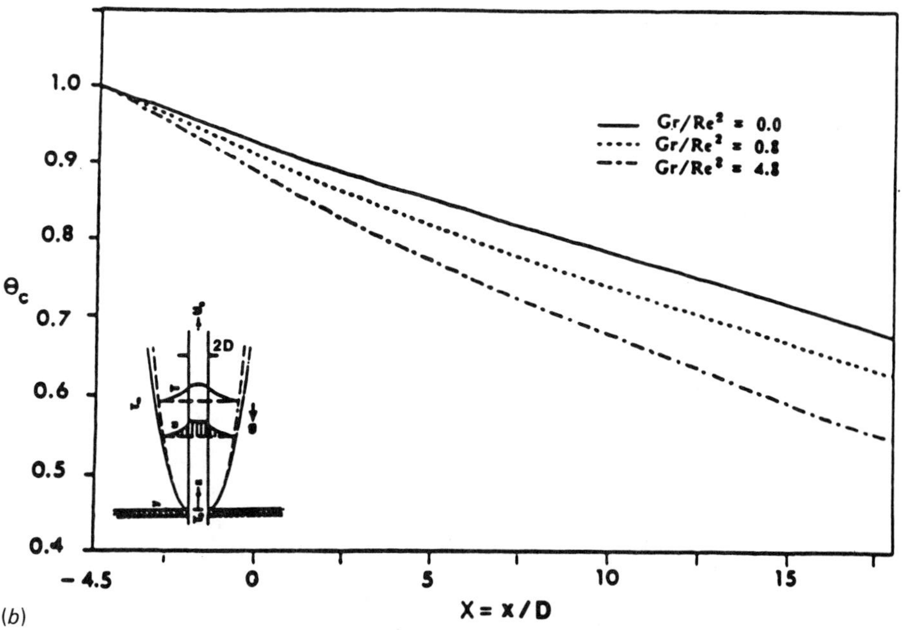

(*b*)

Figure 19 (*a*) Variation of U along Y at X = 15.0, with Pe = 0.05, Pr = 7.0, Re = 25.0, and k_f/k_s = 0.0128, for various values of the mixed-convection parameter Gr/Re^2. Here, the plate is aligned with gravity, moving vertically upward, and the elliptic formulation is employed. (*b*) Downstream temperature variation for various values of Gr/Re^2.

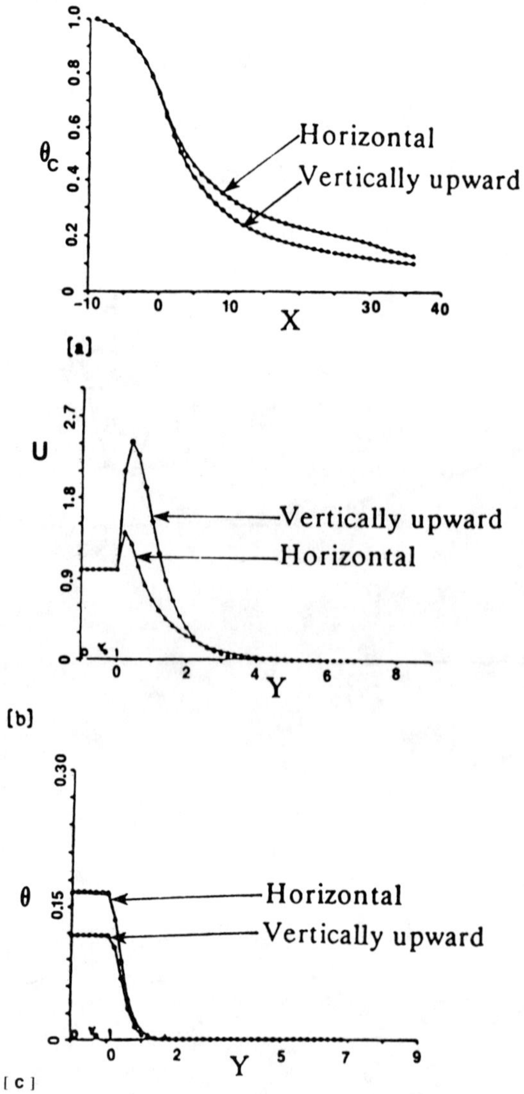

Figure 20 Comparison between the results for a plate moving vertically upward and those for a plate moving horizontally; Pe = 0.25, Re = 25.0, Gr/Re^2 = 8.0, Pr = 7.0, k_f/k_s = 0.0127. The value of X for (*b*) and (*c*) is 31.0.

quently, the computation of $h(x)$ would be in considerable error if $(T_s - T_\infty) \rightarrow 0$. In Fig. 19, the $h(x)$ variation is shown only up to the downstream location were θ drops to about 0.05. This aspect is considered again in the next subsection.

Clearly, there are many aspects that need further work with respect to buoyancy effects. Among those are flows with opposing buoyancy effects, transport from the lower surface of a horizontally moving plate, three-dimensional flow over horizontally moving cylindrical rods, and other orientations. In the opposing-buoyancy case, flow separation may arise, resulting in a very substantial effect on the local heat transfer [44]. The basic mechanisms underlying flow separation due to opposing buoyancy

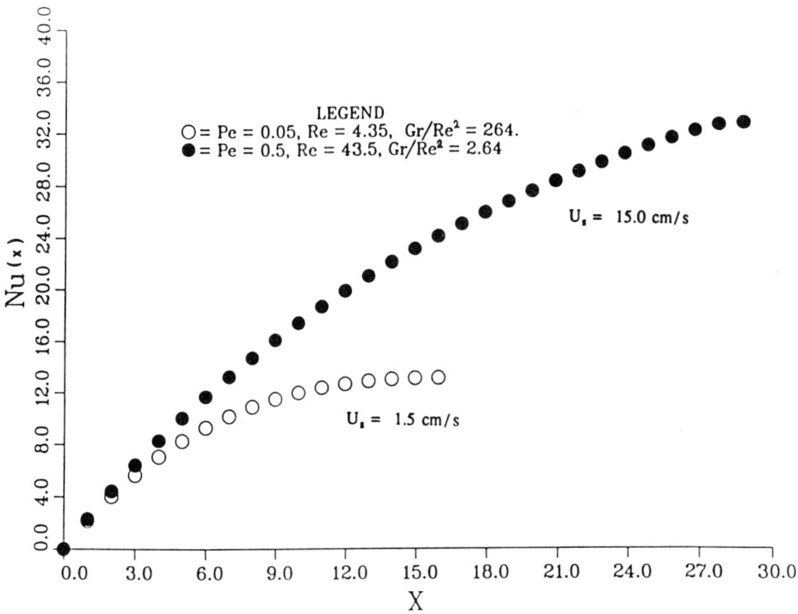

Figure 21 Downstream variation of the local Nusselt number Nu(x) for a 5-mm-thick aluminum plate moving in water at different values of plate velocity U_s.

effects also need detailed investigation. This aspect is considered in terms of experimental results later in this chapter.

5.2 Transient Effects

Very little work has been done on the time-dependent transport that arises following the onset of the process or a change in the operating conditions. The resulting transients are important in determining the time for which the desired thermal processing of the material is not achieved and, thus, finding ways to minimize these effects. These considerations are also important in the operation and control of the process. The work done by Jaluria and Singh [23] was outlined earlier, in terms of the moving leading edge of the material. Khader [46] developed a model for transient mixed convection from a vertically moving, isothermal surface. An explicit finite-difference scheme was employed to solve the transient momentum and energy boundary-layer equations.

The basic formulation and approach outlined in Sec. 4 can easily be extended to study transient effects. In fact, all the steady-state solutions of the elliptic, conjugate problem outlined here were obtained by solving the time-dependent problem, yielding steady-state results at large time. Kang et al. [45] have considered the circumstance of an initially stationary plate, at temperature T_∞, which is set into motion at speed U_∞ at time $\tau = 0$, while imposing the relevant thermal conditions as well at this time. Some of the numerical results obtained by them are discussed here.

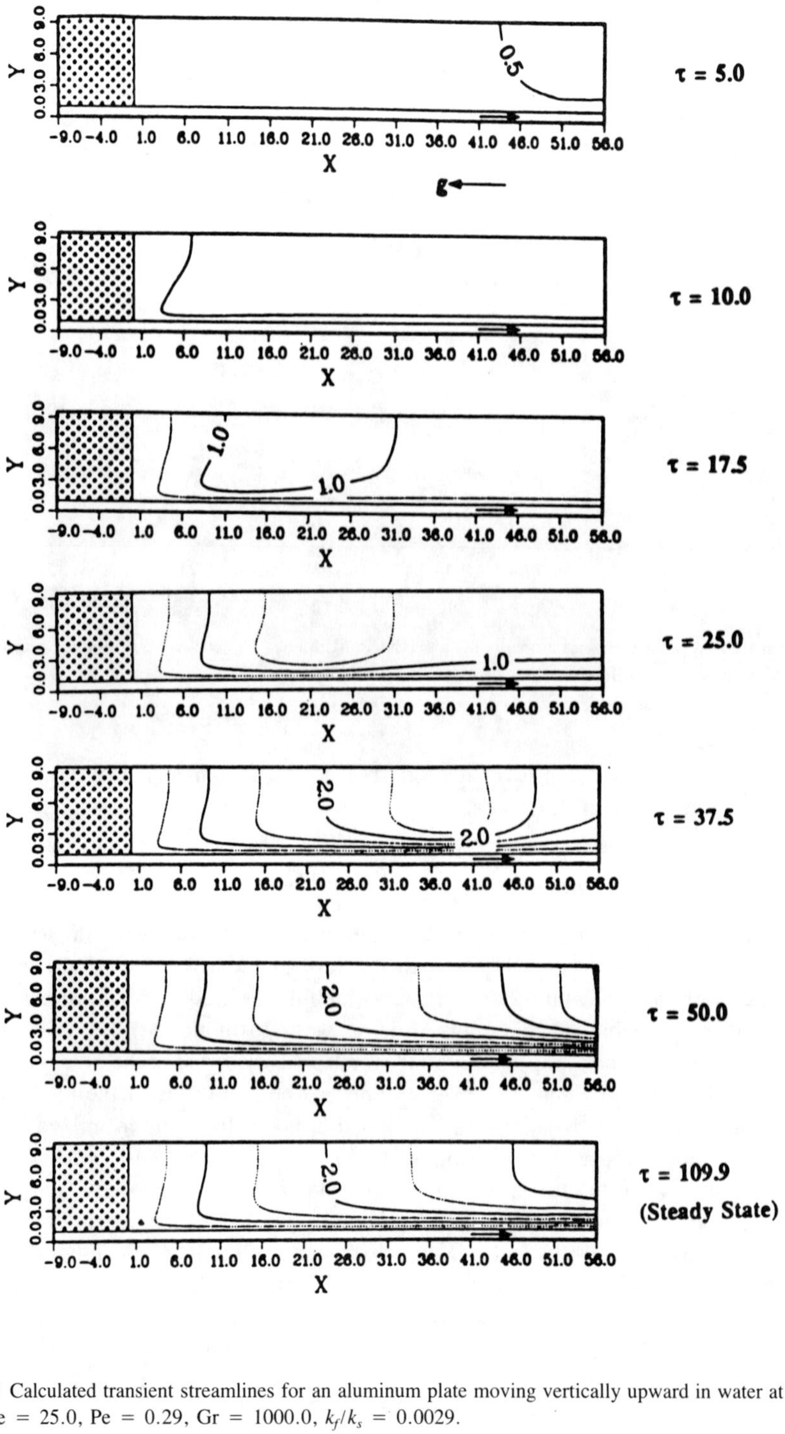

Figure 22 Calculated transient streamlines for an aluminum plate moving vertically upward in water at Pr = 7.0, Re = 25.0, Pe = 0.29, Gr = 1000.0, k_f/k_s = 0.0029.

Figure 22 shows the time-dependent stream lines for an aluminum plate moving vertically upward in water. Here, Y is measured from the plate midplane. Flow starts very close to the moving plate, following the onset of the motion. As time increases, a recirculating flow appears near the slot. This recirculation gradually moves downstream and is finally swept away, yielding the steady-state condition over the flow region under consideration. This recirculating flow region, which arises during the transient phase of the process, is due to the developing nature of the flow and the entrainment required for the moving fluid close to the plate [44]. The heated fluid adjacent to the surface draws in the neighboring fluid, which is unheated at the early stages of the process, leading to recirculation. The isotherms were found to penetrate downstream with time, as expected, leading to steady state at large time. Due to the coupled nature of the problem with buoyancy effects, the flow and temperature fields attain steady state at the same time.

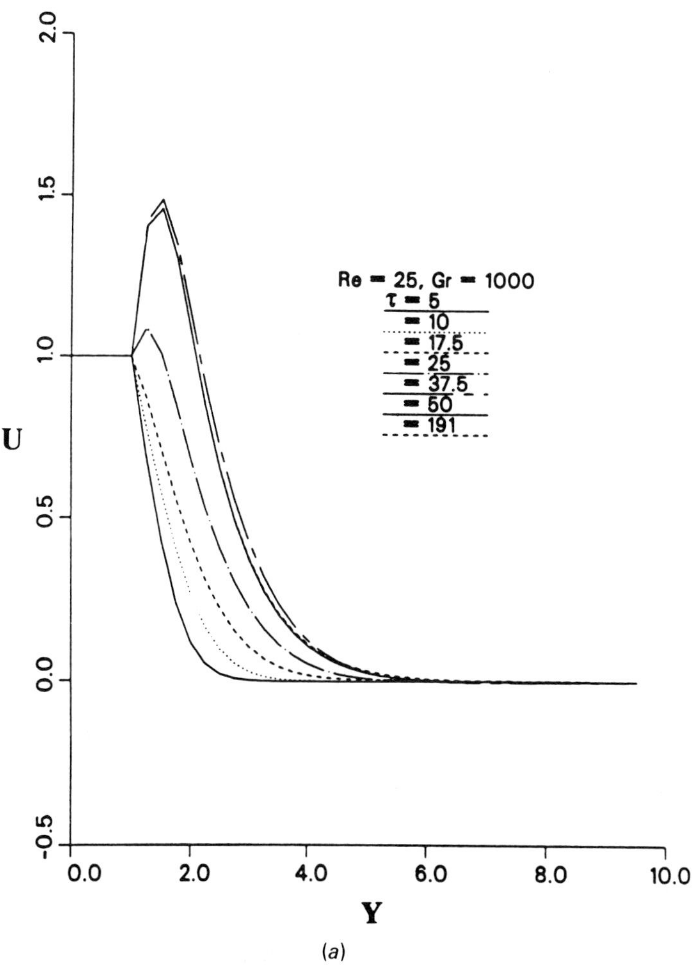

(a)

Figure 23 Transient variation of the calculated (*a*) velocity.

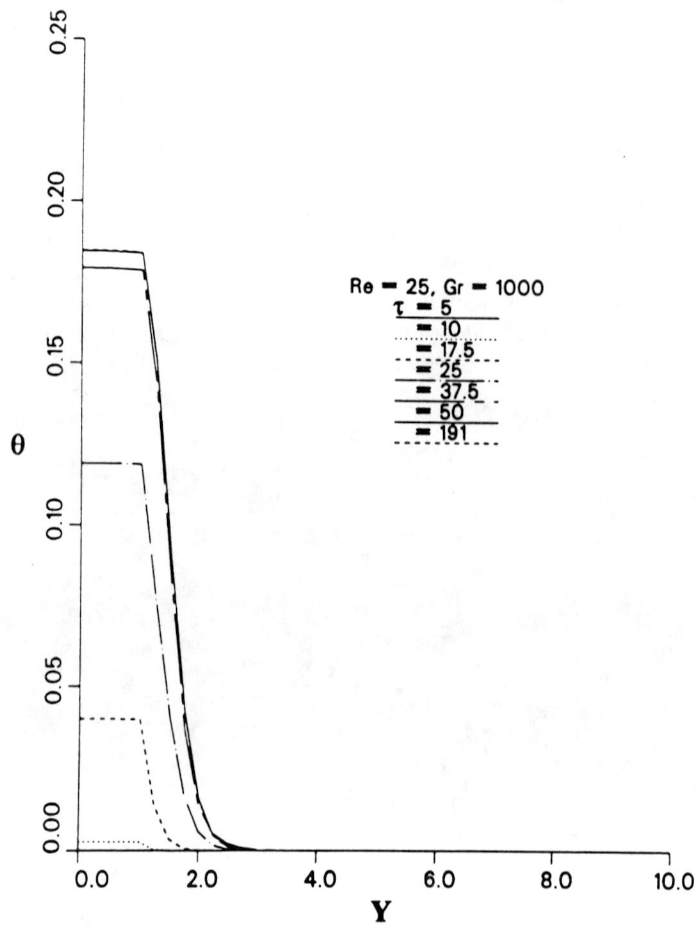

(*b*)

Figure 23 Transient variation of the calculated (*Continued*): (*b*) temperature distributions, at $X = 30.0$, for an aluminum plate moving vertically upward in water at $Pr = 7.0$, $Re = 25.0$, $Pe = 0.29$, $Gr = 1000.0$, $k_f/k_s = 0.0029$.

The corresponding velocity and temperature profiles, shown in Fig. 23, indicate the gradual variation from the no-flow, unheated situation at $X = 30$ to steady state. The maximum velocity is larger than U_s due to the aiding buoyancy effects. The temperature and velocity levels rise with time. A slight overshoot in the maximum velocity, beyond the steady-state value, was found at $\tau = 37.5$ at these conditions. The transient variation of the local Nusselt number Nu(X) with downstream distance X is shown for both the vertical and horizontal cases in Fig. 24. The Nusselt number is calculated only up to the downstream location where the surface temperature $\theta_s = 0.05$, in order to obtain accurate results. Beyond this point, the temperature and the temperature gradient are too small to yield accurate values.

During the transient process, the Nusselt number decreases downstream, reaches a minimum value, and then increases again. It is observed that the minimum arises

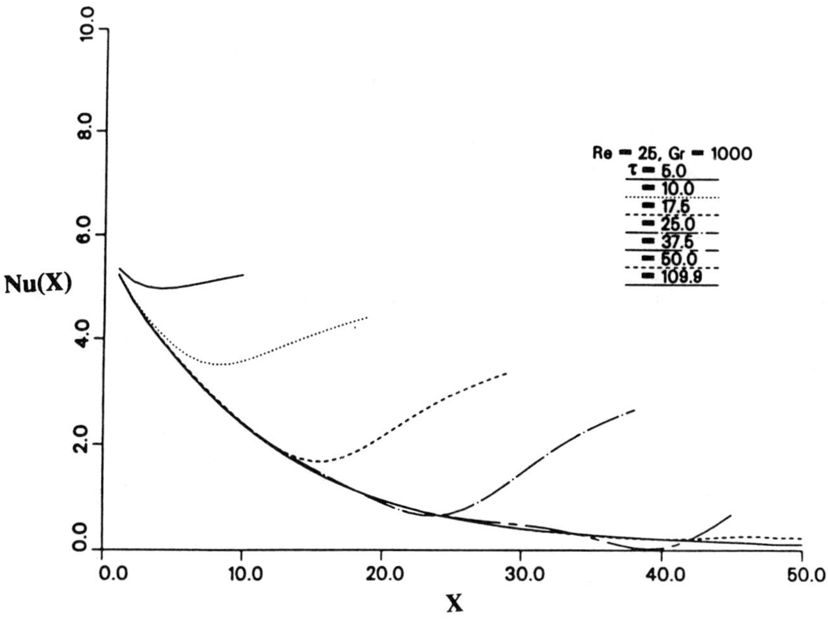

(*a*)

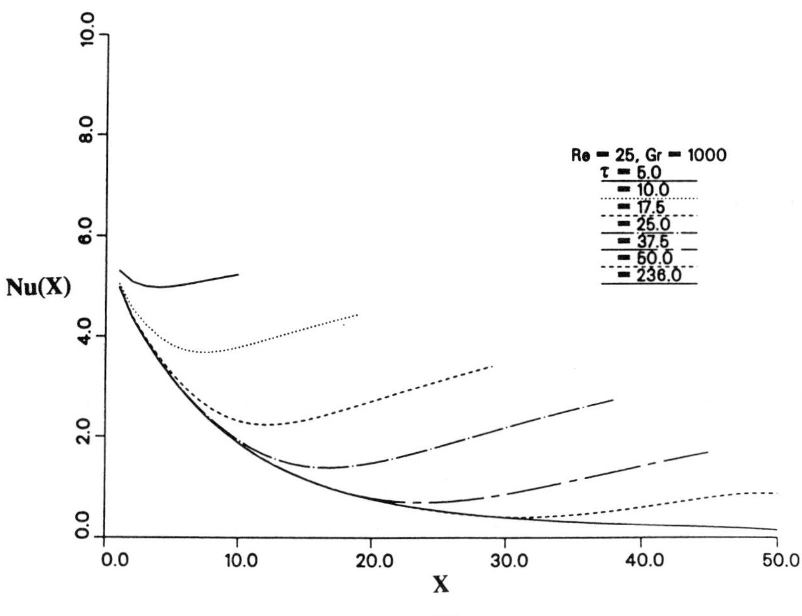

(*b*)

Figure 24 Transient downstream variation of the local Nusselt number for an aluminum plate moving (*a*) vertically upward and (*b*) horizontally in water at Pr = 7.0, Re = 25.0, Pe = 0.29, Gr = 1000.0, k_f/k_s = 0.0029.

at the recirculating flow region, as seen in Fig. 21. Thus, the recirculating flow substantially reduces the heat transfer from the plate, due to the low velocity levels encountered in this region. The heat transfer rate is found to be smaller in the horizontal orientation, as compared to that for the vertical case. This is obviously due to the orientation of the buoyancy force with respect to the plate motion. This is also reflected in lower velocity levels and higher temperature levels for the horizontal orientation, at a given location and time. Also, the transient effects are found to last longer in this case due to the flow being less vigorous than the vertical situation.

The various trends discussed above are important in practical situations. Of particular interest are the time for which the transient process lasts and the temperature distributions that arise in the material, in order to determine the amount of material that has not undergone the desired thermal processing satisfactorily. It is also desirable to determine the dependence of the duration of the transient process on the operating conditions in order to find ways of reducing this time. For instance, a smaller speed U_s at the initial stages, raising it to the desired value after the recirculating region has moved far downstream, may yield better results. Clearly, transient effects have not received much attention, and detailed study of the important basic and applied aspects for different starting circumstances is needed. Of particular interest is the conjugate transport for the case shown in Fig. 2, where the material length increases from zero to large values.

5.3 Additional Forced Flow

We have so far considered the ambient fluid, far from the moving material, as stationary, with the fluid flow being induced by thermal buoyancy and the motion of the solid material undergoing thermal processing. Therefore, the heat transfer rates are determined by these two mechanisms, buoyancy and surface motion. However, in many practical processes, the resulting flow does not provide adequate cooling rates and an additional, externally induced, forced flow is employed to increase the heat transfer rate [47]. Other heat transfer mechanisms, such as boiling with the use of sprays and radiation, are also employed for increasing the heat transfer rates [19, 48]. We shall consider the case of an additional forced-flow velocity U_∞, parallel to the material motion, and the effect it has on the transport processes. Two circumstances are studied, one in which the forced flow is in a channel or duct surrounding the material and the other in which the flow arises in the extensive ambient medium [49]. Obviously, the former is of greater interest in processes such as cooling of glass fibers, where a forced flow of inert gases is blown over the fiber in an enclosed region. If the cooling medium is extensive, as is often the case in cable manufacture for cooling in a water trough, the second circumstance considered is of interest.

The governing equations are still Eqs. (24) and (25) for this problem, with Eq. (24*a*) replaced by Eq. (26) or (27) if buoyancy effects are included. However, the boundary conditions are changed. The velocity component u goes to U_∞, instead of zero, for the stationary ambient medium. Therefore, this boundary condition is written for a moving extensive medium as

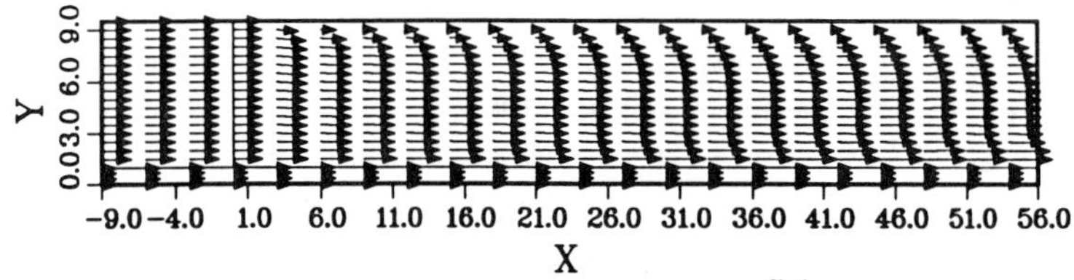

(a)

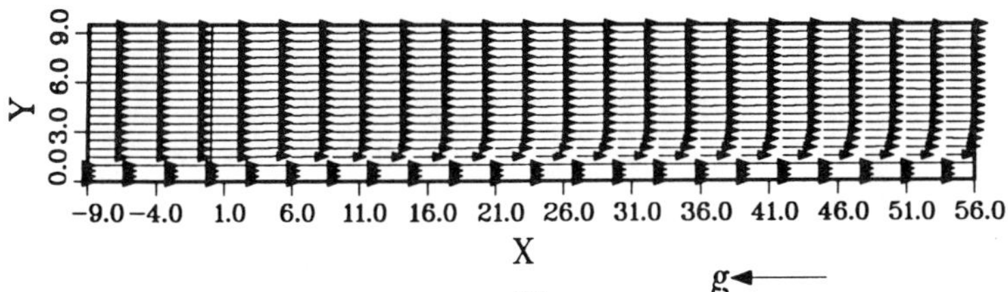

(b)

Figure 25 Computed flow field in terms of the velocity vectors for a plate moving vertically upward, considering (*a*) channel flow, and (*b*) extensive ambient medium, at Re = 25.0, Pe = 0.29, Pr = 7.0, $Gr/Re^2 = 2.4$, $k_f/k_s = 0.0029$, and $U_\infty/U_s = 2.0$.

$$Y \rightarrow \infty \text{ for } X > 0: \qquad \frac{u}{U_s} \rightarrow \frac{U_\infty}{U_s} \tag{29}$$

This implies that an additional parameter U_∞/U_s arises in the problem. Also, if an enclosed region, such as a channel, is considered, no-slip conditions are imposed at the boundary of the enclosure. The temperature there may be taken as uniform at a specific value, such as T_∞, or an adiabatic condition may be employed, depending on the physical process under consideration.

Figure 25 shows the computed velocity field for the vertical orientation, in terms of the velocity vectors. A uniform flow at velocity U_∞ is taken at $x = 0$ and far downstream the parabolic formulation is employed, as outlined earlier. Internal flow in a channel is considered for the results shown in Fig. 25*a*, and flow in an extensive environment for those in Fig. 25*b*. The flow field is seen to develop very rapidly downstream and to yield velocity levels higher than both U_s and U_∞ for the aiding buoyancy circumstance. This aspect is seen more clearly in Fig. 26, where results for the channel flow are shown for varying U_∞/U_s in Fig. 26*a*, at given Gr and Pe, and for varying Gr in Fig. 26*b*, at given U_∞/U_s and Pe. The trends are very much as expected. Here, Y is measured from the material midplane so that $Y = 1.0$ represents the moving surface. Therefore, velocity $U = u/U_s$ is uniform from $Y = 0.0$ to 1.0 and, then, rises sharply for higher values of the Grashof number Gr. At Gr = 5000, the computed velocity U is about 3.7, which is 1.85 times the forced velocity U_∞. A fairly uniform velocity is seen across much of the channel and then U drops to zero at the far wall of the channel, as expected from the no-slip conditions there. The temperature level is highest near the surface, which is moving vertically upward. This leads to the strongest buoyancy effect near the surface, as seen in Fig. 26*b*. At large values of U_∞/U_s, a fairly uniform velocity arises across the channel, at a value higher

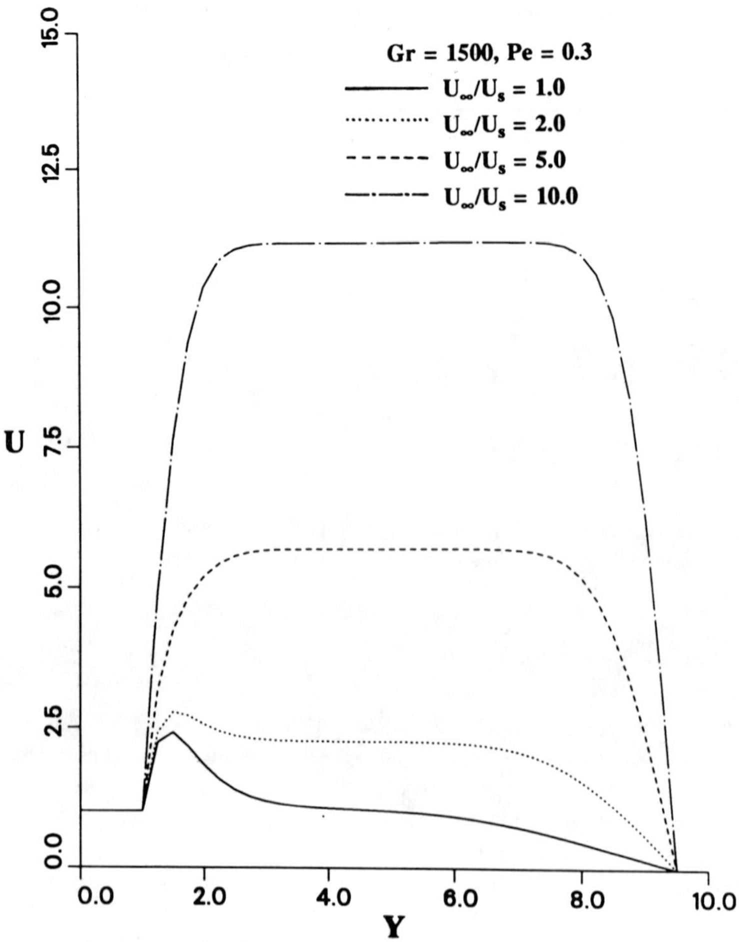

Figure 26 Computed velocity profiles for a plate moving vertically upward in a channel with the forced flow in the direction of motion at Re = 25.0, Pr = 7.0, and k_f/k_s = 0.0029: (*a*) Gr = 1500, Pe = 0.3;

than the forced-flow velocity. Thus, the buoyancy effect is almost uniformly distributed across the channel in this case.

The corresponding heat transfer results for channel flow are shown in Fig. 27, where the downstream variation of the local Nusselt number Nu(X) is plotted for several values of the forced-flow velocity U_∞/U_s. An increase in this parameter increases the Nusselt number, as expected. Thus, the addition of an externally induced forced-flow velocity may be employed to enhance the heat transfer from the moving plate. Figure 28 shows three different materials, Teflon, glass, and aluminum, moving in water in a channel. It is interesting to note that the upstream penetration of thermal effects is small for Teflon, due to its low thermal conductivity. Also, the surface temperature cools very rapidly downstream because of small diffusion in the radial and axial directions. This gives rise to the largest difference between the surface and

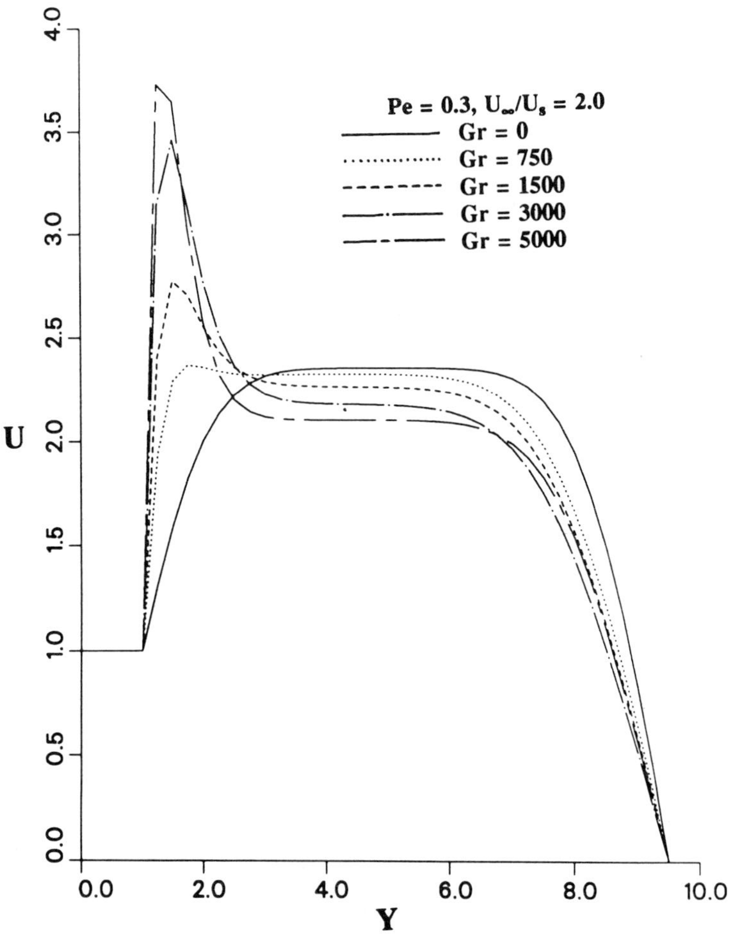

Figure 26 Computed velocity profiles for a plate moving vertically upward in a channel with the forced flow in the direction of motion at Re = 25.0, Pr = 7.0, and k_f/k_s = 0.0029 (*Continued*): (*b*) Pe = 0.3, U_∞/U_s = 2.0.

the midplane temperatures among these three materials. Glass is the intermediate material in terms of the thermal conductivity, and the observed trends lie between those for Teflon and aluminum. The downstream penetration of the thermal field is the largest for aluminum and the temperature difference across the plate is the smallest, as expected from the large thermal conductivity of the material.

Similar trends were observed for the forced flow in an extensive environment. However, the addition of ambient fluid motion in an enclosed region is of greater interest in problems of practical interest. Clearly, this is a very important consideration, and much more needs to be done. For instance, in the cooling of glass fiber for fiber optics applications, an opposing stream of an inert gas is often employed in a channel surrounding the fiber. Similarly, an aiding or opposing stream of water is used in wire

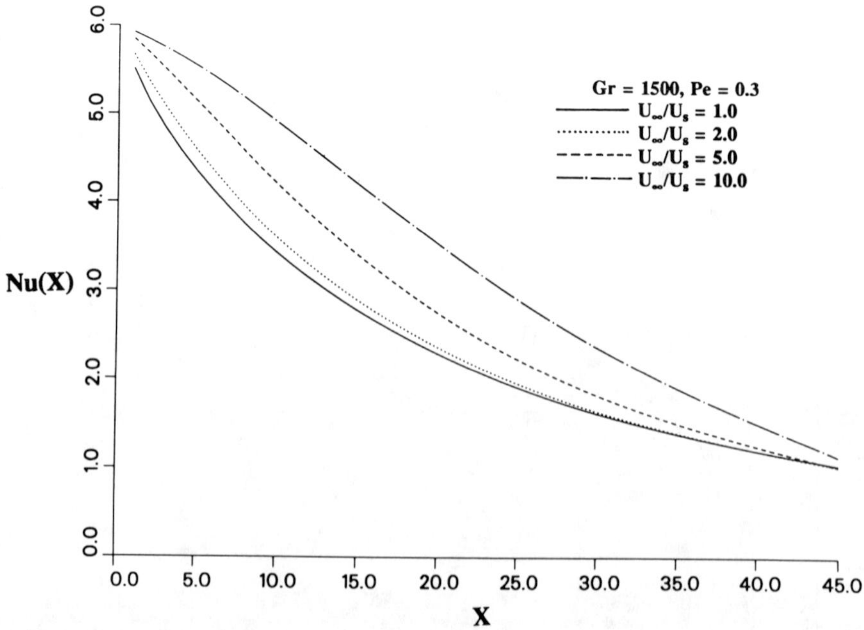

Figure 27 Downstream variation of the local Nusselt number at Re = 25.0, Pr = 7.0, Pe = 0.3, Gr = 1500.0, and k_f/k_s = 0.0029 for vertical-aiding forced flow in a channel at different values of U_∞/U_s.

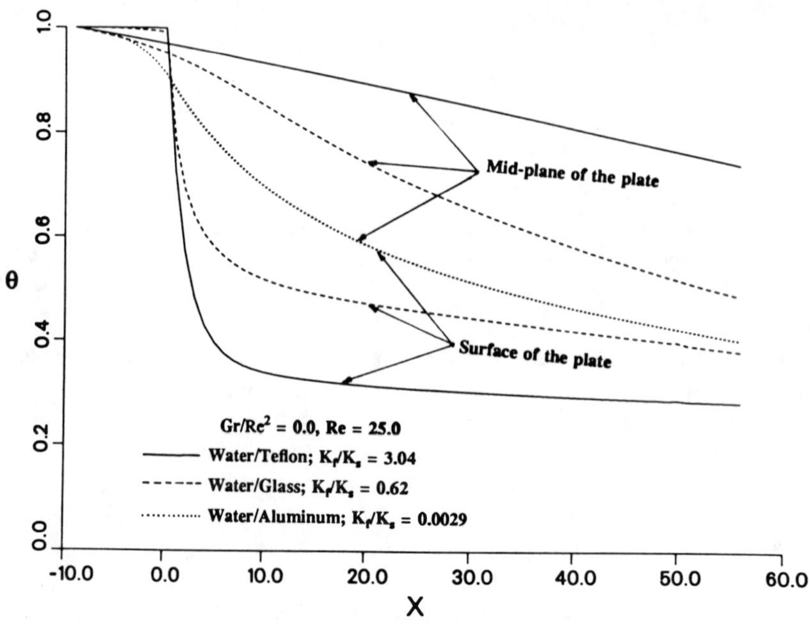

Figure 28 Downstream variation of the surface and midplane temperatures in plates of three different materials moving in water, Pr = 7.0, at U_∞/U_s = 2.0.

drawing during cable manufacture. Therefore, modeling of a forced flow in a direction opposite to material motion, simulation of turbulent flows, transport from materials of circular and other cross sections, three-dimensional transport for cross-flow, and effect of orientation in these cases must be considered in detail to provide the inputs needed for further study and design of these processes.

6 EXPERIMENTAL WORK

An area in which very little work has been done, with respect to thermal transport from a continuously moving material, is on the experimental study of the process. Because of the movement of the material and the conjugate nature of the transport, the experimental arrangement needed for accurate data is quite complicated. However, experimental results are needed for validating the various numerical models outlined in the preceding sections and also for providing the physical insight into many aspects of the process that are not very well understood. These include turbulent transport, separation in circumstances where thermal buoyancy is in a direction opposite to the surface motion, transport near the leading edge of a finite moving surface, three-dimensional effects that arise, for instance, in a horizontally moving cylindrical rod due to thermal buoyancy, transport in the region close to the slot from which the material emerges, transient effects, etc. Only a few experimental studies have been carried out, and the results from some of those are discussed here [17, 31, 50, 51].

Tsou et al. [31] carried out a detailed experimental study of the boundary layer on a continuous moving surface, using a rotating drum with a skimmer plate that fixes

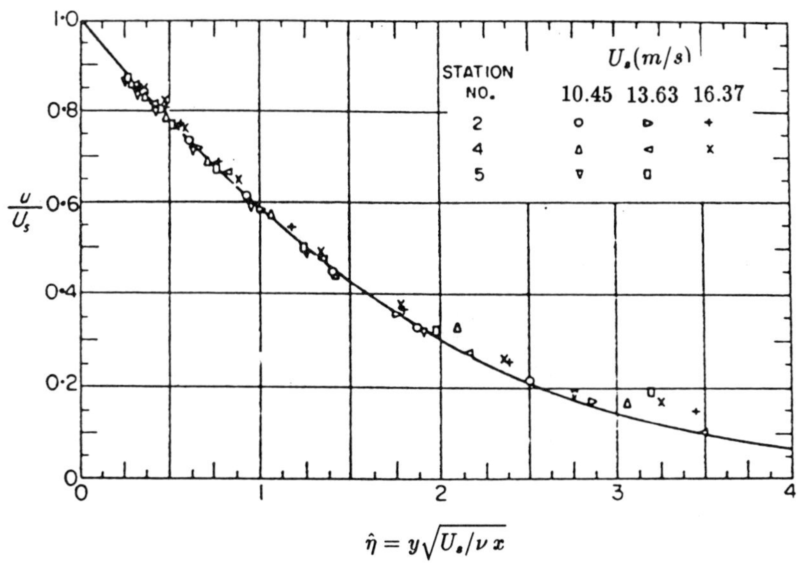

Figure 29 Comparison of experimental and analytical laminar velocity profiles, from the work of Tsou et al. [31].

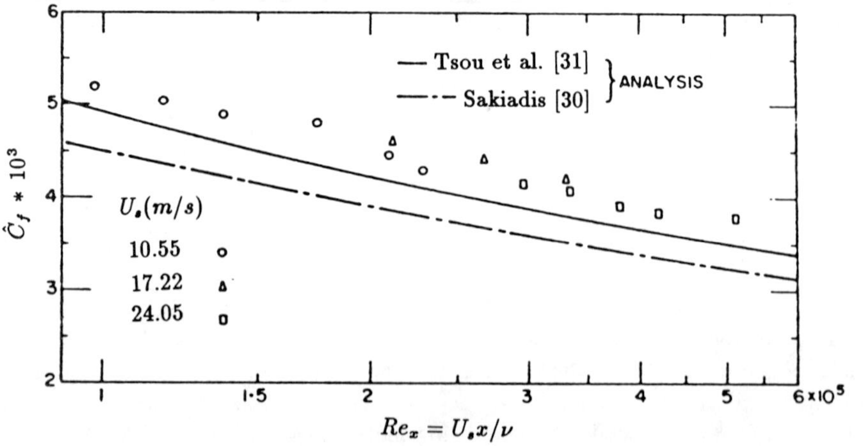

Figure 30 Comparison of experimental and analytical turbulent friction coefficients, from the work of Tsou et al. [31].

the origin of the flow. The radius of the drum was taken as large so that the curvature effects may be neglected. The drum was heated and the isothermal, moving surface condition was simulated. Both laminar- and turbulent-flow circumstances were considered. A comparison between the experimental and analytical laminar velocity profiles is shown in Fig. 29, indicating an excellent agreement between the two. The measurements were taken at various downstream locations, or stations, where 10 equidistant circumferential locations were employed and denoted as 1, 2, 3, . . . , with 0 being the skimmer position. The comparison of experimental and analytical friction coefficients $\hat{C}_f$, where $\hat{C}_f = \bar{\tau}_s/\frac{1}{2}\rho_f U_s^2$, $\hat{\tau}_s$ being the surface shear stress, is shown in Fig. 30. The measured values were about 6% higher than the theoretical values obtained by Tsou et al. [31].

In a study on the conjugate heat transfer from continuously moving surfaces, Chida and Katto [17] carried out experiments on heated flat plates moving in water. Heated, moving, flat belts were used to simulate a continuously moving surface. Velocity and temperature distributions were measured by means of a hot-wire anemometer and a Mach-Zehnder interferometer, respectively. Figure 31*a* shows the results in terms of the measured and predicted temperature distributions in air, along with predicted results at other values of the Prandtl number. A very good agreement between the analytical and experimental results is seen. The results for water are shown in Fig. 31*b*. Here, ξ is a dimensionless coordinate distance in the direction of material motion, with $\xi = x(\rho Ck)_f/(U_s D^2 \rho_s^2 C_s^2)$. The temperature at the slot is determined by extrapolation, and some of these extrapolation points are shown. Good agreement between the experimental and analytical results is seen.

More recently, efforts have been made to simulate a moving plate of finite length experimentally, so that steady-state conditions are obtained at large time following the onset of the transport process [50, 51]. Interest, therefore, lies in the transient as well as in the steady-state circumstance. Figure 32 shows a schematic of the experimental arrangement. A plate is heated by means of electric heaters in a furnace and, when a

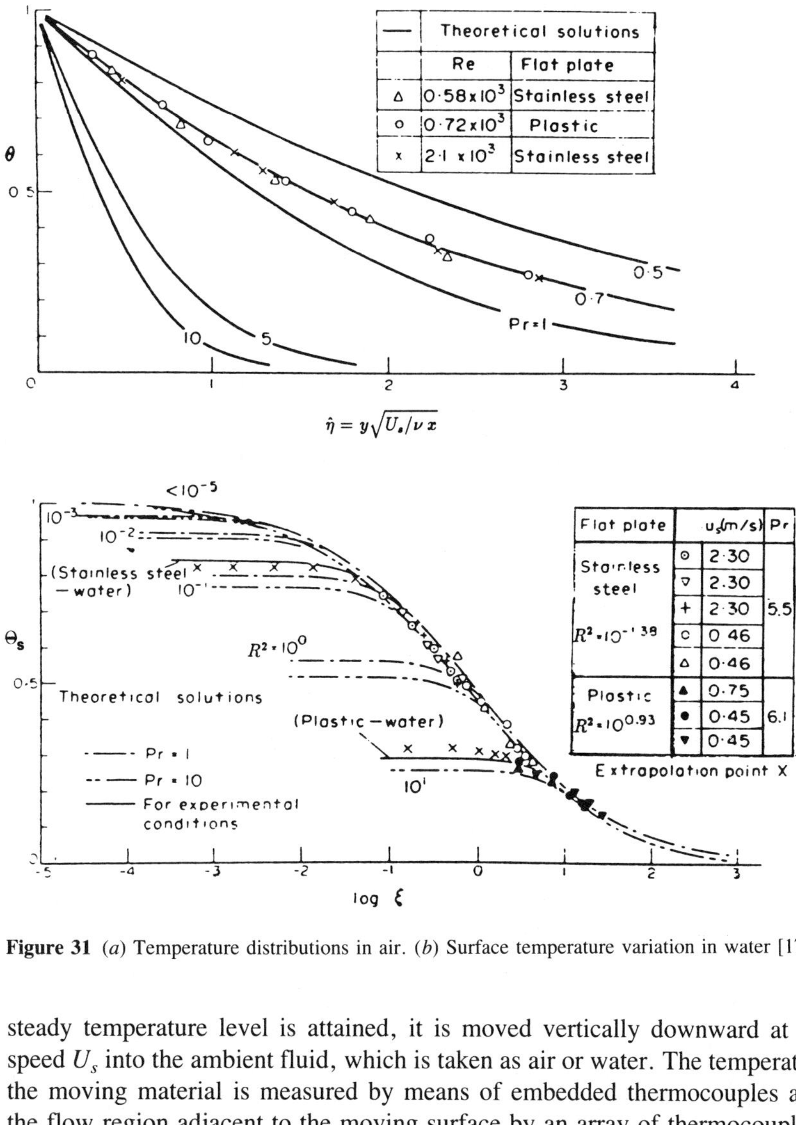

Figure 31 (*a*) Temperature distributions in air. (*b*) Surface temperature variation in water [17].

steady temperature level is attained, it is moved vertically downward at a uniform speed U_s into the ambient fluid, which is taken as air or water. The temperature within the moving material is measured by means of embedded thermocouples and that in the flow region adjacent to the moving surface by an array of thermocouples. A hot-wire anemometer probe is employed to measure the velocity at a given location, which may be varied. For details on the experimental system, the measurement techniques, and the data acquisition system, see [50].

For an aluminum plate of thickness 5 mm, the temperature across the plate thickness is essentially uniform. Figure 33 shows experimentally obtained time-dependent temperature variations in air and water, as indicated by the thermocouples embedded at various vertical locations in the plate. The first thermocouple is closest to the leading edge of the plate. From these results, the temperature distribution within the plate at various dimensionless times τ may be obtained, yielding steady-state conditions at

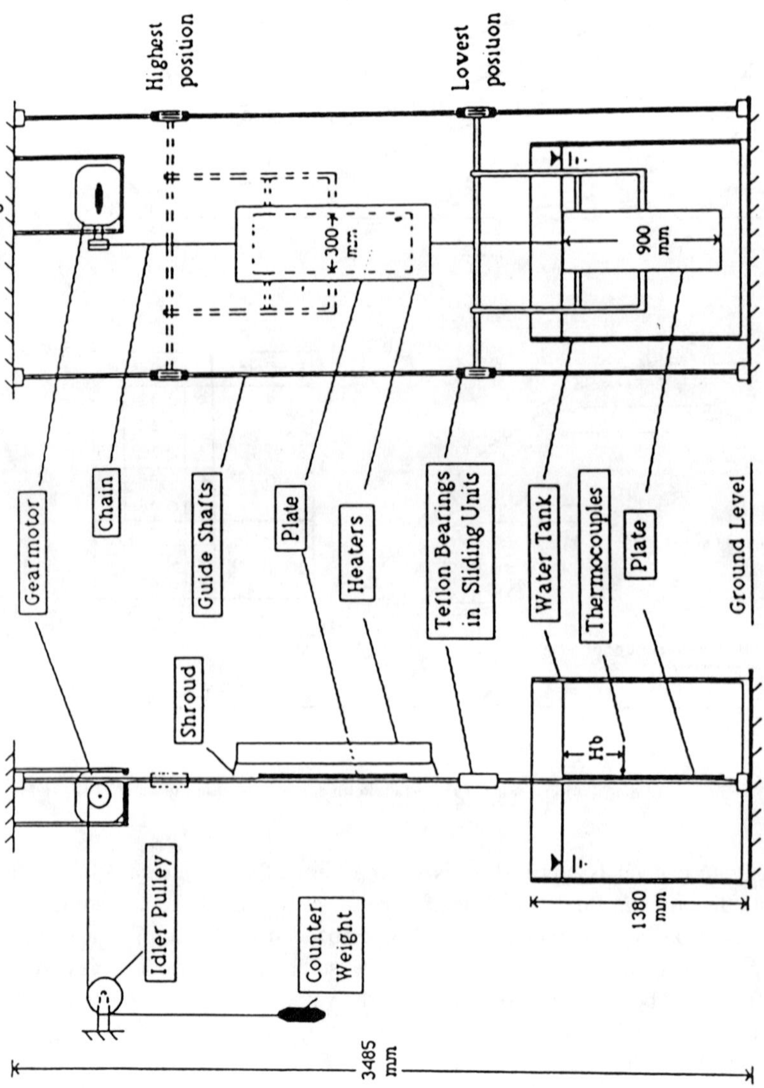

Figure 32 Schematic of the experimental arrangement employed in [50, 51].

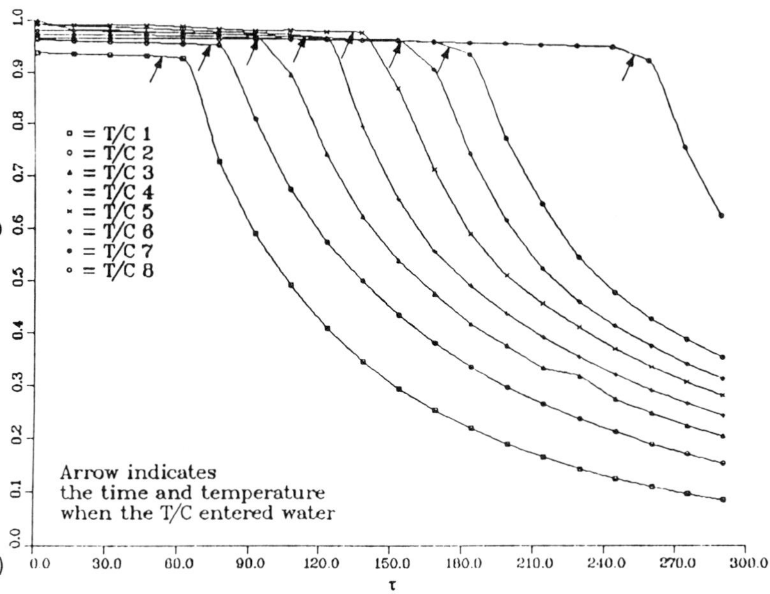

(a)

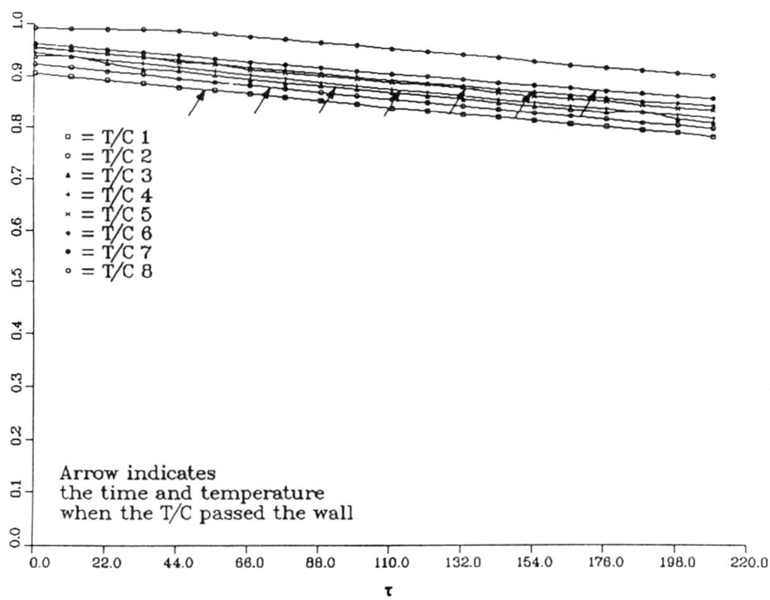

(b)

Figure 33 Measured time-dependent temperatures within a 5-mm-thick heated aluminum plate moving vertically downward in: (*a*) water, at U_s = 2.54 cm/s, Re = 96.4, Pr = 4.31, Pe = 0.73, Gr/Re^2 = 0.85, k_f/k_s = 2.94 × 10^{-3}; (*b*) air, at U_s = 0.44 cm/s, Re = 0.65, Pr = 0.7, Pe = 0.13, Gr/Re^2 = 222.0, k_f/k_s = 1.35 × 10^{-4}.

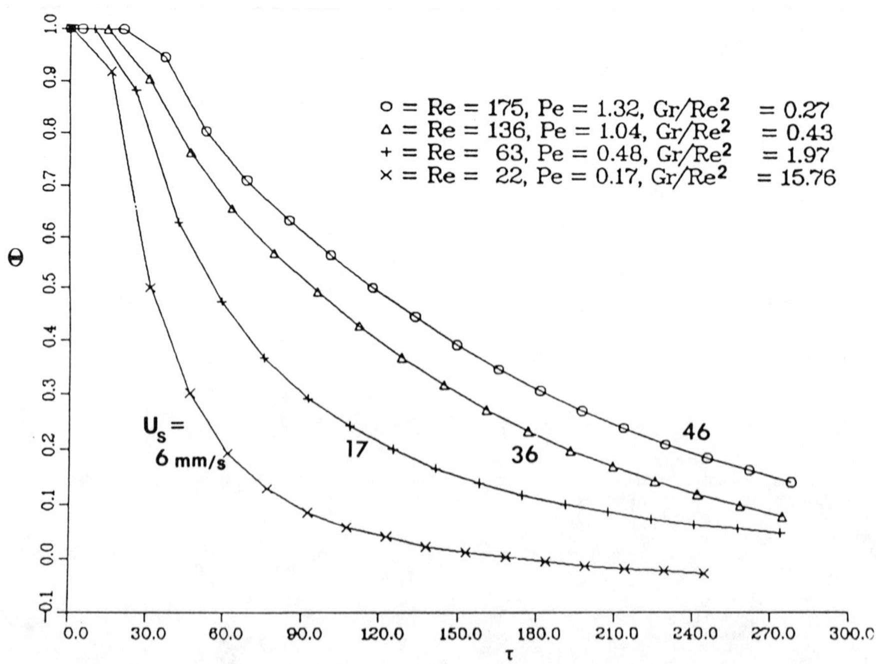

Figure 34 Measured plate temperature variation with time for the aluminum plate moving vertically downward in water at different speeds U_s.

large time, if the plate is long enough. It is seen from this figure that, as expected, the temperature decay with time is much more rapid in water due to the higher heat transfer rates that arise. The effect of the velocity U_s on the temperature decay is seen in Fig. 34, where the time history of a given thermocouple embedded in the plate is plotted for various values of U_s. The corresponding values of the dimensionless governing parameters are also given. As seen earlier from the numerical results, the temperature decay becomes more gradual as the Peclet number Pe increases. Also, time may be expressed in terms of distance x by the equation $x = U_s\tau'$. Here, the Grashof number Gr is essentially constant, varying only about 8%, while Pe is increased to almost eight times its value. The inlet temperature T_0 varies slightly due to the variation in Pe, and this is reflected in a variation in Gr. The corresponding results for air are shown in Fig. 35, and trends similar to those seen earlier are obtained.

A comparison between the experimental results and the theoretically predicted ones, for steady-state conditions, is shown in Fig. 36 for an aluminum plate moving in water. The theoretical results are for the aided-buoyancy circumstance, whereas the experimental results are for the opposed-buoyancy case, since a heated plate moving downward is studied due to the experimental limitations for water. Clearly, the agreement is good, for the small buoyancy levels considered, except near the leading edge. Since the analysis assumes an infinite plate, this disagreement near the leading edge is expected.

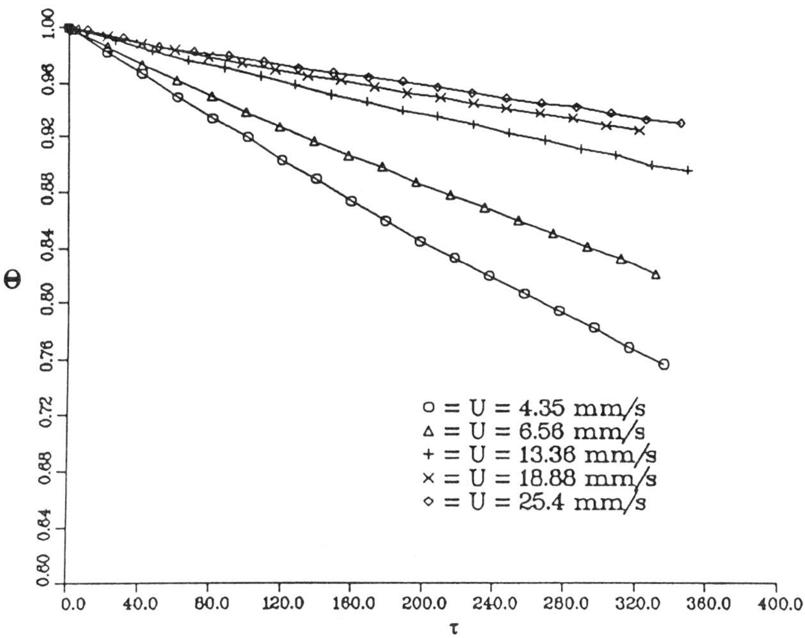

Figure 35 Effect of speed U_s on the time-dependent variation of the plate temperature at a distance 3.5 cm from the leading edge of an aluminum plate moving vertically downward in air.

Figure 37 shows a sequence of photographs from a shadowgraph showing the flow near the surface of the aluminum plate moving vertically downward. As time elapses, a large disturbance leading to flow separation is seen to arise. The flow under consideration is an opposed-buoyancy circumstance, which could lead to flow separation downstream [44]. Some recirculation could also arise due to transient effects, as seen earlier in Sec. 5.2. But the large disturbance and flow separation seen in Fig. 37 is largely due to the opposing-buoyancy effect, since the disturbance was found to remain at one location and not move downstream with time as expected for transient effects. Similar trends were seen at other speeds and inlet temperatures. The flow separation is governed by the local mixed-convection parameters $Gr_x/Re_x^2 = g\beta(T - T_\infty)x/U_s^2$, and the effect was found to be larger at larger values of this parameter.

It is seen from the discussion presented in this section that only a few experimental studies have been carried out on this problem, particularly for the conjugate circumstance which is of interest in practical applications. Much more needs to be done on buoyancy effects, particularly on the opposing-buoyancy situation, transients, turbulence, transport near the leading edge of a plate of finite length, horizontally moving materials, and temperature variation within the material for low-conductivity materials. Some of these aspects are under consideration. But much more needs to be done to provide the inputs needed for the validation of the numerical models and for design of the relevant systems.

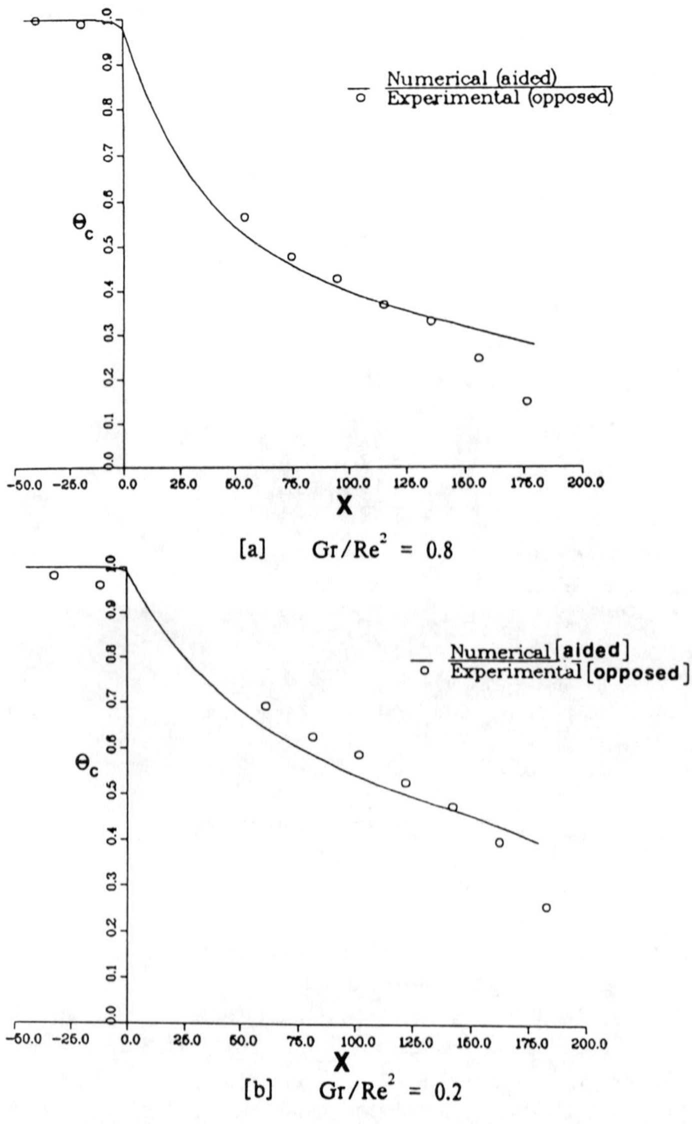

Figure 36 Comparison between numerical and experimental results for the aluminum plate moving vertically downward in water at two different values of Gr/Re^2: (*a*) Re = 103, Pe = 0.727, $k_f/k_s = 2.94 \times 10^{-3}$; (*b*) Re = 175, Pe = 1.32, $k_f/k_s = 2.94 \times 10^{-3}$.

7 ADDITIONAL CONSIDERATIONS, CONCLUSIONS, AND FUTURE WORK

This chapter has presented a review of the research work done on the transport processes associated with a moving material undergoing thermal processing. The various for-

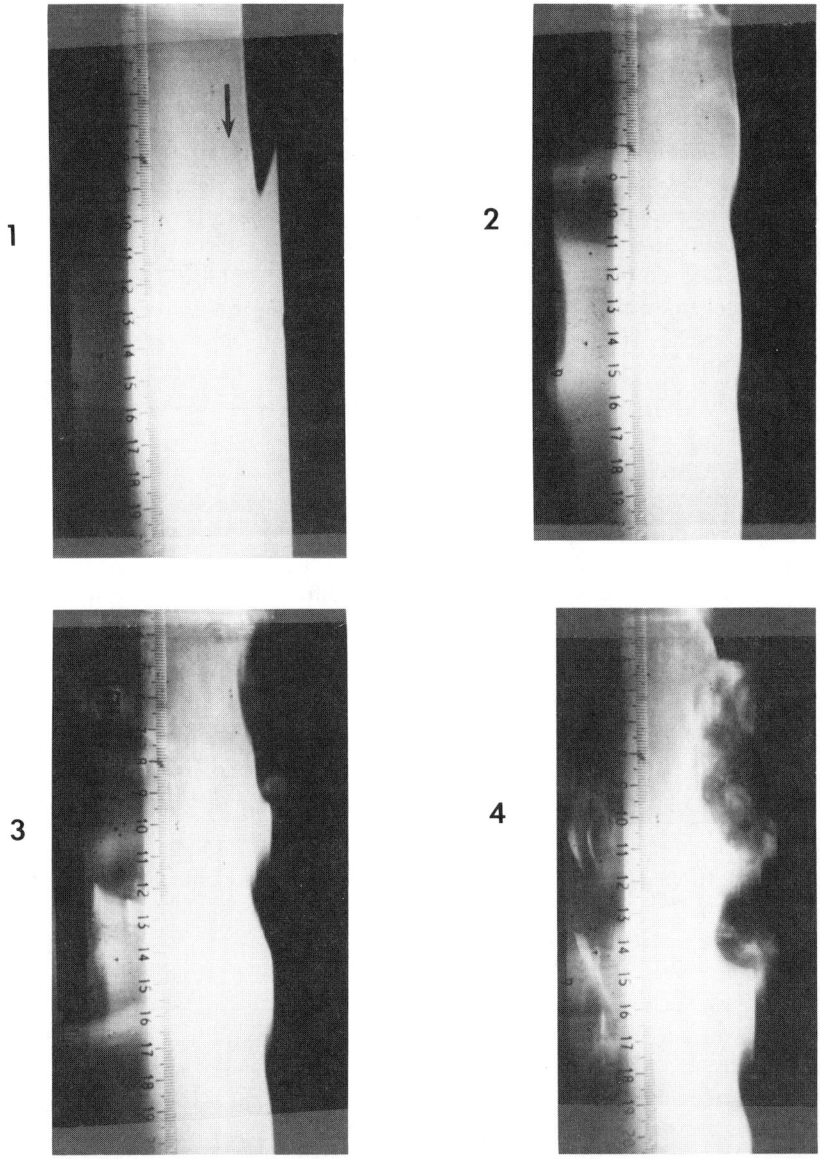

Figure 37 Sequence of photographs showing the flow near the surface of the aluminum plate moving vertically downward at $U_s = 3.7$ cm/s in water at Re $= 140.36$, Pe $= 1.05$, Pr $= 4.31$, and Gr/Re$^2 = 0.45$.

mulations used for simulating the many different aspects of the process and also different circumstances of practical interest are outlined, along with the results obtained. However, the problem is a very vast one, since the underlying transport processes vary with the material and the operating conditions for thermal processing. It applies to processes ranging from crystal growing and continuous casting to extrusion, wire drawing, and fiber drawing. Consequently, it is a very important problem which has received some attention in the past and requires detailed effort in the future on a wide variety of physical aspects associated with the basic process.

We have largely considered the cooling of a continuously moving material in a uniform ambient medium, which may be stationary or moving. Three main approaches to the problem have been discussed. The first is directed at heat transfer within the solid, employing heat transfer coefficients from available empirical correlations at the surface. The second approach considers the flow and heat transfer from a moving surface with an imposed thermal condition, such as isothermal or uniform heat flux. The third approach considers the conjugate problem resulting from a combination of the first two approaches. Thus, the transport in the solid material is coupled with that in the fluid through the boundary conditions and the two are solved for simultaneously. Similar considerations apply for heating of a continuously moving material by convective transport.

Several additional effects and considerations may be included in this problem to simulate many important practical circumstances. Radiative transport is important, and many researchers have included this aspect in the problem; see, for instance, [19, 40]. Two-dimensional laminar flow of absorbing and emitting fluid was considered by Rhodes and Chen [19], and solutions were obtained for the limiting conditions of optically thick and thin boundary layers. Phase change is another important aspect that may be included in the present problem in order to expand the scope of this effort to include processes such as continuous casting and crystal growing. A considerable amount of work has been done on these processes; see, for example, [52–56]. Clearly, this is a very important physical problem, and additional work is needed on solidification and melting as the material moves downstream.

In the preceding sections, the important areas and aspects that need further detailed study were mentioned. The conjugate problem is the one that closely represents many of the important circumstances for the thermal processing of continuously moving materials. Consequently, this is the area where most work is needed. Turbulent flow is an important aspect that needs substantial work in the future. The various turbulence closure models available may be used to compute the resulting transport [27]. Some simple approaches, such as integral equations for a turbulent boundary layer and the law of the wall, were used by some researchers [30, 31], as mentioned earlier. Moutsoglou and Bhattacharya [57] used a mixing-length model for the eddy diffusivities and obtained numerical results on heat transfer, wall shear, and surface temperature. Obviously, turbulent transport from a moving material is an area where future work may be directed in order to obtain better physical insight into the underlying processes and to provide inputs for design. Some work has also been done on the hydrodynamic instability of this flow [58] and, again, further work is needed to characterize the underlying mechanisms.

In addition to the various topics and aspects mentioned above, further work is needed on buoyancy effects, with respect to the effect of the orientation on the transport and on flow separation due to an opposing buoyancy force, three-dimensional effects that arise due to thermal buoyancy on a horizontally moving cylindrical body or due to a transverse forced flow, simulation of the transport processes upstream and downstream in the region near the slot from which the material emerges, flow and heat transfer in the region close to the leading edge of the moving material, and transient effects. Also, detailed experimental work is needed on all these aspects to validate the analytical and numerical models and to provide physical insight into areas that are not well understood at the present time.

ACKNOWLEDGMENTS

The author acknowledges the support of the National Science Foundation, under Grant No. CBT-88-03049, for much of the work reported here. The help of M. V. Karwe and B. H. Kang in obtaining many of the results presented here and in the preparation of the figures is also acknowledged.

REFERENCES

1. J. Szekely, *Fluid Flow Phenomena in Metals Processing*, Academic Press, New York, 1979.
2. A. Ghosh and A. K. Mallik, *Manufacturing Science*, Ellis Harwood (John Wiley), U.K., 1987.
3. L. E. Doyle, C. A. Keyser, J. L. Leach, G. F. Scharder, and M. B. Singer, *Manufacturing Processes and Materials for Engineering*, Prentice-Hall, Englewood Cliffs, N.J., 1987.
4. E. G. Fisher, *Extrusion of Plastics*, John Wiley, New York, 1976.
5. T. Altan, S. Oh, and H. Gegel, *Metal Forming Fundamentals and Applications*, American Society for Metals, Metals Park, Ohio, 1979.
6. Z. Tadmor and C. Gogos, *Principles of Polymer Processing*, John Wiley, New York, 1979.
7. U. A. Jackson, The Present State of the Theory of Crystal Growth from the Melt, *J. Crystal Growth*, vol. 24, pp. 130–136, 1974.
8. F. Hollander, A Model to Calculate the Complete Temperature Distribution in Steel During Hot Rolling, *Iron & Steel Inst.*, vol. 23, pp. 46–74, 1970.
9. L. R. Glicksman, The Cooling of Glass Fibers, *Glass Technol.*, vol. 9, pp. 131–138, 1968.
10. L. R. Glicksman, A Prediction of Upper Temperature Limit for Glass Fiber Spinning, *Glass Technol.*, vol. 15, pp. 16–20, 1974.
11. M. A. Matovich and J. R. A. Pearson, Spinning a Molten Threadline, *Ind. Eng. Chem. Fund.*, vol. 8, pp. 512–520, 1969.
12. W. A. Gifford, A Finite Element Analysis of Isothermal Fiber Formation, *Phys. Fluids*, vol. 25, pp. 219–225, 1982.
13. A. A. Tseng, Finite-Difference Solutions for Heat Transfer in a Roll Rotating at High Speed, *Numer. Heat Transfer*, vol. 7, pp. 113–125, 1984.
14. A. A. Tseng, A Generalized Finite Difference Scheme for Convection-Dominated Metal-Forming Problems, *Numer. Heat Transfer*, vol. 20, pp. 1885–1900, 1984.
15. O. Chopra and P. Niessen, Solidification of Lead and Lead Alloys in Continuous Drum Sheet Casting, *Z. Metallk.*, vol. 66, pp. 10–16, 1975.
16. H. K. Kuiken, Solidification of a Liquid on a Moving Sheet, *Int. J. Heat Mass Transfer*, vol. 20, pp. 309–314, 1977.
17. K. Chida and Y. Katto, Conjugate Heat Transfer of Continuously Moving Surface, Int. J. Heat Mass Transfer, vol. 19, pp. 461–470, 1976. Also, pp. 453–460.
18. V. G. Fox, L. E. Erickson, and L. T. Fan, Methods for Solving Boundary Layer Equations for Moving Continuous Flat Surfaces with Suction and Injection, *AIChE J.*, vol. 14, pp. 726–736, 1968.
19. C. A. Rhodes and C. C. Chen, Thermal Radiation in Laminar Boundary Layers on Continuous Moving Surfaces, *J. Heat Transfer*, vol. 96, pp. 32–36, 1976.

20. K. T. Yangai, Prediction of Strip Temperature for Hot Strip Mills, *Trans. Iron & Steel Inst., Jpn.*, vol. 26, pp. 11–19, 1976.
21. D. A. Peel and A. E. Pengelly, Heat Transfer, Solidification and Metallurgical Structure in the Continuous Casting of Aluminum, *Iron & Steel Inst.*, vol. 23, pp. 186–196, 1970.
22. V. H. S. Kuo and W. R. Wilcox, Influence on Crystal Dimension on the Interfacial Temperature Gradient, *J. Crystal Growth*, vol. 12, pp. 191–194, 1972.
23. Y. Jaluria and A. P. Singh, Temperature Distribution in a Moving Material Subjected to Surface Energy Transfer, *Comp. Meth. Appl. Mech. Eng.*, vol. 41, pp. 145–157, 1983.
24. M. V. Karwe and Y. Jaluria, Thermal Transport from a Heated Moving Surface, *J. Heat Transfer*, vol. 108, pp. 728–733, 1986.
25. E. M. Sparrow, S. V. Patankar, and S. Ramadhyani, Analysis of Melting in the Presence of Natural Convection in the Melt Region, *J. Heat Transfer*, vol. 101, pp. 578–584, 1979.
26. D. W. Peaceman and H. H. Rachford, The Numerical Solution of Parabolic and Elliptic Partial Differential Equations, *J. Soc. Ind. Appl. Math.*, vol. 1, pp. 28–41, 1955.
27. Y. Jaluria and K. E. Torrance, *Computational Heat Transfer*, Hemisphere, New York, 1986.
28. S. V. Patankar, *Numerical Heat Transfer and Fluid Flow*, Hemisphere, New York, 1980.
29. B. C. Sakiadis, Boundary Layer Behavior on Continuous Solid Surfaces: I. Boundary Layer Equations for Two-Dimensional and Axisymmetric Flows, *AIChE J.*, vol. 7, pp. 26–28, 1961.
30. B. C. Sakiadis, Boundary Layer Behavior on Continuous Solid Surfaces: II. The Boundary Layer on Continuous Flat Surface, *AIChE J.*, vol. 7, pp. 221–225, 1961.
31. F. K. Tsou, E. M. Sparrow and R. J. Goldstein, Flow and Heat Transfer in the Boundary Layer on a Continuous Moving Surface, *Int. J. Heat Mass Transfer*, vol. 10, pp. 219–235, 1967.
32. J. F. Griffin and J. L. Thorne, On Thermal Boundary Layer Growth on Continuous Moving Belts, *AIChE J.*, vol. 13, pp. 1210–1211, 1967.
33. L. E. Erickson, L. T. Fan, and V. G. Fox, Heat and Mass Transfer on a Moving Continuous Flat Plate with Suction or Injection, *Ind. Eng. Chem. Fund.*, vol. 5, pp. 19–25, 1966.
34. T. S. Chen and F. A. Strobel, Buoyancy Effects in Boundary Layer Adjacent to a Continuous, Moving Horizontal Flat Plate, *J. Heat Transfer*, vol. 102, pp. 170–172, 1980.
35. A. Moutsoglou and T. S. Chen, Buoyancy Effects in Boundary Layers on Inclined, Continuous, Moving Sheets, *J. Heat Transfer*, vol. 102, pp. 371–372, 1980.
36. E. A. Koldenhof, Laminar Boundary Layer on Continuous Flat and Cylindrical Surfaces, *AIChE J.*, vol. 9, pp. 411–418, 1963.
37. Y. Jaluria, *Computer Methods for Engineering*, Allyn & Bacon (Prentice-Hall), Needham Heights, Mass., 1988.
38. H. K. Kuiken, The Cooling of a Low-Heat-Resistance Sheet Moving Through a Fluid, *Proc. Roy. Soc. London*, vol. A341, pp. 233–252, 1974.
39. H. T. Kuiken, The Cooling of a Low-Heat-Resistance Cylinder Moving Through a Fluid, *Proc. Roy. Soc. London*, vol. A346, pp. 23–35, 1975.
40. M. V. Karwe and Y. Jaluria, Fluid Flow and Mixed Convection Transport from a Moving Plate in Rolling and Extrusion Processes, *J. Heat Transfer*, vol. 110, pp. 655–661, 1988.
41. M. V. Karwe and Y. Jaluria, Numerical Simulation of Thermal Transport Associated with a Continuously Moving Flat Sheet in Rolling or Extrusion, in *Proc. Natl. Heat Transfer Conf.*, Houston, ASME HTD, vol. 96, pp. 37–45, 1988.
42. R. Agarwal and Y. Jaluria, Deflection of a Two-Dimensional Natural Convection Wake Due to the Presence of a Vertical Surface in Close Proximity, *J. Fluid Mech.*, vol. 201, pp. 35–56, 1989.
43. Y. Jaluria, *Natural Convection Heat and Mass Transfer*, Pergamon, Oxford, 1980.
44. B. Gebhart, Y. Jaluria, R. L. Mahajan, and B. Sammakia, *Buoyancy Induced Flows and Transport*, Hemisphere, New York, 1988.
45. B. H. Kang, Y. Jaluria, and M. V. Karwe, Numerical Simulation of Conjugate Transport from a Continuous Moving Plate in Materials Processing, *Num. Heat Transfer*, vol. 19, pp. 151–176, 1991.
46. M. S. Khader, Transient Mixed Convection from a Moving Vertical Surface, ASME Paper No. 81-HT-40, 1981.
47. N. Ramachandran, T. S. Chen, and B. F. Armaly, Correlations for Laminar Mixed Convection in

Boundary Layers Adjacent to Horizontal, Continuous Moving Sheets, *J. Heat Transfer*, vol. 109, pp. 1036–1039, 1987.

48. D. A. Zambrunnen, R. Viskanta, and F. P. Incropera, The Effect of Surface Motion on Forced Convection Film Boiling Heat Transfer, *J. Heat Transfer*, vol. 111, pp. 760–766 1989.
49. B. H. Kang and Y. Jaluria, Flow and Heat Transfer from a Moving Heated Plate in a Uniform Forced Flow, *Num. Heat Transfer*, to appear 1991.
50. M. V. Karwe, Thermal Transport Between a Continuously Moving Heated Plate and a Quiescent Ambient Medium, Ph.D. thesis, Rutgers University, New Brunswick, N.J., 1987.
51. M. V. Karwe and Y. Jaluria, Experimental Investigation of Thermal Transport from a Heated Moving Plate, *Int. J. Heat Mass Transfer*, to appear 1991.
52. L. S. Yao, Natural Convection Effects in Continuous Casting of a Horizontal Cylinder, *Int. J. Heat Mass Transfer*, vol. 27, pp. 697–703, 1984.
53. A. B. Crowley, Mathematical Modeling of Heat Flow in Czochralski Crystal Pulling, *J. Appl. Math.*, vol. 30, pp. 173–189, 1983.
54. S. Toda, H. Sugiyama, H. Owada, M. Kurokawa, and Y. Hori, Laminar Flow Heat Transfer in a Tube with Internal Solidification, *Proc. 8th Int. Heat Transfer Conf.*, Hemisphere, Washington, D.C., vol. 4, pp. 1745–1750, 1986.
55. W. D. Bennon and F. P. Incropera, Developing Laminar Mixed Convection with Solidification in a Vertical Channel, *J. Heat Transfer*, vol. 110, pp. 410–415, 1988.
56. S. S. Wei and S. I. Guceri, Solidification in Developing Pipe Flows, *Int. J. Heat Fluid Flow*, vol. 9, pp. 225–232, 1988.
57. A. Moutsoglou and A. K. Bhattacharya, Laminar and Turbulent Boundary Layers on Moving, Nonisothermal Continuous Flat Surfaces, *J. Heat Transfer*, vol. 104, pp. 707–714, 1982.
58. F. K. Tsou, E. M. Sparrow, and E. F. Kurtz, Hydrodynamic Stability of the Boundary Layer on a Continuous Moving Surface, *J. Fluid Mech.*, vol. 26, pp. 145–161, 1966.

CHAPTER
FIVE

SPECIES-CONTROLLED CONDENSATION IN ENCLOSURES

Per F. Peterson

ABSTRACT

When noncondensable gases or binary mixtures are present in condensing systems, mass diffusion can provide the dominate heat transfer resistance, giving rise to species-controlled condensation. Current technologies typically attempt to minimize mass diffusion effects by isolating and purging noncondensables away from condensing surfaces. Increasingly, future technologies will apply noncondensables for controlling condensing pressures, and apply binary mixtures to obtain condensation at spatially varying temperatures. In all these problems, accurate analysis of species transport and distribution is critical for predicting condenser performance. This chapter focuses on the physical phenomena which occur during species-controlled condensation in enclosures, by first examining experimental measurements, then covering the techniques and simplifying assumptions available for analysis, and finally discussing applications to several areas of current interest.

NOMENCLATURE

A	Margules constant, Eq. (15)
c_p	specific heat at constant pressure
C	heat capacity
C_n	constant, Eq. (17)
D	mass diffusion coefficient
f	fugacity
g	gravitational acceleration

G	constant, Eq. (20)
h	heat transfer coefficient
h_{lv}	latent heat of vaporization
Δh_E	heat of mixing
k	thermal conductivity
l	characteristic enclosure dimension
$\dot{m}$	mass flux
M	molecular weight
Nu	Nusselt number
P	pressure, Pa
$\dot{q}$	heat flux
R	gas constant
Ra	Rayleigh number, Eq. (1)
t	time
T	absolute temperature
T_*	Clausius-Claperyon constant, Eq. (10)
$\mathbf{u}$	mass average velocity vector
$\tilde{\mathbf{u}}$	molar-average velocity vector
v	mass-average velocity, y direction
$\tilde{v}$	molar-average velocity, y direction
x	transverse coordinate
x_n	mass fraction
$\tilde{x}_n$	mole fraction
y	normal coordinate
z	transverse coordinate
γ	activity coefficient
δ	liquid film or wall thickness
μ	dynamic viscosity
ρ	mass density
$\tilde{\rho}$	molar density

Subscripts

b	cooling-medium bulk
g	gas species
i	vapor/liquid interface
L	liquid-phase mixture
n	component number
o	value from previous iteration
r	reference value
T	total
v	vapor species
V	vapor-phase mixture
w	wall
∞	vapor bulk composition

Superscripts

s	saturation

1 INTRODUCTION

Condensation heat transfer offers one of the most effective means of transporting energy, making the study of condensation phenomena important in a broad range of applications. In nuclear, conventional, and alternative Rankine power cycles, nuclear reactor safety assessment, refrigeration and heat pump cycles, aerospace systems, chemical processes, and other applications, condensation in enclosures plays a fundamental heat transport role. In such systems noncondensable gases may be present due to leakage or chemical generation. These gases blanket condenser surfaces, reduce heat transfer efficiency, and induce corrosion. Alternatively, in some systems noncondensables may be added intentionally as a simple, effective means of controlling the system pressure and heat transfer. Binary mixtures can be substituted for pure working fluids to increase heat pump, refrigeration, and power-cycle efficiencies, or to permit other processes such as absorption. For all these condensation problems, it is well worth investigating both the potential positive and negative impacts of the presence of noncondensable gases or the presence of multiple species.

Whether multiple species are present intentionally or unintentionally, mass diffusion dominates a wide variety of condensation problems. Particularly inside enclosures where lower velocities exist, mass diffusion becomes the primary mechanism governing condensation heat transfer. The resulting vapor mixture flow and condensation can be quite complex. The condensation rate can vary strongly and nonlinearly with species concentration. Natural convection, recirculation, and unstable, fluctuating behavior can all occur. Currently, fundamental analysis of such *species-controlled* condensation phenomena requires the adoption of a variety of simplifying assumptions. These assumptions create rather strict limits on the classes of problems which can be considered, making identification and evaluation of the available simplifying assumptions a crucial step in achieving and assessing analytical results.

When a relatively pure vapor or a binary mixture of vapors flow into an enclosure, large spatial variations of condensation rates occur due to progressive changes of species concentration, creating strong effects on the vapor flow field. Under condensation, the noncondensable gas or more volatile species accumulates near the condensing surfaces. Noncondensable or volatile species ultimately tend to accumulate at stagnant locations. Consequently, the resulting concentration and temperature gradients cause mixed double-diffusive convection. To illustrate a small sample of the potential gas recirculation patterns, Fig. 1 shows laser holographic pictures of condensation of an upward-flowing vapor on two vertical, parallel plates. In the first two photographs a light, noncondensable gas, helium, is also present in the enclosure. The resulting fringes show the stable accumulation of the helium at the top of the enclosure and the condensation of the vapor below. In the last two photographs a completely different, unstable, fluctuating flow appears, simply because the noncondensable, argon, has a significantly higher molecular weight than the vapor, water. In between these extreme conditions of stable, stagnant accumulation and unstable, turbulent, and mixed flow, there exists a diverse collection of possible flow patterns linked by bifurcations as functions of the Rayleigh number. Analysis of such systems is not always simple.

Traditionally, investigators of species-controlled condensation problems have considered boundary-layer flows [1–10], where the far-field conditions are well defined

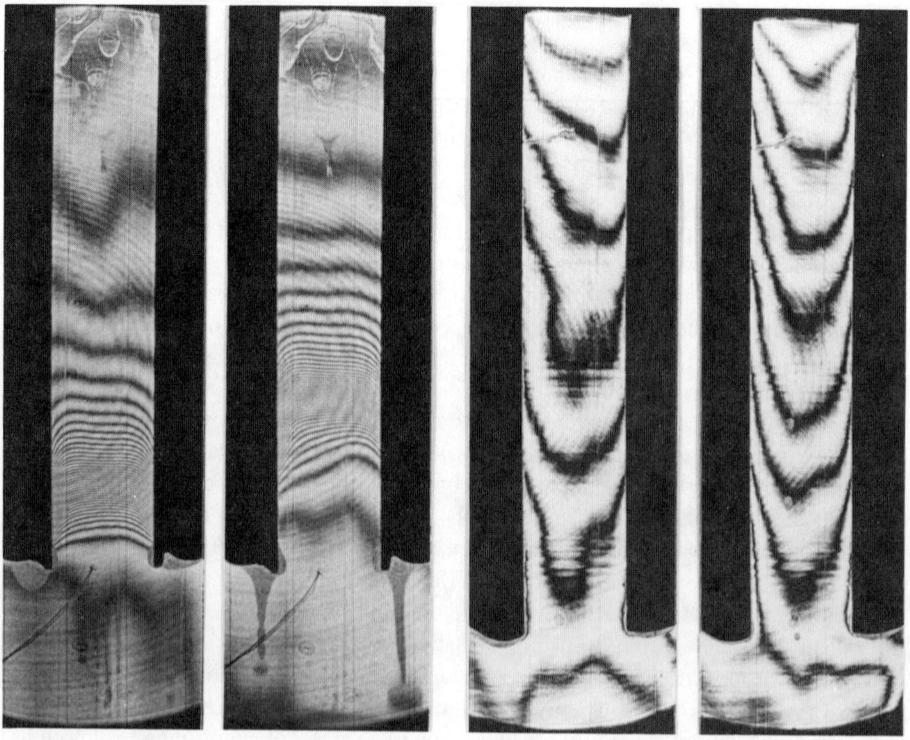

(a)	(b)	(c)	(d)
$q = 115\text{W}$	$q = 220.5\text{W}$	$q = 101.5\text{W}$	$q = 199.1\text{W}$
$P = 40.1\text{kPa}$	$P = 43.4\text{kPa}$	$P = 78.1\text{kPa}$	$P = 95.7\text{kPa}$

Benzene/Helium — Water/Argon

Figure 1 Laser holography photography of the gas distribution in a rectangular enclosure with condensation of a rising vapor at the cooled side walls. (Photos courtesy of Professor Y. Kobayashi, Tsukuba University, Japan [25].)

and relatively simple integration techniques can be applied to the resulting parabolic equations. Additionally, numerical solutions are readily available for the elliptic problems of natural convection in enclosures [11]. Additional solutions exist for evaporation and condensation-induced natural and mixed convection in enclosures with high noncondensable gas concentrations [12–16]. As yet, however, no techniques exist for the synthesis of these areas into exact numerical solutions for mixed-convection condensation of vapors in enclosures with noncondensables, except when noncondensable concentrations are uniformly high. For the large number of enclosure condensation problems to which boundary-layer-type analysis or other simplifying assumptions cannot be applied, empirical and approximate analysis techniques are employed with varying degrees of success and accuracy. At the simplest level, models neglecting transport and assuming local equilibrium are used, while at the next level empirically determined vapor film thicknesses are used to calculate transport rates, as discussed by Webb and Panagoulias [17]. Though film and related approximate analysis techniques are of great importance in the design of complex condenser geometries, it is beyond the scope of this chapter to discuss them adequately. Instead, extensive references are supplied in the sections where applications are discussed.

Solutions to a variety of species-controlled condensation problems can also be obtained without introducing empirical models. It is the purpose of this chapter to systematically identify and evaluate the assumptions available for nonempirical solutions of species-controlled condensation problems, and to relate these assumptions to the broad classes of condensation problems and applications which are of current interest to engineers and researchers. It is hoped that this assumption-based viewpoint will not only provide a useful list of the analytic and experimental tools now available, but will also identify the tools which need to be developed. In this chapter primary attention is paid to condensation of vapors in enclosures containing noncondensable gases, due to the importance of noncondensable effects in applications ranging from nuclear reactor safety analysis to aerospace systems. Attention is also devoted to condensation of nonazeotropic binary mixtures, since such mixtures are now being introduced in the new generation of refrigeration and heat pump equipment, as well as in other heat transfer applications.

Species-controlled condensation phenomena in enclosures is a complex process resulting from the interaction of combined forced and double-diffusive natural convection with highly nonlinear boundary conditions. To establish an understanding of the types of flows which can occur, Sec. 2 presents a description of various experimental results, relying heavily on a series of fundamental investigations by Peterson and Tien [18–22] on condensation in reflux thermosyphons containing noncondensable gases. Since species-controlled condensing flows are governed primarily by boundary effects, Sec. 3 summarizes and discusses the most important boundary conditions for systems with noncondensable gases, as well as for binary mixtures. Convenient phase-equilibrium relationships, which provide optimal fitting of equilibrium data specifically for condensation heat transfer problems, are also given. Section 4 then introduces and discusses a condensed list of the primary assumptions which have been introduced to permit analysis of species-controlled condensation. Section 5 provides motivation for continued study of species-controlled condensation effects by discussing several application areas which benefit from such studies. Finally, conclusions are drawn in Sec. 6, which details the future directions needed to surpass the current limitations in analysis of species-controlled condensation. Since the range of nomenclature which has been applied in mass transfer and condensation problems is wide and diverse, selection of a consistent set of variables provided some difficulty. Ultimately it was decided to follow the nomenclature list developed at the 8th International Heat Transfer Conference, as detailed by Mayhew [23].

2 EXPERIMENTAL OBSERVATIONS OF CONDENSATION IN ENCLOSURES

When noncondensable gases or binary mixtures are present inside an enclosure, the species transport and the resulting condensation on cooled walls or tube banks are complex. Condensation drives a forced flow of vapor from the evaporator to the cooled walls. The combined species and temperature gradients in the vapor phase initiate natural convection, potentially creating recirculation cells or even gross instabilities and periodic fluctuations. Condensation rates at the walls are governed by the combined nonlinear requirements of mass and energy conservation and thermal equilibrium. This creates a highly coupled system where relatively small changes of species concentration at vapor/liquid interfaces can strongly alter the condensation rate and the entire flow structure. Even in a simple geometry such as a reflux thermosyphon condenser, complex flows and heat transfer can be encountered. Perhaps the best way to understand the wide variety of flow patterns which appear during species-controlled condensation is through examination of experimental observations of condensation in thermosyphon condensers. The thermosyphon provides a direct analog for components important in reactor safety analysis and also for aerospace variable-conductance heat pipes. Furthermore, the insight gained from examining of the thermosyphon can be applied to a wide variety of more complex geometries.

Figure 2 illustrates a typical thermosyphon geometry. Vapor flows up from the evaporator and condenses on the walls, forming a liquid film which flows back to the

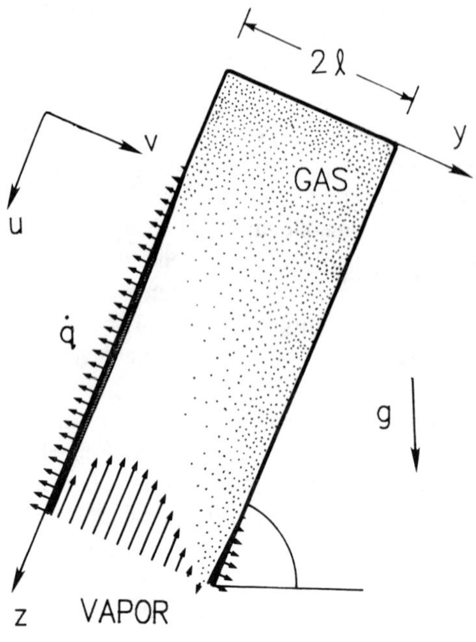

Figure 2 Schematic diagram of a thermosyphon condenser containing noncondensable gas.

evaporator. The noncondensable gas is trapped at the end of the pipe, where it shuts off condensation. In glass thermosyphons this effect is observed visually; mist forms in the interface region, while the glass above remains clear and dry. Between the upper shutoff region and the lower active region, gradients of species concentration and temperature exist. The resulting double-diffusive natural convection can be characterized by the Rayleigh number Ra, which, assuming ideal gas behavior, can be expressed as

$$\begin{aligned}\text{Ra} &= \frac{[\rho(T_b) - \rho_V]gl_w^3}{\mu D} \\ &= \frac{\rho_V g l_w^3}{\mu D}\left\{\frac{T_V}{T_b}\left[1 + \tilde{x}_{g,b}\left(\frac{M_g}{M_v} - 1\right)\right] - 1\right\}\end{aligned} \tag{1}$$

where $\rho(T_b)$ is the equilibrium density of the gas/vapor mixture at the coolant bulk temperature T_b, ρ_V is the density of the pure vapor, μ is the dynamic viscosity, D is the mass diffusivity, $\tilde{x}_{g,b}$ is the equilibrium mole fraction at the cooling-medium temperature, and M is the molecular weight. This Rayleigh number is particularly sensitive to the ratio of gas and vapor species molecular weights M_g and M_v; the total pressure, which determines the density of the pure vapor, ρ_V; and the difference between the coolant and pure vapor saturation temperatures, which influences the maximum concentration of noncondensable $\tilde{x}_{g,b}$. Close temperature approaches, or small differences between the cooling medium and pure vapor saturation temperatures, decrease the Rayleigh number by reducing the noncondensable concentration in the shutoff region, since the maximum gas partial pressure becomes small and thus also the maximum gas concentration ($\tilde{x}_{g,b} = 1 - P^s(T_b)/P_T$). Higher system pressures increase the Rayleigh number by increasing the total system density. This pressure effect permits precise adjustment of the Rayleigh number for studies of transition, bifurcation, and instability.

Since the distribution and transport of noncondensable gas is the most important factor controlling condensation in gas-loaded thermosyphons, experimental techniques which give direct information about the gas concentration are preferable to indirect methods. As illustrated later, the flow and recirculation of the noncondensable can be quite complex, so simple measurements of wall temperature or heat flux provide inadequate information about interior behavior. Laser holography, as seen in Fig. 1, gives excellent snapshots of the gas distribution, even for highly unstable or periodic flows. Laser holography is limited, however, to two-dimensional systems, in this case a planar heat pipe with two cooled side walls and two clear Pyrex walls, heated to prevent condensation. The most important parameter for laser holography is the number of fringes generated, given by the optical path length multipled by the difference between the maximum and minimum refractive indexes and divided by the light wavelength. Obtaining a sufficient number of fringes can be difficult, particularly when a relatively short path length is desired to ensure that the longest wavelength for instability is in the transverse direction. This difficulty is particularly severe at the lower pressures and densities required to obtain the lower Rayleigh-number values where the most interesting transitions and bifurcations occur. The problem can be alleviated somewhat by selecting gas/vapor combinations with widely different refractive indexes, particularly by using noble gases, which have much lower refractive indexes than multiatomic gas molecules of similar molecular weight. Still, it is difficult to obtain sufficient fringes in the low-density, low-positive-Rayleigh-number range where the most interesting bifurcations occur, and thus only negative and higher positive Rayleigh-number values have been studied with laser holography, where the gas is stagnant or fully unstable [24, 25].

The miniature wet-bulb probe provides an alternative technique for measuring noncondensable gas distributions [18]. The device consists of a small (1.5-mm) porous wet bulb with a thermocouple embedded in the center. A second nearby thermocouple measures the dry-bulb temperature. With the miniature wet-bulb probe, gas distributions can be mapped accurately. In air/water systems, condensation occurs on the probe, which eliminates the need for periodic rewetting. This condensation also helps explain the accurate concentration measurement results obtained by Rose [26] in steam condensers, within 0.1 to 0.2% when checked by direct measurements of mass flow rates. In the series of thermosyphon experiments discussed here, a wet-bulb probe was mounted on the end of a 6.4-mm-diameter stainless tube off-center in a 41.3-mm-ID condenser test section. Cooling water circulated in a jacket pipe over the condenser section, causing condensation. The probe could be moved axially and rotated to make measurements in an arc through the condenser, as shown in the scaled diagram in Fig. 3. More complete descriptions of the probe [18] and experimental apparatus [19] are provided elsewhere.

The simplest gas recirculatory flows occur when a thermosyphon is in a horizontal orientation. Figure 4 shows isoconcentration profiles resulting from condensation in the horizontal orientation with water as the working fluid (M_v = 18.0) and with both neon (M_g = 20.2) and a mixture of helium and argon (effective M_g = 20.2) as noncondensable gas [19]. The dashed and dotted curves show the maximum and minimum locations at which a given concentration was measured, while the distance between the curves corresponds to the degree of instability. The right edges of the plots represent the bottom of the horizontal thermosyphon, so it can be seen that the

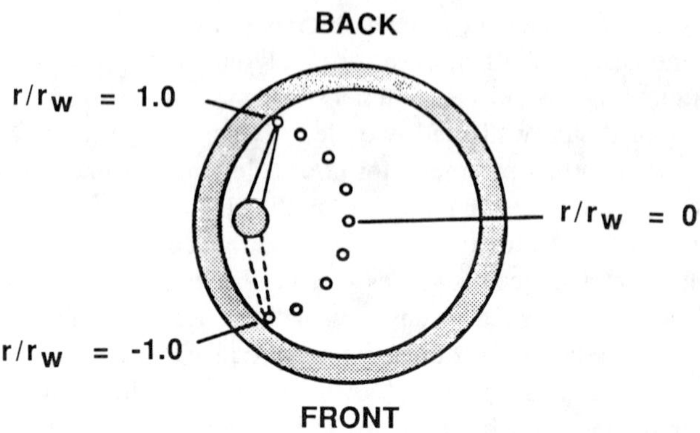

Figure 3 Cross section of thermosyphon condenser test section, showing wet-bulb temperature measurement locations [19].

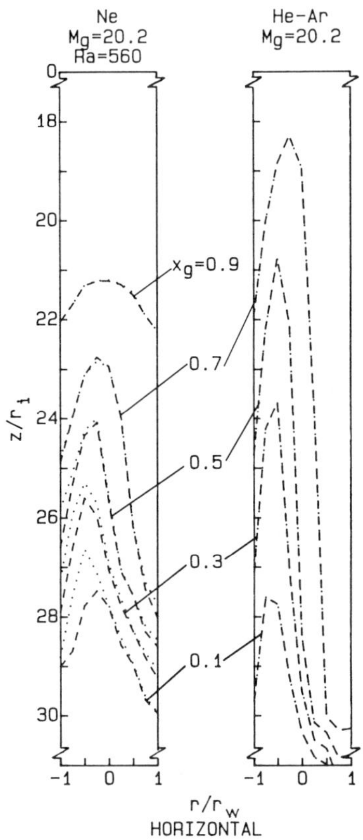

Figure 4 Gas isoconcentration profiles for horizontal thermosyphon (right is bottom) with neon and helium/argon mixture as noncondensable gases [19].

relatively light, warm water vapor tends to flow up over the noncondensable gas and condense on the top side of the tube. The cooler, heavier noncondensable gas flows down, blanketing the bottom side more than the upper side. The concentration gradients cause the noncondensable to diffuse into the counterflowing vapor, which carries the noncondensable and deposits it back near the top of the condenser test section, resulting in recirculation of the gas. It is interesting to note that, with the same average gas molecular weight, gas load, and power level, recirculation with the helium/argon mixture is much stronger than for pure neon. This strong recirculation is due to the much higher diffusion coefficient of helium relative to argon. The helium diffuses preferentially into the vapor, making the vapor significantly lighter than the argon-rich noncondensable gas mixture. The potential augmentation of natural convection by this separation effect should be considered in any system where multiple noncondensable species exist, such as in reactor safety analysis where both hydrogen and nitrogen can be present simultaneously.

In vertical and near vertical thermosyphon orientations, significantly more complex recirculatory behavior can occur, along with gross instability and pressure fluctuations at higher Rayleigh numbers. Figure 5 shows the progression of gas flow patterns observed in the vertical orientation for a water/air system [21]. Here the gas is heavier than the vapor, a situation expected to result in some gas recirculation. At the lowest

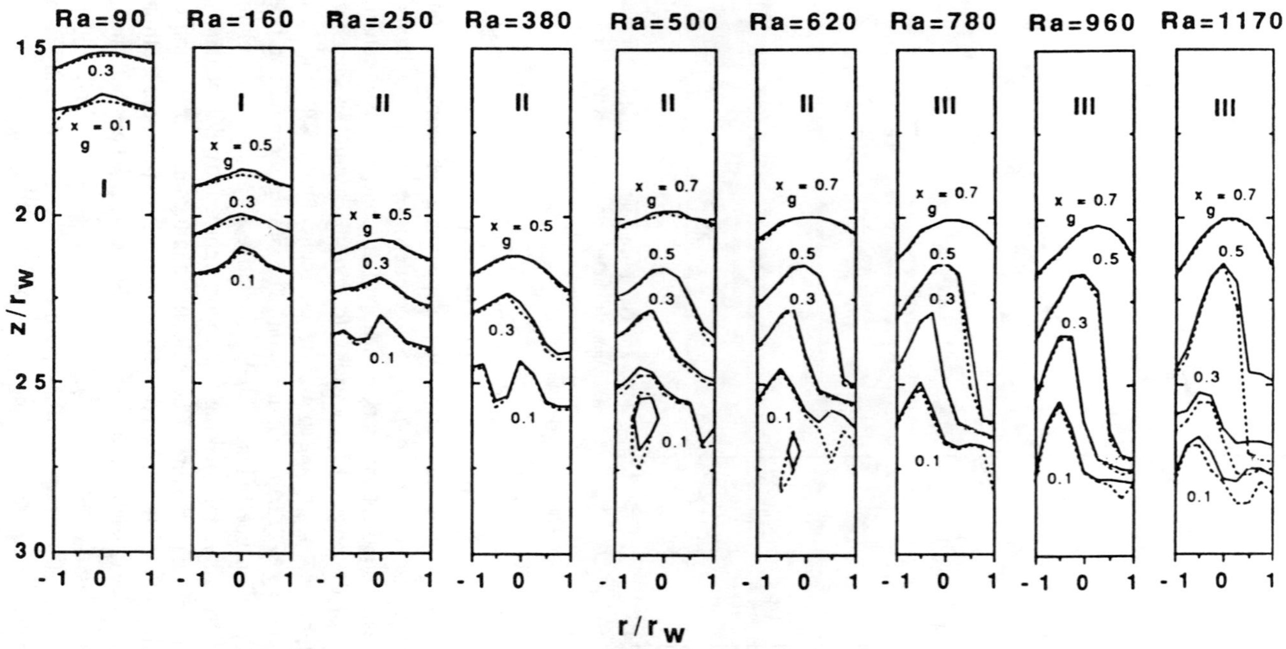

Figure 5 Gas isoconcentration profiles for vertical water/air system at low Ra [21].

Rayleigh number, Ra = 90, however, the gas distribution is virtually axisymmetric, with only slight distortion of the x_g = 0.1 isoconcentration profile due to torroidal recirculation of the gas flowing down along the wall and then being transported back up with the vapor flowing up the centerline. When the Rayleigh number is increased to Ra = 160 by adding a small amount of gas to the pipe, decreasing the area available for condensation and increasing the system pressure and vapor density, the strength of the toroidal gas recirculation cell increases. At Ra = 250 a bifurcation to a three-dimensional structure occurs, characterized by a kink in the x_g = 0.1 isoconcentration profile. With increasing Rayleigh number this kink lengthens and detaches. Since the concentration measurements were made in an arc (Fig. 3), the detached island in the figure is actually a slice through a pocket of noncondensable gas collected against the wall below the primary recirculation region. At Ra = 960 the transition from toroidal gas recirculation to a single roll cell is complete. Some instability of this single roll cell is then observed at Ra = 1170.

For the water/air system, both single roll cells and axisymmetric flow structures can exist in the range from Ra = 1380 to Ra = 2080 [21]. At still higher Rayleigh-number values the single roll cell again predominates. The single roll cell persists, growing in length, until Ra = 7700 is reached, at which point a sudden transition to highly unstable, periodically oscillating flow occurs. Power-density spectra at this transition point show a strong peak at a frequency of 0.008 Hz, as seen in Fig. 6. This frequency, when scaled with the mass diffusion coefficient D, gives a length of

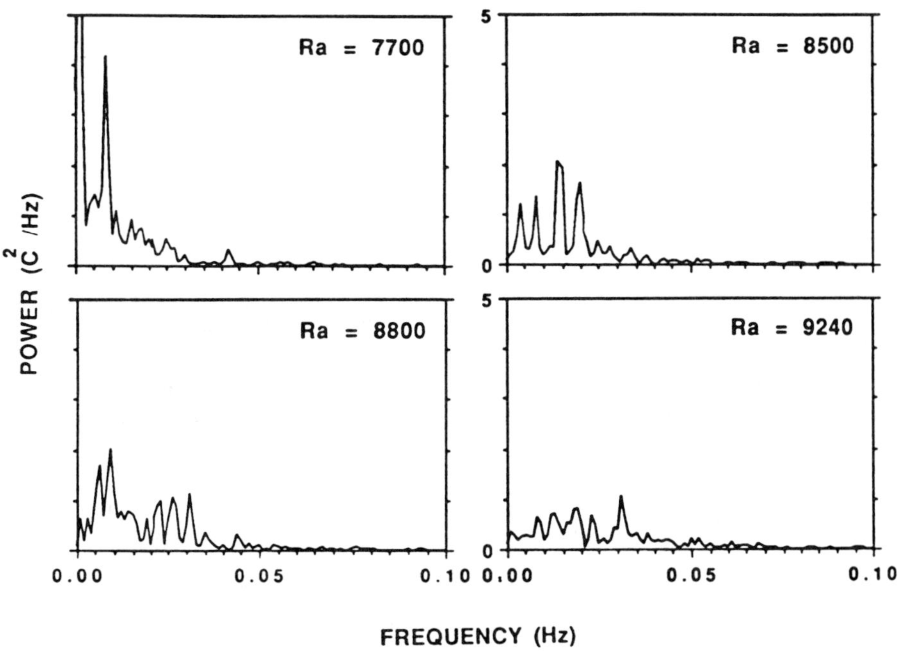

Figure 6 Power-density spectra for the wet-bulb temperature at high Ra [21].

approximately 2.5 pipe diameters, which along with other evidence suggests that the single roll cell is periodically destroyed by a type of Kelvin-Helmholtz instability. At still higher Rayleigh numbers, frequency doubling occurs until finally fully chaotic flow is observed at $\mathrm{Ra} > 9240$.

The potential complexity of the flows induced by natural convection during condensation in the simple thermosyphon condenser geometry helps explain the difficulty of integrating natural-convection effects into models of more complex condensing geometries, such as tube bundles. Even for the case of steady flow in a two-dimensional thermosyphon, exact numerical solutions are not available. Theorists studying bifurcation phenomena now regularly use numerical rather than physical experiments to examine Rayleigh-Bernard convection in rectangular enclosures; difficulties make this option unavailable, as yet, to researchers studying mixed convection in condensing systems. The difficulties do not arise from modeling of the vapor flow, which is straightforward. Rather, the difficulties arise in attempting to satisfy the nonlinear boundary conditions introduced by phase equilibrium and species conservation. Thus the next section examines these vapor/liquid boundary conditions, in order to identify where the difficulties lie.

3 VAPOR/LIQUID INTERFACE RELATIONS

The characteristics which separate multispecies enclosure condensation problems from other condensation and convection problems arise from consideration of phase equilibrium and the conservation of mass, species, and energy at the vapor/liquid interface. These characteristics introduce strong nonlinearity into the boundary conditions for the mass, momentum, species, and energy conservation equations, greatly complicating their solution and thus requiring the introduction of a variety of simplifying assumptions to make analysis tractable. This section discusses at length the specific relationships which govern the vapor/liquid interface for condensation of pure vapors with noncondensable gases present, and for condensation of binary mixtures. For other equations which are not specific to species-controlled problems, such as formulations for the momentum equation and other conservation equations, the reader is referred to the papers referenced here as well as the various texts available on natural convection, such as that of Bejan [11]. Following the discussion of interface conditions here, Sec. 4 introduces the various assumptions which can be introduced to make species-controlled condensation problems tractable, as well as the pitfalls and limitations inherent in those assumptions.

3.1 Application of Fick's Law

The most important relationship used in analysis of binary condensing systems is Fick's law, which may be derived from kinetic theory for ideal gases. The proper formulation for combined diffusive/convective transport is

$$\rho x_g \mathbf{u}_g = \rho x_g \mathbf{u} - \rho D \nabla x_g \tag{2}$$

where ρ is the mass density, x_g is the mass fraction, $\mathbf{u}$ is the mass average velocity vector, $\mathbf{u}_g$ is the absolute velocity vector for the gas or more volatile species, and D is the mass diffusion coefficient. Implicit in this formulation is the assumption that the primary mode of diffusion is due to concentration gradients. Mass diffusion due to temperature gradients, pressure gradients, and external forces is neglected. Reid et al. [28] provide a comprehensive summary of the methods for calculating diffusion coefficients for a variety of species combinations, as well as calculation methods for other transport properties. In molar quantities, Eq. (2) can be rewritten as

$$\tilde{\rho}\tilde{x}_g\mathbf{u}_g = \tilde{\rho}\tilde{x}_g\tilde{\mathbf{u}} - \tilde{\rho}D\,\nabla\tilde{x}_g \tag{3}$$

where $\tilde{\rho}$ is the molar density, $\tilde{x}_g$ is the mole fraction, and $\tilde{\mathbf{u}}$ is the molar average velocity. Often Eq. (2) is written in terms of the species density gradient rather than the mass fraction gradient. This approximate form based on the concentration gradient is given in the earlier heat transfer textbooks, but is valid only in the case of constant mixture mass density. Use of the concentration gradient results in considerable error when the molecular weights of the species are quite different, because the total mass density is far from constant. The molar form given in Eq. (3) is useful in applications where the gas is stagnant (Sec. 4.3), since it can be written in terms of the molar concentration gradient with only small errors, since molar densities of gas mixtures are relatively constant except in systems with large temperature or pressure gradients. In systems where it is necessary to apply the momentum equation, written in terms of the mass-average velocity, Eq. (2) is the most appropriate.

3.2 Pure Vapor and Noncondensable Gas

In most condensing systems, it can be expected that some noncondensable gases will be present in the condensing vapor. Such noncondensables can have particularly large effects in the low-velocity regions of enclosures which are of interest here. At the vapor/liquid interface in these and all other condensing systems, mass continuity requires that

$$\dot{m}_i = \rho_{L,i}v_{L,i} = \rho_{V,i}v_{V,i} \tag{4}$$

where the mass flux $\dot{m}_i$ is related to the interface densities $\rho_{L,i}$ and $\rho_{V,i}$ and normal mass-average interface velocities $v_{L,i}$ and $v_{V,i}$ in the liquid and vapor phases, respectively. For the case of a noncondensable gas, only the vapor species condenses. Thus it is possible to apply the restriction that the absolute normal gas velocity v_g equals zero at the interface. Consequently combination of Eqs. (2) and (4) provides a relationship for the condensing mass flux:

$$\dot{m}_i = \frac{\rho_{V,i}D}{x_{g,i}}\left(\frac{\partial x_g}{\partial y}\right)_i \tag{5}$$

This equation is nonlinear in x_g, and thus it approaches the proper limits as $x_g \to 1$, where the condensation rate is proportional to the slope of the gas fraction, and as $x_g \to 0$, where the condensation rate becomes indeterminate and other heat transfer mechanisms control the condensation rate. Similar forms of Eqs. (4) and (5) can be written in terms of molar quantities.

The energy balance at the interface matches the transport of latent heat and conduction from the vapor to heat removal by the liquid phase, $\dot{q}_{L,i}$,

$$\frac{h_{lV}\rho_V D}{x_{g,i}}\left(\frac{\partial x_g}{\partial y}\right)_i - k_V\left(\frac{\partial T_V}{\partial y}\right)_i = -\dot{q}_{L,i} \tag{6}$$

where h_{lV} is the latent heat of vaporization, k_V is the vapor-phase thermal conductivity, and T_V is the vapor temperature. Since, for condensing systems, the first term (latent heat transport) of Eq. (6) is commonly much larger than the second term (vapor-phase conduction), it is often best to neglect the second term to provide decoupling of the energy and species equations. An added benefit of neglecting the second term is that the constant latent heat value at the saturation temperature T_e^s can be used, which automatically satisfies the overall energy balance between the incoming vapor and the outgoing condensate, if a correction for condensate subcooling is added. There are several ways of treating the right-hand side of Eq. (6) to determine $\dot{q}_{L,i}$. In an exact sense, as is employed by Gerner and Tien [27] for condensation on a liquid sheet in axisymmetric flow, the heat is removed to the liquid by conduction:

$$\dot{q}_{L,i} = k_L\left(\frac{\partial T_L}{\partial y}\right)_i \tag{7}$$

where k_L is the thermal conductivity and T_L is the temperature of the liquid phase. For the case of film condensation on an externally cooled, relatively thin wall, such as for condensation in a thermosyphon or other enclosure, a more convenient approximation can be applied. Considering conjugate heat transfer, and summing the resistances of the liquid film, wall, and external heat transfer,

$$\dot{q}_{L,i} = \left(\frac{\delta_l}{k_L} + \frac{\delta_w}{k_w} + \frac{1}{h_w}\right)^{-1}(T_i^s - T_b) - k_w\delta_w\left(\frac{d^2T_w}{dx^2} + \frac{d^2T_w}{dz^2}\right) + \rho_w C_w \delta_w \frac{\partial T_w}{\partial t} \tag{8}$$

where δ_l is the liquid film thickness, T_i^s is the interface saturation temperature, T_b is the bulk cooling-medium temperature, t is time, x and z are transverse coordinates along the wall, h_w is the external wall heat transfer coefficient, δ_w is the thickness, k_w is the thermal conductivity, ρ_w is the density, and C_w is the specific heat of the wall material. This formulation neglects convective transport in the film, a standard assumption first proposed by Nusselt. The sensible heat of the liquid film is also neglected, but as shown by Rohsenow [29], this effect can be accounted for by adding a sensible heat contribution to the latent heat value in Eq. (6), $h'_{lV} = h_{lV} + 0.68c_{pl}(T_\infty^s - T_w)$, where c_{pl} is the liquid specific heat and T_∞^s is the bulk vapor saturation temperature. In some problems where the exterior heat transfer resistance is large relative to the interior film or capillary structure resistance, such as for many gas-loaded heat pipes [21, 30], it is convenient to assume that the second and first derivatives of the wall temperature are equal to the derivatives of the interface temperature, so that T_i^s can replace T_w in Eq. (8). For falling films a variety of variables can effect the film thickness and heat transport, such as vapor shear and turbulence.

Collier [31] provides a comprehensive discussion of these effects on film condensation. Film waviness effects can also be important, as discussed by Seban and Faghri [32] and by Chu and Dukler [33]. For the case of a wicked surface, δ_l is replaced by the thickness of the capillary structure and k_l by an effective structure thermal conductivity.

In addition to mass and energy conservation at the vapor/liquid interface, a third condition, thermodynamic equilibrium, is required to provide closure for the condensation problem. Since the interface termperature must exist between the limits of the bulk cooling-medium temperature T_b and the saturation temperature of the pure vapor T^s_∞ at the total system pressure P_T, the Clausius-Clapeyron relationship can be used to fit the saturation pressure between these two extremes. Assuming that the partial pressure of the gas obeys Dalton's law, the gas mole fraction at the interface is given by

$$\tilde{x}_{gi} = 1 - \exp\left[T_*\left(1 - \frac{T^s_\infty}{T^s_i}\right)\right] \tag{9}$$

The constant T_* is best given by

$$T_* = \left(1 - \frac{T^s_\infty}{T_b}\right)^{-1} \ln\left(1 - \frac{P^s_b}{P_T}\right) \tag{10}$$

where P^s_b is the saturation pressure at the cooling medium temperature T_b. This formulation is preferable to one based on interpolated values of the latent heat and specific volume, since it exactly matches the desired limit of $T^s_i \rightarrow T^s_\infty$ as $x_g \rightarrow 0$.

Equation (5) provides a normal velocity boundary condition for the momentum equations, coupling the solution of the momentum equations tightly to the species equation solution. Matching the liquid and vapor tangential interface velocities provides the second boundary equation for the momentum equations. For the species equations, Eqs. (6), (8), and (9) must be satisfied simultaneously, giving interface values of $\dot{q}_L$, T^s_i, and $x_{g,i}$. Difficulties can arise in applying the standard numerical techniques used for elliptic natural-convection problems, where the momentum and species equations are solved sequentially and iteratively. Until now numerical solutions for these elliptic problems have been limited to cases of high gas concentration (Sec. 4.2), where the suction velocity is not so sensitive to small changes in the interface concentration, or have required some relaxation of the constraint of interface impermeability to gas (Sec. 4.7). Since vapor conduction usually contributes negligibly to the interface energy balance (Eq. 6), the energy equation is loosely coupled and its primary role becomes determining the vapor temperature distribution and the resulting impact on the buoyancy forces. The equilibrium interface temperature T^s_i from solution of the species equation provides the proper boundary condition for the energy equation.

3.3 Binary Condensing Mixtures

Nonazeotropic binary mixtures have the characteristic property that at vapor/liquid interfaces in equilibrium the vapor and liquid compositions are different. Figure 7 shows how large this difference between vapor and liquid compositions can be for an R-11/R-113 mixture at different temperatures and two total pressures. For condensation

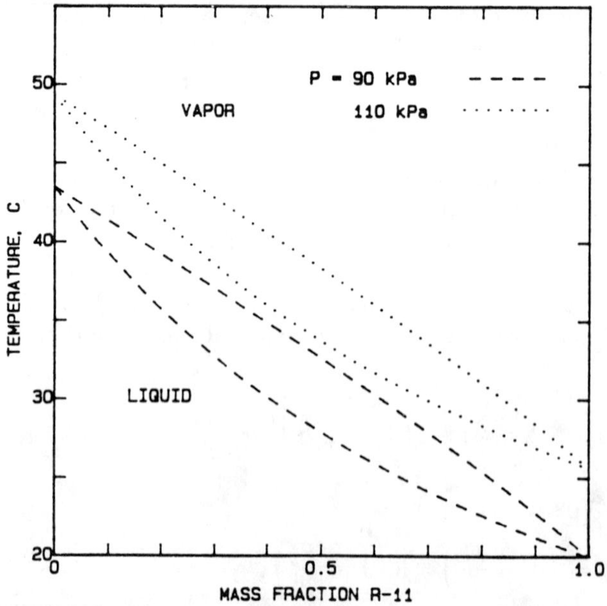

Figure 7 Binary-phase equilibrium diagram for R 11/R 113 [39].

to occur, one species must diffuse away from the interface and the other toward it, so that the equilibrium composition of liquid can condense. The primary problems in analyzing condensation of binary vapor mixtures center on diffusion and mixing in the liquid phase, since liquid diffusion rates are much lower and more difficult to quantify than those for gases. Luckily, as discussed by Denny and Jusionis [7], the same assumptions of Nusselt-type film flow, where convection of energy and momentum can be neglected, can be safely extended to include the convection of mass. For downward boundary-layer vapor flow, the parabolic equation forms make it possible to treat mass transport in laminar films exactly. Unfortunately, for the non-boundary-layer flows which can occur in enclosures, exact treatment of the film becomes difficult. However, two limiting behaviors do exist: no mixing or diffusion in the liquid film, and a well-mixed film.

For the case of a *well-mixed* film, the concentration across the film is uniform, and the liquid concentration at the interface $x_{L,i}$ must equal the average concentration of liquid deposited from the beginning of the film to the point of interest, or

$$x_{L,i} = \frac{-\int_0^x \rho_V(x_{V,i}v_V - D_V\,\partial x_V/\partial y)\,dx'}{\int_0^x \rho_V v_V\,dx'} \tag{11}$$

where v is the vapor interfacial condensation velocity and $x_{V,i}$ is the interfacial vapor concentration. This expression is difficult to apply in practice, since all upstream film conditions must be known to evaluate the integrals. In the case of an enclosure with total reflux, such as a binary thermosyphon [39], or an enclosure where the total mass being withdrawn is known, the average film composition can still be found directly by considering the counterflowing vapor. Given relatively even distribution, mass conservation at any axial location requires that the average liquid concentration flowing down equals the average vapor concentration flowing up, which greatly simplifies Eq. (11).

In the *unmixed* film limit, no mixing, diffusion, or surface renewal occurs in the liquid film. Applying mass conservation at the interface, as given in Eq. (4), coupled with Fick's law as given by Eq. (2) and the assumption of zero liquid diffusion,

$$D_V\left(\frac{\partial x_V}{\partial y}\right)_i = v_{V,i}(x_{V,i} - x_{L,i}) \tag{12}$$

Because the nonmixed model provides a simple boundary condition, it is widely applied in analysis of binary condensation problems. For condensation this unmixed film model works particularly well, since the liquid convection term at the liquid/vapor interface easily dominates the liquid diffusion term. However, under adiabatic conditions with low mass transfer rates, the actual film behavior lies somewhere between the mixed and unmixed models [39], while for evaporators the unmixed model is definitely not suitable and the concentration distribution in the liquid must be considered.

In addition to the species conservation represented in Eq. (11) or (12), an energy balance is required,

$$\frac{\bar{h}_{lV}\rho_V}{x_{V,i} - x_{L,i}}\frac{\partial x_V}{\partial y} - k_V\frac{\partial T_V}{\partial y} = \dot{q}_L \tag{13}$$

where the nonmixed film assumption Eq. (12) has been applied to obtain the condensing mass flux. In general, the latent heat of condensation is related to the pure component latent heats, plus an additional quantity referred to as the heat of mixing or excess enthalpy,

$$\bar{h}_{lV} = \Sigma\, \tilde{x}_n h_{lV,n} + \Delta h_E \tag{14}$$

For ideal liquid mixtures, which the fluorocarbon refrigerants closely approximate, $\Delta h_E = 0$. For other mixtures, particularly polar mixtures with water, the excess enthalpy is not zero. Unfortunately, nonzero excess enthalpy also implies that the phase equilibrium is nonideal, requiring more complicated relationships for evaluating the equilibrium interface concentrations, where Prausnitz et al. [36] provide a comprehensive reference. Christensen et al. [37], as well as others, have tabulated data for heat of mixing for a variety of binary mixtures. For binary combinations such as ammonia/water commonly used in heat transfer applications, tabular and graphical enthalpy and phase equilibrium information are often available from standard sources [38].

For fluorocarbon refrigerants and other near-ideal liquid mixtures, a convenient formulation for phase equilibrium can be developed [39], which has the same advantage of exactly matching the desired limiting behavior under condensing conditions as Eqs. (9) and (10) do for noncondensables. The constants in the relationship are calculated based on tabular data for the pure components. Only four data are required: the pure component saturation temperatures $T^s_{r,1}$ and $T^s_{r,2}$ at a reference pressure P_r, the saturation pressure of component 1, P_1^s, at $T_{r,2}^s$, and the saturation pressure of component 2, P_2^s, at $T_{r,1}^s$. The reference pressure P_r is chosen near the operating pressure P_T. If desired, the accuracy of the equation can then be tuned for small deviations from ideal liquid behavior by introducing experimental data for boiling pressures of liquid mixtures of known composition. This data arises naturally in the course of many binary mixture experiments.

For an equilibrium mixture, the vapor- and liquid-phase fugacities for each component n must be equal, $f_{n,V} = f_{n,L}$. The liquid-phase fugacity can be expressed as $f_{n,L} = \gamma_n x_{n,L} P_n^s$, where γ_n is the activity coefficient. The simplest expression for the activity coefficient, beyond setting it equal to unity, is the two-suffix Margules equation [36],

$$\ln \gamma_1 = \frac{A}{RT} \tilde{x}_{2,L}^2 \qquad \ln \gamma_2 = \frac{A}{RT} \tilde{x}_{1,L}^2 \tag{15}$$

where $\tilde{x}_{n,L}$ is the liquid-phase mole fraction of component n and R is the gas constant. The constant A is determined from experimental data for the mixture and is a weak function of temperature. Here A is assumed to remain constant over the range between the two limiting temperatures $T_{r,1}^s$ and $T_{r,2}^s$. The assumption of ideal-gas behavior permits the use of the Clausius-Clapeyron relation for the saturation pressure and temperature of each component. To provide the optimum fit over the range of temperatures encountered in the enclosure, the Clausius-Clapeyron relationship is written in terms of the exact saturation pressures at the two limiting temperatures. Hence,

$$P_n^s = P_r \exp\left[C_n \left(1 - \frac{T_{r,n}^s}{T} \right) \right] \tag{16}$$

The constants C_n are given by

$$C_1 = \frac{\ln(P_1^s/P_r)}{(1 - T_{r,1}^s/T_{r,2}^s)} \qquad C_2 = \frac{\ln(P_2^s/P_r)}{(1 - T_{r,2}^s/T_{r,1}^s)} \tag{17}$$

where P_1^s is the saturation pressure of component 1 at temperature $T_{r,2}^s$ and P_2^s is the saturation pressure of component 2 at $T_{r,1}^s$.

The ideal-gas assumption also allows the use of a simplified relationship for the vapor-phase fugacity, $f_{n,V} = \tilde{x}_{n,V} P_T$, where $\tilde{x}_{n,V}$ is the vapor mole fraction of component n. Then simple expressions for the liquid- and vapor-phase equilibrium mole concentrations, respectively, are

$$\tilde{x}_{1,L} = \frac{(P_T/P_r) \exp[-G_2(T)] - 1}{\exp[G_1(T) - G_2(T)] - 1} \tag{18}$$

$$\tilde{x}_{1,V} = \frac{(P_r/P_T) \exp[G_2(T)] - 1}{\exp[G_2(T) - G_1(T)] - 1} \tag{19}$$

where

$$G_1(T) = C_1\left(1 - \frac{T_{r,1}}{T}\right) + \frac{A\tilde{x}_{2,L}^2}{RT}$$

$$G_2(T) = C_2\left(1 - \frac{T_{r,2}}{T}\right) + \frac{A\tilde{x}_{2,L}^2}{RT} \qquad (20)$$

Equations (18) and (19) are simple relationships for equilibrium concentrations as functions of temperature and pressure, which can be programmed easily and are useful for binary mixtures which approach ideal behavior, such as fluorocarbon refrigerant mixtures. The Margules constant A can be set equal to zero, or it can be found by fitting a set of experimentally measured values of P_T, T and $\tilde{x}_{n,L}$. Neither Eq. (18) or (19) is explicit in T, but if $\tilde{x}_{1,L}$ or $\tilde{x}_{1,V}$ is specified, a Newton-Raphson iterative scheme converges rapidly to the desired value of T. The binary phase diagram for R-11 and R-113 presented in Fig. 3 was obtained using the above relations. The constant A was found to equal zero, within the accuracy of the experimental measurements, showing the ideality of this fluorocarbon refrigerant mixture [39].

For binary mixture condensation, Eq. (12) can provide a normal velocity boundary condition for the momentum equations, coupling the momentum equation solution tightly to the species equation solution as with noncondensable gases. Matching of the liquid and vapor tangential interface velocities provides the second boundary equation for the momentum equations. For the species equation, Eqs. (8), (13), and (20) must be satisfied simultaneously, giving interface values of $\dot{q}_L$, T_i, and $x_{2,V}$. The equilibrium interface temperature from solving the species equation provides the proper boundary condition for the energy equation. Since vapor conduction is neglected in the interface energy balance, coupling of the energy equation is loose and its solution affects only the vapor temperature distribution and the resulting impact on the buoyancy forces. For binary mixtures at higher driving temperature differences, the mass diffusion resistance decreases as the diffusion layer becomes thinner due to suction, and in the limit of high temperature differences the mixture condenses like a pure vapor, as predicted first by Sparrow and Marschall [6] and later by Hijikata and Himeno [46] and others. There are as yet no available numerical solutions for the elliptic problem of binary mixture condensation in enclosures.

3.4 Enhanced Surfaces for Species-Controlled Condensation

Enhanced surfaces employing a variety of surface geometries such as ribs or fins can greatly increase condensation heat transfer [41–45], often by factors of 10 or more. This exceptional performance is related primarily to increased surface area, capillary effects, and increased film turbulence. In particular, enhanced surfaces are commonly used to increase shell-side condensation. In general, the performance of a surface optimized for condensation of a pure vapor can be greatly diminished by the presence of noncondensable gas or binary mixtures. At the expense of some loss of performance in condensation of pure vapors, surfaces can be optimized to perform well in the presence of small quantities of noncondensable gas. As discussed by Mori et al. [40]

and as illustrated in Fig. 8, when the fin spacing and height are large relative to the thickness of the noncondensable diffusion layer (case A), the heat transfer can be calculated based on the actual surface area including the fin surface area. When the noncondensable gas diffusion layer is thick, relative to the fin spacing and height (case B), a limit is approached where condensation is the same as for a smooth surface. Thus, to obtain acceptable enhancement with noncondensable gases present, larger fin heights and spacing are required.

In general, enhancement effects by fin geometries are similar for binary mixtures and noncondensables [10], and improved binary-mixture condensation enhancement occurs for larger fin spacing and heights. With binary mixtures the diffusion-layer thickness is a strong function of the driving temperature difference and suction rate, so that enhancement effects with binary mixtures become the same as with pure vapors when condensation rates are high.

3.5 Fog and Mist Production

In enclosure problems where thermal diffusion exceeds mass diffusion, the potential exists that subcooling will cause nucleation of mist droplets in the vapor phase. Specifically, when the Schmidt number is greater than the Prandtl number (i.e., steam/air mixtures), mist formation can occur. Such mist formation has been observed experimentally in air/water systems for natural-convection boundary-layer condensation [5] and for condensation on the underside of cooled horizontal surfaces [47]. Mori and Hijikata [5] applied a boundary-layer integral technique to analyze mist formation in a free-convection noncondensable/vapor boundary layer. They determined that, in general, the formation of mist increases the condensation heat flux, by up to a factor of 2 at higher driving temperatures differences. Though mist formation may increase the condensation heat flux, as noted by Steinmeyer [49], mist is normally viewed unfavorably since some means, such as a mist eliminator, is commonly required to remove the mist droplets. Mist formation can also be expected to have a strong effect on binary vapor condensation, where extraction of the less volatile component in the mist droplets will significantly change the composition of the vapor at the condensing

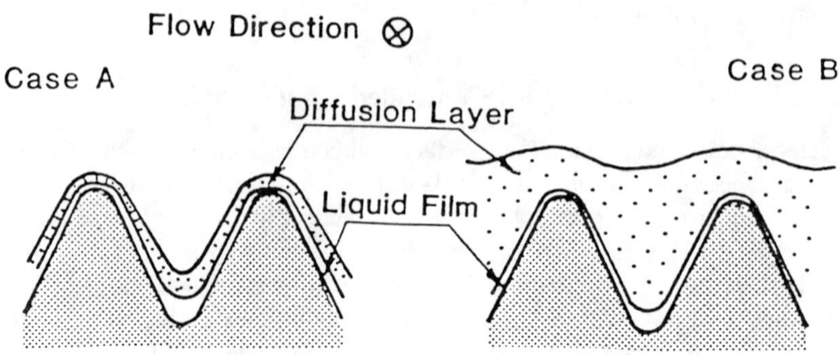

Figure 8 Schematic diagram of the two limiting cases of vapor condensation on a finned plate with noncondensable gas [40].

interface. Studzinski et al. [48] provide criteria that can be applied for the nucleation and growth of droplets in binary vapor systems.

4 SIMPLIFYING ASSUMPTIONS AND APPROXIMATIONS

As noted in the previous section, the nonlinear nature of the relationships governing behavior at the liquid/vapor interface, particularly the nonlinearity introduced by Eqs. (5) and (9) for noncondensable gases, can make analysis of species-controlled enclosure condensation problems particularly difficult. As a result of this difficulty, various workers have introduced a number of simplifying assumptions to make these problems tractable. Some of these assumptions, such as neglecting interfacial resistance, thermal diffusion (Soret), diffusion thermo (Dufour), and thermal radiation effects relative to the mass diffusion resistance [2], can be employed with good results in most enclosure condensation problems. However, there are other common assumptions, discussed in this section, which create clear limits on the range of problems which can be considered. Table 1 illustrates the widespread application of these assumptions by sampling a limited number of papers from the rich literature on various species-controlled condensation problems. This compilation shows where current analytic abilities are lacking, most particularly in dealing with elliptic behavior such as recirculating condensing flows. This table also provides warnings to those who use research results which contain inherent limitations due to assumptions commonly employed. For instance, the results of all the papers which use simplified interface convection relationships (Sec. 4.2) cannot be applied in systems where the noncondensable gas concentration is small, a consideration which limits their practical application to a wide variety of condensation problems.

4.1 Parabolic or Boundary-Layer Behavior

Consideration of boundary-layer flows permits the use of parabolic forms of the conservation equations. These forms can be integrated numerically with relative ease, even when coupled by nonlinear interface equations such as Eqs. (5) and (9). For the case of natural convection of noncondensables in boundary layers, where the gas species is heavier than the vapor, Sparrow and Lin [1] show that similarity solutions are possible. When variable-property and other effects are considered in the natural-convection problem, the computational effort is increased considerably for similarity solutions [2]. By assuming concentration and velocity profiles, Rose [3] applied integral techniques to develop a simple algebraic equation, giving good agreement with the similarity solution of Sparrow and Lin [1] while eliminating the need for a numerical solution. Integral techniques have since been applied to a wide range of condensation problems due to their relative simplicity (i.e., [5, 9]). For the condensation of binary mixtures with downward-developing boundary layers, Sparrow and Marschall [6] also showed similarity solutions are possible. For downward forced convection the concentration profiles are very nonsimilar, and marching finite-difference techniques are required both for noncondensable gases (Denny et al. [4]) and binary mixtures (Denny

Table 1 Compilation of assumptions employed in papers on species-controlled condensation, with section numbers

Author/Year/Ref. #	Boundary-layer behavior, Sec. 4.1	Neglect interfacial convection, Sec. 4.2	Stagnant gas, Sec. 4.3	Constant wall temperature, Sec. 4.4	Simplified natural convection, Sec. 4.5	Constant transport properties, Sec. 4.6	Relax interface impermeability, Sec. 4.7
Noncondensable gases							
Sparrow and Lin, 1964 [1]	X			X		X	
Minkowycz and Sparrow, 1966 [2]	X			X			
Rose, 1969 [3]	X			X		X	
Denny et al., 1971 [4]	X			X	X		
Mori and Hijikata, 1973 [5]	X			X		X	
Hu and El-Wakil, 1974 [12]		X					
Chang et al., 1986 [13]	X	X					
Lin et al., 1988 [14]	X	X		X		X	
Trevisan and Bejan, 1987 [15]						X	X
Close and Sheridan, 1989 [16]		X		X		X	
Gerner and Tien, 1989 [27]					X		
Edwards and Markus, 1972 [30]			X		X		
Hijikata et al., 1984 [51]			X		X	X	
Peterson and Tien, 1988 [19]			X	X	X	X	
Peterson and Tien, 1989 [20]			X		X	X	
Rohani and Tien, 1973 [34]				X		X	X
Galaktionov and Trukhanova, 1985 [35]				X		X	X
Kobayashi and Matsumoto, 1987 [25]				X		X	X
Peterson and Tien, 1990 [22]				X		X	X
Binary mixtures							
Sparrow and Marschall, 1969 [6]	X			X		X	
Denny and Jusionis, 1972 [7]	X			X	X		
Schroppel, 1981 [8]	X				X		
Hijikata et al. 1986 [9]	X			X		X	
Hijikata et al., 1987 [10]	X			X	X	X	
Peterson et al., 1990 [39]	X			X	X	X	

and Jusionis [7]). Similarly, use of the parabolic forms of the conservation equations allows solutions for mass transfer in binary thermosyphons [39], again with a finite-difference marching technique.

The clear advantages of boundary-layer analysis accounts for the large number of papers employing the assumption, a small selection of which are seen in Table 1. Unfortunately, for many problems of practical interest, elliptic system behavior defeats application of parabolic boundary-layer solutions. Elliptical behavior is introduced when boundary layers develop upward, when conjugate conduction exists in the wall, or when the ultimate low-velocity fate of a noncondensable or more volatile component is considered in some stagnant region of an enclosure. For boundary layers and liquid films developing in opposite directions, Hijikata and Himeno [46] showed that the film and vapor resistances can be calculated separately using a reasonable interface temperature value and then summed, giving acceptable results. For many more complex problems the boundary-layer approach must be abandoned.

4.2 Simplified Interfacial Convection Relationships

For condensation with noncondensables present, when the gas concentration is relatively high throughout the enclosure (as in air-conditioning problems), simplified interfacial convection relationships can be applied. The exact application of Fick's law requires consideration of both the diffusive flux and convective flux of each species. For an interface impermeable to gas, this consideration gives rise to the proper form for the condensation mass flux, Eq. (5), as reproduced here:

$$\dot{m}_i = \frac{\rho_{V,i} D}{x_{g,i}} \left(\frac{\partial x_g}{\partial y} \right)_i \tag{5}$$

When $x_{g,i} \rightarrow 1$, this equation has similar form to that for heat transfer, $\dot{q}_i = k_V[\partial T_V/\partial y]_i$. When $\dot{m}_i$ and $\dot{q}_i$ are specified at the boundaries of an enclosure, numerical solutions for the resulting natural convection become straightforward and the double-diffusive effects in enclosures can be analyzed as shown by Trevisan and Bejan [15]. For the wet-bulb thermometer, and many other evaporation problems, the relatively high gas concentration at the vapor/liquid interface permits setting $x_{g,i} = 1$ in the denominator of Eq. (5). When $x_{g,i} \rightarrow 1$, the suction velocity due to condensation can become quite small relative to free-stream velocities. Then the effects of interface suction can be neglected, and correlations available for heat transfer can be applied calculate the mass transfer coefficient. The only requirement is substitution of the Schmidt number for the Prandtl number, because the governing equations and boundary conditions for heat and mass have identical forms. Unfortunately, improper application of the analogy between heat and mass transfer can result in large errors for condensing systems with noncondensable gases, since often gas fractions can vary from zero up to the equilibrium concentration at the cooling-medium temperature, giving rise to large variations in the suction velocity. Heat transfer relationships which include suction are available. When such relationships are used, then the analogy between heat and mass transfer can be employed successfully at lower gas concentrations. Rose [26] provides a more complete discussion of the use of the analogy of heat and mass transfer for condensation with noncondensables present.

When the gas mole fraction is large throughout an enclosure, it is also possible to solve numerically the full elliptic evaporation and condensation problem for an enclosure, fully satisfying the requirements of phase equilibrium. Hu and El-Wakil [12] and Close and Sheridan [16] provide such solutions, but these solutions cannot be extended to cases where any relatively pure vapor exists in the enclosure. In the commonly encountered case where pockets of noncondensable gas exist in an enclosure, the condensation suction velocities found short distances from the pockets can be of the same order as the free-stream velocity. In numerical solutions, small fluctuations of the species concentration in these areas of low gas concentration create large changes in the calculated suction velocity [Eq. (5)], providing a difficult source of numerical stability. Thus current solutions for condensation with locally low gas concentrations exist only for the stagnant gas case, where the momentum equation can be eliminated (Sec. 4.3) and for cases where the requirement of interface impermeability to gas is relaxed (Sec. 4.7).

In contrast to systems with noncondensable gases, the analogy of heat and mass transfer can work much better for binary mixtures, providing rapid solutions for problems such as condensation on enhanced tubes, as shown by Hijikata et al. [10]. For the case of an unmixed film, the condensation rate varies with the difference of the vapor and liquid concentrations, $x_{V,i} - x_{L,i}$, as seen in Eq. (12). For nearly ideal mixtures, such as refrigerants, this difference remains almost constant over the range of condensing temperatures, as seen in Fig. 7.

4.3 Stagnant Noncondensable Gas

The computational difficulties introduced by the coupling of the species and momentum equations can be avoided in a very convenient manner for cases where the noncondensable gas is trapped in some stagnant portion of an enclosure, and can be assumed to remain stationary. The assumption that the gas species velocity is small allows elimination of the momentum equation, simplifying analysis greatly. Because this technique is not presented in any of the standard texts, it is valuable to present the formulation here. One of the earliest applications of this technique can be credited to B. Mikic, as cited by Sakhuja [50]. Since then the stagnant gas technique has been employed by others to determine gas distributions in reflux thermosyphons and heat pipes [19, 20, 22, 51]. For the derivation it is best to start with the molar form of Fick's law, given by Eq. (3). When the divergence of Eq. (3) is taken and continuity for the gas species is applied, the species conservation equation can be derived. However, here a modified form of the species equation is obtained by considering continuity for the mixture instead,

$$\nabla \cdot (\tilde{\rho}\tilde{\mathbf{u}}) = 0 \tag{21}$$

The formulation requires the assumption of constant molar density, $\tilde{\rho}$, which is valid as long as temperature and pressure variations are not too large through the system, so that $\nabla \cdot \tilde{\mathbf{u}} = 0$. The mass diffusion coefficient D is also assumed constant, since no solution for the temperature field is attempted. Dividing Eq. (3) by $\tilde{\rho}\tilde{x}_g$, combining Eqs. (3) and (21), noting that $(\nabla\tilde{x}_g)/\tilde{x}_g = \nabla \ln \tilde{x}_g$, and assuming that the gas remains stationary, $\mathbf{u}_g = 0$, results in

$$\nabla^2(\ln \tilde{x}_g) = 0 \tag{22}$$

This is the modified species equation. It has the same form as the conduction equation, and with appropriate linearization can be solved easily with iterative finite-difference techniques or integral techniques, giving the gas distribution in the enclosure directly. Typical boundary conditions which can be applied include

$$\frac{\partial(\ln \tilde{x}_g)}{\partial y} = 0 \tag{23}$$

for impermeable, adiabatic surfaces. For condensing surfaces,

$$h_{lv} M_v \tilde{\rho}_V D \frac{\partial \ln \tilde{x}_g}{\partial y} = -\dot{q}_L \tag{24}$$

where M_v is the vapor species molecular weight and $\tilde{q}_L$ can be evaluated using Eq. (8). Also required at a condensing surface is an equilibrium condition to relate the mole fraction and interface temperature, as given by Eq. (9). At the entrance plane of the system, the vapor mole average velocity $\tilde{v}$ can be specified by

$$\tilde{v} = \mathrm{D} \frac{\partial(\ln \tilde{x}_g)}{\partial y} \tag{25}$$

Given the form of Eq. (22), solutions can be obtained easily using standard iterative finite-difference techniques for the conduction equation. Iterative numerical solution of Eq. (22) can proceed rapidly if the Clausisus-Clapeyron relation, Eq. (9), is linearized through a Taylor series expansion,

$$\ln \tilde{x}_{g,i} = \ln \tilde{x}_{g,o} + (T_i^s - T_{i,o}^s) \frac{T_* T_\infty^s}{T_{i,o}^{s2}} \left\{ 1 - \exp\left[T_* \left(1 - \frac{T_\infty^s}{T_{i,o}^s} \right) \right] \right\}^{-1} \tag{26}$$

where the previous values of the interface temperature $T_{i,o}^s$ and mole fraction $\ln \tilde{x}_{g,o}$ are updated each iteration. Equations (8), (24), and (26) can be combined algebraically to eliminate T_i^s and $\dot{q}_L$, giving a linear mixed boundary condition for $\ln x_{g,i}$. This formulation has been solved with finite-difference techniques, neglecting wall conduction [22]. In some cases the gas concentration distribution exhibits some degree of similarity along a centerline, as with a thermosyphon or heat pipe. In this case integral techniques can be employed to reduce the problem to coupled ordinary differential equations. Using such integral techniques, wall conduction effects have been included in solutions using the stagnant gas assumption [20].

Figure 9 compares results of the stagnant gas approximation, Eq. (22), for gas isoconcentration profiles and velocity vectors between two infinite parallel plates with results for a full numerical simulation including the momentum equation [22]. Use of the stagnant gas approximation is seen to result in a slug-flow velocity profile, similar to velocity profiles for potential flow, except in this case the driving force is the

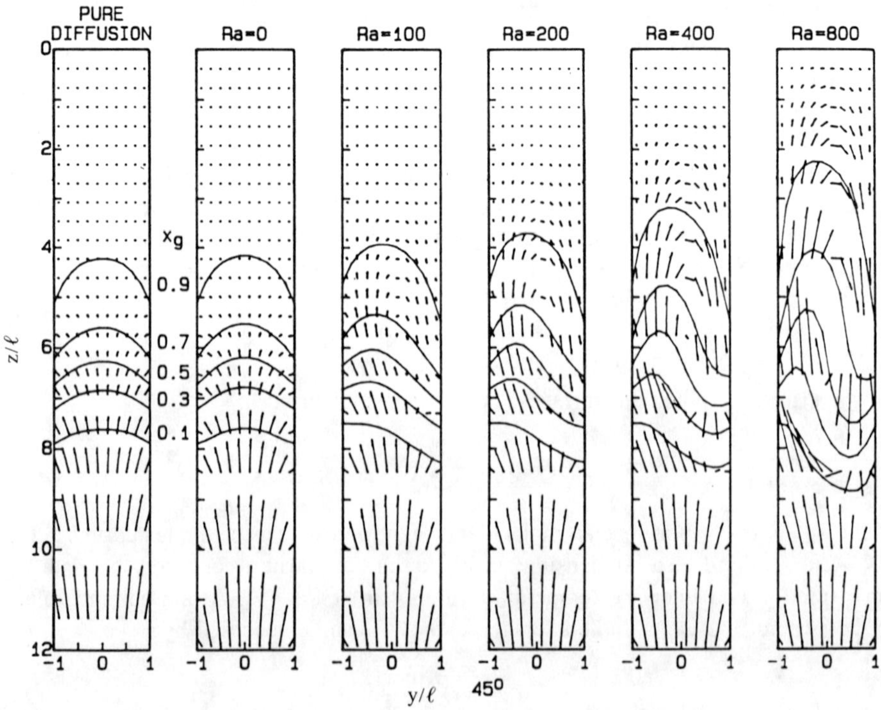

Figure 9 Comparison of approximate [Eq. (22)] and exact numerical isoconcentration profiles and velocity vectors for condensation between infinite plates with noncondensable gas [22].

concentration gradient rather than a pressure gradient. Given the approximate nature of the modified species equation, it is still seen to give excellent agreement for the isoconcentration profiles and the total heat transfer. Good agreement is also obtained with experimental measurements [20].

4.4 Constant Wall Temperature

Most analyses of species-controlled condensation assume a constant wall temperature, or perhaps a linearly varying wall temperature as considered by Schroppel [8]. They neglect the effects of conjugate heat transfer in the wall material, as well as the effects of the finite heat transfer resistance from the wall to the cooling medium. However, in most condensing systems, the exterior resistance from the wall to the cooling medium is of the same order as the film resistance. When the exterior heat transfer resistance has significant magnitude, any spatial variations of the interface temperature due to variations of species concentration will induce conjugate heat transfer in the wall. Particularly in heat pipes and thermosyphons containing noncondensable gases or binary mixtures, steep temperature gradients can exist due to strong variation of the vapor/liquid interface species concentration [19, 39]. Under conditions where significant temperature gradient exist, heat transport along the wall due to the temperature gradient can be of the same order as the condensation heat flux. This effect is particularly strong when the exterior heat transfer resistance is significantly larger than the interior film resistance. For the particularly steep temperature gradients which exist in gas-loaded heat pipes, wall conduction can significantly change the gas distribution, even for walls constructed of low-thermal-conductivity materials such as glass [20]. For unsteady-flow situations, thermal inertia of the wall material becomes an important factor regulating the species concentration at the vapor/liquid interface.

In cases where significant heat transfer gradients exist in the wall, a relationship such as Eq. (8) containing wall conduction terms must be employed to calculate the local condensation flux. Equation (8) implicitly assumes that the wall material is thin, so that conduction in the wall can be assumed to be two- or one-dimensional. Wall conduction is seldom included in boundary-layer calculations, because the elliptic nature of the wall conduction terms eliminates use of the parabolic boundary-layer formulation. Conjugate heat transfer can telegraph information back upstream, making marching numerical techniques ineffective.

Under certain conditions, energy transport via wall conduction can greatly exceed the condensation flux in stagnant regions of an enclosure. Such conditions often occur in gas-loaded heat pipes constructed from aluminum, stainless steel, or metals of similar thermal conductivity. A clearly defined interface exists between the active region, where the vapor is essentially pure and condenses easily, and the stagnant region, where gas significantly decreases the condensation flux. Because the stagnant region at the end of the pipe is at a lower temperature, heat is conducted along the wall and the wall in the stagnant region acts as a fin, transporting significant amounts of heat from the active region. Analysis of such systems can be greatly simplified by considering the wall in the stagnant region as a fin, and solving for the temperature distribution neglecting condensation [52]. From the resulting temperature distribution the quantity of gas in the stagnant region can be calculated by assuming equilibrium at the local wall temperature and integrating the gas concentration over the stagnant volume. This technique makes it possible to analyze the multiple heat pipe radiators now being employed in satellites, where several heat pipe condensers are coupled to a single radiating panel, and interact via conduction.

4.5 Negation of Natural-Convection Effects

In any enclosure where complete condensation occurs, there will necessarily be some regions of low velocity. Given that these are the regions where noncondensable gas or the more volatile component would be expected to accumulate, and given that this accumulated mixture is cooled to a significantly lower temperature and will have a different effective molecular weight than the condensing vapor, some natural convection can be expected to occur. As discussed in Sec. 2, experimental observations show that under some conditions natural-convection effects can be quite strong, and can even result in instable, periodic flows and pressure oscillations.

Because both concentration and temperature gradients drive natural convection in species-controlled condensation, these flows are double-diffusive. Normally double-diffusive effects are considered most interesting when the diffusion rates of thermal energy and mass are significantly different, as occurs for ice melting in salt water. Another double-diffusive effect becomes important in species-controlled condensation problems with noncondensables, making these problems unique. Here double-diffusive effects are augmented by the nonlinear boundary conditions for temperature and concentration. In regions of low gas concentration, small changes in gas concentration along the vapor/liquid interface result in large changes of temperature, making thermal gradients stronger than species gradients. The opposite effect occurs in regions of high gas concentration, where large changes in species concentration result in only modest temperature gradients. This boundary effect provides the most reasonable explanation for the asymmetry of the concentration profiles at the bottom of the gas/vapor interface region in Fig. 1*b*. Here the stabilizing effect of the light gas, helium, is overcome by the much stronger gradient of wall temperature, creating a small region of recirculation. The nonlinear relationship between interface temperature and concentration implies that even for the case of unity Lewis number, the temperature and species concentration profiles are different, as seen in the numerical results shown in Fig. 10 for condensation in a thermosyphon with noncondensable gas. The nonlinear boundary effects make it impossible to employ the assumption of unity Lewis number, an assumption which is convenient in combustion and other gas-phase mass transfer problems.

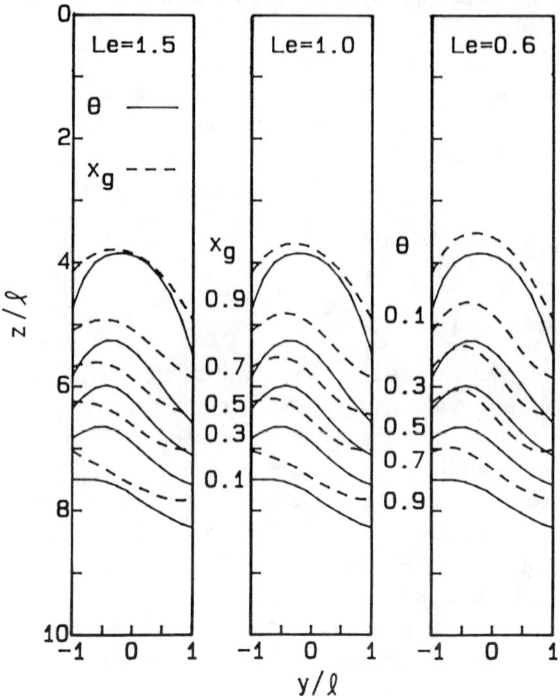

Figure 10 Numerical results for isotherms and isoconcentration profiles in a reflux thermosyphon containing noncondensable gas, for different Lewis numbers [22].

A wide body of literature is available on natural convection in enclosures, and the techniques for dealing with these problems have been summarized by Bejan [11]. A limited amount of literature is available covering double-diffusive convection in enclosures. Hu and El-Wakil [12] provide the earliest experimental and numerical investigation of simultaneous condensation and evaporation in an enclosure with high gas concentrations. Trevisan and Bejan [15] later examined double-diffusive effects under combined constant mass and heat flux boundary conditions. Chang et al. [13], Lin et al. [14], and Close and Sheridan [16] all numerically investigated combined evaporation and condensation in enclosures. The last three works employ the assumptions discussed in Sec. 4.2, limiting application to problems where no condensation of pure vapor occurs anywhere in the system. Other workers have treated the full problem of combined natural and forced convection in two-dimensional thermosyphons [22, 24, 35], but in all cases they relaxed the constraint of gas impermeability at the vapor/liquid interface. Peterson and Tien [22] provide a technique for minimizing the gas flux and thus minimizing the error introduced. As yet, no exact numerical solutions have been obtained for the ostensibly simple problem of condensation of a relatively pure vapor entering an enclosure containing noncondensable gas. Likewise, no numerical solutions exist for natural convection during condensation of binary mixtures in enclosures.

4.6 Constant Transport Properties

The analysis of a wide variety of heat and mass transport problems is eased considerably when constant transport properties are assumed. However, improper application of the assumption of constant transport properties can introduce significant error, so at the very least some valid basis for selecting the constant property values should be employed. Minkowycz and Sparrow [2] provide a thorough investigation of variable-property effects for natural-convection film condensation with noncondensables present. They recommend evaluation of liquid film properties at a reference temperature,

$$T_r = T_w + 0.31(T_\infty^s - T_w) \tag{27}$$

where T_w is the wall temperature and T_∞^s is the saturation temperature of the bulk vapor mixture. Denny and Jusionis [7] note that the transport properties of binary liquid mixtures, particularly viscosity, can vary strongly with composition. They recommend using Eq. (27), based on the assumption that the film composition is constant across the film. This constant composition assumption breaks down if the film is not well mixed and the interface composition changes along the film, since some concentration stratification will result. To provide a reference temperature and concentration for the vapor phase, Lin et al. [14] make use of the recommendations of Hubbard et al. [53] for droplet evaporation,

$$T_r = T_w + \frac{1}{3}(T_\infty^s - T_w) \tag{28}$$

$$x_r = x_w^s + \frac{1}{3}(x_\infty - x_w^s) \tag{29}$$

where x_∞ is the bulk vapor mass fraction and x_w^s is the equilibrium mass fraction at the wall temperature. As yet no detailed assessment of vapor variable property effects exists for enclosure condensation problems. Reid et al. [28] provide techniques for calculating the transport properties of mixtures.

4.7 Relax Interface Impermeability to Noncondensable

One means of obtaining numerical solutions for the problem of condensation of a vapor in an enclosure containing noncondensable gas is to relax the constraint of interface impermeability to the noncondensable. Though not ideal, this technique has permitted various investigators to obtain numerical results. Rohani and Tien [34] and Galaktinov and Trukhanova [35] replaced the requirement of gas impermeability with the assumption of a parabolic temperature profile for the three nodes next to the wall. Kobayashi and Matsumoto [24] used boundary values measured experimentally as inputs for solution of the vapor-space momentum, species, and energy equations. Peterson and Tien [22] used boundary values from a stagnant gas solution (Sec. 4.3) as inputs for a numerical solution with natural convection. These boundary values were then shifted axially to minimize the total quantity of gas passing through the interface. Though these solutions are not exact, they do provide important information about the vapor-space flow dynamics. The most important point is that some quantification of the error introduced by relaxing the impermeability constraint, in terms of a gas mass flux, should be provided.

5 APPLICATIONS

A broad range of applications provides the best motivation for increased understanding of condensation in enclosures containing noncondensables or binary mixtures. These applications range from precision control of high-temperature ovens to highly uniform heating and bending of fine polypropylene filaments for anchoring replacement lenses for cataract patients. This section discusses a few specific areas where species-controlled condensation is particularly important: power-plant surface condensers, reactor safety analysis, geothermal and ocean thermal energy conversion, refrigeration and air conditioning, electronics, and aerospace systems.

5.1 Surface Condensers for Steam Power Plants

Noncondensable gases play a primary role in the design and performance of surface condensers for steam power cycles. For a 1000-MW unit, a typical condenser will contain approximately 600 km of tubing, occupy a volume on the order of 750 m^3, and maintain a vacuum in the range of 40 mbar. Vacuum conditions make some leakage of air into such large systems inevitable, so all such condensers are designed to route noncondensable gases to air-cooling sections, where the concentration of the noncondensable gases can be increased by condensing more vapor, so the gas can be efficiently removed by vacuum pumps. Proper design of such air-removal systems is essential to prevent air blanketing of condenser surfaces and significant reductions in condenser performance, as discussed by Standiford [54]. Rowe et al. [55] note that for older condenser designs it has been found valuable to measure cooling water and vapor temperatures under operating conditions to locate air pockets, and then to remove some tubes to improve flow and entrainment of the air, and to convert other strategically located tubes to air-removal duty.

The design of surface condensers to minimize noncondensable-gas effects and maximize the efficiency of gas removal has been aided by the development of software programs for calculating the flow and heat transfer in condenser tube banks. Rowe et al. [55], in one of the earlier models, found that noncondensable gases create only small thermal resistance as long as the vapor velocities remain above 1 m/s. They concluded that the most important design consideration was to ensure that the steam flow area decreased continuously through the air-cooling section to maintain adequate velocity, and that the section be tightly baffled to prevent steam bypassing into the air pumping system. Chisholm [56] reviewed heat and mass transfer in steam/air mixture, developing a model based on stagnant film models and their modifications. Fujii [57] later recommended use of a mass transfer coefficient based on theoretical analysis of laminar forced-convection condensation of steam-air mixtures on flat plates, and recommended providing individual tube nests for high-speed-steam, low-speed-steam, and air-cooling sections, each to be analyzed independently. Al-Sanea et al. [58] treated the tube nest as a two-dimensional porous medium, solving the equations for continuity, momentum, and species conservation. They concluded that the incorporation of empirical data to calculate heat transfer locally had the most significant effect on the results. These models, and the other models not referenced, typically apply volume averaging or tube-row by tube-row calculations and effective heat and mass transfer coefficients, rather than using exact solutions.

The value of condenser models is determined primarily by how effectively the tube-scale physics of condensation mass and heat transfer are incorporated into the constitutive relationships for the volume-averaged or row-by-row solutions. Of particular importance is the transport of noncondensables. Gas transport mechanisms can include entrainment in boundary layers, as well as the convection and diffusion included in the above models. Andrews et al. [59] found the effect of entrainment of air in air-rich boundary layers surrounding falling condensate to be quite significant, with the potential for falling air bridging the gaps between tubes. They noted that previous models cannot explain the observed sizes of air blankets, except by employing high values of the turbulent diffusion coefficient, beyond what would be expected in stagnant air-blanketed regions. The instability observed in reflux thermosyphons operated with air/water mixtures at pressures typical of surface condensers [21] may provide another explanation for this high effective diffusion coefficient. Such unstability-induced mixing is highly undesirable, since the tubes remain blanketed, while steam is mixed with the vapor between the tubes, increasing the load on the air-pumping system.

A final motivation for improving the design of air-cooling sections centers on the deaeration role the surface condenser plays. As noted by Oliker and Katsman [60], the condenser can potentially serve as an effective deaerator for removing oxygen and carbon dioxide, the primary causes of corrosion in power-plant systems. The most important consideration is minimization of the contact of condensate with noncondensable-rich vapor to reduce the absorption of gas in the condensate.

5.2 Reactor Safety Analysis

Noncondensable gases can have potentially strong effects on condensation during postulated reactor loss-of-coolant accidents (LOCA). In these scenarios, condensation can play an important role in removing reactor decay heat and providing reflux liquid for core cooling. Under certain conditions, condensation in the reactor steam generator tubes is important [61], while direct-contact condensation in the reactor cold legs can also remove decay heat [62]. Condensation in the steam accumulators also affects transient behavior [63].

It is anticipated that noncondensable gases will be present in some quantity during LOCA events. Potential sources include dissolved nitrogen from the accumulators, dissolved air in the emergency core-cooling water, air inleakage if the system experiences subatmospheric pressures, and hydrogen gas generation by fuel cladding-water reaction. The need to quantify the gas transport through the reactor system and the resulting impact on condensation heat transfer has motivated incorporation of gas effects into the thermal-hydraulic system codes used for safety analysis of reactor systems. Nithianandan et al. [63] discuss the methods used to quantify gas effects in the RELAP5/MOD2 code, as well as experiments used to validate the code. Under turbulent flow conditions, a stagnant film model is employed, while for laminar flow the analogy between heat and mass transfer is used. For experiments for downward-condensing flow in a tube with a water seal at the bottom, the model predicted well the tube temperature profile 60 s after a quantity of nitrogen was added. After only 600 s, the nitrogen had disappeared, presumably by absorption into the condensate. With downward flow of nitrogen, the region between the shutoff and active zones was stable. Because of the importance of this type of condensation problem in reactor safety analysis, Hein et al. [61] also performed a thorough investigation. They found similar behavior as Nithianandan for stable systems, but for unstable conditions (nitrogen in upward flow, helium in downward flow) they observed a very long transition region and large fluctuations in the tube-wall temperature. This gross instability is also consistent with observations during the thermosyphon instability studies of Peterson and Tien [21].

Nithianandan et al. [63] also applied RELAP5/MOD2 to condensation onto the water pool in a pressurizer with noncondensable gas present. Here the code performed well for the experiments using nitrogen, but greatly underpredicted the condensation rate for helium experiments, an effect attributed to neglection of buoyancy effects, which tend to isolate helium above and away from the pool surface. Since experimental results were virtually identical for pure steam and for condensation with helium, good agreement was obtained when the helium was assumed to be separated from the pool surface.

A third problem of importance in reactor safety analysis is the condensation of steam on the stream of cold water which may be injected into the cold legs during an accident. Theofanous [62] examined this problem, and obtained much lower condensation rates when he considered the presence of hydrogen gas, with no natural convection. As with pressurizers, cold-leg water injection could also be a situation where natural convection is important, especially considering the large size of the cold legs and the resulting large value of the Rayleigh number.

5.3 Geothermal and Ocean Thermal Energy Conversion

The vertical temperature gradients which can exist in the ocean and the hot water which comes from geothermal wells both offer essentially free if limited supplies of low-grade heat which can be converted to electricity with well-designed thermal cycles. Additionally, such systems can become sources of clean, essentially distilled water. Since the supply temperatures of these sources, particularly ocean sources, are relatively low, cycle efficiencies are much lower than for more common fossil fuel and nuclear power plants. Thus, even though the energy input is free, the system equipment is expensive and must be designed with extreme care, especially considering the highly corrosive nature of ocean and geothermal water. Three primary cycles exist for extracting mechanical work from these fluids. In the first cycle vapor is flashed directly from the warm fluid at one or more pressures, and the resulting steam is expanded through special low-pressure turbines and then condensed. In the second cycle a closed heat exchanger is used to transfer heat to a second working fluid and the vapor generated is expanded through a turbine and then condensed. In the third cycle the warm fluid is expanded to a low pressure directly through a nozzle, so the fluid flashes in the nozzle creating a high-velocity, two-phase jet which can drive a turbine, after which the vapor is condensed. In the second cycle using a second working fluid, binary working fluid mixtures offer the best promise for high cycle efficiency, due to their ability to condense and evaporate at a spatially varying temperature. This varying temperature permits use of the Lorentz cycle, which minimizes the temperature difference across the evaporator and condenser heat transfer surfaces and thus maximizes cycle efficiency. The first and third methods eliminate the need for an evaporator heat exchanger by flashing the warm water directly, but require special considerations for adequate condenser designs. Both ocean and geothermal sources contain large amounts of noncondensable gases, which are liberated with the vapor when the hot water is flashed. The steam condenser designs must deal with these noncondensable gases, which are present in much higher concentrations than are found in standard steam surface condensers (Sec. 5.1). Kreith and Bharathan [64], in their review of ocean thermal energy-conversion systems, give estimates that even with optimal condenser designs, 10% to 15% of the gross plant power output will be required to pump the noncondensables generated by flashing ocean water from the condenser.

5.4 Refrigeration and Air Conditioning

Currently, refrigeration and air-conditioning cycles and equipment designs are undergoing reassessment, since such equipment consumes a major portion of the fluorocarbon refrigerants which deplete ozone and contribute to the greenhouse effect. New refrigerants are being assessed which either do not contain the chlorine which creates these negative effects, or which have relatively short lifespans in the atmosphere. The new refrigerants typically are more expensive, and possess less desirable thermophysical properties, necessitating improved equipment designs. Simultaneously, binary mixtures of refrigerants are receiving serious consideration as working fluids, due to their ability to condense and evaporate at temperatures which vary along the length of the heat exchanger, increasing cycle efficiency. The best examples of successful application of binary working fluids and enhanced surfaces come from the large development project sponsored by MITI (Ministry of International Trade and Industry) in Japan, which has built high-efficiency heat pump/cooling machines with coefficients of performance approaching values double those of conventional systems [65].

Noncondensable gases can be expected to leak into refrigeration systems operating at subatmospheric pressures, such as those using refrigerants R-11 and R-113. Such low-pressure systems are normally equipped with purging devices for removing the noncondensable gases. Noncondensable gases can also collect in high-pressure chillers using refrigerants such as R-12 and R-22 due to insufficient initial evacuation, chemical breakdown of the oil or refrigerant, or leakage of water if water pressures are sufficiently high in the evaporator or condenser. In smaller air-cooled systems, condensation occurs inside tubes. A receiver vessel collects the condensate, and also provides a volume where noncondensables can be isolated away from condensing surfaces. In larger, water-cooled condensers, condensation typically occurs on the shell side of a heat exchanger, and noncondensable gases tend to collect and blanket tube surfaces. In addition to creating a thermal resistance which decreases efficiency, such noncondensable gases are also major contributors to corrosion. For these reasons the effects of noncondensable gases on condenser performance have received considerable attention, as reviewed by Webb and Wanniarachchi [66] and Webb [43].

For low-pressure refrigeration condensers, the purge unit typically consists of a separate small condenser opening at low temperature, where vapor condenses, leaving behind a relatively large concentration of noncondensable gas. When a certain concentration is reached, based on the pressure in the purge condenser, a compressor is activated to expel the vapor/gas mixture, resulting in some discharge of refrigerant to the atmosphere. Such purgers are typically connected by pipe to a single point near the water-inlet end of the condenser. Such designs are not necessarily optimal, since the potential exists that gases could accumulate and blanket surfaces at stagnant points in the condenser away from the purge outlet, without ever being removed. This potential for multiple stagnant regions provides a primary argument for a distributed set of venting points [43]. Purge unit efficiency and capacity are also increased if the vent points are located optimally in regions of high gas concentration. The goal of efficiency enhancement makes accurate evaluation of the transport and distribution of noncondensables critical for optimal purging design. Because the molecular weights of fluorocarbon refrigerants are typically in excess of 100, while typical noncondensables are approximately 30, natural-convection effects can be expected to be stronger than those found for steam/air systems where the vapor/gas molecular weights are both smaller and closer together. With refrigeration condensers, the direction of the buoyancy force is also reversed compared to steam condensers with air, encouraging noncondensable migration to the top of the condenser, rather downward.

5.5 Electronics Cooling

Failure rates of electronic components, particularly the microcircuits used in LSI, VLSI, and ULSI computer chips, increase rapidly with increasing temperature, and also with thermal cycling. Since the power density of chips has increased simultaneously with larger numbers of devices on the chips, cooling techniques capable of removing heat fluxes on the order of 10 W/cm^2 or greater are currently under continuing development. Forced convection of water can provide among the highest heat transfer coefficients, but water is a difficult fluid for electronics cooling applications due to corrosion and relatively high electrical conductivity. Thus nucleate boiling or forced-convection evaporation of fluorocarbon fluids provides the highest heat removal capability, since the dielectric nature of fluorocarbons permits direct immersion of the electronic components in the fluid, allowing direct removal of heat. Design of condensers for these heat transfer systems requires consideration of noncondensable-gas effects, due to the very large solubility of air in typical dielectric fluids and difficulties in completely sealing enclosures. Kraus and Bar-Cohen [67] provide an extensive discussion of the design of condensers for immersion cooling systems. At low power-generation rates cooling occurs by natural convection. As power input increases, nucleate boiling initiates at the component surfaces. Because there are questions about the reliability of separate, vapor-space condensers, due to noncondensable gas blanketing, current attention has been focused on immersion of the condenser cooling coils in the liquid. This arrangement subcools the liquid, so that, initially, vapor bubbles collapse over short distances due to heat transfer to the liquid, while at higher power levels the bubbles reach the condenser surfaces and condense directly. With submerged condensers, consideration must still be made for purging the noncondensables that remain after condensation of vapor bubbles, and of blanketing of condenser surfaces at high power-dissipation rates.

5.6 Aerospace

Convection is not available in satellites for cooling of electronic and other heat-generating components. Thus heat pipes have been employed to transport heat to panels where the energy can be radiated to space. The conceptually simple heat pipe uses evaporation at one end of a tube and condensation at the other end to transport heat, with the condensed liquid returned through a capillary structure on the pipe wall. Dunn and Reay [68] provide a comprehensive discussion of heat pipes. In such heat pipes it is commonly desirable to maintain the evaporator at a constant temperature, regardless of the power input, to reduce thermal cycling and improve the operating characteristics of the cooled components. This thermal control is best accomplished by adding a noncondensable gas to the pipe, along with a reservoir at the end of the condenser to increase the total volume of noncondensable gas. The noncondensable gas blankets a portion of the condenser, shutting it off to condensation. When the power input increases, the system pressure increases, compressing the gas toward the reservoir and exposing more condenser surface. The condenser surface area adjustment helps minimize the evaporator temperature increase. The most important consideration in design of these variable-conductance devices is accurate evaluation of the quantity and distribution of gas in the shutoff region and the resulting heat transfer. This need has prompted development of both one-dimensional [30] and two-dimensional [20] models which include the important effects of conjugate heat transfer. Natural-convection effects can also be of importance [22], not in space, but rather during ground testing, since pipe behavior during ground tests is hoped to be similar to behavior in microgravity. Variable-conductance and constant-evaporator-temperature behavior can also be obtained with binary mixtures of working fluids, without the requirement of a heavy reservoir vessel, making binary heat pipes potential candidates for aerospace applications [39].

6 CONCLUSIONS

A long list of applications exists where species-controlled condensation with noncondensables or binary mixtures must be considered. Additionally, study of such problems requires consideration of a diverse set of heat transfer phenomena of fundamental importance, ranging from condensation to instability and flow bifurcation. In this chapter, the current state of ability in analysis of enclosure condensation with noncondensables or binary mixtures has been presented. The assumptions currently required to make such problems tractable were listed and discussed at length, and the problems which continue to defeat analysis, particularly steady and fluctuating condensation in enclosures with low gas fractions, were identified. A closely related field, the analysis of large, complex condenser systems using volume averaging and other semiempirical techniques, was discussed only briefly. Most of the discussion of the semiempirical models now existing for condenser analysis appeared in the section covering applications, because such models are currently the best tools for condenser design, showing their large importance. What is hoped is that future research will provide a closer wedding of these practical models and the more fundamental mixed-convection and diffusion-controlled flows which are observed and modeled at the thermosyphon scale.

REFERENCES

1. E. M. Sparrow and S. H. Lin, Condensation Heat Transfer in the Presence of Noncondensable Gas, *J. Heat Transfer*, vol. 86, pp. 430–436, 1964.
2. W. J. Minkowycz and E. M. Sparrow, Condensation Heat Transfer in the Presence of Noncondensables, Interfacial Resistance, Superheating, Variable Properties, and Diffusion, *Int. J. Heat Mass Transfer*, vol. 9, pp. 1125–1144, 1966.
3. J. W. Rose, Condensation of a Vapour in the Presence of a Non-Condensing Gas, *Int. J. Heat Mass Transfer*, vol. 12, pp. 233–237, 1969.
4. V. E. Denny, A. F. Mills, and V. J. Jusionis, Laminar Film Condensation from a Steam-Air Mixture Undergoing Forced Flow down a Vertical Surface, *J. Heat Transfer*, vol. 93, pp. 297–304, 1971.
5. Y. Mori and K. Hijikata, Free Convective Condensation Heat Transfer with Noncondensable Gas on a Vertical Surface, *Int. J. Heat Mass Transfer*, vol. 16, pp. 2229–2240, 1973.
6. E. M. Sparrow and E. Marschall, Binary, Gravity-Flow Film Condensation, *J. Heat Transfer*, vol. 91, pp. 205–211, 1969.
7. V. E. Denny and V. J. Jusionis, Effects of Forced Flow and Variable Properties on Binary Film Condensation, *Int. J. Heat Mass Transfer*, vol. 15, pp. 2143–2153, 1972.
8. J. Schroppel, Impact of Wall Temperature Variation on the Film Condensation of Binary Gas-Vapour Mixtures, *Int. J. Heat Mass Transfer*, vol. 24, pp. 165–170, 1981.
9. K. Hijikata, Y. Mori, N. Himeno, M. Inagawa, and K. Takahasi, Free Convective Condensation of a Binary Mixture of Vapors, in *Heat Transfer 1986: Proc. Eighth Int. Heat Transfer Conf.*, San

Francisco, August 1986 (ed. C. L. Tien, V. P. Carey, and J. K. Ferrell), Hemisphere, Washington, D.C., vol. 4, pp. 1621–1626, 1986.
10. K. Hijikata, H. Himeno, and S. Goto, Forced Convection of a Binary Mixture of Vapors, *JSME Int. J.*, vol. 30, pp. 1951–1956, 1987.
11. A. Bejan, *Convection Heat Transfer*, John Wiley, New York, pp. 159–201, 1984.
12. C. Y. Hu and M. M. El-Wakil, Simultaneous Heat and Mass Transfer in a Rectangular Cavity, in *Heat Transfer 1974*, JSME/JSChE, Tokyo, vol. 5, pp. 24–28, 1974.
13. C. J. Chang, T. F. Lin, and W. M. Yan, Natural Convection Flows in a Vertical, Open Tube Resulting from Combined Buoyancy Effects of Thermal and Mass Diffusion, *Int. J. Heat Mass Transfer*, vol. 29, pp. 1543–1552, 1986.
14. T. F. Lin, C. J. Chang, and W. M. Yang, Analysis of Combined Buoyancy Effects of Thermal and Mass Diffusion on Laminar Forced Convection Heat Transfer in a Vertical Tube, *J. Heat Transfer*, vol. 110, pp. 337–344, 1988.
15. O. V. Trevisan and A. Bejan, Combined Heat and Mass Transfer by Natural Convection in a Vertical Enclosure, *J. Heat Transfer*, vol. 109, pp. 104–112, 1987.
16. D. J. Close and J. Sheridan, Natural Convection in Enclosures Filled with a Vapour and a Non-Condensing Gas, *Int. J. Heat Mass Transfer*, vol. 32, pp. 855–862, 1989.
17. D. R. Webb and D. Panagoulias, An Improved Approach to Condenser Design Using Film Models, *Int. J. Heat Mass Transfer*, vol. 30, pp. 373–378, 1987.
18. P. F. Peterson and C. L. Tien, A Miniature Wet-Bulb Technique for Measuring Gas Concentrations in Condensing or Evaporating Systems, *Exp. Heat Transfer*, vol. 1, pp. 1–15, 1987.
19. P. F. Peterson and C. L. Tien, Gas-Concentration Measurements and Analysis for Gas-Loaded Thermosyphons, *J. Heat Transfer*, vol. 110, pp. 743–747, 1988.
20. P. F. Peterson and C. L. Tien, Numerical and Analytical Solutions for Two-Dimensional Gas Distribution in Gas-Loaded Heat Pipes, *J. Heat Transfer*, vol. 111, pp. 598–604, 1989.
21. P. F. Peterson, N. Elkouh, K. W. Lee, and C. L. Tien, Flow Instability and Bifurcation in Variable-Conductance Themosyphons, *J. Heat Transfer*, vol. 113, pp. 158–165, 1991.
22. P. F. Peterson and C. L. Tien, Mixed Double-Diffusive Convection in Gas-Loaded Heat Pipes, *J. Heat Transfer*, vol. 112, pp. 78–83, 1990.
23. Y. R. Mayhew, Use of Physical Quantities, Units, Mathematics and Nomenclature in Heat Transfer Publications, *Exp. Heat Transfer*, vol. 2, pp. 149–172, 1989.
24. Y. Kobayashi and T. Matsumoto, Vapor Condensation in the Presence of Noncondensable Gas in the Gravity Assisted Thermosyphon, *Proc. Sixth Int. Heat Pipe Conf.*, Grenoble, France, May 25–29, Paper No. 3.1.1., 1987.
25. Y. Kobayashi and T. Matsumoto, Two-Dimensional Condensing Vapor Flow on Parallel Flat Plates in an Enclosure, *J. Thermophys. Heat Transfer*, vol. 1, pp. 122–128, 1987.
26. J. W. Rose, Condensation in the Presence of Noncondensing Gases, in *Power Condenser Heat Transfer Technology* (ed. P. J. Marto and R. H. Nunn), Hemisphere, New York, pp. 151–162 (see also comments, p. 185), 1981.
27. F. M. Gerner and C. L. Tien, Axisymmetric Interfacial Condensation Model, *J. Heat Transfer*, vol. 111, pp. 503–510, 1989.
28. R. C. Reid, J. M. Prausnitz, and Bruce E. Poling, *The Properties of Gases and Liquids*, 4th ed., McGraw-Hill, New York, 1988.
29. W. M. Rohsenow, Heat Transfer and Temperature Distribution in Laminar Film Condensation, *Trans. ASME*, vol. 78, pp. 1645–1648, 1956.
30. D. K. Edwards and B. D. Markus, Heat and Mass Transfer in the Vicinity of the Vapor-Gas Front in a Gas-Loaded Heat Pipe, *J. Heat Transfer*, vol. 94, pp. 155–162, 1972.
31. J. G. Collier, *Convective Boiling and Condensation*, 2nd ed., McGraw-Hill, New York, 1981.
32. R. A. Seban and A. Faghri, Wave Effects on the Transport to Falling Laminar Liquid Films, *J. Heat Transfer*, vol. 100, pp. 143–147, 1978.
33. K. J. Chu and A. E. Dukler, Statistical Characteristics of Thin, Wavy Films: Part II. Studies of the Substrate and Its Wave Structure, *AIChE J.*, vol. 20, pp. 695–706, 1974.
34. A. R. Rohani and C. L. Tien, Steady Two-Dimensional Heat and Mass Transfer in the Vapor-Gas Region of a Gas-Loaded Heat Pipe, *J. Heat Transfer*, vol. 95, pp. 377–382, 1973.

35. V. V. Galaktionov and L. P. Trukhanova, Study of the Process of Heat and Mass Transfer in the Region of the Vapor-Gas Front in a Gas-Regulable Heat Pipe, *J. Eng. Phys.*, vol. 48, pp. 296–300, 1985.
36. J. M. Prausnitz, R. N. Lichtenhaler, and E. Gomes de Azevedo, *Molecular Thermodynamics of Fluid-Phase Equilibrium*, Prentice-Hall, Englewood Cliffs, N.J., 1986.
37. J. J. Christensen, L. D. Hansen, and R. M. Izatt, *Handbook of Heats of Mixing*, John Wiley, New York, 1982.
38. *ASHRAE Handbook 1985 Fundamentals*, American Society for Heating, Refrigeration and Air Conditioning, Atlanta, 1985.
39. P. F. Peterson, K. Hijikata, and C. L. Tien, Variable-Conductance Behavior in Two-Phase Binary Thermosyphons, *J. Thermophysics and Heat Transfer*, vol. 4, pp. 325–331, 1990.
40. Y. Mori, K. Hijikata, and P. Wei, Noncondensable Gas Effect on High Performance Vertical Condensing Surface with Parallel Low Fins, ASME-JSME Thermal Engineering Joint Conf., Honolulu, vol. 3, pp. 297–303, 1983.
41. Y. Mori, K. Hijikata, S. Hirasawa, and W. Nakayama, Optimized Performance of Condensers with Outside Condensing Surfaces, *J. Heat Transfer*, vol. 103, pp. 96–102, 1981.
42. P. J. Marto and R. H. Nunn, The Potential of Heat Transfer Enhancement in Surface Condensers, in *Condensers: Theory and Practice*, Inst. Chem. Eng. Symp. Ser. No. 75, Pergamon Press, Oxford, pp. 23–39, 1983.
43. R. L. Webb, Shell-Side Condensation in Refrigerant Condensers, *ASHRAE Trans.*, vol. 90(1B), pp. 5–25, 1984.
44. R. L. Webb, T. M. Rudy, and M. A. Kedzierski, Prediction of the Condensation Coefficient on Horizontal Integral-Fin Tubes, *J. Heat Transfer*, vol. 107, pp. 369–376, 1985.
45. A. E. Bergles, Techniques to Augment Heat Transfer, in *Handbook of Heat Transfer Applications*, McGraw-Hill, New York, pp. 3-1-80, 1985.
46. K. Hijikata and N. Himeno, Condensation of Azeotropic and Nonazeotropic Binary Vapor Mixtures, in *Annual Review of Heat Transfer*, Vol. 3 (ed. C. L. Tien), Hemisphere, Washington, D.C., pp. 39–84, 1990.
47. R. P. Stein, D. H. Cho, and G. A. Lambert, Condensation of Steam on the Underside of a Horizontal Surface in the Presence of Air and Helium, ASME Natl. Heat Transfer Conf., Pittsburgh, Paper 87-HT-31, 1987.
48. R. A. Studzinski, R. A. Zahoransky, S. L. K. Wittig, and D. Barschdorff, Droplet Formation and Growth in Condensing Binary Vapors, *Int. J. Heat Mass Transfer*, vol. 27, pp. 451–461, 1984.
49. D. E. Steinmeyer, Fog Formation in Partial Condensers, *Chem. Eng. Prog.*, vol. 68, no. 7, pp. 64–68, 1972.
50. R. Sakhuja, Effect of Superheat on Film Condensation of Potassium, Ph. D. thesis, Massachusetts Institute of Technology, Cambridge, Mass., 1970.
51. K. Hijikata, S. J. Chen, and C. L. Tien, Non-condensable Gas Effect on Condensation in a Two-Phase Closed Thermosyphon, *Int. J. Heat Mass Transfer*, vol. 27, pp. 1319–1325, 1984.
52. D. Antoniuk, Generalized Modeling of Steady State and Transient Behavior of Variable Conductance Heat Pipes, AIAA Paper 87-1615, June 1987.
53. G. L. Hubbard, V. E. Denny, and A. F. Mills, Droplet Evaporation: Effects of Transients and Variable Properties, *Int. J. Heat Mass Transfer*, vol. 18, pp. 1003–1008, 1975.
54. F. C. Standiford, Effect of Non-condensables on Condenser Design and Heat Transfer, *Chem. Eng. Prog.*, vol. 75, part 2, pp. 59–62, 1979.
55. M. Rowe, B. J. Davidson, E. F. C. Andrews, J. A. Ferrison, and B. J. Taylor, Heat Transfer and Air Blanketing in Steam Condensers, in *Steam Turbines for the 1980's, Design Conference 1979*, I. Mech. E. Conference Publications 1979-12, pp. 153–162, 1979.
56. Chisholm, D., Modern Developments in Marine Condensers: Noncondensable Gases: An Overview, in *Power Condenser Heat Transfer Technology* (ed., P. J. Marto and R. H. Nunn), Hemisphere, New York, pp. 95–142, 1981.
57. T. Fujii, Condensation in Tube Banks, in *Condensers: Theory and Practice*, Inst. Chem. Eng. Symp. Ser. No. 75, Pergamon Press, Oxford, pp. 3–22, 1983.
58. S. Al-Sanea, N. Rhodes, D. G. Tatchell, and T. S. Wilkinson, A Computer Model for Detailed

Calculation of the Flow in Power Station Condensers, in *Condensers: Theory and Practice*, Inst. Chem. Eng. Symp. Ser. No. 75, Pergamon Press, Oxford, pp. 70–88, 1983.
59. J. G. Andrews, R. T. Deam, and J. Smalley, Air Entrainment in Surface Condensers, in *Condensers: Theory and Pactice*, Inst. Chem. Eng. Symp. Ser. No. 75, Pergamon Press, Oxford, pp. 48–69, 1983.
60. I. Oliker and D. Katsman, Aeration and Deaeration in a Condenser, in *Condensers: Theory and Practice*, Inst. Chem. Eng. Symp. Ser. No. 75, Pergamon Press, Oxford, pp. 213–224, 1983.
61. D. Hein, R. Rippel, and P. Weiss, The Distribution of Gas in a U-Tube Heat Exchanger and Its Influence on the Condensation Process, in *Heat Transfer 1982: Proc. Seventh Int. Heat Transfer Conf.*, Munchen, Fed. Rep. of Germany (ed. U. Grigull, E. Hahne, K. Stephan, and J. Straub), Hemisphere, Washington, D.C., vol. 5, pp. 467–473, 1982.
62. T. G. Theofanous, The Cooldown Aspects of the TMI-2 Accident, NUREG/CR-4978, 1987.
63. C. K. Nithianandan, C. D. Morgan, N. H. Shah, and F. R. Miller, RELAP4/MOD2 Model for Surface Condensation in the Presence of Noncondensable Gases, in *Heat Transfer 1986: Proc. Eighth Int. Heat Transfer Conf.*, San Francisco, August 1986 (ed. C. L. Tien, V. P. Carey, and J. K. Ferrell), Hemisphere, Washington, D.C., vol. 4, pp. 1627–1633, 1986.
64. F. Kreith and D. Bharathan, Heat Transfer Research for Ocean Thermal Energy Conversion, *J. Heat Transfer*, vol. 110, pp. 5–22, 1988.
65. M. Akai, A. Yabe, K. Ozaki, N. Endo, and T. Munakata, Qualifications of High Performance Heat Pump—Evaluation of COP for "Super Heat Pump", 3rd Int. Energy Agency Heat Pump Conf., Tokyo, Japan, 12–15 March, 1990.
66. R. L. Webb and A. S. Wanniarachchi, The Effects of Noncondensable Gases in Water Chiller Condensers—Literature Survey and Theoretical Predictions, *ASHRAE Trans.*, vol. 86, part 1, pp. 142–159, 1980.
67. A. D. Kraus and A. Bar-Cohen, *Thermal Analysis and Control of Electronic Equipment*, Hemisphere, Washington, D.C., pp. 377–407, 1983.
68. D. D. Dunn and D. A. Reay, *Heat Pipes*, 3rd ed., Pergamon Press, Oxford, 1982.

CHAPTER

SIX

HEAT TRANSFER IN HETEROGENEOUS PROPELLANT COMBUSTION SYSTEMS

M. Quinn Brewster

ABSTRACT

Heat transfer plays an important role in several critical areas of heterogeneous, solid-propellant combustion systems. These areas include heat feedback to the propellant surface, heat transfer between burning aluminum droplets and their surroundings, heat transfer to internal insulation systems, and heat transfer to aft-end equipment. Gas conduction dominates heat feedback to the propellant surface in conventional ammonium perchlorate (AP) composite propellants, although particle radiative feedback also plays a significant role in combustion of metalized propellants. Particle radiation plays a dominant role in heat transfer to internal insulation, compared with that of convection. However, conduction by impingement of burning aluminum particles, which has not been extensively studied, may also be significant. Radiative heat loss plays an important role in determining the burning rate of molten aluminum particles due to a highly luminous, oxide particle-laden, detached flame envelope. Radiation by aluminum oxide smoke particles also plays a dominant role in heat transfer from the exhaust plume to aft-end equipment. Uncertainties in aluminum oxide particle-size distribution and optical properties still make it difficult to predict radiative plume heat transfer accurately from first principles.

NOMENCLATURE

A_k	ambient oxidizer factor
C	specific heat
d	particle or droplet diameter
G	crossflow mass flux

k	thermal conductivity; rate constant; absorption index
L	motor port length
m	mass regression rate
n	refractive index
N	particle-number density distribution; conduction radiation parameter
p	pressure
q	heat flux
q'''	volumetric heat-generation rate
r	propellant regression rate
R_k	ambient environment factor
t	time, optical thickness
T	temperature
u	velocity component
V	velocity component
x	distance from propellant surface
X	mole fraction
Y	mass fraction
δ	boundary-layer thickness
ε	emissivity
κ	thermal diffusivity
μ	viscosity
ρ	density

Subscripts

0	initial
∞	ambient
c	characteristic; conductive
e	envelope
f	flame
g	gas
L	characteristic length; thickness
m	melt, molten
mp	most probable
O	oxidizer
ox	oxide
r	radiative
rel	relative
s	solid; surface
T	thermal

1 INTRODUCTION

Heat transfer plays an important role in several key areas of heterogeneous solid-propellant combustion systems. These areas include internal rocket motor ballistics as well as thermal protection systems. In the area of internal ballistics, heat transfer plays a critical role in determining both the surface regression rate of the solid propellant as well as the combustion rate of liquid aluminum droplets distributed throughout the motor. In the area of thermal protection, the heat transfer rate to both internal and external thermal protection systems is of critical importance. In particular, radiation is often a strong component of heat transfer, which remains relatively poorly characterized in terms of fundamental properties as well as transport.

Three key areas where heat transfer plays a significant role in solid-propellant rocket systems are labeled in the schematic drawing shown in Fig. 1*a*. First, near the inside surface of the exposed propellant grain, conductive heat feedback from the primary flame zone to the propellant plays a crucial role in determining the regression rate of the propellant. Radiative heating from burning aluminum droplets distributed away from the propellant surface also influences the propellant regression rate. Since the regression rate of the propellant and the corresponding mass production rate of

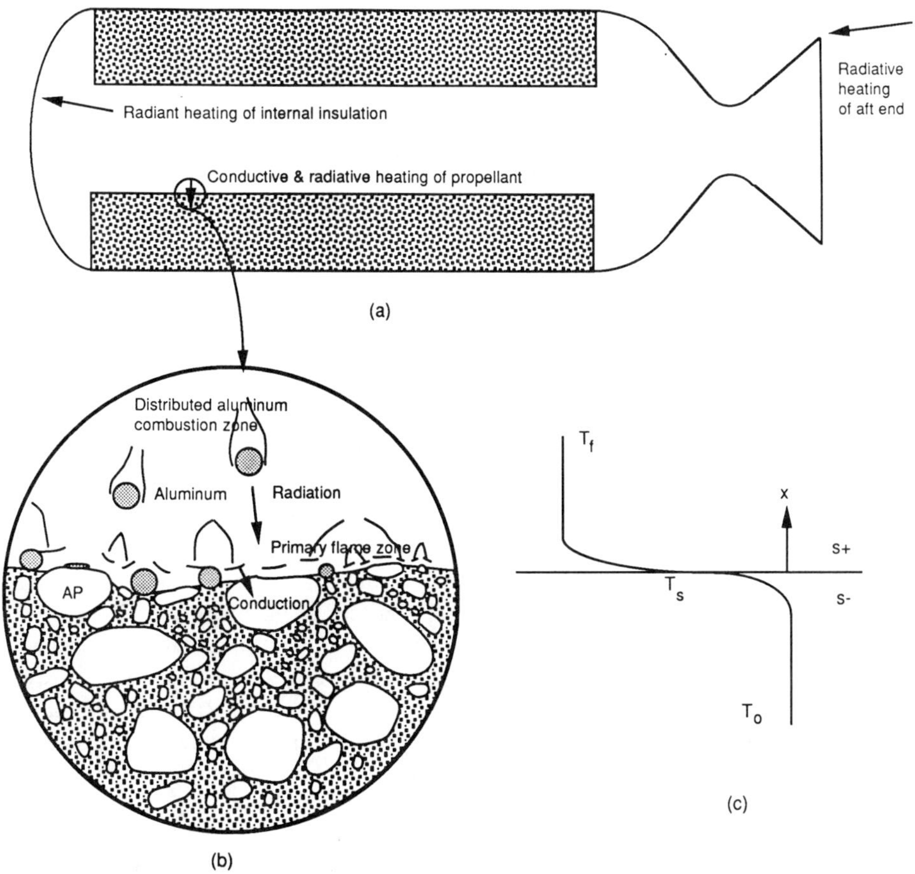

Figure 1 Areas of heat transfer importance in propellant combustion systems.

combustion gases at the propellant surface, together with nozzle discharge characteristics, determines the internal motor pressure and thrust, the rate of heat transfer to the propellant surface is an extremely important factor in internal ballistics considerations. Heat transfer considerations also play a dominant role in determining the size of aluminum agglomerates produced at the surface of the propellant. Large agglomerates are associated with a significant loss in performance. Therefore, the extent of aluminum agglomeration and the size of agglomerates are also key performance parameters determined in part by heat transfer considerations.

The second key area of heat transfer concern is near the head end of the motor or in any other region where propellant does not protect the case from hot combustion products. In these regions, radiation, convection, and particle-impingement heat transfer combine to impose a significant heat load from which the case must be insulated. The primary source of thermal radiation is molten aluminum oxide, Al_2O_3, smoke particles produced by distributed combustion of aluminum. Impingement heat transfer is the result of burning aluminum droplets which are convected by recirculating flow patterns. Since ablative insulation represents inert, dead weight and therefore lost payload, optimal insulation design is essential to competitive performance.

The third area of concern labeled in Fig. 1*a* is the aft end of the rocket. Thermal radiation from the highly luminous exhaust plumes of aluminized propellants constitutes a substantial heat load from which external aft-end equipment must be protected. Aside from heat transfer considerations, thermal radiation in plumes is also a subject of intense interest currently because of observability considerations and remote sensing applications.

2 PROPELLANT HEAT FEEDBACK

Heat feedback to the propellant is a key consideration in solid-propellant systems because it determines the regression rate of the propellant which determines the pressure and thrust of the propulsion system. Conductive, convective, and radiative heat feedback mechanisms have all been recognized as having an influence on the propellant regression rate. Conduction is the principal mechanism of heat feedback to the propellant surface. Conductive heat feedback can originate either with the primary gas-phase flames or, in aluminized systems, with aluminum droplets which often ignite and begin to burn as they move away from the surface of the propellant. Convective heat transfer also plays a role in certain situations. The enhancement of propellant regression rate by flow tangent to the surface of the propellant is known as erosive burning and is a serious concern in systems with high L/D ratios. Radiation also plays a role in burning-rate augmentation, although its importance is generally considered to be limited to low-burning-rate propellants. It should be noted, however, that radiation effects are most likely to be significant in aluminized propellants (which emit a high-intensity continuum) but that aluminum addition changes many features of the propellant combustion. Yet only a few careful, systematic studies with aluminized propellants have been done to delineate the role of radiation. Therefore conclusions about the role of radiative transfer in aluminized propellant combustion should probably be considered as tentative.

2.1 Conductive Heat Feedback in Primary Flame Zone

Conductive heat feedback from the primary gas flames represents the principal "driving force" which establishes the regression rate of most composite solid propellants. The primary flame zone is confined to a very thin region near the surface of the propellant. Figure 1*b* shows a magnified schematic diagram of the surface region of a conventional ammonium percholate (AP) oxidizer, hydrocarbon fuel binder, composite propellant. To achieve the high solids loading necessary for maximum density and optimum stoichiometry, the AP is usually present in a bimodal or trimodal particle-size distribution. Aluminum powder (5 to 60 μm) may also be present in the propellant formulation. The length scale of the heterogeneity of this surface region is approximately 100 μm, which corresponds to the size of the largest AP particles. Because of the heterogeneous nature of the propellant, an intricate flame structure exists above the surface of the propellant. The heat feedback from this complex flame structure is the primary factor in determining the regression rate of the propellant.

2.1.1 One-Dimensional Heat Transfer Models. Although any realistic description of the heat transfer in the primary flame zone must address the heterogeneous, multidimensional nature of the problem, it is instructive first to review the simple, one-dimensional heat transfer models of the system under consideration [1]. Figure 1*c* shows a representation of the average temperature profile near the surface of the propellant. By average is meant an average over the plane parallel to the surface of the propellant. The propellant is assumed to be fed upward at the regression rate r so that the origin at the mean propellant surface stays fixed in space. Under steady-state conditions the mass flux $m = \rho_s r$ is a constant.

The energy balance in the condensed phase is a standard convective-diffusive balance (assuming no chemical reaction or in-depth radiation):

$$k_s \frac{d^2T}{dx^2} - mC_s \frac{dT}{dx} = 0 \tag{1}$$

The solution of Eq. (1) gives an exponential temperature profile in the condensed phase:

$$\frac{T(x) - T_0}{T_s - T_0} = \exp\left(\frac{x}{x_c}\right) \tag{2}$$

The characteristic length scale of the thermal wave in the condensed phase is

$$x_c = \frac{k_s}{mC_s} = \frac{\kappa_s}{r} \tag{3}$$

and its value depends on the regression rate r and the thermal diffusivity κ_s. Typical values for x_c range from 100 μm at low pressures (low regression rates) to 1 μm at high pressures (high regression rates).

The energy balance in the gas phase includes a source term due to conversion of chemical enthalpy to sensible enthalpy q''' as shown in Eq. (4):

$$k_g \frac{d^2T}{dx^2} - mC_{pg}\frac{dT}{dx} + q''' = 0 \tag{4}$$

The solution of Eq. (4) depends on the spatial distribution of the heat release $q'''(x)$, which depends on reaction and mixing rates. Integrating Eq. (4) once, assuming constant properties, gives the heat flux conducted back to the propellant surface:

$$q_{s+} = \int_0^\infty q'''(x)\exp\left(-\frac{mC_{pg}x}{k_g}\right)dx \tag{5}$$

From Eq. (5) it can be seen that the distribution of the average heat release $q'''(x)$ is of prime importance in determining the heat flux to the surface and the regression rate.

Since many negative comments have been made about one-dimensional combustion models, it is worthwhile to point out that a one-dimensional energy balance, as such, is still a perfectly valid concept. While it is not possible to predict $q'''(x)$ adequately without resorting to multidimensional arguments, nevertheless the importance of accurately representing the average spatial distribution of the heat release $q'''(x)$, as illustrated by the one-dimensional analysis, is a lesson to be remembered. In addition, the length scale of the thermal wave in the condensed phase, which can be estimated from the one-dimensional analysis, is an important parameter to keep in mind when expanding the analysis to include propellant heterogeneity.

2.1.2 Heat Feedback from Competing Flames (BDP Model). The first published study to recognize the importance of the multidimensional heat feedback from multiple, competing flames which exist above the surface of composite propellants was that of Beckstead, Derr, and Price [2–4]. As outlined in their original model (the BDP model), there are two principal, competing flames which are of importance in determining the heat feedback and linear regression rate of the propellant: the AP monopropellant, self-deflagration flame, and the primary AP/binder flame (see Fig. 2). The AP self-

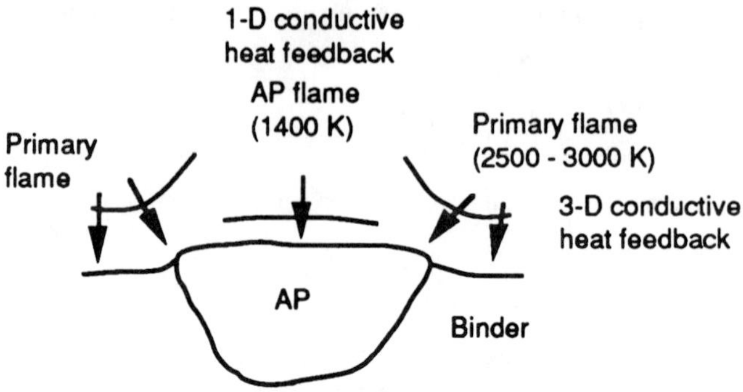

Figure 2 BDP model of competing flame heat feedback.

deflagration flame is a relatively low-temperature, premixed, oxidizer-rich flame (1400 K), which, in the absence of heat from any other source, extinguishes at pressures below 2 MPa. The primary AP/binder flame is a hybrid, premixed-diffusion flame which reaches temperatures on the order of the stoichiometric adiabatic flame temperature (3000 K). The conductive heat feedback from both of these flames provides most of the energy for sustaining the pyrolysis of the condensed propellant constituents. Due to its premixed nature and close proximity to the AP surface, the AP flame conductive feedback is considered to be one-dimensional. On the other hand, the conductive feedback from the primary flames is generally recognized to be three-dimensional, although the BDP model (as well as most subsequent models) assumes one-dimensional heat feedback for the primary flame with an effective flame standoff distance.

Perhaps the most significant contribution of the BDP model is its description of the competing flames, which explains how the heat feedback varies with pressure and AP particle size. Figure 3 shows two sketches of the multiple-flame structure for two different pressures. At low pressures (Fig. 3*a*), the colder AP flame is relatively far from the AP surface. Most of the AP decomposition products are fed into the primary flame, which takes on the nature of a premixed flame over that of a diffusion flame (i.e., it is more kinetically limited than mixing limited). As a result, most of the heat feedback to the propellant comes from the primary flames. Upon examining the propellant surface after extinction of the combustion by rapid depressurization, it is observed that the AP particles are peaked in the middle, which is evidence of the dominant heat feedback from the primary flames. At high pressures (Fig. 3*b*), the AP flame moves closer to the AP surface and most of the AP decomposition products are fed into the AP monopropellant flame. The primary flames, which are somewhat diminished in their overall importance, take on the nature of a diffusion flame over that of a premixed flame and tend toward being mixing limited. The extinguished propellant surface for combustion at high pressures shows that the AP particles are recessed in the middle, which is consistent with the increased importance attributed to the AP flame in the multiple-flame model. The other important aspect of the BDP

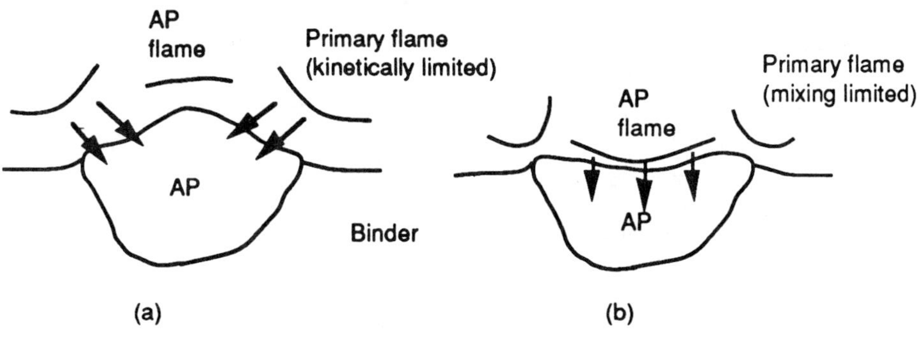

Figure 3 Effect of (*a*) low pressure and (*b*) high pressure on flame heat feedback.

model is that it correctly predicts the general trend of increased heat feedback and burn rate with decreasing AP particle size.

While the BDP model was able successfully to describe the important first-order features of composite propellant combustion behavior, it soon became apparent that the model was not a complete mechanistic description of the heat transfer process. The discrepancies between predicted and measured regression rates for some propellant formulations were still too large for the model to be reliable as a predictive tool in the solid propellant industry. Particularly in propellants with wide AP size distributions, the failure of the model to include some important mechanism was noticeable [5]. Many similar models emerged attempting to improve on the original BDP model, most of which involved evolutionary refinements rather than revolutionary changes. Refinements in terms of oxidizer size distribution statistics [6] and multiple surface temperatures were included [7]. Of these various evolutionary models which followed after BDP, Cohen has written an instructive summary [8].

2.1.3 Sandwich Propellant Studies and KLLEF Model. The next significant advance in understanding of surface combustion behavior and the role of heat transfer came with studies which used a simplified, propellant sandwich geometry [9, 10]. Price and co-workers systematically investigated the combustion behavior of propellant sandwiches which consisted of a thin lamina of binder (25 to 175 μm) between two pressed AP slabs as shown in Fig. 4. From these studies came two significant con-

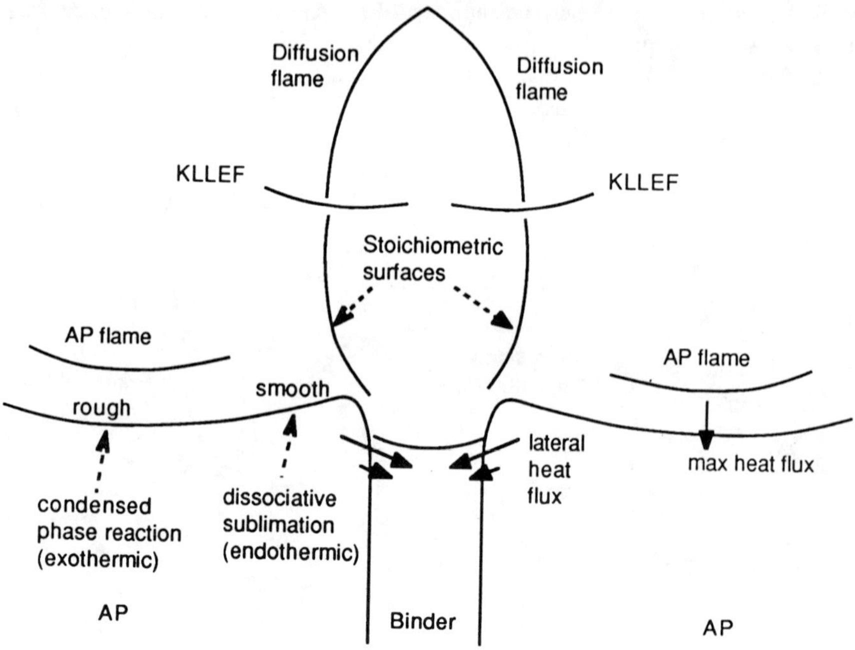

Figure 4 Lateral subsurface heat flux in two-dimensional propellant sandwiches.

clusions: (1) lateral heat transfer from the AP to the endothermic binder was probably altering the normal AP self-deflagration process; and (2) the anomalous AP deflagration was probably affecting the site of heat release in the gas-phase flames which is so important in determining propellant regression rate.

Figure 4 shows a schematic diagram of the profile of an extinguished propellant sandwich surface and the postulated flame structure according to Price. The essential features of the flame structure in Fig. 4 are similar to those of the BDP model. Heat feedback from the AP flame controls the surface regression in regions far away from the binder lamina. Near the plane of symmetry the primary AP/binder flame provides the heat feedback to the surface. In the absence of a better understanding of its structure, the primary flame is postulated to consist of a kinetically limited leading-edge flame (KLLEF) and a trailing diffusion flame. The combination of the KLLEF and trailing diffusion flame can be thought of as playing the same role as the BDP primary flame.

Most of the features of the competing flame heat feedback which were accounted for in the BDP model are also exhibited in the propellant sandwiches. For example, at low pressures the heat feedback from the primary flame is greater than that from the AP flame and the regression rate in the binder/AP interface region is greater than that away from the interface. At high pressures the heat feedback from the AP flame dominates and the regression rate in the AP slabs away from the lamina is greater than that near the binder region. Thus the general correctness of the BDP concepts is verified by the sandwich experiments. In addition to this validation of the general BDP concepts, however, some interesting new insights into the heat transfer near the surface resulted from the sandwich experiments.

Price noticed that in the region adjacent to the binder/AP interface, the AP always protruded above the binder and had a smooth texture when burning was interrupted. This was in contrast to the AP surface away from the binder lamina, which was always rough. Price postulated that this observation could be explained by lateral heat conduction from the AP into the endothermic binder, as indicated in Fig. 4. The drain of energy from the AP by the binder results in a lower AP surface temperature in that region and a lower regression rate normal to the local surface. Lateral heat flow causes a significant change in the AP deflagration mechanism. In the AP region away from the binder lamina, the AP deflagrates in what might be called "augmented" self-deflagration. In this region the flame heat feedback from the gas-phase AP flame combines with some feedback from the primary flame to give a local maximum heat flux and maximum AP surface temperature. (That there is some contribution of heat from the primary flame is known from the fact that the underlying AP regression rate is greater than the AP self-deflagration rate.) This type of deflagration is characterized by an exothermic decomposition reaction in the condensed phase on the surface of the AP followed by further heat release in the gas phase AP flame. The exothermic surface reaction is responsible for the rough surface which remains after extinguishment. In the protruding AP region near the binder, however, the AP probably decomposes at a lower temperature, in an endothermic, dissociative sublimation process. This hypothesis was verified both experimentally and theoretically [9]. Experimental verification was obtained by substituting thermally conductive but nonreactive gold and nonconductive, nonreactive mica for the binder lamina. It was observed that the AP near the lamina still protruded and appeared smooth with gold, but with mica the AP did not protrude and appeared rough, like the AP far from the lamina. Theoretical verification was also obtained by performing a two-dimensional heat conduction analysis [11], which predicted the occurrence of lateral heat transfer.

One important ramification of Price's postulated lateral heat transfer mechanism is that the site of the energy release in the gas phase is probably affected by the siphoning of heat from the AP into the binder. Support for this argument is found in the change in AP decomposition mechanism observed between the AP near the binder and that far from the binder. Knowledge of the site (i.e., distribution) of the energy release is of critical importance in determining the heat feedback to the surface, as demonstrated earlier by the one-dimensional heat transfer analysis. Since the lateral heat transfer mechanism alters the AP decomposition, it indirectly affects $q'''(x)$ and a realistic combustion model must address these issues of multidimensional, coupled heat transfer.

The propellant sandwich studies of Price appear to be very useful for elucidating the multidimensional nature of the heat transfer and combustion in composite propellants. It is recommended that further investigations be conducted with sandwiches, making use of microthermocouples to characterize the temperature profiles near the surface. Extensive experience has been gained using microthermocouples in standard composite propellants, and the results of these studies have been very useful for characterizing the surface temperature of propellants. However, this potentially useful technique has apparently not been tested in propellant sandwiches, which offer a much more well-defined geometry with which to work.

2.2 Aluminum Heating, Agglomeration, Ignition, and Conductive Feedback

The behavior of aluminum in solid propellants has been studied extensively. Most studies have aimed at understanding the agglomeration and ignition mechanisms with the hope of being able to reduce the extent of aluminum agglomeration at the propellant surface. Another aspect of metal combustion in solid propellants which has not received as much attention is the influence of the metal behavior on the burn rate of the propellant. Aluminum can influence the burn rate through a variety of mechanisms, including conductive feedback and inert heating effects.

2.2.1 Aluminum Behavior: Accumulation, Agglomeration, and Ignition. Much of the original work on metal behavior in propellants tried to explain the mechanisms which determine the size of metal agglomerates produced at the surface of the propellant [12–14]. Early theories proposed a correlation between the agglomerate size and the amount of aluminum available in pockets between AP particles [15]. Later work by Price et al. [16] demonstrated that the process of agglomeration involved heat transfer considerations which were not included in the original pocket models.

Price and co-workers showed that powdered aluminum undergoes a series of complicated processes in the region near the surface of a burning solid propellant [17].

These processes include heating as the particles approach the propellant surface, accumulation or concentration of particles on the surface of the propellant, formation of extended filigrees protruding into the hotter gases above the propellant surface, agglomeration of filigrees or surface accumulates into larger molten droplets, and departure of the molten agglomerates from the surface, usually accompanied by ignition. Sambamurthi et al. [18] demonstrated that the size of the agglomerates seemed to be related to their proximity to a hot AP/binder flame. For an agglomerate to become hot enough to ignite and leave the propellant surface, it had to grow large enough to be in close proximity to a high-temperature stoichiometric AP/binder flame. Since local flame conditions depend on the AP particle-size distribution and on pressure, the size distribution of agglomerates leaving the surface also depends on these variables. Generally it has been observed that as pressure increases, agglomerate size decreases, because increasing pressure causes the hot AP/binder flames to move closer to the propellant surface. However, as observed in [18], the decrease in agglomerate size with increasing pressure is distinctly nonlinear. Near a critical value of pressure the agglomerate size decreases rapidly with increasing pressure. The value of the critical pressure depends on the AP particle-size distribution. A recent model developed by Kovalev attempts to incorporate most of what is currently understood about the processes affecting aluminum agglomeration to predict agglomerate size in AP composite propellants [19]. Other details of aluminum behavior near the surface have been described by Babuk and co-workers [20].

One of the key questions which has remained unanswered is what temperature is reached by aluminum before its departure from the surface of a propellant. Particularly puzzling is the fact that window bomb studies often reveal ignited aluminum droplets leaving the surface of a propellant [19], yet the decomposition temperatures of AP and binder are thought to be below typical aluminum agglomerate ignition temperatures. Certainly a lower bound for the temperature of the aluminum at the surface of the propellant would be the binder decomposition temperature (1000–1200 K), assuming negligible surface char-layer formation, and an upper bound would be the melting point of aluminum oxide (2320 K), which is the upper limit for ignition of aluminum droplets. Ignition studies have established that aluminum ignition temperature is a strong function of particle size and oxidizer concentration [21–23]. For very small particles (<10 mm), the ignition temperature decreases with increasing particle size. At intermediate particle sizes (10–20 mm), the ignition temperature goes through a minimum on the order of 1000 K; and for larger particles the ignition temperature increases, approaching the oxide melting point for particles greater than 50–100 mm. The phase of the protective solid oxide film on the surface of the aluminum may also play an important role in aluminum ignition, as pointed out by Gurevich et al. [22]. Crystalline oxide is much more resistive to diffusion of oxidizer than amorphous oxide, and thus crystal formation is thought to raise the ignition temperature.

2.2.2 Aluminum Effect on Regression Rate. While ignition and agglomeration have been rather extensively studied, the influence of metal behavior and heat feedback on regression rate is still not well defined. Nevertheless, some general guidelines have been proposed to suggest what influence metal addition might have on propellant regression rate [17]. One of the propellant properties which metal addition affects is thermal conductivity. Metal staples and wires embedded in propellants have the effect of increasing regression rate, which is usually attributed to higher propellant thermal conductivity [24]. Metal addition can also change the propellant stoichiometry, and thus regression rate, depending on what ingredients the metal replaces in the formulation. Miller et al. [5] systematically varied propellant composition to study the effects of metal addition on regression rate. However, it is generally recognized that in those formulations the dominant mechanism for changing regression rate was replacement of a rate-controlling oxidizer size with metal. Thus it is hard to separate the effect of metal behavior on regression rate from other effects associated with changes in the AP/binder flame structure in those results.

Another factor associated with metal addition which may alter the regression rate is oxidation of the metal. Oxidation of the metal agglomerates as they reside on or near the surface of the propellant will tend to increase the propellant regression rate by transferring heat to the propellant [17, 25]. This oxidative heating can take the form of slow oxidation or fast oxidation (i.e., combustion). Slow oxidation occurs on the surface of the molten metal particles as they accumulate and agglomerate on the surface of the propellant (see Fig. 1*b*). As the agglomerates reach the ignition temperature, thermal runaway occurs, the reaction shifts over into a detached flame in the vapor phase surrounding the droplet, and fast oxidation takes over [19]. As this ignition process is happening, the droplets become detached and begin to move away from the propellant surface. The slow oxidation mechanism has an inherently lower heat transfer rate associated with it than does the fast oxidation mechanism. On the other hand, the agglomerates spend a much shorter period of time in the vicinity of the propellant surface during fast oxidation than during slow oxidation. Thus both oxidative heat transfer mechanisms are potentially significant, and both mechanisms are included in referring to the conductive heat feedback from the metal to the propellant. This conductive heat feedback may depend on the size of agglomerates. For the same mass fraction of agglomerates, smaller agglomerates would have a greater surface area and presumably would transfer more energy to the propellant than would larger agglomerates.

It may be conjectured that the way itself in which agglomerates are carried away from the propellant surface (as observed by high-speed photography) is evidence of the occurrence of oxidative heat feedback from metal agglomerates to the rest of the propellant. It is generally agreed that the drag coefficient on a burning droplet at low Reynolds number is lower than that for a nonburning droplet, for the same Reynolds number. Therefore the acceleration of an agglomerate away from the surface must be due to an increase in the local Reynolds number (i.e., velocity) of the flow around the agglomerate. Such an increase in local velocity could only be due to an increase in the pyrolysis rate of the binder in the immediate vicinity of the agglomerate, which increase must be associated with heat feedback from the metal droplet undergoing thermal runaway and ignition.

Another way that aluminum can affect regression rate is through the inert heating (or heat sink) effect. Until they ignite and move out of range of the hot AP/binder flames near the propellant surface, metal agglomerates can act as a heat sink, siphoning off energy from the primary AP/binder flames which otherwise would have gone to increase the burn rate of the propellant [25, 26]. The higher the ignition temperature of an agglomerate, the more severe the heat sink effect would be. Thus the heat sink effect would depend on particle size, if ignition temperature were a function of agglomerate size.

From the previous discussion it is clear that aluminum behavior and propellant regression rate must be regarded as coupled. Each is influenced by the details of the AP/binder flame structure near the surface of the propellant. It is therefore too simplistic to speak of how the regression rate affects the agglomerate size or how the agglomerate size affects the regression rate. They are both strongly coupled to more fundamental processes such as oxidizer and binder melting, decomposition, and gas-phase reaction.

A recent study [26, 27] sought to make some progress in shedding light on this matter by looking at the correlation between agglomerate size and regression rate in a series of metalized, ammonium perchlorate (AP), composite propellants. An attempt was made to ensure that other burn rate-controlling variables were fixed so that differences in regression rate could be correlated to metal behavior and, in particular, agglomerate size. In order to do this, propellants were formulated so as to keep the AP/binder ratio constant, thus holding constant the primary regression-rate determining influence of the AP/binder flame heat feedback. Two metals, Al and Mg-Al, with similar thermophysical and thermodynamic properties but different oxidative kinetic properties (i.e., ignition temperatures) were used to help distinguish among changes in regression rate due to propellant thermal conductivity, propellant stoichiometry, metal heat feedback, and metal heat sink effects. The results showed a significant correlation between regression rate and agglomerate size (see Table 1 and Fig. 5), which indicate that smaller agglomerates are more conducive to enhanced regression rate. This effect was attributed to more efficient heat transfer from smaller agglomerates to the propellant and a smaller heat sink effect imposed on the AP/binder flames due to lower ignition temperatures for smaller agglomerates. Lower ignition temperatures for Mg-Al alloy compared with pure Al were also found to be more conducive to higher regression rate. The main conclusion and recommendation drawn from that study is still relevant and is therefore reiterated here: metal heat feedback and heat sink effects are important and need to be better understood in order to make sense out

Table 1 Correlation between increase in regression rate and change in volume agglomerate size going from 1.83 to 5.27 MPa (250 to 750 psi)

Metal (%)	Percent increase in burn rate	Percent decrease in agglomerate size
None	82	Not applicable
Al (10)	65	76
Al (20)	43	2
Al (30)	44	9
Mg-Al (10)	53	−9
Mg-Al (20)	54	−24
Mg-Al (30)	51	7

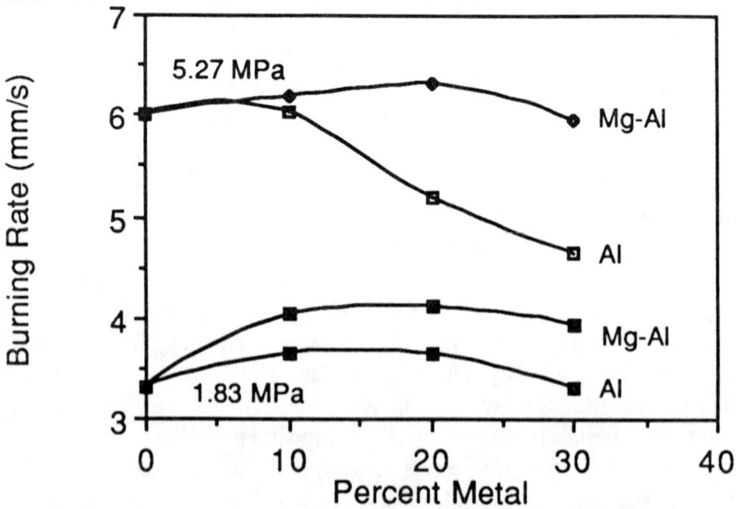

Figure 5 Regression rate for metalized composite propellants.

of composite-propellant regression-rate data. Future and ongoing studies may shed additional light in this area.

2.3 Radiative Heat Feedback

Radiative heat feedback has also been considered as a mechanism for enhancing propellant regression rate. For purposes of considering radiative feedback it is convenient to discuss aluminized and nonaluminized systems separately. For several reasons, these two types of systems behave quite differently. In terms of their absorptivities they are very different. Nonaluminized propellants are generally translucent at visible and near-infrared wavelengths unless an opacifier or opaque catalyst has been added to the propellant. Aluminized propellants are very opaque at all wavelengths due to the aluminum powder [28]. Thermal emission in nonaluminized systems originates with molecular gas combustion products and soot [29]. In aluminum-containing propellants, burning aluminum droplets and molten, submicrometer aluminum oxide particles emit a broad continuum of high-temperature radiation.

2.3.1 Steady Radiative Heat Feedback in Nonaluminized Propellants. Several investigators [30–34] have studied the effect of a constant radiant heat flux (q_r) on the steady mass regression rate (m) in nonaluminized propellants. The principal conclusion of these studies is that at low to moderate flux levels (<400 W/cm^2), radiant energy absorbed in-depth, below the location of the rate-controlling chemical reactions, is equivalent to an increase of $\Delta T = q_r/mC_s$ in the initial temperature of the propellant T_0 (the so-called equivalence principle). This conclusion has both positive and negative consequences, as pointed out by Ibiricu and Williams and by Caveny, Ohlemiller, and

Summerfield. On the positive side, it means that in-depth, radiation absorption effects can be relatively easily incorporated into the theoretical analyses of both steady and unsteady combustion (assuming that the absorption takes place beneath the rate-controlling reaction layers). On the negative side, the equivalence principle means that no new information about the chemical or physical reaction mechanisms can be obtained from externally supplied radiant heat flux experiments that could not be obtained from regression-rate temperature sensitivity studies.

The radiant flux emitted by nonaluminized-propellant combustion gases and soot has been measured by Zenin et al. using calorimeters [25]. The magnitude of the flux is typically of the order of 20 W/cm^2, which is not enough to influence the regression rate appreciably. Other studies of propellants burning in electrically heated tubes have also shown that, except for low-regression-rate propellants, the radiant flux present in a nonaluminized rocket is not large enough to increase the regression rate significantly. Hence, radiative feedback is generally ignored in nonaluminized systems, except for propellants with very low regression rates or in laboratory studies where an external flux is supplied to study the combustion behavior.

2.3.2 Unsteady External Radiative Flux in Nonaluminized Propellants. Many investigators have also studied the effect of a time-varying, external radiant flux incident on nonaluminized solid propellants for the purpose of understanding the unsteady combustion response of propellants [35–44]. One of the first studies in this area was conducted by Mihlfeith et al. [35]. They used a piezoelectric, microforce transducer to measure the combustion recoil induced by an oscillatory incident radiant heat flux. Recently workers at the Naval Weapons Center have revived the original Mihlfeith experiment [43], while Strand at the Jet Propulsion Laboratory [42] has developed a microwave technique for measuring the instantaneous propellant response. In all these efforts a combustion model is needed to relate the measured heat flux-coupled response function to the desired pressure-coupled response function. A key assumption in the combustion model has to do with the absorption of the radiant flux. It is analytically more convenient to assume that the radiant flux is absorbed uniformly at the surface of the propellant rather than deal with in-depth absorption and selective absorption by different propellant constituents. However, it is still uncertain if the assumption of uniform absorption is valid (in light of the fact that the constituents have markedly different optical properties [28, 45]). It is also uncertain if the effect of in-depth absorption is important. Although AP is known to be absorptive at 10.6 μm [45] (the CO_2 laser is commonly used as a radiation source), a recent study [44] indicates that there would still be significant in-depth absorption in AP which covers a substantial fraction of the propellant surface area. Concerns such as these need to be investigated further before the oscillatory radiant heat flux experiment can be used as a reliable tool for propellant characterization and development.

2.3.3 Radiative Feedback in Aluminized Propellants. It is safe to say that radiative feedback in aluminized propellants is definitely more significant than that in nonaluminized propellants, due to the broad-band, high-temperature radiant emission from burning aluminum droplets and aluminum oxide particles. Whether or not the radiative feedback is large enough relative to the conductive feedback to influence the regression rate, however, is still not certain. In any event, radiative feedback is still usually ignored in aluminized-propellant combustion studies. One reason for the neglect of radiation even in aluminized systems is that studies comparing regression rates of aluminized and nonaluminized propellants have been inconclusive. Sometimes aluminum enhances the regression rate and sometimes the opposite result is obtained. This behavior at first is puzzling in light of evidence that external radiation at flux levels typical of that emitted by burning aluminum droplets (at least 100 W/cm^2) increases the regression rate of nonmetalized propellants significantly [32, 33].

The question of whether or not radiative heat feedback is important is complicated by the fact that, as mentioned earlier, addition of aluminum changes several properties of the propellant and flame simultaneously (in addition to increasing the radiant heat flux). The thermal conductivity of the propellant is increased by adding aluminum, which, like the radiant flux, would also tend to increase the regression rate. However, addition of aluminum also introduces a heat sink effect which tends to decrease the regression rate [27]. Thus conclusions drawn from regression rate studies about the importance of radiative feedback are subject to considerable error.

A more satisfactory method of judging the importance of radiative feedback, rather than inference from regression-rate studies, is to measure the radiative flux directly. Zenin et al. [25] measured radiative heat flux and conductive heat flux using microthermocouples and calorimeters embedded in stoichiometric, AP/polyformaldehyde mixtures and concluded that radiative feedback played a minor role in determining regression rate relative to that of conductive feedback from oxidizing aluminum particles. The radiant heat flux was measured as 80 to 100 W/cm^2. Felton and Hitchcock [46] measured radiative feedback in aluminized propellants using light pipes and obtained total radiant flux levels on the same order as those of Zenin et al., 80 to 100 W/cm^2 at 2 to 5 MPa. In both of these studies, however, there is some question as to whether smoke and aluminum agglomerates may have reduced the measured signal.

Analytical studies have also been performed to predict the radiant flux back to the surface of an aluminized propellant [47] and its effect on aluminum agglomeration via in-depth absorption [48]. From these studies it is estimated that the radiant flux incident on the propellant surface is between 300 and 400 W/cm^2, which is a factor of 4 larger than reported measured values. However, it should be noted that the emissive properties of burning aluminum droplets play a strong role in such predictions, and these properties are not very well characterized, though some attempts have been made [49].

Another important property which influences the radiative feedback is the propellant absorptivity. The absorptivity of metalized propellants has been estimated in some cases [50] to be negligibly small based on the fact that metals have intrinsically low bulk absorptivities. However, this view is too simplistic and overlooks the phenomenon of multiple scattering. More realistic estimates based on the solution of the radiative transport with multiple scattering in the propellant [28, 48] indicate that the propellant total absorptivity is at least 0.4. Hemispherical, spectral reflectance measurements [25, 51] give propellant spectral absorptivities of 0.8 in the visible and near-infrared for aluminized propellants. Binder optical properties can also influence the absorptivity, particularly in the infrared region. Measurements of transmissivity of composite propellants by Mihlfeith [36] indicate that binder absorption coefficients are temperature dependent and that certain binders have positive temperature coefficients while others have negative temperature coefficients. Observations of the influence of external CO_2 laser radiation on propellant regression rate reported in a recent paper by Yin et al. [34] also seem to confirm the influence of temperature-dependent (both positive and negative) binder absorption coefficients, although that interpretation was not given in the paper.

To summarize the influence of radiative feedback on propellant regression rate, it must be said that it is still uncertain. Further careful measurements are needed. In undertaking such measurements it is important to deal effectively with the problem of obscuration of the radiation detector by smoke, soot, and aluminum agglomerates. If the radiation signal is measured when the detector or collection optics are located at the bottom of a cavity (such as in [25] and [46]), there is the uncertainty of obscuration by particles which is very difficult to account for unless the propellant is extinguished and examined. This precaution has been taken in a current study [51] using optical fibers embedded in the propellant, and initial results indicate radiant fluxes of 350 W/cm^2 with 20% aluminum loading at 3.5 MPa, which is more than double the previously reported value of Zenin at 4.5 MPa of 160 W/cm^2.

2.4 Convective Heat Feedback

Convective effects also play a role in altering the heat feedback to the propellant surface. There is always a mean component of velocity tangent to the propellant surface, which is usually small near the head end of the motor but can become rather large (hundreds of meters per second) near the aft end of the motor (Fig. 1*a*). This crossflow usually results in an increase in the heat feedback to the propellant, which causes an increase in the regression rate. The enhancement of the regression rate by the mean flow tangent to the surface of the propellant is known as erosive burning and is a significant factor which must be dealt with in systems with high L/D ratios. Usually the effect of flow tangent to the propellant is to increase the heat feedback and thereby increase the propellant regression rate as shown in Fig. 6*a*. However, with certain propellants, negative erosive burning (a reduction in regression rate) has also been observed at low crossflow velocities, as indicated by Fig. 6*b*.

The mechanism of enhanced regression rate due to crossflow is associated with an increase in heat feedback due to the mean convective crossflow, much like a boundary-layer flow. Various levels of sophistication have been used to describe this mechanism, ranging from the simple analysis of Lenoir and Robillard, which uses the Chilton-Colburn correlation with a correction for transpiration [52], to the turbulent, chemically reacting boundary-layer analyses of Beddini and Razdan [53, 54]. Because of its simplicity and widespread use, the Lenoir-Robillard result is repeated here. The regression rate is given by the following:

$$r = ap^n + \alpha \frac{G^{0.8}}{L^{0.2}} \exp\left(-\frac{53m}{G}\right) \qquad (6)$$

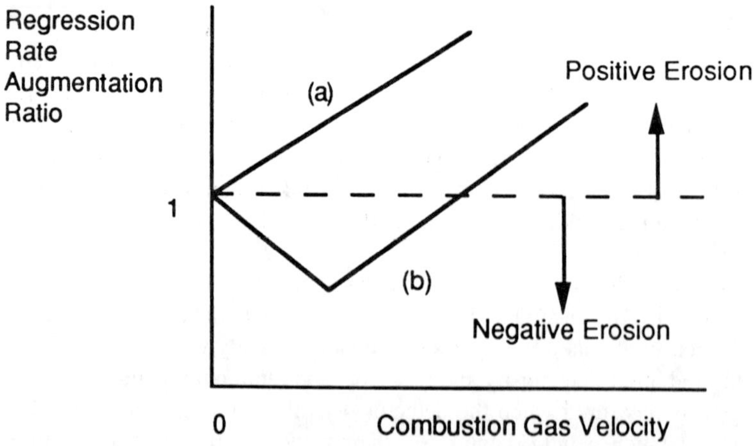

Figure 6 Regression rate augmentation due to crossflow: (*a*) positive only; (*b*) mixed.

where

$$\alpha = \frac{0.0288 C_{pg} \mu_g^{0.2} \mathrm{Pr}^{-2/3}}{\rho_s C_s} \left(\frac{T_f - T_s}{T_s - T_0} \right) \tag{7}$$

The first term in Eq. (6) is the normal pressure-dependent regression rate law and the second term accounts for the erosive burning effect. In Eq. (6), G is the mass flux of the crossflow and L is the port length of the motor. King [55] took a slightly different approach to modeling erosive burning effects. King's model focussed on the BDP model and tried to simulate the way in which a convective crossflow might distort the AP/binder flames and push them closer to the propellant surface. The net result of pushing the primary gas flames closer to the surface is to increase the conductive heat feedback and regression rate of the propellant.

While convective heat transfer analyses, such as those just mentioned, appear to be adequate for describing positive erosive burning, the observation of negative erosive burning (Fig. 6*b*) indicates that the actual mechanism of erosive burning may be more complicated in some situations. In negative erosive burning, an increase in crossflow velocity is accompanied by a decrease in regression rate. One explanation which has been offered is that negative erosive burning is caused by competing mass transfer and convective heat transfer effects [56]. At low crossflow velocities, it is argued, a decrease in heat feedback associated with mass blowing causes a decrease in regression rate, while at high velocities convective heat transfer dominates. This argument sounds reasonable because of the familiar concept of reduced heat transfer that is associated with mass blowing in a convective boundary layer. However, it must be remembered that the reduction in heat transfer which occurs in boundary-layer flows occurs between a boundary-layer flow *without* blowing and a similar boundary-layer flow *with* blowing.

There is no reason to expect a decrease in heat transfer due to blowing when crossflow is added to situation where blowing already existed. Only an increase in heat transfer would be expected. A more plausible explanation for negative erosive burning is that partial coverage of AP surfaces by melted binder flowing under the effect of shear stresses is responsible for reduced regression rate. This argument would explain why some propellants (perhaps those with high-melting binders) exhibit negative erosive burning and others do not. An experimental effort should be conducted using extinguishment of propellants burning under crossflow conditions to test this hypothesis. By examining the propellant surface after extinguishment it should be possible to determine if binder melting and flowing plays a role in erosive burning. Much more could be said on the topic of erosive burning, but to keep the topic of convective feedback in balance with the rest of this chapter the reader is referred to a recent review on erosive burning by Razdan and Kuo [57].

Often there is also an unsteady, oscillatory component of velocity tangent to the propellant surface arising from longitudinal acoustic flow oscillations. The enhancement in regression rate which accompanies the oscillatory tangential flow is referred to as velocity coupling. (Ed Price, the originator of the term "velocity coupling," has suggested that "convective coupling" might be a more accurate description of the phenomenon.) Velocity coupling is an abnormal situation which is to be avoided. The topic of velocity coupling involves interactions between the mean and unsteady flow fields and is complicated by distinctly nonlinear behavior. A recent workshop was held on the topic of velocity coupling and can be consulted for more details [58, 59].

3 HEAT AND MASS TRANSFER IN ALUMINUM DROPLET COMBUSTION

Aluminum agglomerates formed at the surface of an aluminized solid propellant do not burn to completion in the vicinity of the surface of the propellant, like the rest of the propellant constituents, but burn in a distributed fashion away from the surface. Whether or not the droplets burn completely before exiting the nozzle depends on their initial size, their burning rate, and their residence time in the motor. Because of the negative performance consequences of exhausting unburned aluminum from the nozzle, the complete combustion of aluminum is of great interest. Hence the combustion rate of aluminum particles, which is very dependent on heat and mass transfer rates, is of considerable interest in aluminized propulsion systems.

The subject of heat and mass transfer in metal combustion (particularly aluminum combustion) was studied extensively 20 to 30 years ago [60–81]. Following that very prolific period there was a period of relative inactivity (at least as far as the open literature is concerned) as metals found wide application in propulsion systems and explosives, while fundamental interest waned [82–84]. In the past few years, however, fundamental interest in metal combustion has been revitalized both for propulsion and explosives applications [85–90], as well as for other applications such as materials synthesis [91–93].

3.1 General Features of Aluminum Combustion

A brief review of the essential features of aluminum combustion is given, with an emphasis on the heat and mass transfer aspects of the problem. A model of a burning aluminum droplet is shown in Fig. 7. There are two distinct reaction zones, an inner zone and an outer zone. These two zones are distinguished from each other by luminosity (continuum emission), temperature, and submicrometer Al_2O_3 particle population. The inner zone is relatively nonluminous (in continuum emission), achieves the highest temperature in the system (T_f), and is rather devoid of particles (at least particles large enough to emit significant continuum radiation). The outer zone, on the other hand, is highly luminous, somewhat cooler than the inner zone, and is populated by particles which are sufficient in size and number density to emit a strong continuum of radiation. Heat release occurs throughout the inner zone, primarily due to condensative recombination of suboxide vapors to form Al_2O_3(liq). The condensed-phase Al_2O_3(liq) particles thus formed in the inner zone are rapidly transported both outward away from the droplet (by bulk convection and thermophoresis), forming a luminous trail, and inward toward the droplet (by thermophoresis), where they are deposited on the droplet surface and often coalesce into a large cap (see Fig. 7). In the inner flame zone the combustion-generated, submicrometer particles are too small

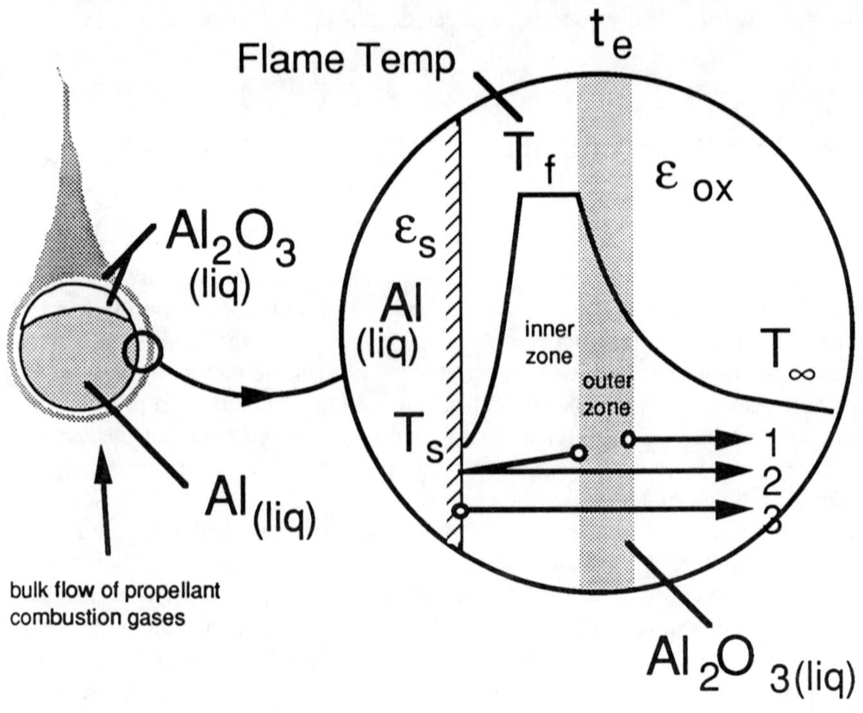

Figure 7 Model of burning aluminum droplet showing radiative transfer contributions: (1) flame emission; (2) reflected flame emission; (3) droplet emission.

(and possible too few in number density) to emit a significant continuum of radiation. In the outer zone, where conditions are more favorable for particle growth and the bulk velocity component away from the droplet is much lower, the Al_2O_3 smoke particles achieve sufficient size and number density to emit a substantial continuum of thermal radiation. While the above features are based on observations from many studies, there is disagreement over various aspects of this description. For example, the continuum radiation in the flame zone has also been attributed to chemiluminescence [81], and plausible reaction mechanisms have been set forth.

3.2 Radiative Heat Transfer Considerations in Aluminum Droplet Combustion

Glassman, who did much of the pioneering work in metal combustion, suggested that radiative transport plays an important role in metal combustion [60]. Brzustowski and Glassman even went so far as to suggest that Glassman's original vapor-phase burning criterion for metal combustion needed to be modified to include the effect of radiative heat loss [61]. The original vapor-phase combustion criterion was that if the boiling point of the metal oxide was greater than that of the metal, combustion would occur in the vapor phase. Otherwise surface combustion would occur. According to this criterion (and thermodynamic data available at that time), however, it was predicted that Ti and Zr should burn in the vapor phase, which they do not. Hence, Brzustowski and Glassman suggested that the boiling-point criterion was only a necessary but not a sufficient condition, subject to radiant heat loss considerations. It should be noted that more recent, and therefore presumably more accurate, estimates of TiO_2 and ZrO_2 boiling points were found to be less than the corresponding metal boiling points [94], which revalidates the original vapor-phase criterion. Nevertheless, the issue of the influence of radiative transfer in aluminum droplet combustion has never been adequately addressed.

Following Glassman's original suggestion about the importance of radiative transfer, several studies tried to include the effects of radiation in combustion models. Kuehl [64] included the term $\varepsilon_{ox}\varepsilon_s\sigma(T_f^4 - T_s^4)$ in the droplet energy equation to represent the net radiative flux to the droplet. Micheli and Schmidt [83] used the term,

$$\frac{\sigma(T_0^4 - T_s^4)}{(1/\varepsilon_s) + (d_s/d_f)^2[(1/\varepsilon_{ox}) - 1]}$$

which correctly accounts for curvature and multiple reflection effects between the droplet and flame, but still assumes that the flame is opaque. Both of these studies included the term $\varepsilon_{ox}\sigma(T_f^4 - T_\infty^4)$ to represent the net radiative flux from the droplet to the surroundings. Neither study mentions the values of ε_{ox} and ε_s that were assumed. A serious problem with the radiative representations of these models is that they treat the detached flame as an opaque, nontransmitting surface. There is no allowance made for emission loss from the droplet surface to the surroundings through the flame envelope. Furthermore, the contribution to the emission loss from the flame which is reflected from the droplet is not included.

A more realistic model of the radiative transfer in a burning aluminum droplet is shown in Fig. 7 [49]. Three important components of radiant emission are designated in Fig. 7. Component 1 is the direct emission from the luminous detached flame envelope to the surroundings. Component 2 is the flame emission which is reflected by the droplet and transmitted through the flame. Component 3 is the emission by the droplet surface which is transmitted through the flame envelope. Based on this more rigorous representation, it can be seen that the optical thickness of the envelope t_e becomes a very important parameter in determining the radiative properties of burning aluminum droplets. While it is virtually impossible to predict the optical thickness of the flame envelope from first principles, it has been suggested that the envelope is probably optically thin [49]. This conclusion is drawn from comparisons between theoretical predictions and measurements of spectral intensity emitted by burning aluminum droplets.

3.3 Convective and Radiative Heat Transfer Effects on Droplet Burning Rate

In most metal combustion applications the most important consideration is the combustion rate or droplet burnout time. The droplet combustion rate has been widely estimated using Hermsen's correlation [95, 96]:

$$\frac{d_s}{d_{s0}} = \left(1 - \frac{kt}{d_{s0}^{1.8}}\right)^{1/1.8} \tag{8}$$

where

$$k = 8.3144 \times 10^{-5} R_k A_k^{0.9} p(\text{psi})^{0.27} (\text{cm}^{1.8}/\text{s}) \tag{9}$$

$$A_k = \sum_i X_i \qquad i = H_2O,\ CO_2,\ OH,\ O,\ O_2 \tag{10}$$

$$R_k = \begin{cases} 1 & \text{laboratory} \\ 2.7 & \text{motor} \end{cases} \tag{11}$$

Equation (8) gives the droplet diameter as a function of time in the form of a classical "d^2" law, where the diffusion-limited exponent 2 has been modified to 1.8. The influence of ambient oxidizer concentration is included in the factor A_k.

Two important effects that can influence the burning rate of liquid droplets (metal or otherwise) are convective heat transfer and radiative transfer with the ambient environment. Burning-rate data obtained from motor firings and from laboratory tests indicate that combustion rates are substantially higher in motors than in the laboratory, and this difference has been attributed to the difference in convective and radiative environments. In Hermsen's correlation the influence of convective and radiative heat transfer are lumped together and represented by the adjustable constant R_k. A value of $R_k = 1$ corresponds to burning-rate data that have been obtained in a laboratory environment with a single burning droplet where there is no convective effect (other than natural convection) and the radiative environment is a blackbody at room temperature. A factor of $R_k = 2.7$ has been recommended for motor environments.

In an effort to improve on Hermsen's correlation and to separate the effects of convective and radiative heat transfer, a comparative study was done between Hermsen's correlation and the classical theory of diffusion-limited droplet combustion including convective heat and mass transfer [90]. Stoichiometric, single-step reaction between aluminum and water vapor (the primary oxidizer in most solid-propellant combustion gases) was assumed. A typical result of that comparison is shown in Fig. 8. Figure 8 shows four cases of droplet diameter as a function of time. Two cases represent the situation with no convective heat transfer ($V_{rel} = 0$ in the diffusion model and $R_k = 1$ in Hermsen's correlation). The other two cases represent the situation with convective heat transfer ($V_{rel} = 1$ m/s in the diffusion model and $R_k = 2.7$ in Hermsen's correlation). A velocity of 1 m/s is a reasonable estimate of the average relative velocity between burning aluminum droplets and the surrounding propellant gases at typical rocket pressures. As can be seen in Fig. 8, a slight increase in droplet burning rate (decrease in combustion time) due to convective heat transfer is predicted by the diffusion model. At the same time, Hermsen's correlation predicts a much larger increase in burning rate. Assuming that the increase in burning rate predicted by the diffusion model corresponds to the convective heat transfer effect, then the much larger increase in burning rate predicted by Hermsen's correlation would suggest that radiation is the dominant heat transfer mechanism acting to increase the droplet combustion rate in motors.

The situation just described is the reverse of the "usual" situation encountered with hydrocarbon droplets, where convective effects are much more important than radiative effects. However, the reversal in roles of radiative and convective heat transfer is quite understandable. In hydrocarbon droplet combustion, the droplets are typically injected into the combustor at high velocities (hundreds of meters per second). Thus relative Reynolds numbers are high and convective effects are important in hydrocarbon combustion. In solid-propellant metal droplet combustion relative velocities are low, thus convective heat transfer effects are small. However, the oxide-smoke combustion

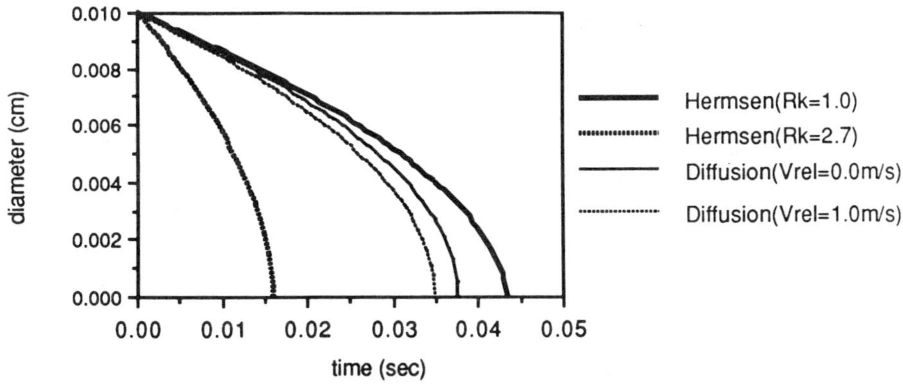

Figure 8 Comparison between Hermsen's correlation and diffusion model showing effect of convective environment ($Y_{0\infty} = 0.5$, $T_\infty = 2400$ K, $p = 0.1$ MPa).

product of metals as well as the burning droplets themselves are more luminous and hotter than their hydrocarbon counterparts. Thus radiative transfer plays a more important role than convection in typical metal combustion applications.

Based on the foregoing discussion, some suggestions can be made of ways in which future studies might contribute to the understanding of heat transfer effects in aluminum droplet combustion: (1) include a more realistic model of radiative transfer in detailed droplet combustion models; (2) develop a simple engineering combustion correlation which delineates convective and radiative effects; and (3) do experimental studies to characterize the radiative properties of burning droplets, such as envelope optical thickness and temperature.

4 INTERNAL INSULATOR HEAT TRANSFER

Near the head end of many rocket motors and in other regions there are exposed surfaces where propellant does not protect the case from hot combustion products (Fig. 1*a*). In these regions, radiation, convection, and particle-impingement heat transfer combine to impose a significant heat load from which the case must be insulated. Typically, an ablative, rubber insulation is used as thermal protection. Since ablative insulation represents inert, dead weight and therefore lost payload, optimal insulation design is essential to competitive and reliable performance.

4.1 Heat Transfer Mechanisms and Relative Importance

The three principal mechanisms of heat transfer to internal insulating surfaces are gas convection, radiation, and particle impingement. Gas convection is usually considered to be of importance as a heat transfer mechanism near the nozzle throat, where the combination of high velocities and high temperatures results in maximum convective heat transfer. Inside the motor, gas temperatures are high (3600 K) but velocities are usually low enough (<1 m/s near the head end) that convective heat transfer is not a serious problem. At the exit of the supersonic nozzle, velocities reach a maximum but the gas temperature is reduced such that convective heat transfer is also reduced. Thus convective heat transfer is of primary concern in the nozzle throat region. However, it should be pointed out that the gas flow field plays a significant role in heat transfer to forward dome internal insulation in the way that it modifies the radiative heat transfer. The gas flow has a significant effect on the aluminum particle trajectories and on the optical thickness of the thermal boundary layer and thus can strongly influence the heat transfer to the head end.

Radiative heat transfer is usually considered to be of importance inside the motor, especially in aluminized propellant systems [29]. The primary source of thermal radiation is molten aluminum oxide Al_2O_3 smoke particles produced by distributed combustion of aluminum. Pearce [97] modeled radiative heat transfer in aluminized solid rockets using an isothermal cylinder and concluded that radiation was of minimal importance at the nozzle throat (relative to convection) but of major importance at the nozzle entrance and inside the motor. Near the head end, where velocities are low, radiation is the dominant mechanism of heat transfer to forward dome internal insulation surfaces.

Particle impingement is also a potentially significant mechanism of heat transfer to insulation surfaces which has not been extensively studied. Impingement heat transfer is the result of burning aluminum droplets which are convected by recirculating flow patterns near the head end. Because of the very efficient heat transfer between burning, molten metals and any surface they impact, particle impingement is a very important heat transfer mechanism. The effects of impinging, molten, burning metal particles are often evidenced in catastrophic ways. For example, severe damage to forward motor enclosures has been observed in high-pressure, unstable combustion of highly aluminized propellants, which is attributable to impingement heat transfer [98]. In another impressive demonstration of the tremendous heat transfer capacity of burning metals, Grosse and Conway [99] reported using a premixed flame of aluminum and pure oxygen to cut a 3-in. (7.5-cm) hole in a 30-in. (75-cm)-thick concrete slab at a rate of more than 1 in. (2.5 cm) per minute.

4.2 Buoyancy-Induced Recirculating Flow

Buoyancy-induced recirculation has been suggested as a posssible mechanism for enhancing heat transfer to forward dome internal insulators [100]. Buoyancy-induced recirculation is driven by acceleration of the flight vehicle. In flight, the acceleration vector is aligned with the axis of the motor, which imposes a body force on fluid particles in the direction of the nozzle. Acceleration in solid rockets may range from 3 g (space shuttle solid rocket boosters) to tens and hundreds of g (missile systems). In the presence of acceleration body forces, colder gas near the head end insulator is unstable relative to the warmer gas in the chamber interior. Thus the potential exists for a recirculating flow pattern to be established near the head end, as shown in Fig. 9. Purcell and Daines [100] have simulated the nonreacting, nonaluminized flow field near the head end of a typical motor and concluded that buoyancy could play a significant role in determining the head-end flow pattern.

The buoyancy-induced recirculating flow has a primary stagnation point (or locus of points) and a secondary stagnation point as shown in Fig. 9. Since the primary stagnation point is the region of greatest heat transfer, this region is of particular interest. A study which considered the local stagnation flow in the primary stagnation region, including radiation (see Fig. 10), indicated the following interesting features of the flow: (1) for typical aluminized propellant compositions the thermal boundary layer is optically thick with aluminum oxide smoke; (2) radiation is the dominant mode of heat transfer (assuming negligible particle impingement); (3) the flow behaves like that of a low-Prandtl-number fluid ($\delta \ll \delta_T$ even though fluid Pr is of order 1) due to the strong radiation influence; and (4) the important dimensionless parameters which determine dimensionless heat flux are wall-to-bulk fluid temperature ratio T_0/T_∞, conduction-radiation parameter N_0, boundary-layer optical thickness t_δ, wall emissivity ε_w, and fluid Prandtl number Pr (see Fig. 11).

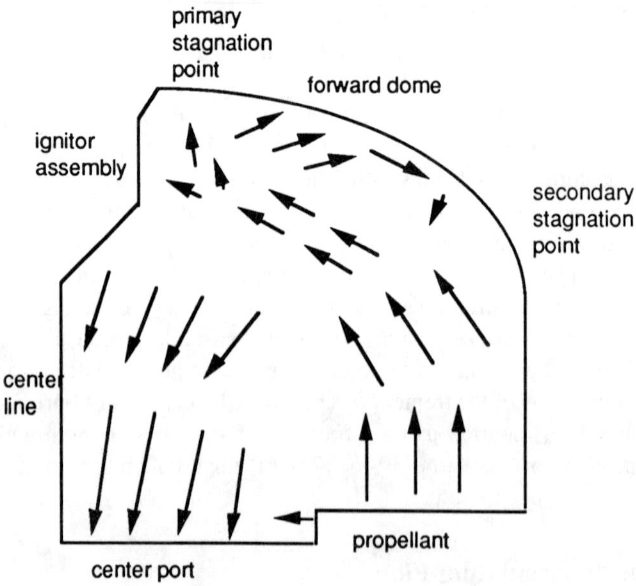

Figure 9 Buoyancy-induced flow pattern near head end.

4.3 Flow Field Simulation with Distributed Aluminum Combustion

As might be expected, multidimensional flow-field simulation which includes distributed aluminum combustion is not as advanced as nonreacting flow simulation. Studies which include distributed aluminum combustion typically do not include buoyancy effects [101, 102], for example, and often utilize the assumption that complete combustion of aluminum occurs at the propellant surface [103]. The inclusion of aluminum introduces the added complexity of multiphase flow as well as chemical reaction. In order to develop a reliable tool for predicting aluminum particle impingement, it will be necessary to include the effects of buoyancy and distributed aluminum combustion in a multidimensional, viscous, turbulent internal flow-field code. A fully coupled, turbulent, multiphase, multidimensional flow simulation with chemical equilibrium calculated at each grid point is computationally prohibitive and probably unwarranted. A reasonable first approximation is to assume a one-step, global reaction in which the restricted equilibrium composition consisting of AP/binder products and aluminum are converted to the final equilibrium-state products [102]. Salita and Levine have given recent workshop summary reports on the status of internal flow code development [104, 105].

Recommendations for future development in this area are as follows. Better models for predicting particle impingement heat transfer are needed. There is also a need for a better capability to predict aluminum oxide particle flow fields and concentrations. To accomplish these goals, a multidimensional, body-fitted flow-field model should

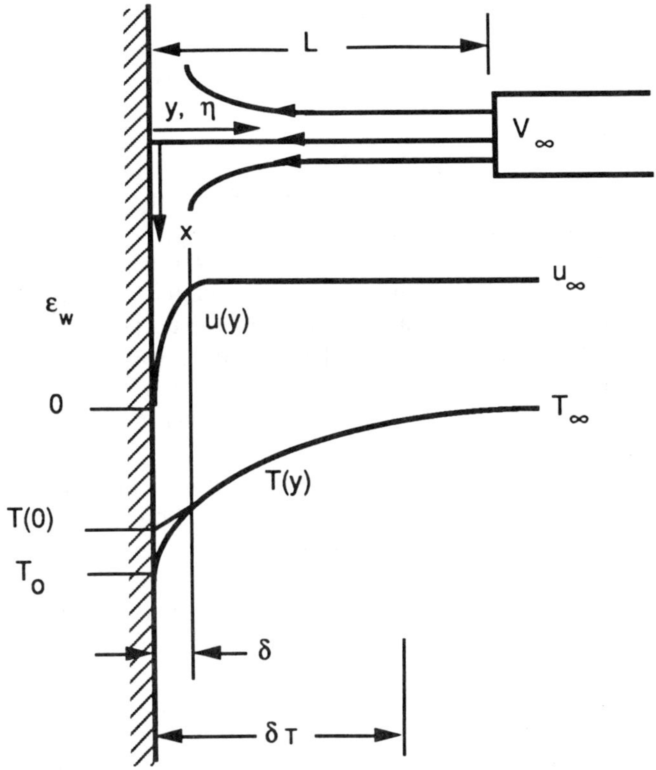

Figure 10 Optically thick, radiating stagnation flow.

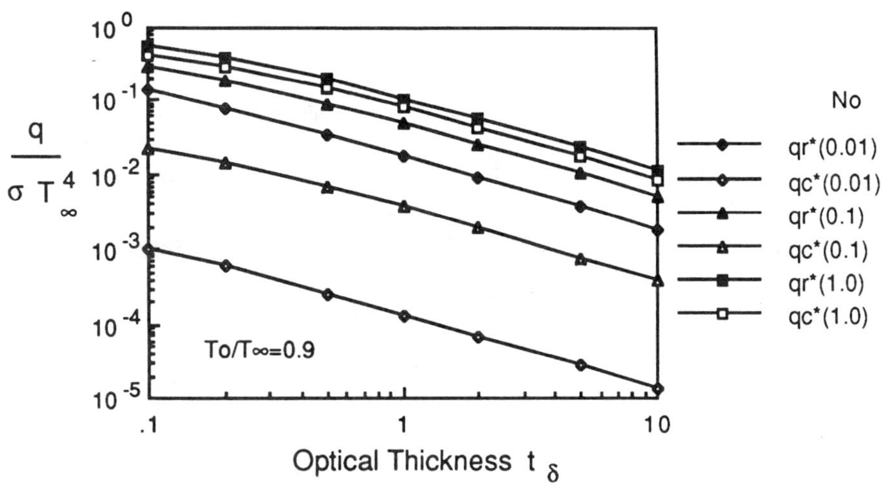

Figure 11 Nondimensional radiative (q_r^*) and conductive (q_c^*) heat flux components in optically thick, two-dimensional planar stagnation flow (Pr = 0.5, ε_w = 1).

be developed which includes distributed aluminum combustion, buoyancy, and radiative transfer.

5 PLUME RADIATION

The final area of heat transfer interest to be considered here is the aft end of the motor (Fig. 1*a*), where radiative heating of aft-end equipment by the exhaust plume is of concern. During the development of the Saturn launch vehicles, which began in the late 1950s, concerns over radiative base heating became the driving force behind much of the progress made in the understanding of gas radiation heat transfer [106]. Nowadays, with more widespread use of solid rockets and aluminum fuel, plume radiation is an even bigger concern because of the intense continuum radiation emitted by condensed aluminum oxide particles [107]. Hence, this discussion will focus on radiation by aluminum oxide particles.

Initial calculations of aluminized plume radiosity used Mie theory to predict the emission by aluminum oxide–laden plumes. Optical constants based on pure Al_2O_3 (sapphire) were used, and particle-size distribution was assumed. Most of these predictions resulted in estimates which grossly underpredicted the measured values of plume radiosity. In hindsight it can be seen that there were two reasons for the disagreement: (1) the absorption index was severely underestimated; and (2) the aluminum oxide particle-size distribution was not well known. Unfortunately, the predicted radiosity is very sensitive to these two sets of data, which, even today, are known with very little confidence. Thus, much research in plume radiation is focusing on understanding the physics and chemistry of the processes which determine the optical constant and size distribution of aluminum oxide, and trying to measure these properties with better accuracy.

5.1 Aluminum Oxide Particle-Size Distribution

Al_2O_3 particle-size distribution in aluminized plumes and motors has been studied extensively over the years [108]. It is generally recognized that the size distribution is bimodal due to the two competing mechanisms of aluminum oxidation: (1) detached "vapor-phase" oxidation, which produces submicrometer smoke particles (0.1 to 2 μm); and (2) surface accumulation, which produces large residual caps (10 to 100 μm). However, several mechanisms act which change the size distribution, including:

1. Distributed combustion of aluminum
2. Growth by collision due to velocity slip-induced coalescence
3. Breakup of large droplets due to viscous drag forces
4. Deposition and shedding of molten Al_2O_3 in nozzle
5. Density change due to solidification and melting
6. Condensation of $AlCl_3$

Efforts to measure particle size have proceeded along two different lines of approach. One approach is to capture particles and then measure the size distribution.

The other is to use nonintrusive optical techniques. These two techniques and their relative merits are discussed in the following sections.

5.1.1 Particle-Capturing Techniques. Most of the early experimental studies of aluminum oxide particle size were based on capturing particles at the exit of the nozzle and measuring the size from the captured samples. A critical review of the work up to 1980 has been given by Hermsen [109]. This review forms the basis for particle-size prediction used in the industry standard SPP code [96]. It is based on empirical correlation of particle-size data with motor and propellant variables. The most important variables for correlating the size data are taken to be chamber pressure, chamber aluminum oxide concentration, nozzle throat diameter, and residence time. Correlations of this type are usually given in terms of d_{43}, which has shown to be the appropriate moment for predicting two-phase flow loss:

$$d_{nm} = \left[\frac{\int_0^\infty d^n \, N(d) \, d(d)}{\int_0^\infty d^m \, N(d) \, d(d)} \right]^{1/(n-m)} \tag{12}$$

In Eq. (12), $N(d)$ is the number density distribution as a function of diameter d. Typically, the value of d_{43} reported in [109] ranges between 2 and 10 μm.

Experimental determination of the size distribution has been discussed in detail by Kraeutle [110]. By breaking the distribution up into subfractions, analyzing the subfractions using appropriate microscopy (i.e., optical or electron), and recombining the subfractions, Kraeutle demonstrated that the size distribution is bimodal, with the transition between lobes occurring between 2 and 20 μm. That study pointed out that although microscopic analysis could be done accurately, it was very tedious. For example, for a combustion product sample obtained in a T-burner, analysis showed that by number 98% of the particles were below 1 μm, while by mass only 15% were smaller than 1 μm. That is, 10^3 particles would have to be counted to count one 2.5-μm particle, and 3×10^8 particles would have to be counted to count one 40-μm particle. This points out the tedious nature of measuring size distribution and why often the form of the size distribution is assumed (e.g., log-normal, Rosin-Rammler, Gamma function, etc.) and attention is focused on determining just the appropriate averages (e.g., d_{32}, d_{43}, d_{63}, etc.). Unfortunately, when seven different commercial, automated size analyses were compared with the "exact" distribution, only one came close, illustrating the caution with which such techniques should be viewed [110].

Another study of particle size using a quench collection bomb was reported by Salita [111]. Burning particles of aluminum, aluminum oxide caps, and aluminum oxide smoke particles were collected in a buffered isopropyl alcohol bath. A mass balance was conducted to account for all of the original aluminum in the propellant. Lost mass was attributed to aluminum oxide smoke, which is difficult to capture in a quench bath. A bimodal size distribution was obtained with the small mode attributed to aluminum oxide smoke particles and the large mode to conglomerates of aluminum and aluminum oxide. The size and mass fraction of the large mode varied depending on the distance of the quench bath from the propellant and pressure. However, the small mode (smoke) had a mass mean diameter of 1.5 μm ($d_{32} = 1$ μm), which was independent of pressure and collection distance. The actual particle sizing was done in that study with a Microtrac optical sizing system.

5.1.2 Optical Techniques. Attention has also been given to in-situ optical techniques for measuring particle size. A review of the various optical techniques for measuring particle size in propellant combustion systems has been given recently by Koo [112]. Optical techniques have the advantage that they are generally nonintrusive. However, usually the form of the size distribution must be assumed and the optical constants must usually be known. In addition, most of the techniques reported (except multiwavelength extinction) are limited to optically thin, single-scattering media. Many techniques are limited to measuring only a certain portion of the size distribution. For example, diffraction-based techniques can usually measure particles only above 2 or 3 μm such that $\pi d/\lambda > 5$. Many different techniques have been reported, including extinction, emission, and scattering measurements.

One of the first to apply optical techniques to rocket particle-size measurements was Dobbins [113]. He used a two-wavelength extinction technique (0.365 and 1.01 μm) and obtained d_{32} values of 0.4 to 0.6 μm, which was significantly smaller than sizes indicated by particle-collection techniques. In a follow-up study to try to explain the discrepancy, Dobbins and Strand [114] compared results from particle collection and optical extinction measurements (three wavelengths), assuming various monomodal size distributions. Again, the d_{32} sizes obtained by the optical technique were significantly smaller than the collected sizes (less than 1 μm for optical compared with 1 to 3 μm for collected samples). They also noted that the optically determined sizes were internally inconsistent in that the magnitude of d_{32} differed by a factor of 2 depending on which combination of wavelengths was used. They concluded that self-consistent optical results could be obtained only if a bimodal size distribution was assumed; however, such an attempt was not reported.

Powell et al. [115] developed an optical technique for determining unknown size d_{32}, refractive index n, and volume fraction f_v, simultaneously from forward scattering and extinction measurements. Drawbacks of this technique which limit its usefulness for aluminized propellant flames and plumes are the assumption of diffraction-dominated scattering ($\pi d/\lambda > 5$), the assumption of nonabsorbing particles, the assumption of single scattering, and the assumption of a monomodal size distribution.

Konopka et al. [116] used in-situ, multiwavelength extinction, SEM photographic analysis, and a Coulter counter to determine size distribution for samples from two rocket exhausts. They presented both log-normal and Gamma-function correlations for size distribution. The agreement between the various methods was quite reasonable, with a d_{32} value of 1.15 μm and most probable size of $d_{\mathrm{mp}} = 0.153$ μm for the Gamma-function distribution (rocket 2).

Recently, efforts have been reported [117] to measure particle size both inside a small motor as well as outside by using purged windows on the motor. Two techniques have been used: (1) diffraction scattering (similar to the Malvern system using a linear array detector instead of a ring array) and (2) holography. While this work represents an important step in trying to look inside the motor, both of these techniques require optically thin, single-scattering conditions. This limits the application to aluminum loadings of about 2% or less. Also, the diffraction-based Malvern type of approach is again limited to larger particles (>3 μm in [117]) so that only a portion of the size distribution can be characterized, and the submicrometer smoke particles are missed.

A two-dimensional, digital imaging technique has also been reported [118]. This technique offers the benefit of an instantaneous and spatially resolved measurement as opposed to a time- and line-of-sight-averaged measurement obtained by the previously discussed optical methods. However, this method (which uses 90°-scattered light) is also limited to single scattering and particles larger than about 10 μm.

A recent design study [119] has also been reported which examines the advantages and disadvantages of various optical techniques (e.g., emission, extinction, and scattering) for determining particle size and optical constants over various ranges. The main conclusion of that study is that extinction and emission measurements (using a shock tube for heating) are the most generally suitable. They also conclude that scattering measurements are difficult and therefore unsuitable due to the highly structured angular dependence of scattered intensity for moderate to large particle-size parameters ($\pi d/\lambda$). This conclusion is basically true for single-scattering, monodisperse systems (although a broad-band laser can be used to overcome this difficulty [118]). However, polydispersity and multiple scattering quickly smooth out this structure. Thus, the only remaining valid objections to scattering measurements from [119] are the need to have many lasers to cover the spectrum (which is not a serious fundamental problem) and the difficulty of calibrating for absolute scattered intensity.

In an effort to overcome some of the difficulties of the previous studies, an experimental investigation was conducted using two-wavelength, extinction and absolute scattering measurements (forward and backward) in optically intermediate propellant flames [120]. Multiple scattering was included in the data reduction by incorporating an inverse solution of the radiative transfer equation. Optical constants were not assumed but were measured simultaneously. A monomodal Gamma-function size distribution was assumed. The resulting optical mean size was $d_{32} = 1$ μm, in agreement with the reported size of the smoke fraction from [111].

From the work on particle-size analysis done to date, several observations can be made:

1. Particle collection techniques are subject to questions as to what changes may have occurred in size during collection and separation.
2. Particle collection inside a motor is not feasible at present.
3. The physical separation and size analysis which must follow particle collection techniques are tedious but accurate.
4. Optical techniques are subject to questions of the assumed size distribution.
5. Optical measurements often indicate sizes smaller than those from collection techniques.
6. Optical techniques are useful for obtaining mean sizes but not the full size distribution.

5.1.3 Evolution of Particle-Size Distribution. Based on the results of these past investigations, a scenario for the evolution of aluminum oxide particle-size distribution is presented. Figure 12 shows the bimodal aluminum oxide particle-size distribution which exists inside the motor after aluminum combustion has reached near-completion. The submicrometer smoke phase from the detached flame envelopes constitutes 85–90% (by mass) of the oxide product. The surface residual caps constitute the remaining 10–15%. As the flow proceeds through the motor and especially through the nozzle, the smoke fraction grows due to velocity slip-induced coalescence (smaller, faster particles colliding with larger, slower particles). This growth is depicted by an arrow in Fig. 12, indicating a shift of the smoke fraction to larger sizes. At the same time the small particles are coalescing with the larger particles, the large fraction is breaking up due to shear forces in the nozzle. At Weber numbers ranging from 4 to 8 [121], viscous drag forces overcome surface tension forces holding the droplet together and the droplet breaks up. This analysis indicates that particles larger than about 20 μm cannot survive passage through the nozzle. Thus there is a shift in the large mode to smaller sizes. Since coalescence and breakup mechanisms depend on residence time and velocity lag in the nozzle, the size distribution in the plume will vary considerably from motor to motor. Further systematic study is needed before particle sizes can be predicted with enough confidence for accurate plume radiation simulation.

5.2 Aluminum Oxide Optical Constants

Initial attempts to predict plume emission from first principles gave results which underestimated plume radiosity. It is now apparent that this was due to a lack of information about the particle-size distribution and optical constants [122–124]. It was soon recognized that the absorption index of molten aluminum oxide must be much bigger than that of solid aluminum oxide [122], and that impurities such as carbon [125] and aluminum [125, 126] might be playing a significant role. Other factors which have been mentioned as influencing the optical constants are supercooling and

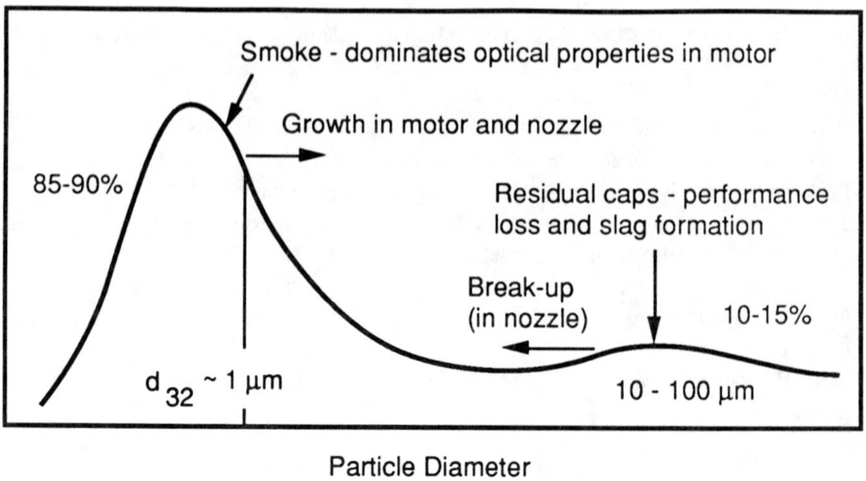

Figure 12 Bimodal size distribution of aluminum oxide produced by aluminum combustion.

crystalline phase; lately the possible influence of condensation of aluminum trichloride on the outside of aluminum oxide particles has come to light as a possible mechanism for changing optical properties [127].

5.2.1 Solid Phase. The optical constants of Al_2O_3 in the pure solid state have been relatively well established by many previous investigations [116, 128–142]. Figures 13 and 14 show that, at room temperature, pure Al_2O_3 is essentially nonabsorbing between 0.5 μm and 5 μm. Below 0.2 μm and above 6 μm, electronic transitions and lattice vibrations, respectively, result in photon absorption. The refractive index, n_s, varies between 1.8 and 1.6 [128], and the absorption index, k_s, varies between 10^{-7} and 10^{-5}, depending on wavelength [129]. Measurements at higher temperatures [130] indicate that n_s increases as temperature increases with a nearly constant coefficient ($2.9 \times 10^{-5}\ K^{-1}$) [131], as shown in Fig. 13. The absorption index also increases with increasing temperature [116, 130, 139], as shown in Fig. 14. However, the data of Konopka et al. [116], which were taken from measurements using actual rocket particles heated in a shock tube, as compared with Gryvnak's data for pure Al_2O_3, indicate that composition is an important factor in determining the absorption index as well as temperature.

Polycrystalline structure and porosity also play a potentially significant role in determining aluminum oxide optical properties. Lee and Kingery [138] studied the transmission of both single-crystal (sapphire) and sintered ceramic aluminum oxide samples. The ceramic samples were porous and polycrystalline. Their experimental results and theoretical interpretations showed that the effects of porosity (pore size and volume fraction) are much stronger than the effects of polycrystallinity (grain size). Their results pointed out that although both pores and grain boundaries increase photon scattering, the porosity effect dominates for normal sintered ceramics which usually have at least 2% porosity with a typical pore size of 1–5 μm. The presence of pores does not alter the intrinsic optical constants of the material (n and k), but the

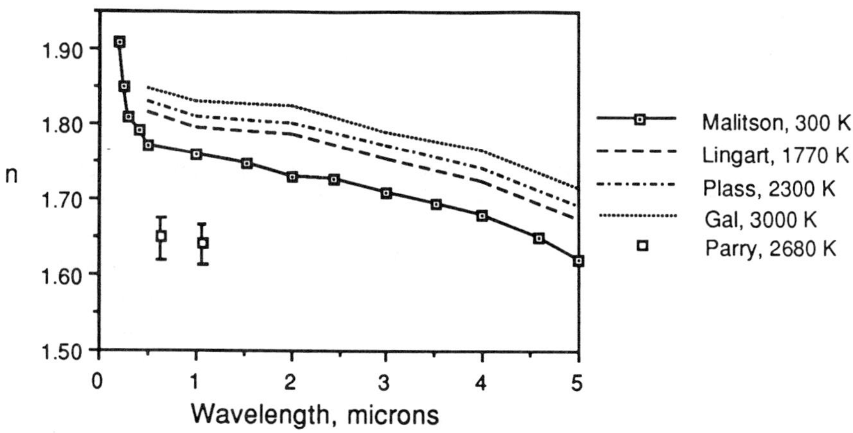

Figure 13 Survey of refractive index of aluminum oxide.

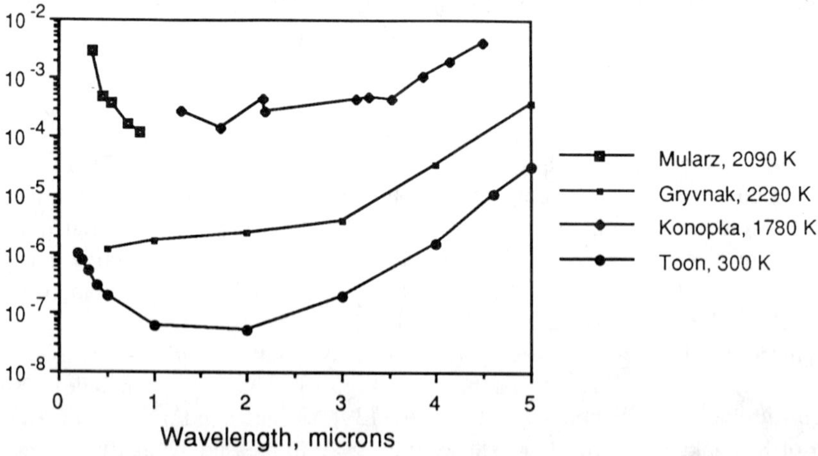

Figure 14 Survey of absorption index of solid aluminum oxide.

multiple-scattering effect does significantly increase the apparent emissivity of a given sample. The same enhancement of apparent emissivity by multiple scattering has also been noticed in MgO [143]. This multiple-scattering enhancement effect could be significant for particles of the order of tens or hundreds of micrometers but not for micrometer-sized or submicrometer particles.

Other studies [144, 145] have indicated that stoichiometry can also significantly influence the value of the absorption index. Substoichiometric aluminum oxide (Al_2O_{3-x}) can appear gray or black even at room temperature, in contrast to the normal white color of stoichiometric Al_2O_3. In oxygen-deficient aluminum oxide, free aluminum atoms act as free charge carriers to increase the absorption index.

5.2.2 Liquid Phase. The optical constant of Al_2O_3 in the molten state ($T_m = 2320$ K) have not been as well characterized as those for the solid state. Most studies indicate that k increases substantially upon melting, with a bigger increase occurring in relatively pure and stoichiometric samples [116, 130, 133, 135, 136]. Figure 15 shows that most of the reported values for k_m between 0.5 and 5 μm are in the range of 10^{-3} to 10^{-2}. Recently, Reed [137] suggested a relation for estimating k_m between 1.7 and 4.5 μm, based on extinction and emission measurements using particles that were collected from a rocket plume and heated in a shock tube:

$$k_m = 3.7 \times 10^{-4} \cdot T(\mathrm{K})^{1.5} \cdot \lambda(\mu\mathrm{m}) \cdot 10^{-13500/T(\mathrm{K})} \tag{13}$$

Predictions for k_m based on Eq. (13) are in reasonable agreement with other data plotted in Fig. 15, however Eq. (13) does not extend to short enough wavelengths to predict the upswing in k_m due to electronic transitions in the long-wavelength tail of the fundamental electronic absorption edge.

Conspicuously absent from the literature are data for direct measurements of the refractive index of molten Al_2O_3. The data of Gal and Kirch [132] plotted in Fig. 13 at 3000 K are apparently a simple extrapolation based on solid-phase data and not

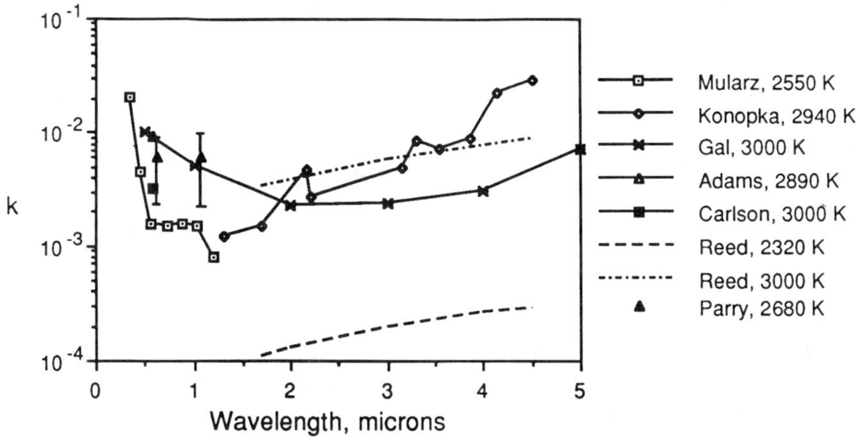

Figure 15 Survey of absorption index of molten aluminum oxide.

actual measurements. While this type of extrapolation has been widely used to estimate n_m [116, 132, 135, 136, 140], it ignores the change in density which occurs upon melting and the accompanying change in refractive index which takes place. Ignoring the change in refractive index due to melting is probably reasonable for predicting emission and absorption properties, but it is not reasonable for predicting scattering properties.

Recently, Parry and Brewster [120] reported the development of an in-situ light-scattering and extinction technique for determining the optical constants and mean optical size (d_{32}) of molten Al_2O_3 smoke particles in propellant flames. Direct transmittance and bidirectional transmittance and reflectance measurements were made using scattered laser light on aluminized solid-propellant flames at visible and near-infrared wavelengths ($\lambda_1 = 0.6328$ μm and $\lambda_2 = 1.064$ μm). The optical properties of the molten Al_2O_3 smoke combustion product were obtained from the light-scattering and extinction measurements by inverse solution of the radiative transfer equation. The values of n obtained for molten Al_2O_3 at 2680 K ($n_{m,\lambda 1} = 1.65$ and $n_{m,\lambda 2} = 1.64$) were significantly less than the values which have been reported for solid Al_2O_3 at temperatures just below the melting point of 2320 K ($n_{s,\lambda 1} = 1.82$ and $n_{s,\lambda 2} = 1.81$), indicating that a substantial decrease in n occurs upon melting. This decrease in n can be attributed to the expansion which takes place upon melting and is in good qualitative agreement with the predictions of the Lorentz-Lorenz equation. The value of k obtained for molten Al_2O_3 at 2680 K was 0.006 ± 0.004 (at both wavelengths), which is in reasonable agreement with other reported values. A dispersion analysis was also performed to fit this and other data over the spectral region from 0.5 to 5.0 μm and for temperatures from 2320 K to 3000 K.

A summary of aluminum oxide optical properties is given in Table 2 and Fig. 16. Molten aluminum oxide can be classified as a lossy dielectric in terms of its optical constants ($n \approx 1.65$ and $k \approx 10^{-3}$ to 10^{-2}). Because the particle size varies so widely, the scattering regime ranges from Rayleigh scattering to geometric optics scattering.

Table 2 Summary of liquid aluminum oxide particle radiative properties

Properties Size, d (μm)	Detached flame envelope 10^{-2} to 10^{-1}	Semiequilibrated smoke trails 1	Caps from surface accumulation 10 to 100
Size parameter $\pi d/\lambda$	$\ll 1$	~ 1	$\gg 1$
Radiative scattering regime	Rayleigh	Mie	Geometric optics
Particle emissivity	0.5–1	0.03–0.06	$\frac{4n}{(n+1)^2} \sim 0.94$
Extinction efficiency	10^{-5}–10^{-3}	2–4	2(1)

In the detached flame envelopes, the particle size is submicrometer, the size parameter is small, and Rayleigh scattering prevails. Because the refractive index is finite while the particle size is small, the particle "emissivity" (1 minus albedo) is of order 1. However, the extinction efficiency is very small. In the semiequilibrated smoke trails the particles have grown in size to the order of a micrometer, and Mie scattering predominates. Now the size is large enough to give extinction efficiency of order 1 but not large enough for the weakly absorbing particle to be semi-infinite and thus the particle "emissivity" is small. The residual caps, however, are large enough to be in the geometric optics regime, with extinction efficiency and emissivity of order 1. Thus a wide range of aluminum oxide particle sizes is present through a heterogeneous-propellant combustion system, and a corresponding wide range of scattering regimes is also evident.

5.3 Plume Radiation Models

While it is always a desirable goal to be able to predict everything including heat transfer from first principles, reality dictates that models based on empirical coefficients are usually necessary. The Bobco engineering model [146] for calculating plume radiation to aft-end equipment has been employed in the aerospace industry for the past 23 years and represents a reasonable balance between design needs and accurate

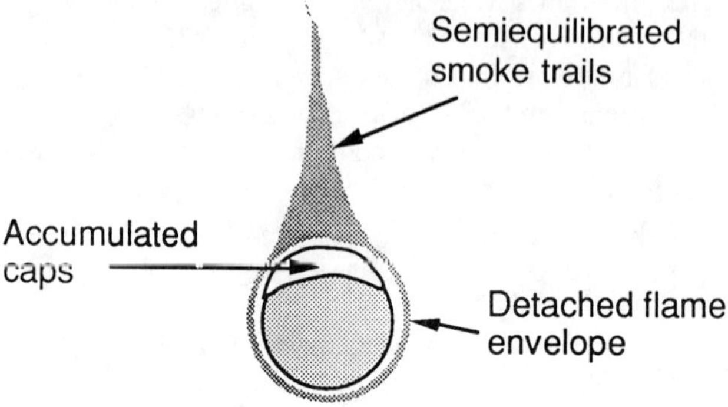

Figure 16 Liquid aluminum oxide particle.

physical representation. Many improvements have also been made to the Bobco model. Edwards et al. [147] included a two-particle model to represent cold, small, scattering particles and hot, large, emitting/absorbing particles in overlapping conical clouds. The clouds have axially varying radiative properties, with the hot particles imagined to lose emissive power and particle emissivity with increasing axial distance. Parametric calculation were made using a hybrid Monte Carlo, radiosity/irradiation technique to show which parameters are critical and how they may be inferred from engineering tests.

Edwards and Babikian [148] also considered the effects of nongray gas radiation and soot radiation. They concluded that gas radiation and soot radiation are negligible compared with emission by aluminum oxide. In that study the limiting anomalous diffraction results [149] were used to predict particle scattering and extinction efficiencies. These results assume that $n - 1 \ll 1$ and $\pi d/\lambda \gg 1$, which is not necessarily satisfied for aluminum oxide particles ($n = 1.6$ to 1.8; $\pi d/\lambda = 3$ to 5). Also, the limiting results of Rayleigh Gans theory [149] were used to predict the particle-scattering asymmetry factor. Rayleigh Gans theory also assumes that $n - 1 \ll 1$, in addition to assuming that $(n - 1)\pi d/\lambda \ll 1$. Again, these conditions are not necessarily satisfied and the asymmetry factor is overpredicted by using the Rayleigh Gans results. Comparing exact Mie calculations for extinction efficiency, albedo, and asymmetry factor with corresponding anomalous diffraction and Rayleigh Gans results for $n = 1.62$, $k = 0.001$, and $d = 1\ \mu m$ gives significant differences, but it is not certain how much these differences would influence the conclusions of [148]. It is not likely that the conclusions of the study would change significantly, although the numbers may be altered somewhat.

6 SUMMARY

Heat transfer plays an important role in several areas of heterogeneous-propellant combustion systems. Heat feedback to the propellant surface determines the regression rate of the propellant, which helps determine the thrust and action time of the motor. Heat transfer to internal and external surfaces creates the need for insulation systems which must be optimally designed for maximum performance. Gas and particle conduction, gas convection, and radiation by aluminum oxide particles all play significant roles at various locations through the propulsion system.

In the past, design of heterogeneous propellant combustion systems has relied on extensive testing and empiricism. This is partly because understanding of the fundamental processes has been limited and partly because extensive research has not been viewed as being cost-effective when compared with testing. This trend will change as testing becomes prohibitively expensive and confidence in numerical simulation increases. In the future, improvements in performance will come in small increments. Emphasis will be given to improving system reliability, safety, and environmental compatibility. At the same time, the need for accurate fundamental property data will always be present.

ACKNOWLEDGMENT

Support for the author's research from the National Science Foundation (CBT 86-96162), the office of Naval Research (N00014-87-K-0547), and Thiokol Corporation is gratefully acknowledged.

REFERENCES

1. N. Kubota, Survey of Rocket Propellants and Their Combustion Characteristics, in *Fundamentals of Solid Propellant Combustion* (ed. K. K. Kuo and M. Summerfield), Prog. in Astronautics and Aeronautics, vol. 90, AIAA, New York, chap. 1, 1984.
2. M. W. Beckstead, R. L. Derr, and C. F. Price, A Model of Composite Solid Propellant Combustion Based on Multiple Flames, *AIAA J.*, vol. 8, pp. 2200–2207, 1970.
3. M. W. Beckstead, Combustion Calculations for Composite Solid Propellants, *13th JANNAF Combustion Mtg.*, CPIA Publications, Laurel, MD, pp. 299–312, 1976.
4. M. W. Beckstead, A Model for Solid Propellant Combustion, *14th JANNAF Combustion Mtg.*, CPIA Publication 292, vol. 1, pp. 281–306, 1977.
5. R. R. Miller, M. T. Donohue, R. A. Yount, and J. R. Martin, Control of Solids Distribution in HTPB Propellants, AFRPL-TR-78-14, Hercules, Inc., Allegheny Ballistics Laboratory, Cumberland, Md., April 1978.
6. R. L. Glick and J. A. Condon, Statistical Combustion Modeling—The Effect of Additives, *14th JANNAF Combustion Mtg.*, CPIA Publication 261, vol. 1, pp. 341–378, December 1977.
7. N. S. Cohen and L. D. Strand, An Improved Model for the Combustion of AP Composite Propellants, *AIAA J.*, vol. 20, pp. 1739–1746, 1982.
8. N. S. Cohen, Review of Composite Propellant Burn Rate Modeling, *AIAA J.*, vol. 18, pp. 277–293, 1980.
9. E. W. Price, J. K. Sambamurthi, R. R. Panyam, and R. K. Sigman, Combustion of Ammonium Perchlorate-Polymer Sandwiches, *Combustion and Flame*, vol. 63, pp. 381–413, 1986.
10. E. W. Price, C. Markou, and R. K. Sigman, Effect of Types of Binder and Burning Rate Catalyst on Edge Burning of AP-Binder-AP Sandwiches, *Proc. 26th JANNAF Combustion Mtg.*, CPIA Publication #529, Vol. II, pp. 93–106.
11. R. R. Panyam, Combustion at the Interface of an Oxidizer-Fuel Slab System, Ph.D. thesis, Georgia Institute of Technology, Atlanta, Ga., 1983.
12. J. E. Crump, Aluminum Combustion in Composite Propellants, *ICRPG Combustion Conf.*, CPIA Publication No. 105, vol. I, pp. 321–329, May 1966.
13. H. L. Churchill, R. W. Fleming, and N. S. Cohen, Aluminum Behavior in Solid Composite Propellant Combustion, AFRPL-TR-74-13, May 1974.
14. V. G. Grigoyverm, V. A. Arko, and K. P. Kutsenogy, Experimental Study of Aluminum Particle Agglomeration by Combustion of Condensed Systems, *Fizika Goreniya i Vzryva*, vol. 17, no. 3, pp. 3–10, May–June 1981.
15. N. S. Cohen, A Pocket Model for Aluminum Agglomeration in Composite Propellants, *AIAA J.*, vol. 21, no. 5, pp. 720–725, May 1983.
16. E. W. Price, R. K. Sigman, J. K. Sambamurthi, and C. J. Park, Behavior of Aluminum in Solid Propellant Combustion, AFOSR-TR-77-0050, November 1976.
17. E. W. Price, Combustion of Metalized Propellants, in *Fundamentals of Solid Propellant Combustion* (ed. K. K. Kuo and M. Summerfield), Prog. in Astronautics and Aeronautics, vol. 90, AIAA, New York, chap. 9, 1984.
18. J. K. Sambamurthi, E. W. Price, and R. K. Sigman, Aluminum Agglomeration in Solid-Propellant Combustion, *AIAA J.*, vol. 22, pp. 1132–1138, August 1984.
19. O. B. Kovalev, Physicomathematical Model of the Agglomeration of Aluminum in the Combustion of Mixed Condensed Systems, *Fizika Goreniya i Vzryva*, vol. 25, no. 1, pp. 39–48, January–February 1989.

20. V. A. Babuk, V. P. Belov, V. V. Khodosov, and G. G. Shelukhin, Study of the Structure of Agglomerates with Combustion of Aluminized Mixed Condensed Systems, *Fizika Goreniya i Vzryva*, vol. 24, no. 5, pp. 52–57, September–October 1989.
21. M. A. Gurevich, K. I. Lapkina, and E. S. Ozerov, Ignition Limits of Aluminum Particles, *Fizika Goreniya i Vzryva*, vol. 6, no. 2, pp. 172–175, April–June 1970.
22. M. A. Gurevich, G. E. Ozerova, and A. M. Stepanov, Heterogeneous Ignition of an Aluminum Particle in Oxygen and Water Vapor, *Fizika Goreniya i Vzryva*, vol. 6, no. 3, pp. 326–335, July–September 1970.
23. A. F. Belyaev, Y. V. Frolov, and A. I. Korotkov, Combustion and Ignition of Particles of Finely Dispersed Aluminum, *Fizika Goreniya i Vzryva*, vol. 4, no. 3, pp. 323–329, 1968.
24. M. K. King, A Model of the Effect of Wires on Solid Propellant Burning Rate, *25th JANNAF Combustion Mtg.*, CPIA Publication No. 498, pp. 95–108, 1988.
25. A. A. Zenin, A. P. Glaskova, O. I. Leipunskyi, and V. K. Bobolev, Effects of Metallic Additives on the Deflagration of Condensed Systems, *12th Symp (Int.) on Combustion*, The Combustion Institute, Holt, Rhinehart and Winston, New York, pp. 27–35, 1968.
26. B. E. Hardt and M. Q. Brewster, Investigation of Al and Mg/Al Alloy Behavior in Composite Solid Propellant Combustion, *25th JANNAF Combustion Mtg.*, CPIA Publication No. 498, pp. 199–206, 1988.
27. M. Q. Brewster and B. E. Hardt, Influence of Metal Agglomeration and Heat Feedback on Composite Solid Propellant Burn Rate, *J. Propulsion and Power*, to appear 1991.
28. M. Q. Brewster and B. E. Hardt, Selective Radiation Absorption in Composite Propellant Combustion, *24th JANNAF Combustion Mtg.*, vol. 1, CPIA Publication No. 476, pp. 157–163, 1987.
29. M. Q. Brewster and K. W. Gross, Radiation Effects on Flow Characteristics in Combustion Chambers, *Proc. 26th JANNAF Combustion Mtg.*, CPIA Publications, #529, Vol. II, pp. 343–354, 1989.
30. M. D. Horton and L. Z. Youngberg, Effect of Radiant Energy on the Burning Rate of a Composite Solid Propellant, *AIAA J.*, vol. 8, pp. 1738–1741, 1970.
31. R. L. Coates and S. Kwak, Effect of External Radiation on the Burning Rates of Solid Propellants, *J. Spacecraft and Rockets*, pp. 742–744, 1972.
32. L. H. Caveny, T. J. Ohlemiller, and M. Summerfield, Influence of Thermal Radiation on Solid Propellant Burning Rate, *AIAA J.*, vol. 13, pp. 202–205, 1975.
33. M. M. Ibiricu and F. A. Williams, Influence of Thermal Radiation on the Burning Rates of Homogeneous Solid Propellants, *Combustion and Flame*, vol. 24, pp. 185–198, 1975.
34. J. Yin, B. Chen, Q. Luo, J. Ma, and K. Wang, Effect of External Radiation on the Burning Rates of Composite Solid Propellants, AIAA 89-2533, 25th Joint Propulsion Conf., 1989.
35. C. M. Mihlfeith, A. D. Baer, and N. W. Ryan, Propellant Combustion Instability as Measured by Combustion Recoil, *AIAA J.*, vol. 10, pp. 1280–1285, 1972.
36. C. M. Mihlfeith, Some Experiments on the Effect of Thermal Radiation on Composite Solid Rocket Propellants, Ph.D. thesis, University of Utah, Salt Lake City, Utah, 1971.
37. V. F. Mikheev, V. E. Zarko, S. M. Borin, K. P. Kutsenogii, and V. N. Simonenko, Measurement of Burning Rates in Transient Combustion Processes Under the Influence of External Radiation, in *AIAA Prog. Astro. and Aero.* (ed. T. Boggs and B. Zinn), vol. 63, pp. 173–187, 1978.
38. V. E. Zarko and A. B. Kiskin, Numerical Modeling of Nonsteady Powder Combustion Under the Action of a Light Flux, *Combustion, Explosion and Shock Wave*, vol. 16, pp. 650–654, 1980.
39. V. E. Zarko, V. N. Simonenko, and K. P. Kutsenogii, Effects of Light Intensity on the Nonstationary Combustion Rate in a Transient Process, *Combustion, Explosion and Shock Wave*, vol. 11, pp. 459–464, 1975.
40. T. J. Ohlemiller, L. H. Caveny, L. DeLuca, and M. Summerfield, Dynamic Effects on Ignitability Limits of Solid Propellants Subjected to Radiative Heating, *Fourteenth Symp. (Int.) on Combustion*, The Combustion Institute, Holt, Rhinehart and Winston, New York, pp. 1297–1307, 1972.
41. L. DeLuca, Frequency Response Function of Burning Solid Propellants, *Meccanica*, pp. 195–205, December 1980.
42. L. Strand, M. Weil, and N. S. Cohen, Solid Propellant Combustion Response to Oscillatory Radiant Heat Flux, AIAA 89-2667, 25th Joint Propulsion Conf., 1989.
43. J. C. Finlinson, T. P. Parr, and R. Stalnaker, Measurement of Propellant Combustion Response to

Oscillatory Radiant Heat Flux, *Proc. 26th JANNAF Combustion Mtg.*, CPIA Publications, to appear, 1989.

44. T. D. Prevish and M. Q. Brewster, Combustion Response of Solid Rocket Propellants to Oscillatory Radiant Heat Flux, *Proc. 26th JANNAF Combustion Mtg.*, CPIA Publications, #529, vol. II, pp. 127–142, 1989.
45. R. S. Patel and M. Q. Brewster, Optical Constants of Propellant Grade Ammonium Perchlorate, *AIAA J.*, vol. 24, pp. 1878–1880, 1986.
46. R. F. Felton and J. E. Hitchcock, An Optical Technique to Measure Energy Incident on a Burning Solid Propellant Surface, AIAA 71-469, 6th Thermophysics Conf., 1971.
47. M. Q. Brewster and D. L. Parry, Radiative Heat Feedback in Aluminized Solid Propellant Combustion, *J. Thermophys. Heat Transfer*, vol. 2, pp. 123–130, 1988.
48. M. Q. Brewster and R. S. Patel, Selective Radiative Preheating of Aluminum in Composite Solid Propellant Combustion, *J. Heat Transfer*, vol. 109, pp. 179–184, 1987.
49. M. Q. Brewster and D. M. Taylor, Radiative Properties of Burning Aluminum Droplets, *Combustion and Flame*, vol. 72, pp. 287–299, 1988.
50. D. M. Chen, W. H. Hsieh, W. H. Snyder, T. S. Yang, and K. K. Kuo, Study of the Thermophysical Properties and Combustion Behavior of Metal-Based Solid Fuels, AIAA 88-3041, 1988.
51. A. Ishihara, M. Q. Brewster, T. A. Sheridan, and H. Krier, The Influence of Radiative Heat Feedback on Burning Rate in Aluminized Propellants, *Combustion and Flame*, vol. 84, pp. 141–153, 1991.
52. J. M. Lenior and G. Robillard, A Mathematical Method to Predict the Effects of Erosive Burning in Solid Propellant Rocket, *Sixth Symp. (Int.) on Combustion*, Reinhold, New York, pp. 663–667, 1957.
53. M. K. Razdan and K. K. Kuo, Turbulent Flow Analysis of Erosive Burning of Cylindrical Composite Solid Propellant, *AIAA J.*, vol. 20, pp. 122–128, 1982.
54. R. A. Beddini, Aerothermochemical Analysis of Erosive Burning in a Laboratory Solid Rocket Motor, *AIAA J.*, vol. 18, pp. 1346–1353, 1980.
55. M. K. King, A Model of Erosive Burning of Composite Propellants, *J. Spacecraft and Rockets*, vol. 15, pp. 139–146, 1978.
56. M. J. Zucrow, J. R. Osborn, and J. M. Murphy, An Experimental Investigation of the Erosive Burning Characteristics of a Nonhomogeneous Solid Propellant, *AIAA J.*, vol. 3, pp. 523–525, 1965.
57. M. K. Razdan and K. K. Kuo, Erosive Burning of Solid Propellants, in *Fundamentals of Solid Propellant Combustion* (ed. K. K. Kuo and M. Summerfield), Prog. in Astronautics and Aeronautics, vol. 90, AIAA, New York, chap. 10, 1984.
58. M. K. King, One Person's View of Velocity Coupled Instability, *24th JANNAF Combustion Mtg.*, CPIA Publication 476, pp. 115–130.
59. G. L. Vogt, Workshop Report: Velocity Coupling, *24th JANNAF Combustion Mtg.*, CPIA Publication 476, pp. 131–140.
60. I. Glassman, Combustion of Metals, Physical Considerations, in *Solid Propellant Rocket Research* (ed. M. Summerfield), Prog. in Astronautics and Rocketry, vol. 1, Academic Press, New York, pp. 253–258, 1960.
61. T. A. Brzustowski and I. Glassman, Spectroscopic Investigation of Metal Combustion, in *Heterogeneous Combustion* (ed. H. G. Wolfhard, I. Glassman, and L. Green), Prog. in Astronautics and Rocketry, vol. 15, Academic Press, New York, pp. 41–74, 1964.
62. A. M. Mellor and I. Glassman, Vapor Phase Diffusion Flames in the Combustion of Magnesium and Aluminum: III. Experimental Observations in Carbon Dioxide Atmospheres, in *Heterogeneous Combustion* (ed. H. G. Wolfhard, I. Glassman, and L. Green), Prog. in Astronautics and Rocketry, vol. 15, Academic Press, New York, pp. 159–176, 1964.
63. R. Friedman and A. Macek, Ignition and Combustion of Aluminum Particles in Hot Ambient Gases, *Combustion and Flame*, vol. 6, pp. 9–19, 1965.
64. D. K. Kuehl, Ignition and Combustion of Aluminum and Beryllium, *AIAA J.*, vol. 3, pp. 2239–2247, 1965.
65. G. H. Markstein, Combustion of Metals, *AIAA J.*, vol. 1, pp. 550–562, 1963.
66. A. M. Mellor and I. Glassman, A Physical Criterion for Metal Ignition, *Pyrodynamics*, vol. 3, pp. 43–64, 1965.

67. W. M. Fassell, C. A. Papp, D. L. Hildebrand, and R. P. Sernka, The Experimental Nature of the Combustion of Metallic Powders, in *Solid Propellant Rocket Research* (ed. M. Summerfield), Prog. in Astronautics and Rocketry, vol. 1, Academic Press, New York, pp. 259–260, 1960.
68. R. W. Bartlett, J. N. Ong, W. M. Fassell, and C. A. Papp, Estimating Aluminum Particle Combustion Kinetics, *Combustion and Flame*, vol. 7, pp. 227–234, 1963.
69. G. H. Markstein, Heterogeneous Reaction Processes in Metal Combustion, *Eleventh Symp. (Int.) on Combustion*, The Combustion Institute, Holt, Rhinehart and Winston, New York, 1967.
70. A. Macek, Fundamentals of Combustion of Single Aluminum and Beryllium Particles, *Eleventh Symp. (Int.) on Combustion*, The Combustion Institute, Holt, Rhinehart and Winston, New York, pp. 203–217, 1967.
71. A. Macek, R. Friedman, and J. M. Semple, Techniques for the Study of Combustion of Beryllium and Aluminum Particles, in *Heterogeneous Combustion* (ed. H. G. Wolfhard, I. Glassman, and L. Green), Prog. in Astronautics and Rocketry, vol. 15, Academic Press, New York, pp. 3–16, 1964.
72. A. Davis, Solid Propellants: The Combustion of Particles of Metal Ingredients, *Combustion and Flame*, vol. 7, pp. 395–367, 1963.
73. L. S. Nelson, N. L. Richardson, and J. L. Prentice, Apparatus for the Production and Ignition of Metal Droplets with a Pulsed Laser, *Rev. Sci. Instruments*, vol. 39, pp. 744–747, 1968.
74. R. P. Wilson and F. A. Williams, Experimental Study of the Combustion of Single Aluminum Particles in O_2/Ar, *Thirteenth Symp. (Int.) on Combustion*, The Combustion Institute, Philadelphia, pp. 833–845, 1971.
75. J. L. Prentice and L. S. Nelson, Differences Between the Combustion of Aluminum Droplets in Air and in an Oxygen-Argon Mixture, *J. Electrochem. Soc.*, vol. 115, pp. 809–812, 1968.
76. C. M. Drew, A. S. Gordon, and R. H. Knipe, Study of Quenched Aluminum Particle Combustion, in *Heterogeneous Combustion*, (ed. H. G. Wolfhard, I. Glassman, and L. Green), Prog. in Astronautics and Rocketry, vol. 15, Academic Press, New York, pp. 17–39, 1964.
77. C. K. Law, A Simplified Theoretical Model for the Vapor Phase Combustion of Metal Particles, *Combustion Sci. Technol.*, vol. 7, pp. 197–212, 1973.
78. W. G. Courtney, Condensation During Heterogeneous Combustion, *Eleventh Symp. (Int.) on Combustion*, The Combustion Institute, Philadelphia, pp. 237–244, 1967.
79. T. H. Rautenberg and P. D. Johnson, Light Production in the Aluminum-Oxygen Reaction, *J. Opt. Soc. Am.*, vol. 50, pp. 602–605, 1960.
80. I. Glassman, A. M. Mellor, H. F. Sullivan, and N. M. Laurendeau, A Review of Metal Ignition and Flame Models, AGARD-CP52, pp. (19)1–(19)29, 1970.
81. C. E. Kolb, M. E. Gersh, and D. R. Herschbach, A Suggested Mechanism for the Visible Chemiluminescence Observed in Gas Phase Aluminum Oxidation, *Combustion and Flame*, vol. 25, pp. 31–41, 1975.
82. P. L. Micheli, Prediction of Burning Rate of Aluminum in Solid Rocket Motors, *13th JANNAF Combustion Mtg.*, CPIA Publications, Philadelphia, pp. 487–499, 1976.
83. P. L. Micheli and W. G. Schmidt, Behavior of Aluminum in Solid Rocket Motors, AFRPL-TR-77-29, 1977.
84. W. G. Schmidt, R. L. Lovine, and R. C. Poynter, Zirconium/Aluminum Combustion, AFRPL-TR-81-19, 1981.
85. A. Fontijn, Temperature Dependence of the Kinetics of Elementary Combustion Reactions of Refractory Species, *Combustion Sci. Technol.*, vol. 50, pp. 151–162, 1986.
86. R. A. Ogle, J. K. Beddow, L. D. Chen, and P. B. Butler, An Investigation of Aluminum Dust Explosions, *Combustion Sci. Technol.*, vol. 61, pp. 75–99, 1988.
87. S. R. Turns, S. C. Wong and E. Ryba, Combustion of Aluminum-Based Slurry Agglomerates, *Combustion Sci. Technol.*, vol. 54, pp. 299–318, 1987.
88. S. C. Wong and S. R. Turns, Ignition of Aluminum Slurry Droplets, *Combustion Sci. Technol.*, vol. 52, pp. 221–242, 1987.
89. A. G. Merzhanov, Y. M. Grivorjev, and Y. A. Galchenko, Aluminum Ignition, *Combustion and Flame*, vol. 29, pp. 1–14, 1977.

90. M. R. Jones, K. C. Tang, and M. Q. Brewster, An Investigation of the Combustion of Aluminum in Water, *Proc. 26th JANNAF Combustion Mtg.*, CPIA Publication #529, Vol. II, pp. 213–222, 1989.
91. K. Brezinsky, K. Davis, and I. Glassman, Gas Phase Combustion Synthesis of TiN, paper 89-6, Western States Section/The Combustion Institute, Philadelphia, 1989.
92. W. Felder and H. F. Calcote, A Sodium Flame Process for Synthesis of Pure Metals, Alloys and Ceramics, paper 89-8, Western States Section/The Combustion Institute, 1989.
93. E. V. Chernenko, L. F. Afanaseva, V. A. Lebedeva, and V. I. Rozenband, Inflammability of Mixtures of Metal Oxides with Aluminum, *Fizika Goreniya i Vzryva*, vol. 24, pp. 3–11, 1988.
94. I. Glassman, *Combustion*, 2nd ed., Academic Press, New York, p. 388, 1987.
95. R. W. Hermsen, Improved Technical Elements for SPP: Combustion Efficiency Models, CPIA Publication 321, Philadelphia, pp. 53–60.
96. G. R. Nickerson, D. E. Coats, R. W. Hermsen, and J. T. Lamberty, A Computer Program for the Prediction of Solid Propellant Rocket Motor Performance (SPP), vols. 1–5, AFRPL-TR-83-036, 1984.
97. B. E. Pearce, Radiative Heat Transfer With a Solid Propellant Rocket Motor, *J. Spacecraft and Rockets*, vol. 15, pp. 125–128, 1978.
98. T. P. Rudy, R. S. Brown, E. J. Shanabrook, H. H. Weyland, and L. S. Bain, Unstable Combustion of HTPB/AP Propellants at High Pressure, *24th JANNAF Combustion Mtg.*, vol. III, CPIA Publication 476, pp. 1–14, 1987.
99. A. V. Grosse and J. B. Conway, Combustion of Metals in Oxygen, *Ind. Eng. Chem.*, vol. 50, pp. 663–672, 1958.
100. S. P. Purcell and W. L. Daines, Buoyant Flow and Its Effect on Internal Insulator Performance-Numerical Approach, *Proc. 1985 JANNAF Propulsion Mtg.*, CPIA Publications, pp. 1–10, 1985.
101. R. S. Larson, Prediction of Aluminum Combustion Efficiency in Solid Propellant Rocket Motors, *AIAA J.*, vol. 25, pp. 92–101, 1987.
102. E. G. Bradley and M. Q. Brewster, A Multi-dimensional Combustion Model for Solid Propellant Flowfields, with Burning Aluminum Droplets, *Proc. 26th JANNAF Combustion Mtg.*, CPIA Publications, #529, Vol. II, pp. 57–68, 1989.
103. R. Madabhushi, J. Sabnis, F. DeJong, and H. Gibeling, Navier-Stokes Analysis of Aft Dome Flow Field in Solid Motors with Submerged Nozzle, AIAA 89-2781, 1989.
104. M. Salita, Solid Rocket Motor Performance: A Workshop Report, *24th JANNAF Combustion Mtg.*, CPIA Publication 476, pp. 253–276, 1987.
105. J. N. Levine, Application of Internal Flow Field Modeling to Solid Rocket Motor Design: A Workshop Report, *25th JANNAF Combustion Mtg.*, CPIA Publication 498, pp. 213–226, 1988.
106. C. B. Ludwig, W. Malkmus, J. E. Reardon, and J. A. L. Thomson, Handbook of Infrared Radiation from Combustion Gases, NASA SP-3080, NTIS N73-27807, 1973.
107. J. Walker, W. Malkmus, and C. B. Ludwig, Handbook of the Standardized Infrared Radiation Model (SIRRM), AFRPL-TR-81-61, 1981.
108. D. K. Edwards and R. P. Bobco, Effect of Particle Size Distribution on the Radiosity of Solid Propellant Rocket Motor Plumes, AIAA-81-1032, 16th Thermophysics Conf., 1981.
109. R. W. Hermsen, Aluminum Oxide Particle Size for Solid Rocket Motor Performance Prediction, *J. Spacecraft and Rockets*, vol. 18, pp. 483–490, 1981.
110. K. J. Kraeutle, Particle Size Analysis in Solid Propellant Combustion, *Prog. in Aeronautics and Astronautics*, AIAA, vol. 63, pp. 76–108, 1978.
111. M. Salita, Quench Bomb Investigation of Al_2O_3 Formation from Solid Rocket Propellants (Part II): Analysis of Data, *25th JANNAF Combustion Mtg.*, CPIA Publication 498, pp. 185–198, 1988.
112. J. Koo, A Review of Particle Sizing Methods in Rocket Propulsion, *24th JANNAF Combustion Mtg.*, CPIA Publication 476, pp. 141–156, 1987.
113. R. A. Dobbins, Remote Size Measurements of Particulate Products of Heterogeneous Combustion, *Eleventh Symp. (Int.) on Combustion*, The Combustion Institute, Philadelphia, pp. 921–926, 1967.
114. R. A. Dobbins and L. D. Strand, A Comparison of Two Methods of Measuring Particle Size of Al_2O_3 Produced by a Small Rocket Motor, *AIAA J.*, vol. 8, pp. 1544–1550, 1970.
115. E. A. Powell, R. A. Cassanova, C. P. Bankston, and B. T. Zinn, Combustion Generated Smoke

Diagnostics by Means of Optical Measurement Techniques, *Prog. in Astronautics and Aeronatics*, AIAA, vol. 53, pp. 449–463, 1977.

116. W. L. Konopka, R. A. Reed, and V. S. Calia, Measurements of Infrared Optical Properties of Al_2O_3 Rocket Particles, *Prog. in Aeronautics and Astronautics*, AIAA, vol. 91, pp. 180–197, 1984.
117. S. Orguc, T. E. Pruitt, T. D. Edwards, E. D. Youngborg, J. P. Powers, and D. W. Netzer, Measurement of Particulate Size in Solid Propellant Rocket Motors, *24th JANNAF Combustion Mtg.*, CPIA Publication 476, pp. 165–182, 1987.
118. D. L. Hofeldt, M. G. Allen and R. K. Hanson, Instantaneous Two-Dimensional Multiple Particle Sizing Diagnostic, *24th JANNAF Combustion Mtg.*, CPIA Publication 476, pp. 183–192, 1987.
119. T. E. Parker, D. I. Rosen, and W. T. Rawlins, Infrared Optical Property Measurement Techniques: Definition and Preliminary Design, AFRPL-TR-86-034, 1986.
120. D. L. Parry and M. Q. Brewster, Optical Constants of Al_2O_3 Smoke in Propellant Flames, to appear in *J. Thermophysics and Heat Transfer*, Vol. 5, No. 2, 1991.
121. M. Salita, Implementation and Validation of the One-Dimensional Gas/Particle Flow Code OD3P, *Proc. 26th JANNAF Combustion Mtg.*, CPIA Publication #529, Vol. III, pp. 69–82, 1989.
122. C. D. Bartky and E. Bauer, Predicting the Emittance of a Homogeneous Plume Containing Alumina Particles, *J. Spacecraft and Rockets*, vol. 3, pp. 1523–1527, 1966.
123. S. J. Morizumi and H. J. Carpenter, Thermal Radiation from the Exhaust Plume of an Aluminized Composite Propellant Rocket, *J. Spacecraft and Rockets*, vol. 1, pp. 501–507, 1964.
124. B. W. Worster, Particulate Infrared Radiation in Aluminized Solid Fuel Rocket Plumes, *J. Spacecraft and Rockets*, vol. 11, pp. 260–262, 1974.
125. A. B. Pluchino and D. E. Masturzo, Emissivity of Al_2O_3 Particles in Rocket Plume, *AIAA J.*, vol. 19, pp. 1234–1237, 1981.
126. T. J. Rieger, On the Emissivity of Alumina/Aluminum Composite Particles, *J. Spacecraft and Rockets*, vol. 16, pp. 438–439, 1979.
127. W. R. Cofer III, E. L. Winstead, and L. E. Key, Surface Composition of Solid Rocket Exhausted Aluminum Oxide Particles, *J. Propulsion and Power*, vol. 5, pp. 674–677, 1989.
128. I. H. Malitson, Refraction and Dispersion of Synthetic Sapphire, *J. Opt. Soc. Am.*, vol. 52, no. 12, pp. 1377–1379, 1962.
129. O. B. Toon and J. B. Pollack, The Optical Constants of Several Atmospheric Aerosol Species: Ammonium Sulfate, Aluminum Oxide, and Sodium Chloride, *J. Geophys. Res.*, vol. 81, no. 33, pp. 5733–5748, November 1976.
130. D. A. Gryvnak and D. E. Burch, Optical and Infrared Properties of Al_2O_3 at Elevated Temperatures, *J. Opt. Soc. Am.*, vol. 55, no. 6, pp. 625–629, 1965.
131. G. N. Plass, Temperature Dependence of the Mie Scattering and Absorption Cross Sections for Aluminum Oxide, *Appl. Optics*, vol. 4, no. 12, pp. 1616–1619, 1965.
132. G. Gal and H. Kirch, Particulate Optical Properties in Rocket Plumes, AFRPL-TR-73-99, pp. 28–29, 1973.
133. D. J. Carlson, Emittance of Condensed Oxides in Solid Propellant Combustion Products, *Tenth (Int.) Symp. on Combustion*, The Combustion Institute, Philadelphia, pp. 1413–1424, 1965.
134. E. Bauer and D. J. Carlson, Mie Scattering Calculations for Micron Size Alumina and Magnesia Spheres, *J. Quant. Spectro. Rad. Transfer.*, vol. 4, pp. 363–374, 1964.
135. J. M. Adams, A Determination of the Emissive Properties of a Cloud of Molten Alumina Particles, *J. Quant. Spectro. Rad. Transfer*, vol. 7, pp. 273–277, 1967.
136. E. J. Mularz and M. C. Yuen, An Experimental Investigation of Radiative Properties of Aluminum Oxide Particles, *J. Quant. Spectro. Rad. Transfer*, vol. 12, pp. 1553–1568, 1972.
137. R. Reed, Optical Properties for Liquid Al_2O_3, Private communication, Arnold AFB, February 29, 1988.
138. D. W. Lee and W. D. Kingery, Radiation Energy Transfer and Thermal Conductivity of Ceramic Oxides, *J. Am. Ceramic Soc.*, vol. 43, no. 11, pp. 595–607, 1960.
139. V. H. Myers, A. Oro, and D. P. DeWitt, A Method for Measuring Optical Properties of Semi-Transparent Materials at High Temperatures, *AIAA J.*, vol. 24, no. 2, pp. 321–326, 1986.

140. Y. K. Lingart, V. A. Petrov and A. A. Tikhonova, Optical Properties of Leucosapphire at High Temperatures, *Teplofizika Vysokikh Temperatur*, vol. 20, no. 5, pp. 872–880, 1982.
141. A. S. Barker Jr., Infrared Lattice Vibrations and Dielectric Dispersion in Corundum, *Phys. Rev.*, vol. 132, no. 4, pp. 1474–1481, 1963.
142. D. G. Goodwin and M. Mitchner, Infrared Optical Constants of Coal Slags: Dependence on Chemical Composition, *J. Thermophys. and Heat Transfer*, vol. 3, no. 1, pp. 53–60, 1989.
143. V. G. Skogarev and A. V. Florko, Effect of the Temperature on the Integrated Emissivity of Magnesium Oxide, *Fizika Goreniya i Vzryva*, vol. 23, pp. 92–94, 1987.
144. L. S. Nelson and N. L. Richardson, Effects of Oxygen and Argon Atmospheres on Pendant Drops of Aluminum Oxide Melted with Carbon Dioxide Laser Radiation, *High Temp. Sci.*, vol. 5, pp. 138–154, 1973.
145. K. J. Kraeutle and H. H. Bradley, Jr., Combustion of Aluminized Propellants: The Influence of Pressure and Propellant Composition on Formation of Aluminum Combustion Residue, *14th JANNAF Combustion Mtg.*, CPIA Publication 292, vol. 1, pp. 209–219, 1977.
146. R. P. Bobco, Radiation from Conical Surfaces with Nonuniform Radiosity, *AIAA J.*, vol. 4, pp. 544–546, 1966.
147. D. K. Edwards, Y. Sakurai, and D. S. Babikian, A Two Particle Model for Rocket Plume Radiation, *J. Thermophys. and Heat Transfer*, vol. 1, pp. 13–20, 1987.
148. D. K. Edwards and D. S. Babikian, Radiation from a Nongray Scattering Emitting and Absorbing SRM Plume, AIAA-89-1721, 24th Thermophysics Conf., 1989.
149. H. C. van de Hulst, *Light Scattering by Small Particles*, Dover, New York, 1981.

CHAPTER
SEVEN

HEAT TRANSFER IN VORTICALLY ENHANCED MIXING OF VAPORIZING DROPLET SPRAYS

Roger H. Rangel

ABSTRACT

A discussion of a methodology for the study of heat transfer in the vortically enhanced mixing of heating and vaporizing sprays is provided. A brief description of work in related areas, particularly vortical mixing of inert particles and droplet vaporization theories, is provided. The fluid mechanics of vortical mixing and its implications for the efficient entrainment and vaporization of droplets, and the mixing of the vaporized species in some idealized configurations, are described with the use of streamlines and particle paths. Recent droplet vaporization models are discussed, and the numerical shortcomings of current gas-liquid coupling models are addressed. Some discussion on the distinction between dense and dilute sprays, the current status of gas-phase coupling, and the importance of radiation heat transfer in some applications are given.

NOMENCLATURE

B	transfer number
c_p	gas specific heat
C_D	drag coefficient
C_{Re}	convective correction
D	drag force
$\mathcal{D}$	diffusion coefficient
e	error
$F_{\Delta r}$	correction factor

$F(B_{\text{eff}})$	function of the transfer number
$\mathbf{g}$	gravity vector
h_{fg}	latent heat of vaporization
i	imaginary unit
m	droplet mass
$\dot{m}$	droplet vaporization rate
n	number of vortex elements
N	number of droplets or droplet groups
$\mathcal{N}$	number droplet density
Nu	Nusselt number
p	pressure
Pe	Peclet number
Pr	Prandtl number
$\dot{q}$	heat flux
$\dot{Q}_r$	thermal radiation energy
r	radial coordinate
r'	grid size-to-droplet radius ratio
$\mathbf{r}$	position vector
R	droplet radius
Re	Reynolds number
R^0	gas constant
$\tilde{R}$	droplet-cluster radius
R^*	equivalent radius
s	interface coordinate
Sc	Schmidt number
St	Stanton number
t	time
T	temperature
u	velocity component
ΔU	velocity jump across vortex sheet or layer
v	velocity component
V_{rel}	magnitude of gas-droplet relative velocity
$\mathbf{v}$	velocity vector
W	molecular weight
x	cartesian coordinate
X	molar fraction
y	cartesian coordinate
Y	mass fraction
z	$x + iy$, complex variable
α	thermal diffusivity
γ	vortex strength
δ	vortex size
$\delta(\mathbf{r})$	delta function
Δ	differential or incremental
ε	fractional vaporization rate

μ	gas viscosity
ρ	density
τ	characteristic time
ψ	streamfunction
ω	vorticity

Subscripts

bn	normal boiling point
eff	effective
F	F species
g	gas
i	vortex index
j	vortex index
ℓ	liquid
ref	reference
s	at the droplet surface
T	temperature based
∞	free stream

1 INTRODUCTION

Spray mixing by vortical flows occurs in many applications involving two-phase flows, particularly liquid droplet sprays. These applications include propulsion systems such as liquid rockets and gas turbines, pollutant dispersion in the air or in liquid streams, flame propagation in fires, and fire safety devices such as sprinkling systems, among others. In these systems, the formation and growth of large-scale vortical structures play an important role in enhancing the rates at which entrainment of the dispersed phase, and subsequent mixing within the continuous phase, take place. When the dispersed phase is a condensed material, capable of vaporization, then a combination of heat and mass transfer phenomena take place under the effect of the vortical field. The vortical structures may be initially absent in the flow field, but the existence of shearing flows will promote the growth of instabilities by the Kelvin-Helmholtz mechanism. The structures so formed will evolve into large-scale vortical eddies by pairing of the small ones.

This chapter outlines a theoretical research methodology to investigate the development of vortically enhanced mixing in two-phase flows. The configuration chosen for this investigation is one in which an initially condensed phase, such as a fuel or other liquid spray, is introduced into a gaseous shearing flow. The conditions under which the shearing flow can evolve into a number of large-scale vortical structures may be controlled, and the behavior of the spray, in terms of its spacial distribution, entrainment rates, and local and overall mixing, can be investigated so that the important controlling parameters are identified. In the following sections, the background of the problem is outlined, and some suggestions concerning future directions are given in detail.

2 BACKGROUND

Rather than a comprehensive compilation of all the research efforts into the areas of vortical flows and spray systems, this section contains a selection of some investigations which provide a particular foundation for the study of vortically enhanced heat transfer, vaporization, and mixing of liquid sprays. We begin with a presentation of the subject of vortical flows, continue with some insights into spray modeling, and close with a brief description of some experimental studies.

2.1 Vortical Flows

Vortical flows occur naturally or are induced in most energy applications involving turbulent mixing of two or more species. Thus, many propulsion systems involving separate injection of the reactants typically rely on the formation of a vortical field to enhance the rates of mixing and heat and mass transfer. The existence of coherent structures in turbulent flows, particularly in turbulent mixing layers, was observed and described in the experiments of Brown and Roshko [1] and Browand and Weidman [2]. Over the past 15 years, an increasing number of studies focusing on the behavior of these coherent structures, their effects on the flow field, and their interaction with the scalar fields have appeared. A few examples are the works of Christiansen and Zabusky [3], Acton [4], Ho and Huang [5], and Corcos and Sherman [6]. One of the fundamental motivations for the study of these flows is their direct relevance in the understanding and modeling of turbulent reactive flows. Some investigations that have explored this phenomenon are contained in the works of Marble [7], Masutani and Bowman [8], Givi et al. [9], Ghoniem and Givi [10], and Cetegen and Sirignano [11]. A very complete overview of the current status of research on the interaction of particles with shear layers has been given recently by Crowe et al. [12]. An investigation of the effect of large-scale vortex structures on particle dispersion was performed by Crowe et al. [13], who introduced a time scaling ratio, the Stokes number, to quantify the effect of the large-scale structures on the dispersion of particles. The Stokes number is defined as the ratio of the aerodynamic response time of the particle $\tau_A = 2\rho R^2/(9\mu)$ to the time scale associated with the large-scale structure $\tau_F = \delta/\Delta U$. Thus:

$$\mathrm{St} = \frac{\tau_A}{\tau_F} = \frac{2\rho R^2\,\Delta U}{9\mu\delta} \tag{1}$$

For small values of the Stokes number, the particles will respond very rapidly to gas velocity changes and will disperse as gas particles. For intermediate Stokes numbers, the particles may be entrained by the vortices and centrifuged away from the vortices. For large Stokes numbers, the large-scale coherent structures have insufficient time to influence the particle motion.

The above studies have not considered heat transfer effects (in the gas or in the particles) during the vortical mixing of particles. Neither have the effects of droplet vaporization been considered to any significant extent. In this chapter, the interaction between the fluid mechanics and the mixing of a two-phase flow, including vaporization of the condensed phase, will be examined with the objective of determining the conditions that lead to a more efficient entrainment and mixing. Some preliminary results will be presented, and some indications of future work will be made.

2.2 Spray Modeling Theory

Spray systems are encountered in many engineering applications that include liquid-fueled combustors, irrigation, drying, gas absorption, and fire control. Extensive reviews of the spray combustion literature can be found in the works of Faeth [14], Law [15], and Sirignano [16]. The complexities involved in the coupled exchange of energy, momentum, and mass between the gas and liquid phase renders the analysis of such systems very complicated. Analytical treatment of the equations governing typical spray phenomena is usually impossible due to the characteristic nonlinearities associated, for example, with the convective motion, the coupling between the phases, and the temperature dependence of the physical properties, to name a few reasons. In some cases, the formulation of certain simplifying assumptions permits an analytical or semianalytical solution of the equations. The combined insight gained by the use of both theoretical and numerical results is always more beneficial than the use of either one alone when explaining existing experimental behaviors.

In a number of recent papers, a model of heat and mass transfer, vaporization, and combustion of a spray system consisting of a number of parallel streams injected in an oxidizing gas flow has been developed. A two-dimensional analysis of steady channel flow with two streams of heating and vaporizing droplets was first considered [17]. An analytical procedure based on the Green's function approach was possible. Since the nonlinear vaporization terms were retained in the gas-phase energy equations, the final solution, after a suitable transformation to integrodifferential equations, was obtained by numerical integration. The liquid-phase equations were also integrated numerically because of the nonlinearities introduced by the vaporization process. The model has been more recently extended to include the chemically reactive case by considering the one-step Arrhenius chemical reaction of Westbrook and Dryer [18]. Transient effects in the gas phase were later introduced [19, 20] so that the development of the reaction zone could be investigated. In addition, a second transient effect, more properly described as an inherent intermittency that results from the discrete character of the droplets, was described. Because of the increased complexity of the physical model, the solution procedure employed was based on a finite-difference scheme.

The behavior of a group of droplets in a turbulent flow field generally depends on the concentration of the droplets and on the size of the droplets as compared with the scale of the turbulence [21]. At high droplet concentrations, the turbulence may be completely damped. Also, droplet agglomeration and collision can occur. At low concentrations, each particle may be considered isolated except for the indirect interaction which occurs if high vaporization rates exist. Theoretical studies of particle-turbulence interaction were performed by Tchen [22], Corrsin and Lumley [23], and Friedlander [24]. Some theoretical modeling of droplet-containing eddies has been initiated by Bellan and Harstad [25]. In this work, the behavior of a cluster of droplets inside an already existing eddy is modeled in a global manner. The droplet cluster is embedded into a cylindrical vortex which is infinite in the longitudinal direction. Since the droplet cluster is assumed to be within the eddy at the beginning of the calculation, no information can be inferred as to whether or not the droplets are entrained during the formation of the eddy and what mechanisms control this rate of entrainment. Those studies do provide interesting information on the subsequent behavior of the droplet cluster. In most cases it is shown that the droplets tend to form a shell near the outer rim of the vortex due to the action of the centrifugal force.

2.3 Experimental Studies

The interaction between discrete particles and grid generated turbulence was studied by Snyder and Lumley [26]. Experiments with small oil droplets in a plane air jet were performed by Ahmadi and Goldschmidt [27]. A complete review of the most relevant experimental investigations of particle dispersion by jets and other flow structures is reported by Crowe et al. [12], Kobayashi et al. [28] injected solid particles of 50 μm in diameter into one stream of a planar mixing layer. Their results indicate that dense particles appear to have velocities directed away from the mixing plane, toward the free stream. The turbulent interaction between a shear layer and a well-characterized spray under isothermal conditions has been investigated by Lazaro and Lasheras [29, 30]. In their two-dimensional experiments, they looked at the turbulent mixing region behind a flat plate separating two fluid streams, one of which contained a water spray. They examined the effect of the coherent vortical structures on the entrainment rates, as well as other characteristics of the two-phase interaction. Flow visualization, as well as instantaneous particle concentration measurements, revealed the presence of a highly organized particle dispersion layer [30]. Their results also indicated that the spectral response of the particles to the coherent velocity field leads to a selective dispersion of the droplets along the mixing layer, and a nonuniform droplet-size probability density functions (pdf) distribution across the dispersion layer was obtained.

3 THEORETICAL APPROACH

The theoretical modeling of complex flow systems such as the ones described in this chapter typically require the solution of a number of nonlinear, unsteady, and multi-dimensional partial and ordinary differential equations. Only in very limited cases and after a number of simplifications are performed by neglecting some terms in the equations is it possible to obtain analytical solutions of any validity over a wide range of parameters. The solution of the complete set of equations and even the solution of many simplified physical models requires complex numerical algorithms. We propose two basic configurations for the study of vortically enhanced mixing of two-phase flows. The first involves the temporally evolving shear layer and the second involves the spacially developing shear layer. In both cases, constant and variable gas density should be investigated. In the temporally evolving shear layer, periodic boundary conditions are employed upstream and downstream of the computational domain. This is equivalent to assuming that one is moving with the average velocity between the two streams. In the spacially developing shear layer, inflow and outflow conditions are required. The following sections describe a methodology to be used in the proposed investigations to solve the condensed-phase (spray) equations and the continuous-phase equations.

3.1 Liquid Spray Modeling

Our approach is based on a deterministic model. In it, the Lagrangian equations governing the behavior of a few representative elements (droplets or droplet groups) are solved and, in this way, conclusions pertaining to the behavior of large quantities of droplets are obtained. This methodology is based on the fact that the statistical aspect of these flows is due to the uncertainty on the exact initial spacial distribution and the exact initial values of other parameters, such as velocities and sizes, related to the spray. Once these quantities are known, the history of the spray may be obtained in an exact manner if other aspects of the flow are also described in a deterministic fashion. Clearly, this does not rule out the validity of the statistical approaches but merely complements it.

The Lagrangian equations governing the momentum dynamics of a droplet in a certain droplet group are the equations for the trajectory and the momentum exchange with the gas phase:

$$\frac{d\mathbf{x}}{dt} = \mathbf{v} \qquad \frac{d\mathbf{v}}{dt} = \frac{C_D}{2}\frac{\rho\pi R^2}{m} v_{\text{rel}}(\mathbf{v}_{\mathbf{g}} - \mathbf{v}) + \mathbf{g} \tag{2}$$

where $\mathbf{x}$ and $\mathbf{v}$ are the droplet position and droplet velocity vectors, respectively. The gas velocity and density are $\mathbf{v}_g$ and ρ, respectively, and the absolute value of the droplet-gas relative velocity is v_{rel}. The droplet mass is m and the droplet radius is R. The drag coefficient is obtained from available numerical correlations or from experimental ones. The gravitational field $\mathbf{g}$ can play an important role, depending on the orientation, in modifying the rates of entrainment of the droplets by the vortical flow. Equation (2) is strictly valid for $R \to 0$. Other terms that should be included in Eq. (2) are discussed in Lazaro and Lasheras [30].

Droplet drag calculations that account for both the Stefan flow and for the influence of neighboring droplets are inconclusive. Consequently, modelers usually use values of the drag coefficient corresponding to a solid sphere, with a correction for the effect of vaporization. Results of these numerical studies suggest that neighboring droplets affect the droplet drag significantly. Recently, Nguyen et al. [31] compiled several droplet-drag correlations for use in spray calculations. The drag coefficient is defined as

$$C_D = \frac{D}{\tfrac{1}{2}\rho v^2 \pi R^2} \tag{3}$$

where D is the magnitude of the drag force acting on the droplet. Most drag-coefficient correlations used in the spray literature account for a Reynolds-number effect and a vaporization effect through a dependence on the transfer number B. No correlation for the drag coefficient accounting for the combined dependence on the Reynolds and Transfer numbers as well as on droplet-droplet interaction is available in the literature, although the work of Chiang and Sirignano [32] may produce correlations for two droplets in tandem. A common droplet correlation is that of Yuen and Chen [33], who showed experimentally that for low to moderate vaporization rates ($B = c_p(T_\infty - T_s)/h_{fg} \le 3$), the drag coefficient of an evaporating droplet may be approximated by the standard drag curve for a solid sphere, provided that the gas viscosity μ is evaluated at a reference temperature and concentration obtained by using the one-third rule:

$$T_{\text{ref}} = T_s + \frac{1}{3}(T_\infty - T_s), \qquad Y_{F,\text{ref}} = Y_{Fs} + \frac{1}{3}(Y_{F\infty} - Y_{Fs}) \tag{4}$$

Yuen and Chen argued that the decrease in viscous drag due to blowing is accompanied by an increase in pressure drag of similar magnitude. The solid-sphere drag curve is approximated within 10% by the curve-fit formula [34]:

$$C_D = \frac{24}{\text{Re}} + \frac{6}{1 + \text{Re}^{0.5}} + 0.4 \tag{5}$$

in the range $0 \le \text{Re} \le 2 \times 10^5$. However, the solid-sphere correlation is invalid for high vaporization rates and for droplets with wake flames.

Renksizbulut and Yuen [35] conducted experiments with droplets in air streams up to 1059 K, and proposed the following correlation for the drag coefficient:

$$C_D = \frac{24}{\text{Re}}(1 + 0.2\,\text{Re}^{0.63})(1 + B)^{-0.2} \tag{6}$$

for $10 < \text{Re} < 300$. In this equation, all thermophysical properties are evaluated using the mean between the surface and ambient values of the temperature and mass fraction; the density in Re, however, is based on the free-stream value. The $(1 + B)^{0.2}$ factor accounts for the reduction in drag due to the blowing effect of evaporation. Chiang et al. [36] propose the following drag correlation, which agrees within 4% with results from their variable-property Navier-Stokes numerical calculations:

$$C_D = \frac{24}{\text{Re}}(1 + 0.41\,\text{Re}^{0.405})(1 + B)^{-0.32} \tag{7}$$

for $30 < \text{Re} < 250$, $0.6 < B < 6$. The thermophysical properties are evaluated as in the Renksizbulut-Yuen correlation. The Chiang et al. correlation indicates a larger reduction of drag due to blowing than does the Renksizbulut-Yuen correlation.

For a vaporizing droplet there is also an equation governing the rate of change of the droplet mass:

$$\frac{4}{3}\pi\rho_\ell \frac{dR^3}{dt} = -\dot{m} \tag{8}$$

Traditionally, the vaporization rate $\dot{m}$ is obtained from a quasi-steady analysis of the gas film around the droplet [37] as

$$\dot{m} = 4\pi R \rho \mathcal{D} C_{\text{Re}} \ln(1 + B_{\text{eff}}) \tag{9}$$

In this expression, B_{eff} is an effective transfer number that takes into account the heat of vaporization and the heat conducted into the droplet [37, 38]. The effective transfer number, B_{eff}, is defined by the mass fractions of fuel vapor at the droplet surface and in the surrounding air:

$$B_{\text{eff}} = \frac{Y_{Fs} - Y_{F\infty}}{1 - Y_{Fs}} \tag{10}$$

An energy balance, including the latent heat of vaporization, heat conduction into or out of the liquid, and conduction in the vapor phase, serves to determine B_{eff}. The convective correction C_{Re}, appearing in the equation for the vaporization rate, modifies the spherically symmetric result to account for the convective effects. Several correlations are available for the convective correction, including the one suggested by Clift et al. [39]:

$$\begin{aligned} &\text{Re} \geq 1 \qquad C_{\text{Re}} = 0.5[1 + (1 + \text{Re Pr})^{1/3}\,\text{Re}^{0.077}] \\ &\text{Re} < 1 \qquad C_{\text{Re}} = 0.5[1 + (1 + \text{Re Pr})^{1/3}] \end{aligned} \tag{11}$$

The appropriateness of Eq. (9) for high-Reynolds-number vaporization has been questioned by several investigations [16]. Recently, Abramzon and Sirignano [40] have proposed an expression for the vaporization rate that yields the logarithmic dependence on the transfer number for low Reynolds number, but which results in the more appropriate power-law dependence for high-Reynolds-number vaporization. The vaporization rate is given by

$$\dot{m} = 4\pi R\rho\mathcal{D}\left[1 + \frac{C_{\text{Re}} - 1}{F(B_{\text{eff}})}\right]\ln(1 + B_{\text{eff}}) \tag{12}$$

where

$$F(B_{\text{eff}}) = (1 + B_{\text{eff}})^{0.7}\,\frac{\ln(1 + B_{\text{eff}})}{B_{\text{eff}}} \tag{13}$$

All thermophysical properties are evaluated with the one-third rule except the density appearing in the Reynolds number, which is evaluated at the free-stream value. The function $F(B_{\text{eff}})$ increases from 1 to 1.285 as B_{eff} increases from 0 to 8. For $8 < B_{\text{eff}} < 20$, $F(B_{\text{eff}})$ remains practically constant. Renksizbulut and Haywood [41] provide an expression for the vaporization rate as

$$\dot{m} = 4\pi R\rho\mathcal{D}\,\frac{\text{Nu}}{2}\,B_T \tag{14}$$

where the Nusselt number Nu is given by

$$\text{Nu} = (2 + 0.57\,\text{Re}^{0.5}\,\text{Pr}^{0.333})\,(1 + B_T)^{-0.7} \tag{15}$$

and the transfer number $B_T = c_p(T_\infty - T_s)/h_{fg,\text{eff}}$ takes into account the heat transferred into the droplet. The one-half rule is used to evaluate the thermophysical properties except for the density in the Reynolds number, which should be the free-stream density. This correlation is valid for $10 < \text{Re} < 2000$.

If transient internal heating is considered, there is an equation for the conservation of energy inside the droplet. Employing a spherically symmetric model,

$$\frac{\partial T}{\partial t} = \alpha_{\text{eff}} \frac{1}{r^2} \frac{\partial}{\partial r} \left(r^2 \frac{\partial T}{\partial r} \right) \tag{16}$$

where T is the liquid temperature and α_{eff} is an effective thermal diffusivity that takes into account the internal circulation in the droplet due to the convective gas flow surrounding it [40]:

$$\alpha_{\text{eff}} = \alpha \left\{ 1.86 + 0.86 \tanh \left[2.245 \log \left(\frac{\text{Pe}_\ell}{30} \right) \right] \right\} \tag{17}$$

The energy conducted into the droplet is obtained from the definition of the effective transfer number as

$$\dot{q}_\ell = \dot{m} \left[\frac{c_p (T_\infty - T_s)}{B_{\text{eff}}} - h_{fg} \right] \tag{18}$$

Other equations needed to complete the film model are the thermodynamic equilibrium equations relating the droplet surface temperature to the gas-phase fuel mass fraction at the surface. If the Clausius-Clapeyron relation is employed, the molar fraction at the surface is given by

$$X_{Fs} = \frac{1}{p} \exp \left[\frac{h_{fg} W_\ell}{R^0} \left(\frac{1}{T_{\text{bn}}} - \frac{1}{T_s} \right) \right] \tag{19}$$

and the mass fraction at the surface is obtained from

$$Y_{Fs} = \left[1 + \frac{W_g}{W_\ell} \left(\frac{1}{X_{Fs}} - 1 \right) \right]^{-1} \tag{20}$$

3.2 Constant-Density Flows

When the gas phase is at a constant, uniform temperature, or if the temperature difference between the two streams in a shear flow configuration is sufficiently small, the density of the gas phase may be taken constant. In addition to the considerable simplification it implies in the solution of the flow field, the constant-density case is fundamentally interesting because it provides a description of the entrainment and mixing rates due solely to the action of the flow dynamics in the absence of temperature gradients. In the two-dimensional, inviscid case and assuming that the discrete phase has no effect on the continuous flow field [42], the only nonzero component of vorticity, ω_z, is convected by the flow field according to the equation:

$$\frac{D\omega_z}{Dt} = \frac{\partial \omega_z}{\partial t} + u \frac{\partial \omega_z}{\partial x} + v \frac{\partial \omega_z}{\partial y} = 0 \tag{21}$$

In order to solve these equations, one introduces a streamfunction, ψ, so that:

$$u = \frac{\partial \psi}{\partial y} \qquad v = -\frac{\partial \psi}{\partial x} \tag{22}$$

The streamfunction satisfies the Poisson equation:

$$\nabla^2 \psi = -\omega_z \tag{23}$$

The solution of Eq. (23) for the case in which vorticity is confined to an infinitesimally thin sheet of arbitrary shape [43] yields the velocity components at any point (x, y) as

$$u(x, y) = -\frac{1}{2\pi} \int \frac{y - y'}{r^2} \gamma(s)\, ds \tag{24}$$

$$v(x, y) = \frac{1}{2\pi} \int \frac{x - x'}{d^2} \gamma(s)\, ds \tag{25}$$

where $\gamma(s)$ is the circulation per unit length (strength) of the vortex sheet at the point (x', y'), $r^2 = (x - x')^2 + (y - y')^2$, and s is a coordinate running along the vortex sheet. The integration is along the interface from $-\infty$ to $+\infty$. As we are interested in disturbances which are periodic in the horizontal direction, we need only consider the integration along one wavelength of the disturbance. Using complex notation with $z = x + iy$, Eqs. (24) and (25) become

$$u(z) - iv(z) = \frac{i}{2} \int \gamma(s) \cot[\pi(z - z')\, ds \tag{26}$$

where the integration is over one period of the disturbance (one unit of length). Separating Eq. (26) into its real and imaginary parts results in

$$u = -\frac{1}{2} \int_0^1 \frac{\gamma \sinh 2\pi(y - y')}{\cosh 2\pi(y - y') - \cos 2\pi(x - x')}\, ds \tag{27}$$

$$v = \frac{1}{2} \int_0^1 \frac{\gamma \sin 2\pi(x - x')}{\cosh 2\pi(y - y') - \cos 2\pi(x - x')}\, ds \tag{28}$$

When the continuous distribution of vorticity represented by a vortex sheet whose velocity field is described by Eqs. (27) and (28) for the case of periodic disturbances of wavelength λ is approximated by a discrete distribution of vorticity with N line vortices per wavelength, these equations become

$$u = -\frac{1}{2} \sum_{j=1}^{n} \frac{\Delta\Gamma_j \sinh 2\pi(y - y_j)}{\cosh 2\pi(y - y_j) - \cos 2\pi(x - x_j)} \tag{29}$$

$$v = \frac{1}{2\lambda} \sum_{j=1}^{n} \frac{\Delta\Gamma_j \sin 2\pi(x - x_j)}{\cosh 2\pi(y - y_j) - \cos 2\pi(x - x_j)} \tag{30}$$

where $\Delta\Gamma_j = \gamma_j\, \Delta s_j$ is the circulation associated with the segment of sheet Δs_j represented by vortex j.

Since the vorticity is simply convected by the flow field, the motion of the various vortex elements is obtained by integrating the equations

$$\frac{dx_i}{dt} = u_i \qquad \frac{dy_i}{dt} = v_i \tag{31}$$

Figure 1, from [42], illustrates a situation in which a sinusoidal disturbance and its first subharmonic, both of dimensionless amplitude 0.025, are simultaneously imposed on the interface at $t = 0$. Since the vortices that define the interface are allowed to move closer to or farther away from each other, the initial resolution does not remain fixed in time. In order to prevent the resolution from deteriorating in regions where vorticity is being stretched, and in order to eliminate unnecessary calculations in regions where the vorticity agglomerates, addition of new vortices and combination of vortices is accordingly performed. Whenever the separation between any two consecutive vortices decreases below a certain value (typically one-half of the initial separation), these two vortices are combined and replaced by an equivalent single vortex. The number of vortex elements n is indicated at each time frame.

Figure 1 shows how the presence of the first subharmonic forces the pairing of every two structures in the manner described in the experiments of Ho and Huang [5]. No complete merging occurs because there is no diffusion mechanism present to spread the vorticity away from the interface.

Figure 2, also from [42], shows the results for a case in which the second subharmonic is present in addition to the fundamental. In this case, grouping of every three structures occurs and, most important, the transverse thickening of the layer occurs more rapidly since there are now three structures rolling up together. (Note that the x and y scales in Fig. 2 are different from those in Fig. 1.) Figure 3 shows the amplitude of the disturbance as a function of time for the case of a single disturbance, a disturbance plus the first subharmonic, and a disturbance plus the second subharmonic. The exponential growth rate predicted by the linear theory is valid until a dimensionless time of approximately 0.8. From then on, the single disturbance grows slowly to a maximum and then actually decreases due to the strong shear generated by the resulting vortex row and the absence of viscous diffusion. The cases containing the first and second subharmonics continue to grow nearly linearly until a dimensionless time of 3. From this point on, the case containing the first subharmonic reaches a maximum amplitude and then decreases, also due to the absence of merger in the presence of strong shear. The amplitude of the case containing the second subharmonic continues to grow linearly, although at a faster rate induced by the three structures that begin to roll up together. Eventually, as a result of shearing and the absence of viscosity, the amplitude of this case should also reach a maximum and start decreasing, although the calculations have not been extended beyond the stage shown.

Some calculations for the two-dimensional, constant-density case have been performed [44]. The configuration studied corresponds to the temporal growth of a periodic shear layer. Two flow configurations have been investigated. In the first one, the evolution of the shear layer under the presence of the fundamental alone is considered. In the second case, we consider the formation and growth of two vortical structures

Figure 1 Rollup and pairing of two vortical structures induced by the presence of the first subharmonic. The characteristic time is the wavelength divided by the velocity jump. The number of vortex elements is *n*. (From [42].)

in the presence of the fundamental and the first subharmonic as illustrated in Fig. 1. In both cases we investigated the effect of the vortical structures on a few droplet streams initially located in the upper gas stream.

Figure 4 shows the evolution of the gas-phase streamlines with time during the formation of one vortex singularity induced by a single disturbance (the fundamental).

Figure 2 Rollup and grouping of three vortical structures induced by the presence of the second subharmonic. The characteristic time is the wavelength divided by the velocity jump. The number of vortex elements is *n*. (From [42].)

Realize that the fluid particle paths are different from the streamlines in this unsteady flow. The steady state is reached after about 2 units of dimensionless time; after this time, the fluid particle paths and the streamlines coincide. While the fluid particle paths may cross the streamlines during the transient evolution, thus making entrainment possible, no entrainment of fluid particles occurs during steady state. Further entrainment is possible only by the process of continuous pairing. Figure 5 shows the gas-phase streamlines at various stages in the formation and pairing of two structures.

In this one-way coupling approach, the effect of the droplets on the gas-phase fluid dynamics is ignored. Future studies should remove this limitation. Consequently, the gas-phase scalar equations for conservation of energy and species can be solved once the flow velocity is known. Similarly, the droplet equations are solved with knowledge of the gas velocity. In this two-dimensional representation, the equations of conservation of energy and species in the gas phase are, assuming constant properties,

$$\frac{\partial \rho T}{\partial t} + \frac{\partial \rho u T}{\partial x} + \frac{\partial \rho v T}{\partial y} = \frac{1}{\mathrm{Re\ Pr}} \frac{\partial^2 T}{\partial x^2} + \frac{1}{\mathrm{Re\ Pr}} \frac{\partial^2 T}{\partial y^2} + S_{T,\mathrm{vap}} \tag{32}$$

$$\frac{\partial \rho Y_i}{\partial t} + \frac{\partial \rho u Y_i}{\partial x} + \frac{\partial \rho v Y_i}{\partial y} = \frac{1}{\mathrm{Re\ Sc}} \frac{\partial^2 Y_i}{\partial x^2} + \frac{1}{\mathrm{Re\ Sc}} \frac{\partial^2 Y_i}{\partial y^2} + S_{Y_i,\mathrm{vap}} \tag{33}$$

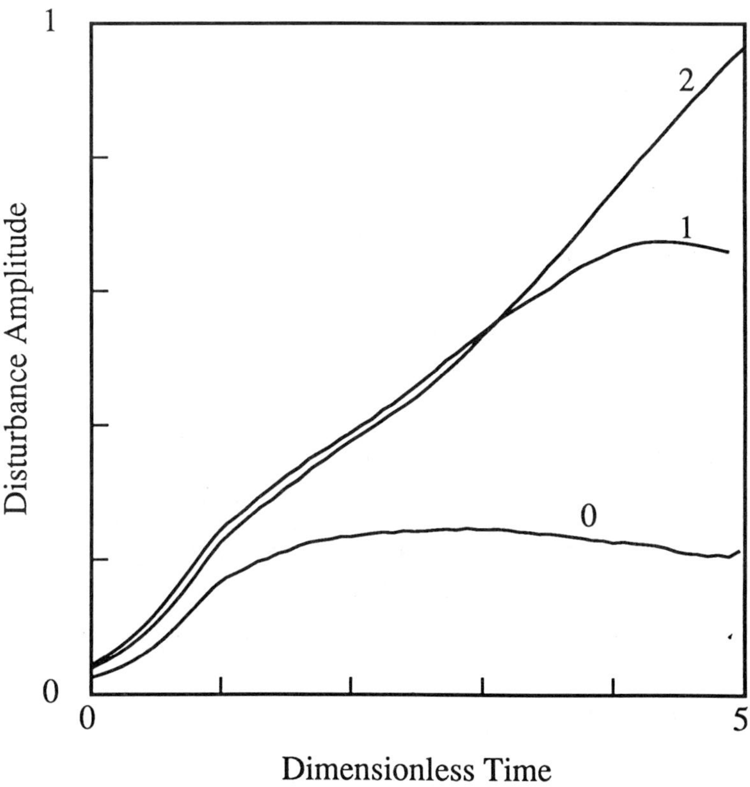

Figure 3 Amplitude of the disturbance as a function of time for the fundamental disturbance alone, the fundamental plus the first subharmonic, and the fundamental plus the second subharmonic.

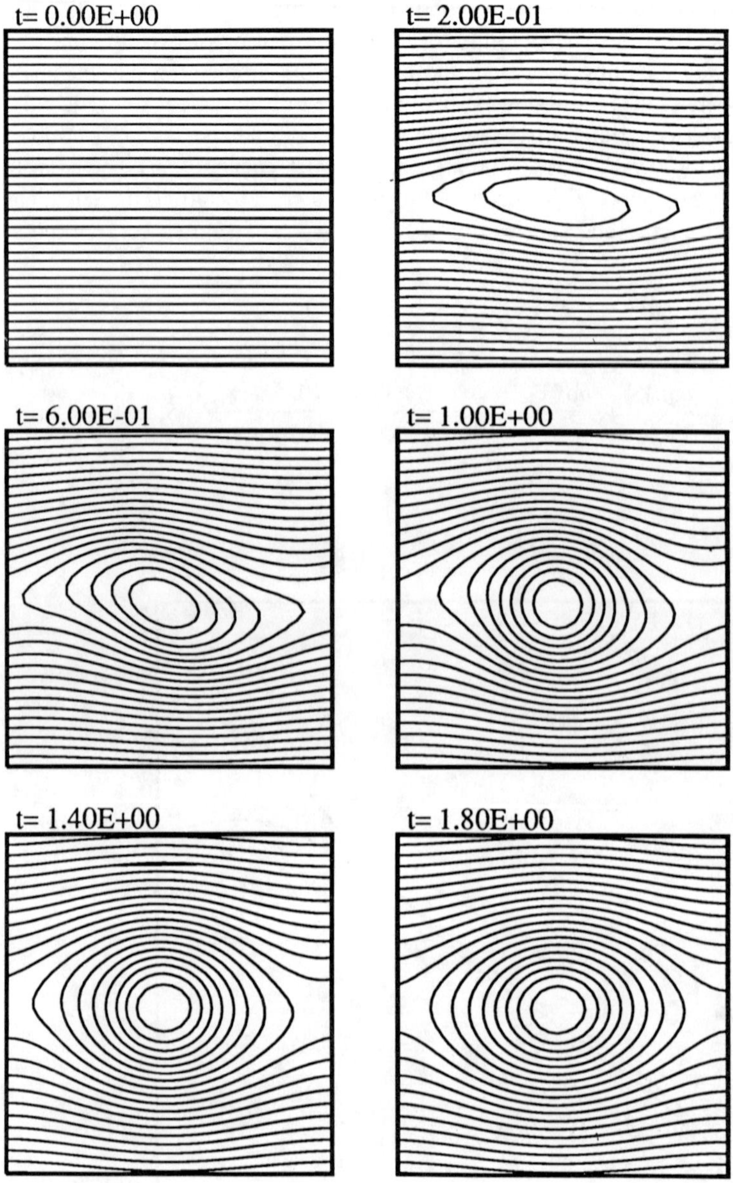

Figure 4 Evolution of the gas-phase streamlines during the formation of one vortex structure from a vortex sheet.

where the vaporization source term is given by

$$S_{\phi,\mathrm{vap}} = \sum_{j=1}^{N} \zeta_\phi \dot{m}_j \mathcal{N} \delta(\mathbf{r} - \mathbf{r}_j) \tag{34}$$

and where the delta functions are nonzero only at the locations of the droplets. The function ζ_ϕ generally depends on both gas and temperature properties [45]. The solution procedure involves rewriting the governing equations in finite-difference form using a control-volume approach and solving a set of algebraic linearized equations [46].

Figures 6–9 present results for what we will consider the base-case calculation. Here the flow field corresponds to that of two vortical structures pairing under the

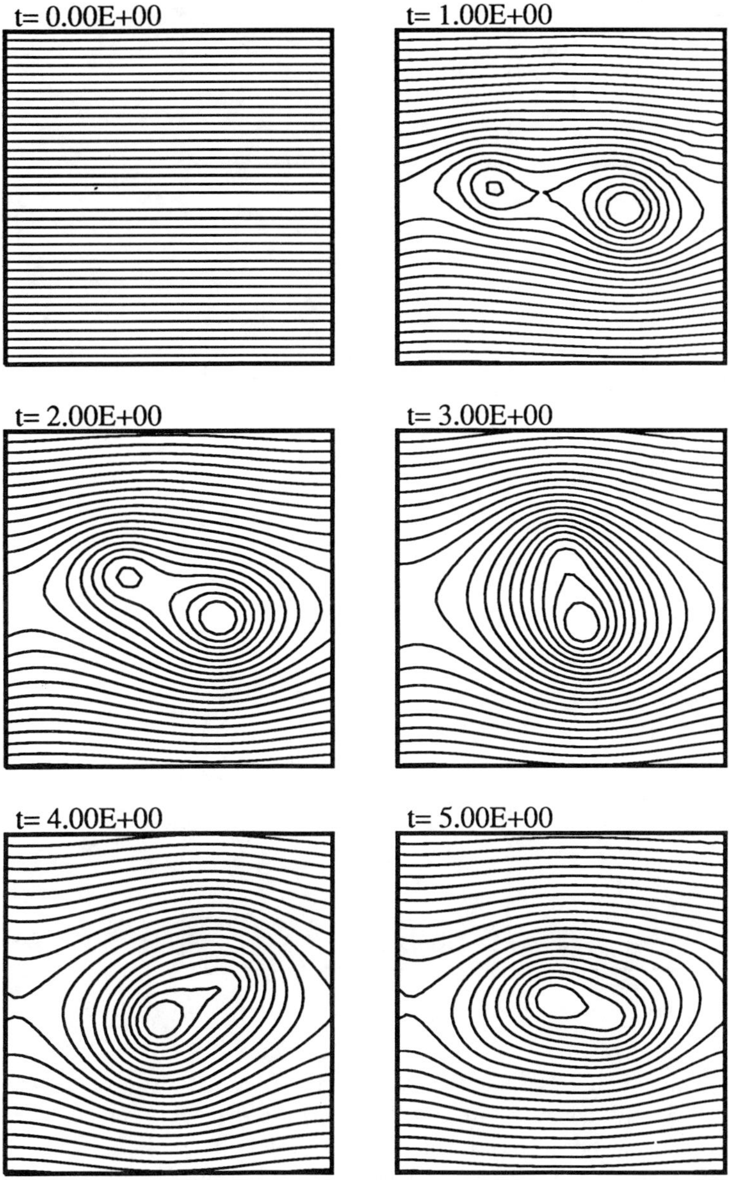

Figure 5 Evolution of the gas-phase streamlines during the formation of two vortex structures from a vortex sheet, and their pairing induced by the first subharmonic.

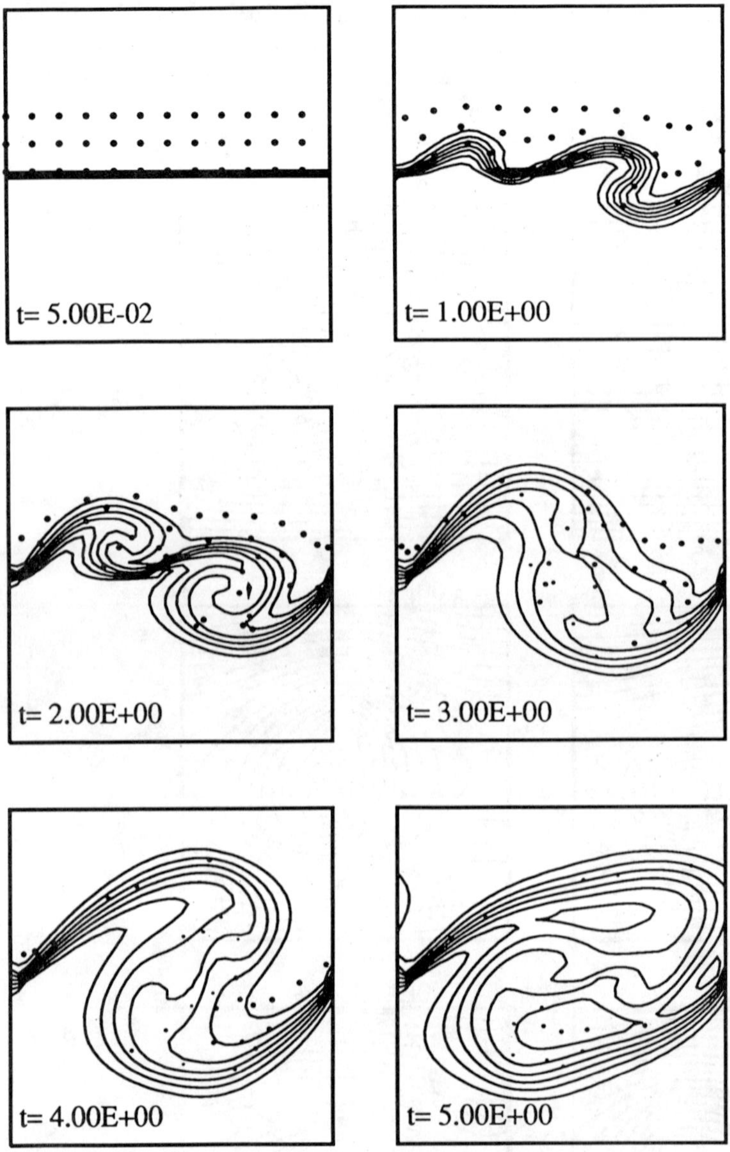

Figure 6 Development of the gas-phase isotherms in the base-case calculation. Droplet size: 100 μm.

effect of the first subharmonic. The interface evolution and the streamlines are those of Figs. 1 and 5. We consider three streams of 100 μm in diameter droplets of *n*-octane. The initial spacing between adjacent streams as well as the initial spacing between two adjacent drops in a stream is 2.5 cm. The Peclet number based on the disturbance wavelength and the velocity jump across the vortex sheet is 2500. The initial velocity of the droplets is that of the upper gas stream. The initial distance from the vortex sheet to the stream closest to it is 2 mm. In the base-case calculation the inertia of the droplets is neglected so that the droplets move with the flow and the droplet paths coincide with the fluid particle paths. In Fig. 6, the contour lines of constant temperature are shown. Initially, the upper gas stream, which contains the droplets, is at 500 K, while the lower one is at 1500 K. The droplet locations are indicated in Figs. 6 and 7 by the black circles. The size of the circles decreases as

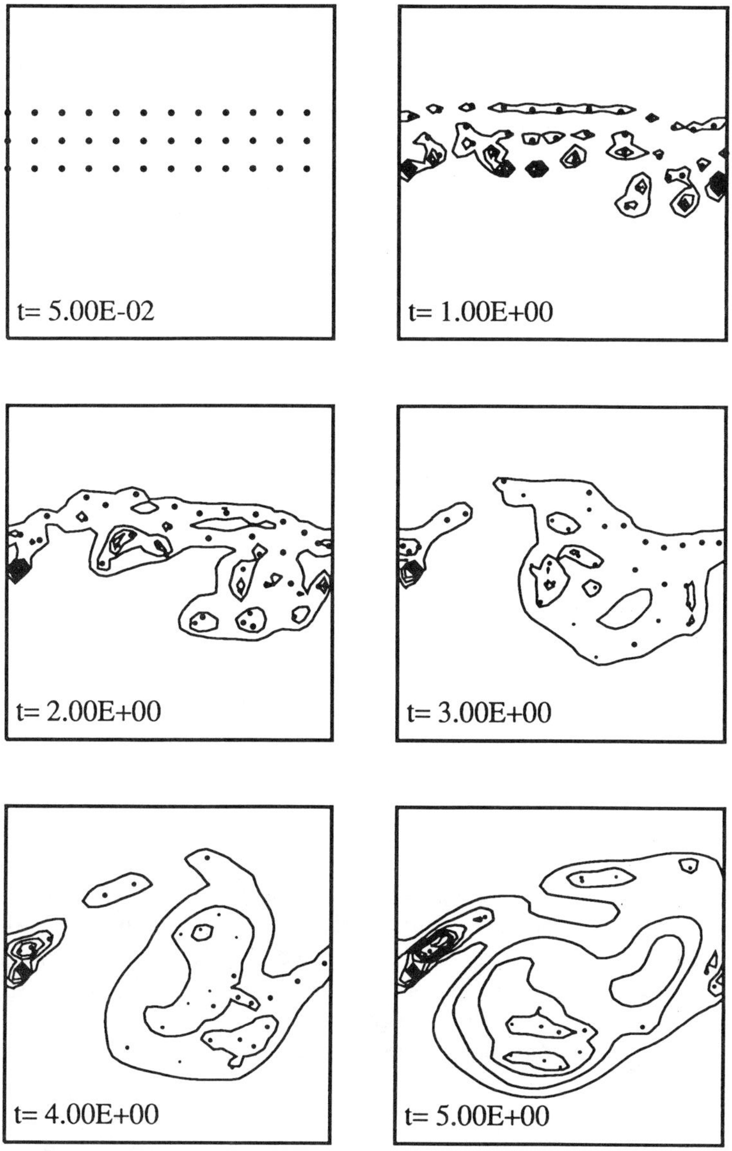

Figure 7 Development of the gas-phase constant mass fraction contours in the base-case calculation. Droplet size: 100 μm. No inertia.

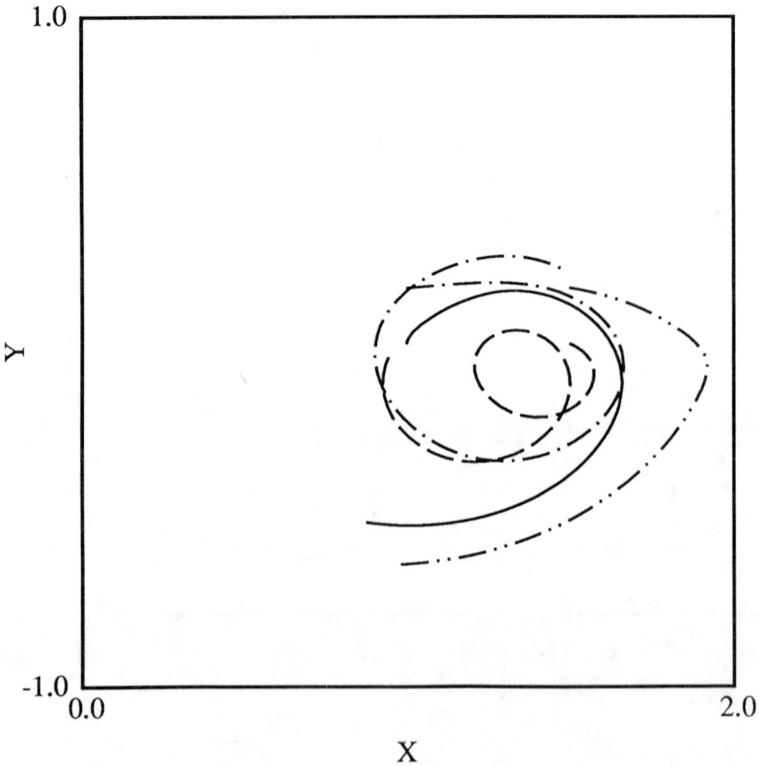

Figure 8 Droplet trajectories of selected droplets without inertia.

the droplets vaporize. The efficiency of the mixing process is better understood with the aid of Fig. 7, which shows the contour lines of constant mass fraction of the vaporized species. Initially, the mass fraction of the vaporizing species is zero. Since in the base-case calculation the droplets move as the gas particles, the only sensible measure of mixing is related to how well the gassified fuel is mixed with the initially fresh gas.

The droplet trajectories for droplets number 7, 10, 19, and 22 in the base-case calculation are shown in Fig. 8. With reference to the first frame of Fig. 6, the droplets are numbered from left to right starting with the stream closest to the initial vortex sheet. There are 12 droplets in each stream, thus, droplets 7 and 10 are in the stream closest to the vortex sheet, and droplets 19 and 22 are in the second stream, above droplets 7 and 10, respectively. In this case, in which the inertia of the droplets is neglected, these four droplets are entrained by the first vortex structure they encounter. The droplet temperature and size history are shown in Fig. 9. As this figure indicates, the droplets in the first stream are rapidly reached by the thermal diffusion wave propagating from the interface. These two droplets are heated up more rapidly independently of the formation of the structures. Droplets 19 and 22 rely on the formation of the vortex structures to reach the hot gas.

The results obtained by accounting for the inertia of the droplets are contained in Figs. 10–13. Figures 10 and 11 show the contour lines of constant gas temperature and constant mass fraction of the vaporizing species. The main feature observed in Figs. 10 and 11 is the fact that the droplets, on the average, tend to remain on the upper side of the shear layer. This is due to their inertia, which makes them fairly insensitive to the changing gas velocity field. Figure 12 shows the paths of the same selected droplets of the base case. The droplets closest to the shear layer are entrained very easily. The droplets on the second droplet stream are not entrained by the first vortical structure, and by the time they reach the second structure (reentering on the left side of Fig. 12), that structure has already reached a quasi-steady state (closed streamlines). Thus, in order to proceed with the entrainment process, further pairing of vortical structures must occur. Figure 13 shows the evolution of the droplet surface temperature and of the droplet mass versus time for the same selected droplets of Fig. 12. Droplet number 7, represented by the solid line, which is entrained very easily, is exposed to the lower temperature fluid that is entrained with it. This droplet remains in the relatively cooler core of the structure. Other droplets that remain outside the core are exposed to higher temperatures.

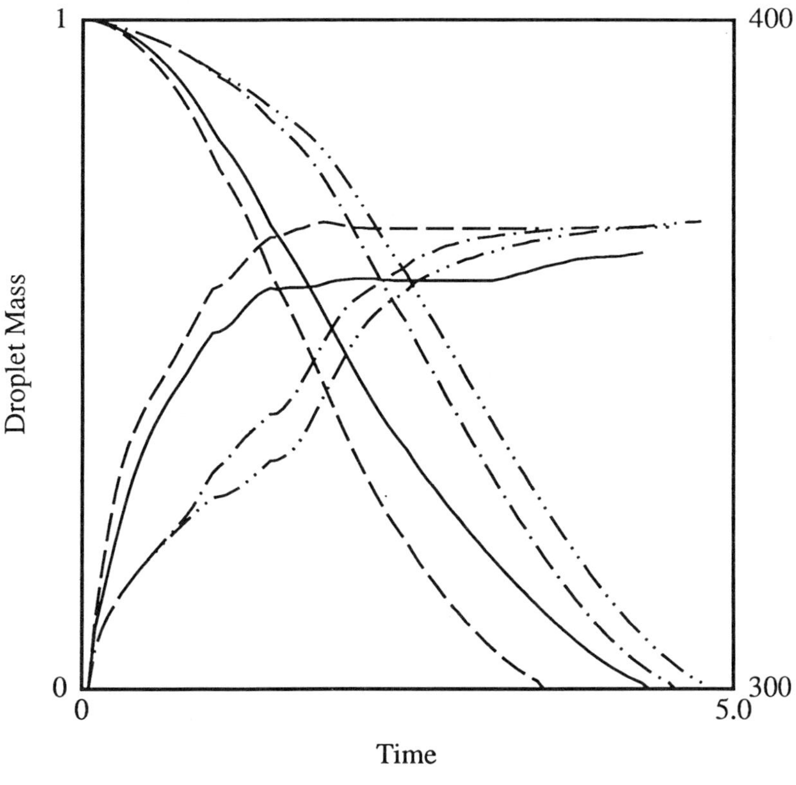

Figure 9 Droplet size and surface temperature histories in the base case. No inertia.

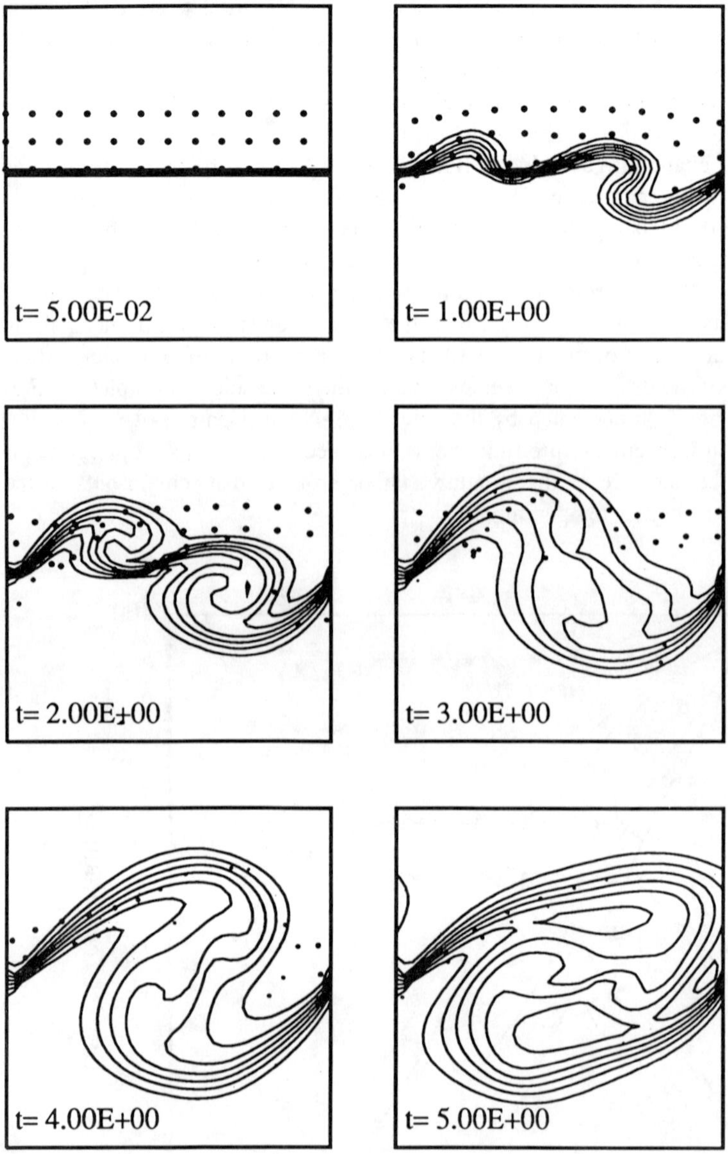

Figure 10 Development of the gas-phase isotherms accounting for droplet inertia. Droplet size: 100 μm.

The effect of reducing the droplet size from 100 μm to 50 μm is illustrated with the aid of Fig. 14. This figure shows the contour lines of constant mass fraction of the vaporizing species in a calculation accounting for the inertia of the droplets. The smaller droplets respond more easily to the gas motion, but their lifetimes are shorter than those of the larger droplets. Consequently, by a dimensionless time of 3, the droplets are almost fully vaporized and further entrainment and mixing occurs in the gas phase only.

Further investigations should explore the conditions under which the entrainment is most or least efficient as a function of various controlling parameters such as droplet sizes, gas temperature, presence of subharmonics in the disturbance, and so on. The

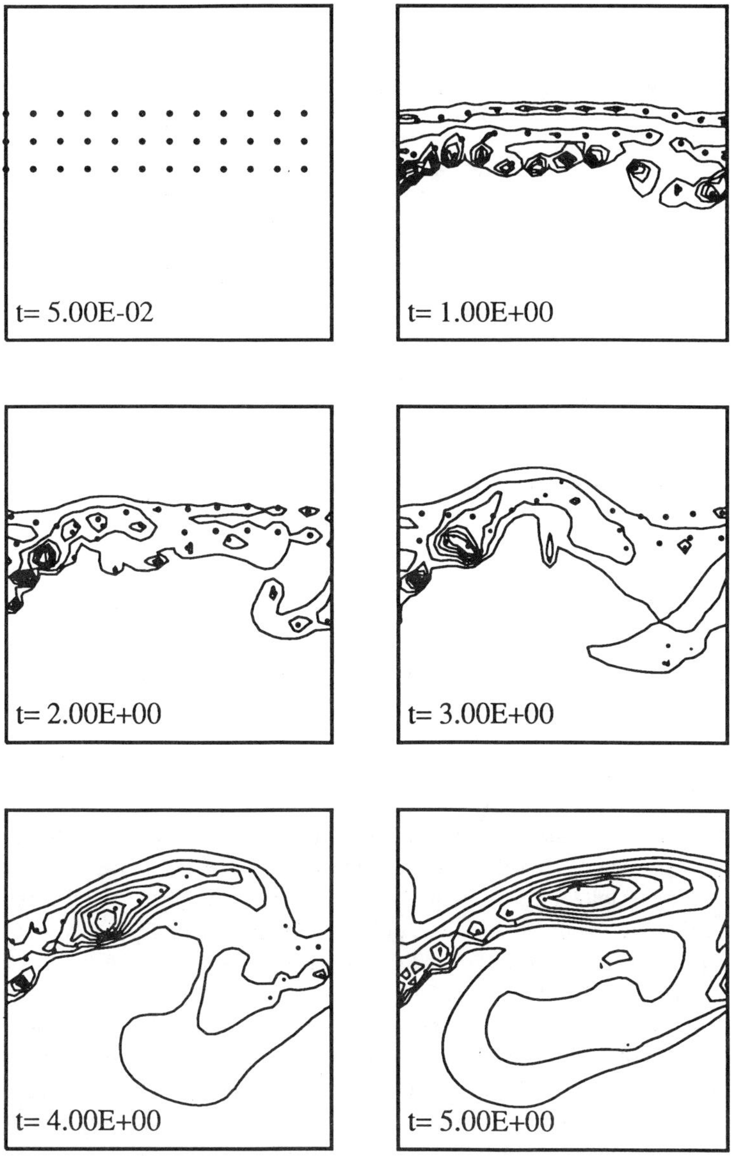

Figure 11 Development of the gas-phase constant mass fraction contours accounting for droplet inertia. Droplet size: 100 μm.

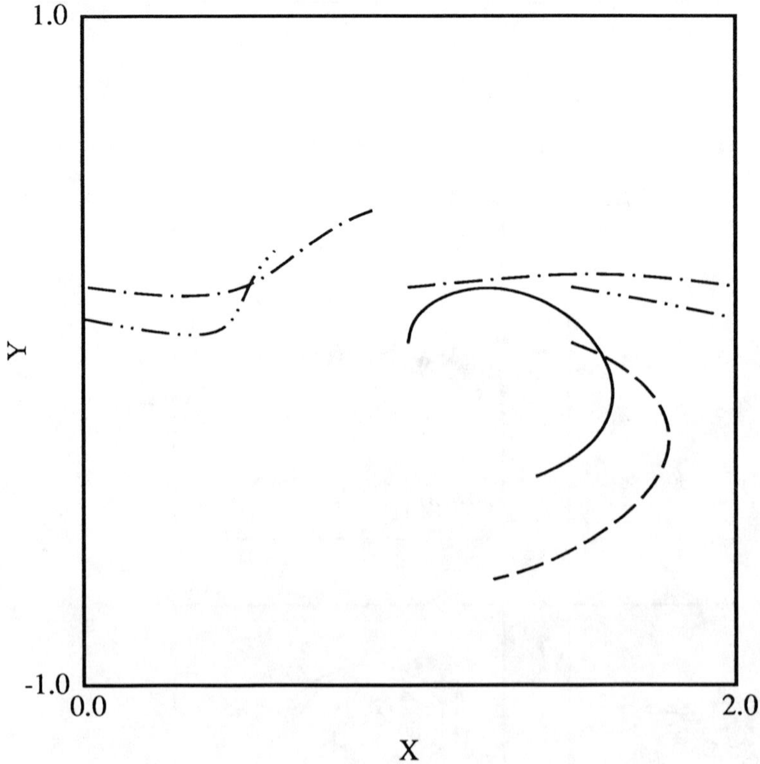

Figure 12 Droplet trajectories of selected droplets with inertia.

two-way heat transfer coupling between the gas and the droplets should also be investigated. This point is described further in the next section.

3.3 Variable-Density Flows

The coupling between the gas phase and the liquid droplets is a two-way coupling in the sense that heat transferred from the gas into the droplets affects the heating and vaporization rates of the droplets and, simultaneously, affects the gas phase by reducing its temperature and thus increasing its density. Further, the vaporization process changes the gas-phase composition, which also affects the gas density. These interactions occur in addition to the fluid dynamic interactions of momentum exchange between the gas and the droplets. The variable-density case requires numerical solution of the more complete equations of motion. In this case, large density ratios may exist in the flow field, and the fluid dynamics and the energy flow field are coupled. Also, there is a two-way interaction between the continuous phase and the dispersed phase by means of momentum, mass, and energy exchanges. The dispersed-phase equations are those described earlier. The gas-phase equations are those of conservation of mass, momentum, energy, and species:

$$\frac{\partial \rho}{\partial t} + \nabla \cdot (\rho \mathbf{v}) = \sum_{j=1}^{J} \dot{m}_j \delta(\mathbf{r} - \mathbf{r}_j) \tag{35}$$

$$\frac{\partial \rho \mathbf{v}}{\partial t} + \nabla \cdot \rho \mathbf{v}\mathbf{v} = -\nabla p - \nabla \cdot \tau + \sum_{j=1}^{J} S_{mj} \delta(\mathbf{r} - \mathbf{r}_j) \tag{36}$$

$$\frac{\partial \rho c_p T}{\partial t} + \nabla \cdot \rho c_p \mathbf{v} T - \nabla \cdot \rho \alpha c_p \nabla T + \left(\frac{\partial \ln \rho}{\partial \ln T}\right)_p \frac{Dp}{Dt} = \sum_{j=1}^{J} \dot{m}_j \left(c_p T_{sj} - \sum_{m=1}^{M_j} \varepsilon_m h_{fg,m}\right) \delta(\mathbf{r} - \mathbf{r}_j) \tag{37}$$

$$\frac{\partial \rho Y_i}{\partial t} + \nabla \cdot \rho \mathbf{v} Y_i - \nabla \cdot \rho \mathcal{D} \nabla Y_i = \sum_{j=1}^{J} \delta_{ij_i} \varepsilon_i \dot{m}_j \delta(\mathbf{r} - \mathbf{r}_j) \tag{38}$$

where the source terms on the right-hand side correspond to the mass, momentum, energy, and species interactions with J heating and vaporizing droplets. In the vaporization source term, $\delta_{ij_i} = 1$ only if species i is present in droplet j. Otherwise it

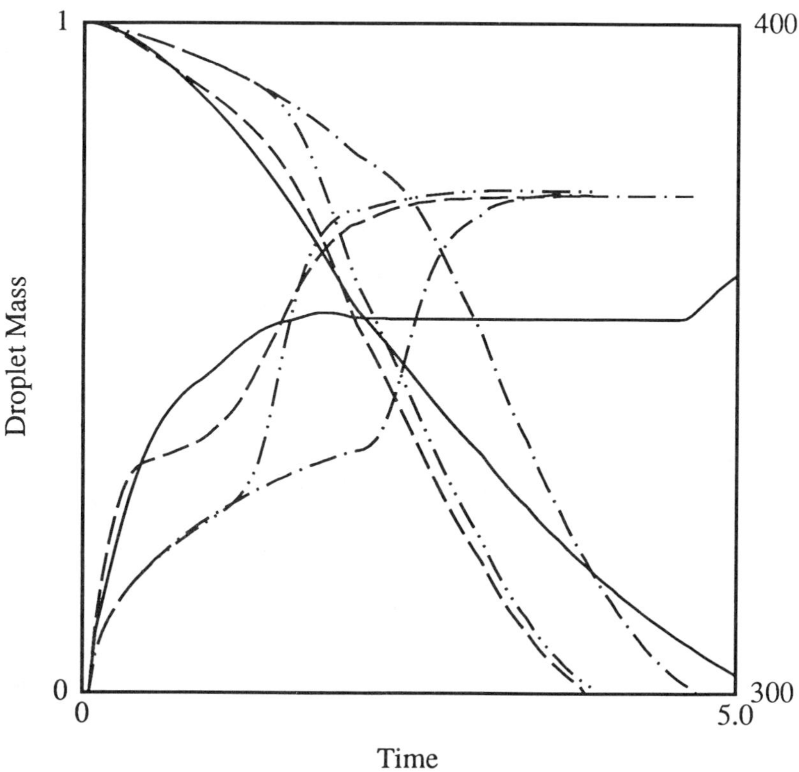

Figure 13 Droplet size and surface temperature histories·accounting for droplet inertia.

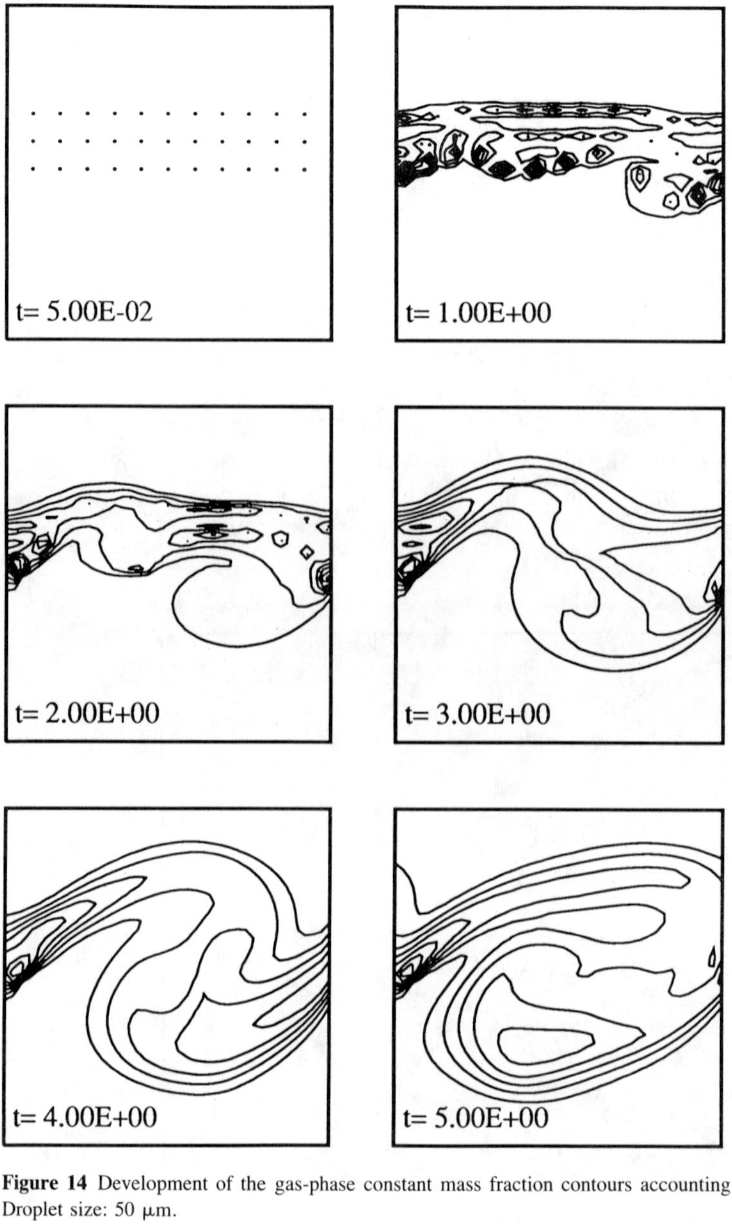

Figure 14 Development of the gas-phase constant mass fraction contours accounting for droplet inertia. Droplet size: 50 μm.

is zero. In Eq. (36), the source term S_{mj} contains the effect of droplet j on the gas linear momentum.

A major portion of future research into the problem of vortically enhanced heat transfer in sprays should be devoted to solving the above equations in a systematic manner and identifying the important controlling parameters. Both two- and three-dimensional studies should be conducted. Additional issues that may play an important role in some situations are described below.

3.4 Gas-Liquid Coupling

The traditional approach to numerical coupling between the gas and liquid phases in spray calculations has relied on the use of the point-source approximation. Recently, Rangel and Sirignano [47] have discussed the shortcomings of this approximation when one is interested in resolving on a scale of a few droplet diameters. The assumption implied in such a representation is that the droplets are negligibly small compared to any scale of interest in the gas phase. Rangel and Sirignano provide the following expressions for the numerical vaporization rate and the error in the spherically symmetric case:

$$\dot{m}^* = 4\pi\rho\mathcal{D}RF_{\Delta r}\ln(1 + B_F) \tag{39}$$

$$e = -\frac{1}{j\,\Delta r'} + \frac{1}{j^2\,\Delta r'^2} - \cdots = \sum_{n=1}^{\infty}\left(-\frac{1}{j\,\Delta r'}\right)^n = -\frac{1}{1 + j\,\Delta r'} \tag{40}$$

As this last equation indicates, the error is dependent only on the ratio of the grid size to the droplet size, r', times the index number of the node used to represent the ambient conditions. This product is simply the dimensionless distance from the selected node to the droplet center. The numerical factor modifying the vaporization rate is found to be

$$F_{\Delta r} = 1 - \frac{1}{j\,\Delta r'} + \frac{1}{j^2\,\Delta r'^2} = \sum_{n=0}^{\infty}\left(-\frac{1}{j\,\Delta r'}\right)^n = \frac{j\,\Delta r'}{1 + j\,\Delta r'} \tag{41}$$

where the last expression is valid if $j\,\Delta r'$ is greater than 1, as in their analysis. The vaporization rate is always underpredicted. The most important conclusion is that a reduction in the gas-phase mesh size, presumably in order to improve the gas-phase resolution, results in a larger error in the calculation of the vaporization rate.

In the convective axisymmetric case, a doublet should be added to the point source in order to obtain a more appropriate representation of the flow field. When the transfer number is calculated by employing a fuel mass fraction averaged at the location of the droplet, substantial errors can be involved. It is suggested that the transfer number be evaluated with the fuel mass fraction of the approaching flow, beyond the fuel mass fraction layer in the front of the droplet. This method reduces the error to less than 10% in all the cases investigated in the range of Reynolds numbers between 2.5 and 100 and for grid-to-drop ratios as low as 2.

3.5 Dilute and Dense Sprays

The vast majority of research into the behavior of droplet spray systems has centered around the so-called dilute regime. In this regime, many droplets or droplet groups are considered but their concentration in the carrier gas is low enough so that the spray may be considered a collection of relatively isolated droplets. This occurs when the typical separation distance between droplets is of the order of 10 droplet diameters or more. Droplet interaction still occurs, since the droplet heating and vaporization processes have an effect on the surrounding ambiance by modifying the mass fraction of gasified species and the temperature field. Whether these modifications increase or decrease the above parameters and to what extent depends on whether heating and vaporization or cooling and condensation are occurring.

Bellan and Harstad [48] have examined the behavior of evaporating sprays in the so-called dense or nondilute regime. In these regime, droplet separation distances are less than 10 droplet diameters, typically three or four from center to center, and thus, direct droplet interaction occurs easily. Direct droplet interaction includes, in addition to the phenomena listed for the dilute sprays, a modification of the droplet Nusselt number and drag coefficient due to the presence of other droplets, droplet collisions, and possibly droplet coalescense. Bellan and Harstad addressed the effects of initial relative velocity and initial droplet and gas temperature on the evaporation times of nondilute clusters of drops [48]. In this approach, the details of the flow and temperature field around each individual droplet are not considered. Instead, three relevant aspects of the interaction between the droplet cluster and the surroundings are considered: (a) the extent of the flow penetration inside the cluster volume; (b) the magnitude of the instantaneous relative velocity between the cluster and the flow; and (c) the relationship between the evaporation rate with and without flow.

The extent of the flow penetration was summarized by the general momentum equation:

$$\frac{dv_{\text{rel}}}{dt} = -\frac{3}{8}\frac{\rho}{\rho_\ell}\frac{1}{R^*}C_D v_{\text{rel}}^2 \tag{42}$$

where v_{rel} is the relative velocity between the cluster and the gas and R^* is an equivalent radius defined by

$$R^* = R + (\tilde{R} - R)\min(1, \sqrt{\chi}) \tag{43}$$

with $\chi = u_s/(u_\infty - u_c)$.

When there is no penetration of the external flow, the average evaporation rate of the cluster is defined by the sum of the evaporation rates of the drops in the core of the cluster, the droplets in the outermost shell of the sphere of radius R^*, and the droplets in the spherical shell between the radius R^* and the cluster radius $\mathcal{R}_h$.

Bellan and Harstad have examined the dispersion of evaporating, dilute and dense, clusters of particles in large coherent eddies [25]. In their model, a cylindrical cluster of single-component fuel, uniformly distributed, monosized droplets of known initial size, is contained by a cylindrical vortical domain of infinite radial dimension. Variations in the direction of the vorticity vector and the azimuthal direction are neglected. Both droplets and gas possess radial and azimuthal velocities that are functions of the radial coordinate and time only. The cluster of droplets is bounded on the inside by an inner surface that evolves in time as the centrifugal force pushes the droplets outward, and on the outside by one defined by the outermost droplets in the cluster. In this simple, yet very interesting idealization, the domain is divided into three distinct regions. Region I is defined as the core, which contains gas with no droplets and also contains the line vortex at the center. Region II is the one containing both gas and the cluster of droplets bounded by the inner and outer surfaces described earlier. Finally, region III comprises the outer gas field, beyond the droplet cluster.

Bellan and Harstad employ the concept of ''spheres of influence'' [49] to address the phenomenon of droplet interaction. Each droplet in the cluster is surrounded by a fictitious sphere of influence whose radius is half the distance between the centers of two adjacent droplets. The volume of the cluster is that of the spheres of influence and the space between the spheres of influence. The droplet temperature is a function of both time and radial coordinate. The density, composition, and temperature of the gas contained by each sphere of influence are also functions of time and the same radial coordinate. The gas outside the spheres of influence is assumed spacially uniform and, thus, its thermodynamic properties are functions only of time. Bellan and Harstad solve the equations of continuity, momentum, and energy for the gas and liquid using a Eulerian approach. Details are given in [25]. Among the various results obtained by this approach, it is interesting to note the importance of the initial tangential velocity of both gas and droplets on the evaporation in the dense cluster regime. In practical situations, the droplets are entrained by the vortical structures during the formation of the latter, and consequently the idea of an initial tangential velocity is somewhat idealized. Results obtained for dense clusters of droplets indicate that the evaporation time is not well correlated with the initial Stokes number, while the volume ratio, defined as the ratio of the final cluster volume to the initial cluster volume, and the thickness ratio (final shell thickness to initial shell thickness) correlate very well with the initial Stokes number. Some comparisons with existing experimental data on non-evaporative, dilute regimes show good qualitative agreement. As Bellan and Harstad indicate, further experimental research is required in the nondilute, evaporative regime.

3.6 Radiation Absorption by Droplets

In applications involving high temperatures, as occurs for example in spray combustion systems, thermal radiation absorption by the droplets may affect significantly the transient heating behavior of the liquid, and in some cases, may result in substantial differences in the overall dynamic behavior of the droplets. This scenario goes all the way from shortened droplet lifetimes to droplet disruption induced by overheating of one or more of its components. In the most simplified approach to thermal radiation absorption by the droplets, one may consider modifying the droplet energy equation to include a volumetric source term representing a uniform or spacially dependent heating source resulting from the thermal energy effectively penetrating the droplet interior. Assuming a spherically symmetric droplet model, the energy equation becomes

$$\frac{\partial T}{\partial t} = \alpha_{\mathrm{eff}} \frac{1}{r^2} \frac{\partial}{\partial r}\left(r^2 \frac{\partial T}{\partial r} \right) + \frac{\dot{Q}_r}{mc_p} \tag{44}$$

The radiative heating of a droplet within the volume, $\dot{Q}_r$, may be computed from summing the radiant interchange between all the volume elements in the flow domain. Edwards [50] has shown that the radiant transfer factor matrix may be found from transforming the volume shape factor matrix via the plating algorithm [51], which accounts for optical scattering and wall reflection. In this approach, one would calculate the number and size of the droplets in each volume element as well as their optical properties. Under optically thin conditions, prevailing at all photon wavelengths except in the 4.3-μm CO_2 band, the shape-factor matrix can be updated readily using the mean-square separation distance concept [51]. The 4.3-μm region is treated using the effective transmissivity concept. In this way, a transient calculation can be carried out, including the radiation interactions.

The previous approach is the so-called external-field approach [52], in which attention is concentrated on the external field and in which the droplet internal field receives little or no attention as far as the thermal radiation is concerned. The local volumetric interaction of thermal irradiation with an absorbing particle has been investigated by Tuntomo et al. [53]. They considered a homogeneous spherical particle, irradiated by a monochromatic plane wave, using classical electromagnetic wave theory and the geometric optics approach. The Monte Carlo ray-tracing technique is used to determine the two-dimensional spacial dependence of radiative absorption. The results of Tuntomo et al. [53] are valid for the range of size parameter, $2\pi R/\lambda_e$, of 0.1 to 100, for a range of real part of the complex refractive index of 1.179 to 2, and for a range of imaginary part of the complex refractive index of 0.001 to 1. The pattern of the local absorption is divided into three different regimes, namely, uniform absorption, front-concentrated absorption, and back-concentrated absorption. In fuel-spray combustion applications, the radiant feedback from the flame is absorbed by the droplets. As Tuntomo et al. recommend, it is desirable to have front-concentrated absorption in order to enhance vaporization on the illuminated side of the droplet. Since typically the higher-temperature gas will also be on this side of the droplet, the result will likely be a reduction in the ignition delay.

4 CONCLUSIONS

This chapter has described some of the phenomenology affecting heat transfer in the vortically enhanced mixing of vaporizing sprays. This problem has applications in many engineering systems in which it is desirable to effect efficient mixing of a condensed substance within a carrier fluid. Propulsion systems in which the fuel is injected in liquid form into an oxidizing air stream is an example of such applications. Fire safety sprinkler devices constitute additional examples. An idealized configuration and a methodology for the study of this phenomenon has been proposed. To some extent, the most current related literature in vortical flow analyses and spray theories have been presented, and a consistent way to approach the combined problem is proposed. Some preliminary results for the heat transfer in constant-density vortical flows are presented to illustrate the richness of this problem. Some of the unresolved questions, including the difference between dense and dilute sprays, the coupling between the gas and liquid phases in numerical calculations, and the potential importance of radiation, are discussed in some detail.

REFERENCES

1. G. L. Brown and A. Roshko, On the Density Effects and Large Structure in Turbulent Mixing Layers, *J. Fluid Mech.*, vol. 64, pp. 775–816, 1974.
2. F. K. Browand and P. D. Weidman, Large Scales in the Developing Mixing Layer, *J. Fluid Mech.*, vol. 76, pp. 127–144, 1976.
3. J. P. Christiansen and N. J. Zabusky, Instability, Coalescence and Fision of Finite-Area Vortex Structures, *J. Fluid Mech.*, vol. 61, pp. 219–243, 1973.
4. E. Acton, The Modeling of Large-Scales Eddies in a Two-Dimensional Shear Layer, *J. Fluid Mech.*, vol. 76, pp. 561–592, 1976.
5. C-M. Ho and L.-S. Huang, Subharmonics and Vortex Merging in Mixing Layers, *J. Fluid Mech.*, vol. 119, pp. 443–473, 1982.
6. G. M. Corcos and F. S. Sherman, The Mixing Layer: Deterministic Models of a Turbulent Flow. Part 1. Introduction and the Two-dimensional Flow, *J. Fluid Mech.*, vol. 139, pp. 29–65, 1984.
7. F. E. Marble, Growth of a Diffusion Flame in the Field of a Vortex, in *Recent Advances in Aerospace Sciences* (ed. C. Casci), 1985.
8. S. M. Masutani and C. T. Bowman, The Structure of a Chemically Reacting Plane Mixing Layer, *J. Fluid Mech.*, vol. 172, pp. 93–126, 1986.
9. P. Givi, W.-H. Jou, and R. W. Metcalfe, Flame Extinction in a Temporally Evolving Mixing Layer, *21st Symp. (Int.) on Combustion*, pp. 1251–1261, 1986.
10. A. F. Ghoniem and P. Givi, Vortex-Scalar Element Calculations of a Diffusion Flame, 25th AIAA Aerospace Sciences Mtg., Paper 87-0225, 1987.
11. B. M. Cetegen and W. A. Sirignano, Analysis of Molecular Mixing and Chemical Reaction in a Mixing Layer, 26th AIAA Aerospace Sciences Mtg., Paper 88-0730, 1988.
12. C. T. Crowe, J. N. Chung, and T. R. Troutt, Particle Mixing in Free Shear Flows, *Prog. Energy Combustion Sci.*, vol. 14, pp. 171–194, 1988.
13. C. T. Crowe, R. A. Gore, and T. R. Troutt, *Particle Sci. Technol. J.*, vol. 3, p. 149, 1985.
14. G. M. Faeth, Evaporation and Combustion of Sprays, *Prog. Energy Combustion Sci.*, vol. 9, pp. 1–76, 1983.
15. C. K. Law and W. A. Sirignano, Unsteady Droplet Combustion with Droplet Heating—II: Conduction Limit, *Combustion and Flame*, vol. 28, pp. 175–186, 1977.
16. W. A. Sirignano, Fuel Droplet Vaporization and Spray Combustion, *Prog. Energy Combustion Sci.*, vol. 9, pp. 291–322, 1983.
17. R. H. Rangel and W. A. Sirignano, Rapid Vaporization and Heating of Two Parallel Fuel Droplet Streams, *21st Symp. (Int.) on Combustion*, The Combustion Institute, pp. 617–624, 1986.
18. C. K. Westbrook and F. L. Dryer, Chemical Kinetic Modeling of Hydrocarbon Combustion, *Prog. Energy Combustion Sci.*, vol. 10, pp. 1–57, 1984.
19. R. H. Rangel and W. A. Sirignano, Unsteady Flame Propagation in a Spray with Transient Droplet Vaporization, *22nd Symp. (Int.) on Combustion*, The Combustion Institute, pp. 1931–1939, 1988.
20. R. H. Rangel and W. A. Sirignano, Combustion of Parallel Droplet Streams, *Combustion and Flame*, vol. 75, pp. 241–254, 1989.
21. J. O. Hinze, *Turbulence*, 2nd ed., McGraw-Hill, New York, 1975.
22. C. M. Tchen, Mean Value and Correlation Problems Connected with the Motion of Small Particles Suspended in a Turbulent Fluid, Ph.D. thesis, University of Delft, the Netherlands, 1947.
23. S. Corrsin and J. Lumley, *Appl. Sci. Res.*, vol. 6A, p. 114, 1956.
24. S. K. Friedlander, *AIChE J.*, vol. 3, p. 381, 1957.
25. J. Bellan and K. Harstad, Some Aspects of Liquid Dispersion and Evaporation in Swirling Flows, Fall Meeting of the WSS/CI, Livermore, Calif., 1989.
26. W. H. Snyder and J. L. Lumley, *J. Fluid Mech.*, vol. 48, p. 41, 1971.
27. G. Ahmadi and V. W. Goldschmidt, *Tech. Rep. FMTR 70-3*, Purdue University, West Lafayette, Ind., 1970.
28. H. Kobayashi, S. M. Masutani, S. Azuhata, N. Arashi, and Y. Hishinuma, *Second Int. Symp. on Transport Phenomena in Turbulent Flows*, Tokyo, 1987.

29. B. J. Lazaro and J. C. Lasheras, Droplet Dispersion and Transport Mechanisms in a Turbulent, Free Shear Layer, *22nd Symp. (Int.) on Combustion*, The Combustion Institute, pp. 1991–1998, 1988.
30. B. J. Lazaro and J. C. Lasheras, Particle Dispersion in a Turbulent, Plane, Free Shear Layer, *Phys. Fluids*, vol. 1, pp. 1035–1044, 1989.
31. Q.-V. Nguyen, R. H. Rangel, and D. Dunn-Rankin, Measurement and Prediction of Trajectories for Vaporizing Droplet Pairs, Fall Meeting of the WSS/CI, Paper No. 89-62, Livermore, Calif., 1989.
32. C. Chiang and W. Sirignano, Numerical Analysis of Interacting, Convecting, Vaporizing Fuel Droplets with Variable Properties, 28th AIAA Aerospace Sciences Mtg., Paper 90-0357, 1990.
33. M. C. Yuen and L. W. Chen, On Drag of Evaporating Liquid Droplets, *Combustion Sci. Technol.*, vol. 14, pp. 144–154, 1976.
34. F. M. White, *Viscous Fluid Flow*, McGraw-Hill, New York, 1974.
35. M. Renksizbulut and M. C. Yuen, Experimental Study of Droplet Evaporation in a High Temperature Air Stream, *J. Heat Transfer*, vol. 105, pp. 384–388, 1983.
36. C.-H. Chiang, M. S. Raju, and W. A. Sirignano, Numerical Analysis of Convecting, Vaporizing Fuel Droplets with Variable Properties, *Int. J. Heat Mass Transfer*, in press, 1991.
37. C. K. Law, Recent Advances in Droplet Vaporization and Combustion, *Prog. Energy Combustion Sci.*, vol. 8, pp. 169–199, 1983.
38. F. A. Williams, *Combustion Theory*, 2nd ed., Addison-Wesley, Reading, Mass., 1985.
39. R. Clift, J. R. Grace, and M. E. Weber, *Bubbles, Drops, and Particles*, Academic Press, New York, 1978.
40. B. Abramzon and W. A. Sirignano, Droplet Vaporization Model for Spray Combustion Calculations. *Int. J. Heat Mass Transfer*, vol. 32, pp. 1605–1618, 1989.
41. M. Renksizbulut and R. J. Haywood, Transient Droplet Evaporation with Variable Properties and Internal Circulation at Intermediate Reynolds Numbers, *Int. J. Multiphase Flows*, vol. 14, pp. 189–202, 1988.
42. R. H. Rangel and W. A. Sirignano, The Dynamics of Vortex Pairing and Merging, 27th AIAA Aerospace Sciences Mtg., Reno, Nev., Paper No. 89-0128, 1989.
43. G. K. Batchelor, *An Introduction to Fluid Dynamics*, Cambridge University Press, Cambridge, 1970.
44. R. H. Rangel, Vortex Entrainment and Mixing in a Temporally-Growing Shear Layer, ASME Winter Annual Meeting, HTD-Vol. 148, pp. 11–18, Dallas, Texas, 1990.
45. J.-P. Delplanque and R. H. Rangel, Thermal Expansion Effects on the Behavior of Vaporizing Fuel Droplet Streams, 1989 Fall Meeting of the Western States Section of The Combustion Institute, Livermore, Calif., Paper 89-63, 1989.
46. S. V. Patankar, *Numerical Heat Transfer and Fluid Flow*, McGraw-Hill, New York, 1980.
47. R. H. Rangel and W. A. Sirignano, An Evaluation of the Point Source Approximation in Spray Calculations, *Numer. Heat Transfer*, vol. 16, pp. 37–57, 1989.
48. J. Bellan and K. Harstad, Analysis of the Convective Evaporation of Nondilute Clusters of Drops, *Int. J. Heat Mass Transfer*, vol. 30, pp. 125–136, 1987.
49. J. Bellan and R. Cuffel, A Theory on Nondilute Spray Evaporation Based upon Multiple Drop Interactions, *Combustion and Flame*, vol. 51, pp. 55–67, 1983.
50. D. K. Edwards, Hybrid Monte-Carlo Matrix-Inversion Formulation of Radiation Heat Transfer with Volume Scattering. In *Heat Transfer in Fire and Combustion Systems* (ed. C. K. Law, Y. Jaluria, W. W. Yuen, and K. Miyasaka), ASME HTD-Vol. 45, New York, pp. 273–278, 1985.
51. D. K. Edwards, The Plating Algorithm for Radiation Script-F Transfer Factor, *J. Heat Transfer*, vol. 108, pp. 237–238, 1986.
52. C.-L. Tien, Thermal Radiation in Packed and Fluidized Beds, *J. Heat Transfer*, vol. 110, pp. 1230–1242, 1988.
53. A. Tuntomo, S. H. Park, and C.-L. Tien, Internal Field of Radiant Absorption in a Spherical Particle, submitted for publication, 1989.

CHAPTER
EIGHT

HEAT AND MASS TRANSFER IN FOULING

S. H. Chan

ABSTRACT

Recent advances in analytical modeling of silica and calcium carbonate depositions on heat exchanger surfaces are reviewed. General physical and chemical processes leading to $CaCO_3$ fouling are described. The $CaCO_3$ fouling model assumes that the deposition process is controlled by two processes: the mass transport of ions and molecular species to and from heat transfer surface, and the crystallization reaction on the same surface. Analytical predictions of the $CaCO_3$ deposition flux, the mean, and the profile of the local fouling-layer thickness along a heated plate by a laminar falling film are presented. They are found to be in good agreement with experimental data. Good agreement is also shown when applied to a turbulent annulus flow system. Similarly, physical and chemical processes pertinent to silica fouling are reviewed. Silica deposition models that account for supersaturation, pH factor, salinity, and thermal-hydraulic effects are presented. Analytical predictions are made and results are then compared to the field and laboratory data.

NOMENCLATURE

a_j	activity of species j, molality (m)
A	kinetic constant, preexponential factor
A_{flow}	flow area in the annulus
A_i	area, m^2
C_b	bulk concentration of amorphous silica
C_e	equilibrium concentration, molality

C_j or $[j]$	concentration of species j
C_j^*	nondimensional concentration
C_0	equilibrium concentration of amorphous silica in pure water, molality
$\hat{C}_p$	heat capacity of solution
d_1, d_2	inner and outer annulus diameter
d_e	equivalent diameter of annulus
D^+	nondimensional diffusivity
D_{eff}	effective diffusivity
D_i, D	mass diffusivity of species i in water
D_0	diameter of tube
D_{turb}	turbulent mass diffusivity
E_a	reaction activation energy
f_1, f_2	friction factors for inner and outer annulus
g	gravitational constant
h	heat transfer coefficient
H_1, H_2	quantities defined in Eqs. (17) and (18)
I	ionic strength, molality
$j_i^{(x)}$	mass flux due to concentration gradient
k	thermal conductivity of solution
k_{des}	desorption coefficient
k_f	thermal conductivity of fouling scale
k_r	crystallization rate constant of $CaCO_3$ scale
K_1, K_2	dissociation constant of CO_2
K	ratio of K_2 to K_1
K_R	crystallization rate constant of silica scale
K_{sp}	solubility product of $CaCO_3$
l_m	mixing length for turbulence model
L	length of heated section
LMTD	log mean temperature difference
$\dot{m}$	fluid mass flow rate
$\dot{m}_p$	mass flow rate per unit parameter
M	molarity
M_i	molecular weight of species i
N_{CaCO_3}	calcium carbonate deposition flux
$N^*_{CaCO_3}$	nondimensional deposition flux
NaAc	sodium acetate
Nu	Nusselt number
p	correlation constant in Eq. (1)
Pr	Prandtl number
q	correlation constant in Eq. (1)
q_0	heat flux
r	radial coordinate
r_i	rate of consumption/generation by reaction
R	outer annulus radius
Re	flow Reynolds number

R_f	fouling resistance
R_g	gas constant
t	time
T_s	temperate at scale-brine interface, **K**
T_{sc}, T_{blk}	surface and bulk temperature
U_i	overall heat transfer coefficient, W/m^2 °C
$\overline{V}_z$	average or mean velocity in z direction
$\dot{V}$	volume flow rate, m^3/h
y, z	rectangular coordinates
y^+, z^+	dimensionless coordinate
Z_i	valence of species i
γ_i	molal activity coefficient of species i
Γ	nondimensional reaction rate
δ	film thickness in laminar flow
ε	dielectric constant of pure water
κ	ratio of inner to outer annulus radii
κ_1	von Karman constant, 0.418
λ	quantity defined in Eq. (9)
μ	liquid viscosity for water
ρ	liquid density for water
ρ_f	scale density for $CaCO_3$
τ	scale thickness
τ_0	shear stress at the wall
ϕ	osmotic coefficient

Subscripts

aq	aqueous
b	bulk
c	coolant
cl	clean
e	equilibrium
g	gas phase
liq	liquid phase
s	solid phase
w	wall

Superscripts

0	initial condition

1 INTRODUCTION

Fouling is the process in which unwanted materials are deposited onto a surface. The process can occur either in the presence or absence of temperature gradients. Fouling on a heat transfer surface causes a decrease in the efficiency of the heat exchanger as well as impeding fluid flow. It "can typically degrade the performance of heat exchangers by as much as 80%, and can sometimes cause complete failure" [1].

With few exceptions all heat transfer equipment is subject to fouling to some degree. It is important for designers of heat transfer equipment to understand the fundamental processes and parameters that affect the rate of fouling in heat exchangers, if effective design criteria are to be obtained. Such understanding would, for example, allow the operators of such equipment to prepare cleaning schedules having the least impact in terms of costly plant or unit shutdowns. The main motivation for a better understanding of fouling, however, is not the design of elaborate heat exchangers, but rather the increasing cost of wasted energy and raw materials used in the construction of such equipment.

Until a decade ago the topic of fouling was little understood and was termed "the major unresolved problem in heat transfer" [2]. In recent years, however, an increased effort on the part of the scientific and engineering community has been directed toward understanding the fundamental physical and chemical process of fouling [3–6]. This by no means implies that a complete treatment of fouling processes has been accomplished. In fact the topic is still in "an uncertain state of affairs with respect to prediction" [7], and much work remains in developing comprehensive models, consistent with experiments. Still further effort will be required to efficiently integrate theoretical models of fouling into the actual design of heat exchangers.

Fouling in heat exchangers has been regarded as an additional resistance to heat transfer. Currently its effect in the design of heat exchangers is accounted for by use of an assumed fouling resistance, R_f, obtained from TEMA tabulations, and adding this to the clean surface resistance in order to arrive at the required heat transfer surface area:

$$A = \frac{Q}{U\,\Delta T}$$

where

$$\frac{1}{U} = R_f + \frac{1}{h}$$

This approach, though widespread, has been observed to be inadequate in several respects [7]. The available values for R_f are restricted "mainly to water and hydrocarbon process streams flowing through shell-and-tube heat exchangers" [7]. They fail to account for the various heat exchanger geometries and the variety of process fluids. The effect of pertinent operating conditions, such as fluid velocity, surface temperature, and bulk temperature and composition, on R_f are neglected. However, the most serious objection to the use of R_f data is the failure to recognize the time-dependent behavior of the fouling process. An instantaneous steady state with a constant value of R_f is assumed. This has serious consequences for equipment design. An initially clean heat transfer surface implies an overdesigned exchanger at startup, requiring the use of lower initial fluid velocities, which causes increased surface temperatures, thus resulting in greater fouling than would otherwise be expected [7].

McCabe and Robinson [8] are reported to have performed the initial analytical analysis of deposition rates in evaporators. Since then most work on the analytical treatment of fouling has had as its starting point the attributing of the process to the two simultaneous processes of deposition and removal. The difference between the two rates is then taken as the net rate of fouling:

$$\frac{dm}{dt} = \dot{m}_{\text{dep}} - \dot{m}_{\text{rem}}$$

The focus, then, of fouling research has been directed toward establishing the functional relationship of $\dot{m}_{\text{dep}}$ and $\dot{m}_{\text{rem}}$ on the various process parameters. One of the first such models was proposed by Kern and Seaton [9]. They assumed a constant deposition rate with time, and a removal rate proportional to the mass of the deposit, thus increasing with time and reaching $\dot{m}_{\text{dep}}$ asymptotically. Epstein [10] reviewed other alternatives proposed for the evaluation of the deposition and removal terms. It has been noted [10] that the deposition term is the result of several sequential steps: (1) processes in the bulk; (2) transport to the liquid-solid interface; and (3) attachment/formation. Similarly, the removal term may involve a combination of the following steps: (1) dissolution; (2) erosion/reentrainment; and (3) spalling/sloughing. In cases where the deposition is the only significant process, a linear fouling growth rate behavior is often observed; if both the deposition and a time-dependent removal term are present, the behavior is asymptotic [2]. An initial delay in the growth of fouling layer is observed due to the nucleation processes on the surface and is referrred to as the induction time.

Due to the extremely complicated processes of real-life fouling, a classification scheme has been devised by workers in this area. This has allowed the breakdown of these processes into independent modes, rendering their study more manageable. Various types of fouling at liquid-solid interfaces have been grouped into six main categories [7]:

1. Precipitation fouling—crystallization of dissolved salt(s) onto the heat transfer surface, also referred to as scaling
2. Particulate fouling—deposits due to accumulation of suspended solids in the solution onto the heat transfer surface
3. Chemical reaction fouling—deposits formed onto the heat transfer surface as the product of a chemical reaction at the surface
4. Corrosion fouling—fouling due to in-situ corrosion of the heat transfer surface
5. Biological fouling—attachment of micro/macroorganisms onto the heat transfer surface
6. Solidification fouling—freezing of a melt or its constituents onto the heat transfer surface

In each category the successive events leading to the formation of deposit on the surface are initiation, transport, attachment, removal, and aging [7].

Fouling is a broad and complex subject. The present review is by no means intended to be comprehensive. Rather, it is concerned mainly with precipitation fouling. Specifically, recent advances in modeling two types of scales, silica and calcium carbonate, are reviewed below. They are selected partly because they exhibit exactly opposite solubility characteristics, namely, silica will form on cooled surfaces while calcium carbonate will form on heated surfaces. Due to their abundance as components of the earth's crust, both scales are found everywhere, in valves, pipelines, evaporators, heat exchanger tubes, or wherever equipment is in contact with underground water (brine). Finally, considerable progress has been made on modeling calcium carbonate scale, so its deposition rate can be successfully predicted from first principles. However, this is true to a much less extent for silica scale. Thus, by reviewing these two types of scales, it may provide readers a better perspective about fouling in general.

2 CALCIUM CARBONATE SCALING

Since calcium carbonate solubility decreases with increasing temperature (retrograde solubility), calcium carbonate scale is found on heated surfaces. It can also form in isothermal systems. In geothermal energy conversion plants, for example, as a geothermal brine passes through fluid transmission devices such as values, constrictions, or flow-measuring instruments, its pressure reduces rapidly. When the pressure of the brine solution decreases rapidly, carbon dioxide gas is evolved from the brine due to its lower solubility limit. This causes an increase in the pH of the brine, which in turn increases the carbonate concentration and, therefore, produces the deposition and crystal growth of calcium carbonate. The kinetics of this reaction are very fast, causing scale formation immediately downstream of such pressure drops [11]. For scaling of $CaCO_3$, transport and formation in the deposition term are most significant. Due to the hardness and adhesion characteristics of $CaCO_3$, it is agreed that any removal term would be insignificant. In most models the deposition term is given as the product of a concentration difference times a mass transfer coefficient. It is noted in passing that, compared to the deposition, the removal term is far less understood; thus, fewer models for this process are available [12]. One such model, proposed by Taborek et al. [12], indicates that the rate of removal is proportional to the wall shear stress and inversely proportional to the adhesion strength of the deposit. A more recent explanation of the removal process has been given by Cleaver and Yates [13], who utilize the concept of "turbulent bursts" for near-wall turbulent flows. In their opinion, such turbulent bursts would act "like a miniature tornado, lifting deposited material from the surrounding surface and dispersing it in the fluid stream" [13].

Hasson et al. [14, 15], Sheikholeslami and Watkinson [16], and Watkinson and Martinez [17] have considered the deposition of $CaCO_3$ in laminar falling film or turbulent shell-and-tube flows. In the absence of removal processes, the deposition of $CaCO_3$ increases linearly with time under constant heat flux condition. Hasson et al. [14, 15] considered the fouling of this inverse-solubility salt to be the result of the mass transport of chemical species from the bulk to the liquid-solid interface, followed by a crystallization reaction to form $CaCO_3$ scale. They used values of mass transfer coefficients based on nonfouling flow conditions for their studies. Although Hasson's model has been considered to be "the most successful model for predicting $CaCO_3$ scaling rates" [16], recent experimental work [16] aimed at determining the applicability of this model in predicting scaling on various materials has shown that serious shortcomings in terms of accuracy, as well as predicted fouling behavior, are present in this analytical formulation.

A mechanistic approach has recently been proposed [18, 19] whereby the fundamental equations of heat and mass transfer are solved with appropriate boundary conditions. This eliminates the use of mass transfer coefficients calculated for nonfouling conditions. The approach is applicable to laminar- and turbulent-flow conditions, both of which are reviewed next.

3 DEPOSITION OF $CaCO_3$ IN A LAMINAR FALLING FILM

3.1 Physical and Chemical Aspects

To facilitate the description and modeling of $CaCO_3$ scale, the physical system similar to the scaling experiments performed by Hasson and Perl [15] is illustrated in Fig. 1, in which a vertically falling film of scale-producing water is evaporated on the outside surface of a cylindrical tube. The diameter-to-film thickness ratio is assumed to be large so that the problem geometry can be approximated in rectangular coordinates as shown. The surface is heated at constant flux, q_0, while a constant flow of scale-producing water of known inlet temperature and chemical composition is maintained over it. The heat flux is low enough so that evaporation is in the nonboiling region. Due to the inverse solubility characteristic of the salt, precipitation of $CaCO_3$ from the solution creates a scale of increasing thickness on the heat transfer surface. The problem, then, is predicting the local rate of deposition of $CaCO_3$ given the known operating conditions, i.e., velocity and inlet temperature and composition.

The mechanistic model assumes that calcium, carbonate, and bicarbonate ions (i.e., Ca^+, CO_3^{2-}, and HCO_3^-, respectively) must diffuse to a heated surface to react and form a $CaCO_3$ scale and dissolved CO_2. At the liquid-scale interface, the equilibrium condition is assumed such that the excess CO_2 produced must diffuse through the falling film and desorb at the liquid-gas interface. Since the reaction rate constant and diffusion coefficients are, respectively, surface and bulk temperature dependent, both heat and mass transfer transports should be considered in the model. The required chemical kinetics and equilibrium reactions are presented next.

Consider the equilibrium of the H_2O-CO_2 system [20–22]. Dissolved CO_2 gas in water exists as both aqueous CO_2 as well as carbonic acid, H_2CO_3. It can also be found in the form of bicarbonate and carbonate ions, HCO_3^- and CO_3^{2-}, respectively. The following equilibrium reactions can be identified for the H_2O-CO_2 system:

$$CO_{2(g)} \overset{K_H}{\rightleftarrows} CO_{2(aq)} \tag{1}$$

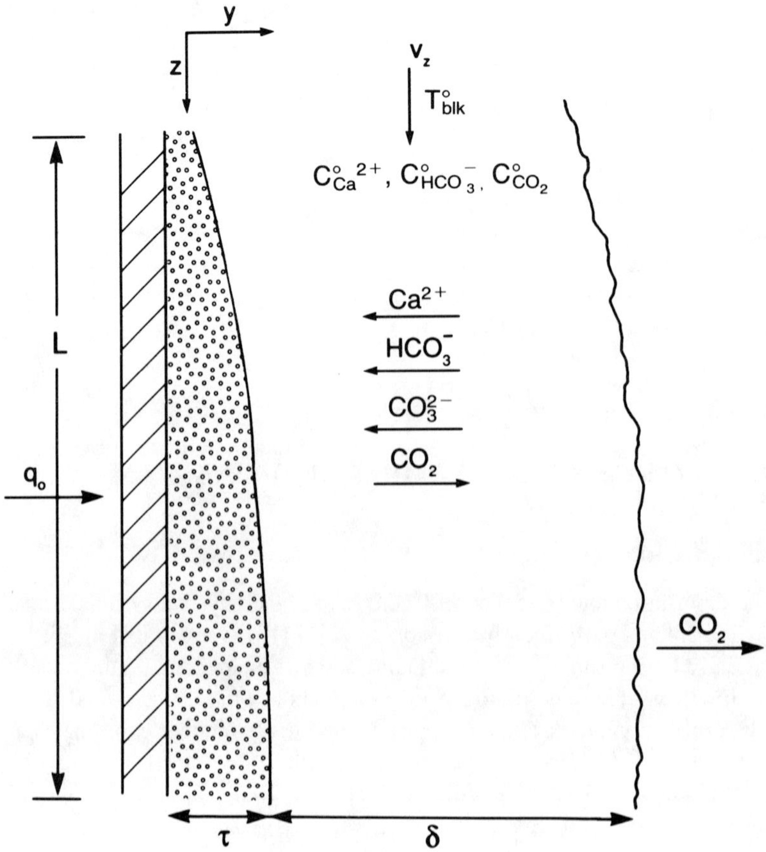

Figure 1 Geometry for falling film deposition.

$$CO_{2(aq)} + H_2O \overset{K_1}{\rightleftarrows} H^+ + HCO_3^- \tag{2}$$

$$HCO_3^- \overset{K_2}{\rightleftarrows} H^+ + CO_3^{2-} \tag{3}$$

Reactions in Eqs. (2) and (3) are the first and second dissociations of CO_2 with dissociation constants K_1 and K_2, respectively. These equilibrium constants are temperature dependent, and this functionality has been expressed by Plummer and Busenberg [20] as

$$\log(K_1) = -356.3094 - 0.06091964*T + 21834.37/T + 126.8339*\log T - 1684915./T^2 \tag{4a}$$

$$\log(K_2) = -107.8871 - 0.03252849*T + 5151.79/T + 38.92561*\log T - 563713.9/T^2 \tag{4b}$$

where K_1 and K_2 are on a molar basis and T is the temperature in kelvins. The dissociation of the various ionic species in the H_2O-CO_2 system is quite sensitive to the pH of the solution. This distribution can be seen as a function of pH in Fig. 2. In the present analysis, a low pH solution (i.e., $6 < pH < 8.5$) is assumed so that most of the dissolved CO_2 in water is in the form of the bicarbonate species, HCO_3^- [11].

Calcium carbonate crystals exist in three forms: aragonite, calcite, and vaterite. The salt in its aragonite form is considered here, since this is the form found in experimental studies of Hasson and Perl [15]. Calcium carbonate is an inverse-solubility salt; therefore, its concentration in solution will tend to decrease as the temperature increases. Under equilibrium conditions the extent of the dissolution of $CaCO_3$ can be represented by the solubility product, K_{sp}:

$$CaCO_{3(s)} \overset{K_{sp}}{\rightleftarrows} Ca^{2+} + CO_3^{2-} \tag{5}$$

where the subscript s denotes the solid phase. The solubility product is normally sensitive to temperature variations. For $CaCO_3$ this dependency has been given by Plummer and Busenberg [20] as follows.

$$\text{Aragonite:} \quad \log(K_{sp}) = -171.9773 - 0.077993*T + 2903.293/T + 71.595*\log T \tag{6}$$

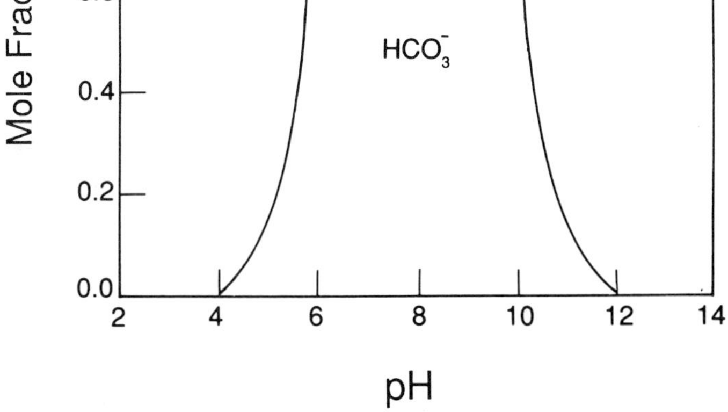

Figure 2 Effect of pH on CO_2 distribution (in terms of mole fraction of carbonate, bicarbonate, and aqueous carbon dioxide) in water at 100°C [11].

$$\text{Calcite:} \quad \log(K_{sp}) = -171.9065 - 0.077993*T + 2839.319/T + 71.595*\log T \quad (7)$$

$$\text{Vaterite:} \quad \log(K_{sp}) = -172.1295 - 0.077993*T + 3074.688/T + 71.595*\log T \quad (8)$$

where K_{sp} is in molar units and T is the absolute temperature.

The inlet compositions, i.e., concentrations, of Ca^{2+}, CO_2, and HCO_3^- are assumed to be known. Since $CaCO_3$ has an inverse solubility characteristic, its equilibrium concentration will tend to decrease in the solution adjacent to the heated surface. Due to the concentration gradient, Ca^{2+} and HCO_3^- will be transported from the bulk of the film toward the surface. At the liquid-solid interface, the HCO_3^- undergoes the following equilibrium reaction [14]:

$$2HCO_3^- \xrightarrow{K} H_2O + CO_2 + CO_3^{2-} \quad (9)$$

It should be noted that the reaction in Eq. (9) is actually the net of the two dissociation reactions in Eqs. (2) and (3); therefore, the equilibrium constant, K, is given by

$$K = \frac{K_2}{K_1} \quad (10)$$

in which K_1 and K_2 are the dissociation constants given by Eq. (4). The CO_2 formed from the reaction given by Eq. (9) is transported through the liquid film and desorbed at the liquid-gas interface. The CO_3^{2-} species produced in the same reaction reacts with Ca^{2+}, which has been transported to the surface, to form $CaCO_{3(s)}$:

$$Ca^{2+} + CO_3^{2-} \xrightarrow{k_r} CaCO_{3(s)} \quad (11)$$

where k_r, the reaction rate constant, is given later [see Eq. (27)]. This completes the precipitation of $CaCO_3$ scale on the heat transfer surface. Reactions in Eqs. (9) and (11) are sometimes combined in the form [15]

$$Ca^{2+} + 2HCO_3^- \rightarrow CaCO_{3(s)} + CO_{2(g)} + H_2O \quad (12)$$

3.2 Governing Transport Equations

3.2.1 Fluid Flow. For a laminar film flowing under the force of gravity, the only velocity component is in the downward or z direction. Assuming a fully developed velocity profile $[v_z = v_z(y)]$, the z component of the momentum equation for the coordinate system shown in Fig. 1 can be readily solved to yield the velocity profile as [23]

$$v_z(y) = \frac{3\overline{v}_z(2y/\delta - y^2/\delta^2)}{2} \quad (13)$$

where $\overline{v}_z = \rho g \delta^2 / 3\mu$ is the mean velocity.

3.2.2 Heat Transfer. As will be seen later, it is necessary to have expressions for the surface and bulk temperatures along the length of the heated section. A constant heat flux supplied to the surface and the bulk inlet temperature and mass flow rate are assumed to be given. A convective film heat transfer coefficient having negligible variation in the z direction is assumed. Then, a simple heat transfer analysis yields the bulk and surface temperature profiles as

$$T_{\text{blk}}(z) = \frac{q_0 z}{\dot{m}_p \hat{C}_p} + T^0_{\text{blk}} \tag{14}$$

and

$$T_{\text{sc}}(z) = \frac{q_0 z}{\dot{m}_p \hat{C}_p} + T^0_{\text{blk}} + \frac{q_0}{h} \tag{15}$$

The least justified assumption in developing the expression for the scale surface temperature is that of the existence of a constant convective heat transfer coefficient, which is inevitably calculated based on correlations for nonfouling conditions. However, this assumption does not seem significant enough to seriously affect the results of the present analysis. In fact, it provides at least a means of improvement by not using the mass transfer coefficient based on nonfouling condition, as will be seen later. The heat transfer coefficient used in the computations is calculated from the correlation developed by Chun and Seban [24] in the form

$$h = 0.606\left(\frac{k^3\rho^2 g}{\mu^2}\right)^{1/3}\left(\frac{\dot{m}_p}{\mu}\right)^{-0.22} \tag{16}$$

which accounts for the effect of rippling at the liquid-gas interface.

3.2.3 Mass Transfer. The equation of continuity of species i of a multicomponent system, assuming the flux due to concentration differences to be the only significant mass flux, is given as

$$\frac{\partial \rho_i}{\partial t} = -[\nabla \cdot (\rho_i \mathbf{v} + j_i^{(x)})] + r_i \tag{17}$$

For the present study, a steady state is assumed, since the experimental data to be compared with are obtained as such, and little would be gained by considering the unsteady-state behavior except for greatly complicating the solution. Assuming that no net generation or consumption occurs in the bulk, r_i can be neglected. A crystallization reaction is assumed to occur at the liquid-solid interface, and so is the equilibrium reaction as described by Eq. (9). The mass flux of species i in a dilute aqueous solution due to concentration gradients is

$$j_i^{(x)} = -M_i D_i \nabla C_i \tag{18}$$

Therefore, for steady state and no bulk reactions, the species equation reduces to

$$0 = -\nabla \cdot (M_i C_i \mathbf{v} - M_i D_i \nabla C_i) \tag{19}$$

For a laminar falling film, with $v_z(y)$ as the only velocity component and assuming negligible diffusive mass transfer in the z direction relative to convective transport, the above equations are combined to give

$$\frac{3}{2}\bar{V}_z\left(\frac{2y}{\delta}-\frac{y^2}{\delta^2}\right)\frac{\partial C_j}{\partial z}=D_j\frac{\partial^2 C_j}{\partial y^2} \tag{20}$$

This is the governing mass transport equation for the four chemical species, Ca^{2+}, CO_3^{2-}, HCO_3^-, and CO_2, in the falling film system. The inlet concentration of all species is assumed to be known, i.e.,

$$\text{at } z = 0,\ C_j = C_j^0 \tag{21}$$

where j denotes any one of the four species. Except for CO_2, transport of the other species occurs only in the liquid phase. This implies that at the gas-liquid interface no mass flux exists for these components, which is written as

$$\text{B.C.1:} \qquad \text{at } y = \delta,\ \frac{\partial C_j}{\partial y} = 0 \tag{22}$$

where j excludes CO_2. The carbon dioxide produced at the heated surface diffuses through the liquid phase and is desorbed at the gas-liquid interface. The film thickness is assumed to be constant. Depending on the extent of the desorption of CO_2 in water, two types of boundary conditions can be written for CO_2 at the gas-liquid interface:

$$\text{B.C.1 } (CO_2): \qquad \text{at } y = \delta,\ C_{CO_2} = 0 \tag{23a}$$

$$-D_{CO_2}\frac{\partial C_{CO_2}}{\partial y} = k_{des}(C_{CO_2,liq} - C_{CO_2,g}) \tag{23b}$$

At the liquid-solid interface various species are being consumed and/or generated according to the reaction given in Eqs. (11) and (12). Under steady-state conditions, mass flux to/from the surface must equal the consumption/generation by net chemical reactions. Let the rate of generation of $CaCO_3$ scale be denoted by r. Then, from the stoichiometry in Eqs. (11) and (12), a surface-reaction condition can be written for the various species:

$$\text{B.C.2:} \qquad \text{at } y = 0,\ D_{Ca^{2+}}\frac{\partial C_{Ca^{2+}}}{\partial y} = r$$

$$D_{CO_3^{2-}}\frac{\partial C_{CO_3^{2-}}}{\partial y} = r$$

$$D_{HCO_3^-}\frac{\partial C_{HCO_3^-}}{\partial y} = 2r$$

$$D_{CO_2}\frac{\partial C_{CO_2}}{\partial y} = -r \tag{24}$$

In these expressions the mass diffusivities are temperature-dependent quantities [18].

Knowledge of the form of the crystallization reaction rate, r, though crucial to any predictive model of scaling, has been quite limited. Several studies in this area have been conducted [25], and it seems that the most likely candidates for the form of the reaction rate are

$$r = k_r([Ca^{2+}] - [Ca^{2+}]_e)^2 \tag{25}$$

$$r = k_r([Ca^{2+}][CO_3^{2-}] - K_{sp}) \tag{26}$$

where the concentrations and other properties are evaluated at the surface conditions. The equilibrium concentration of calcium, Ca_e^{2+}, is assumed to be given by the square root of the solubility product, K_{sp}, which is given by Eq. (6). However, a recent study [18] has showed that the second form of the reaction rate yields results in better agreement with experimental data. It has been proposed that k_r, the crystallization rate coefficient, would have an Arrhenius form [15] such that its dependence on surface temperature is given as

$$k_r = A \exp\left(\frac{-E_a}{R_g T_{sc}}\right) \tag{27}$$

Attempts to determine the values of the preexponential factor and activation energy in Eq. (27) have had limited success. One reason for the lack of a consistent set of values is the sensitivity of the two constants to the experimental measurements, since the constants are specific to the crystallizing salt and extremely sensitive to errors introduced by the presence of impurities that affect the crystal growth rate. Assuming Eq. (9) to be an equilibrium reaction at the heated surface, the concentration of CO_3^{2-} at the interface can be written as

$$C_{CO_3^{2-}} = K \frac{C^2_{HCO_3^-}}{C_{CO_2}} \tag{28}$$

where K is given in Eq. (10). This eliminates C^{2-}_{CO3} from the reaction rate expression given in Eq. (26). The resulting reaction rate expression is to be used in the boundary condition at the scale deposition surface [see Eq. (32) later]. Since it is in terms of concentrations of Ca^{2+}, HCO_3^-, and CO_2 species, three conservation species equations will be needed to describe the mass transport processes. By defining the following groupings,

$$y^+ \equiv \frac{y}{\delta} \qquad z^+ = \frac{z}{L} \qquad D_j^+(z^+) \equiv \frac{2LD_j(z^+)}{3\bar{v}_z\delta^2} \qquad K_{R,j} \equiv \frac{k_r(z)\delta}{D_j(z)} \tag{29}$$

where Ca^{2+}, HCO_3^-, and CO_2 species are represented by the subscripts $j = 1, 2$, and 3, respectively, the three transport equations with appropriate conditions are summarized as

$$(2y^+ - y^{+2})\frac{\partial C_j}{\partial z^+} = D_j^+(z^+)\frac{\partial^2 C_j}{\partial y^{+2}} \qquad j = 1, 2, \text{ and } 3 \tag{30}$$

$$\text{At } z^+ = 0: \qquad C_j = C_j^0 \qquad j = 1, 2, \text{ and } 3 \tag{31}$$

At $y^+ = 0$: $\dfrac{\partial C_j}{\partial y^+}$

$$= n_j K_{R,j}(z^+)\left[K(z^+)\frac{C_1 C_2^2}{C_3} - K_{sp}(z^+)\right] \qquad j = 1, 2, \text{ and } 3 \qquad (32)$$

In Eq. (32), the constant $n_j = 1, 2$ and (-1) for $j = 1, 2,$ and 3, respectively, in view of the combined chemical reaction in Eq. (12). Also, it is noted that the reaction rate constant, k_r, of an Arrhenius form [see Eq. (27)] is built in the $K_{R,j}$ group. The other conditions, at $y^+ = 1$, are

$$\text{For } j = 1, 2: \qquad \frac{\partial C_j}{\partial y^+} = 0 \qquad (33)$$

$$\text{For } j = 3: \qquad (i)\ C_3 = 0$$

or

$$(ii)\ \frac{\partial C_3}{\partial y^+} = \frac{K_{des}(C_{3,liq} - C_{3,g})}{(-D_3)} \qquad (34)$$

In Eq. (34), two possible forms of the boundary condition of CO_2 at the film-air interface are listed. The first assumes negligible CO_2 concentration at the interface, while the second assumes that the diffusion rate is equal to the desorption rate. These assumptions are tested in the present study.

3.2.4 Solutions and Discussion. The three species equations and their conditions are solved numerically using a finite-difference technique [19]. Once the concentration profiles are obtained at a position z^+, the concentration gradient at the wall is obtained using a quadratic polynomial fit,

$$\frac{\partial C_j}{\partial y^+} = \frac{-9C_{j,1}^{K+1} + 10C_{j,2}^{K+1} - C_{j,3}^{K+1}}{8\delta\ \Delta y^+}$$

at $y^+ = 0$. From the above, the local deposition rate can be calculated as

$$N_{CaCO_3} = D_1 \left.\frac{\partial C_{Ca^+}}{\partial y}\right|_{y=0} \qquad (35)$$

and the local scale thickness can also be determined from a simple mass balance,

$$\tau = \frac{-D_0 + (D_0^2 + 4D_0 t N_{CaCO_3} M_{CaCO_3}/\rho_f)^{1/2}}{2} \qquad (36)$$

To compare the model with experimental data of Hasson and Perl [15], the physical properties and initial conditions used in the numerical computations are given in Table 1. The diffusivities of Ca^{2+}, HCO_3^-, and CO_3^{2-} species are assumed to be equal to (D). Since accurate a-priori values of the activation energy and preexponential constant in Eq. (27) were not available, a trial-and-error procedure was used to obtain

Table 1 Physical properties and initial conditions used in calculations for deposition from a falling film

Run no.	pH	Initial concentrations ($M \times 10^3$) $C^0_{Ca^{2+}}$	$C^0_{HCO_3^-}$	$C^0_{CO_2}$	$C^0_{CO_3^{2-}}$	T^0_{blk}(K)	Length of run (h)
E-3	7.75	6.30	4.60	0.119	0.030	318.5	52.5
E-2	7.55	4.25	8.00	0.330	0.033	319.0	65.4
E-9	7.25	8.15	4.40	0.343	0.010	320.0	84.5
E-11	7.10	14.60	4.40	0.431	0.009	320.5	82.5
E-10	7.20	10.25	4.40	0.370	0.010	321.0	83.0
E-5	7.55	6.20	4.40	0.178	0.020	326.0	78.75
E-7	7.20	6.20	4.40	0.411	0.009	335.5	84.5
E-4	7.25	6.30	4.50	0.374	0.011	336.0	32.2
E-8	7.25	6.05	4.20	0.359	0.010	339.5	61.0

δ	2.3×10^{-4} m	M_{CaCO_3}	0.1 kg/mol
$\overline{V}_z$	0.32609 m/s	g	9.8 m/s^2
q_0	420.16 W/m^2	k	0.628 w/m K
$\dot{m}_p$	0.075 kg/ms	k_f	2.25 w/m K
μ	1×10^{-3} kg/ms	ρ	10^3 kg/m^3
$\hat{C}_p$	4186 J/kg K	ρ_f	2.45×10^3 kg/m^3
R_g	8.33 J/mol K		

agreement for run E-3 of Hasson and Perl [15]. Run E-3 was chosen because it is the only run for which the complete experimental thickness (and hence deposition) profile was given. The best values found [18] for E_a/R_g and ln(A) are 8800 and 38.5, respectively.

The mass deposition flux and scale-layer thickness using two different boundary conditions for CO_2 [see (i) and (ii) in Eq. (34)] were calculated for the experiments of Hasson and Perl [15] and are tabulated in Table 2. The experimental and calculated thickness profile for run E-3 is shown in Fig. 3. Figure 4 presents dimensionless mass flux of calcium carbonate deposition versus dimensionless inlet Ca^{2+} concentration, with the dimensionless heat flux and inlet CO_3^{2-} concentration as varying parameters. The nondimensional species concentration, molar deposition flux, and heat flux were defined as follows:

$$N^*_{CaCO_3} = \frac{N_{CaCO_3}\delta}{D_1 K_{sp}^{1/2}} \qquad q^* = \frac{q_0\delta}{k_f(T^0_{blk} - T_{ref})} \qquad C^*_j = \frac{C^0_j}{K_{sp}^{1/2}}$$

where the temperature dependent quantities such as K_{sp} and D are evaluated at the inlet temperature, T^0_{blk}. The reference temperature, T_{ref}, is chosen for convenience at 273 K. The plots are for constant Schmidt and Reynolds numbers, the values of which are $Sc = \mu/\rho D = 436$ and $Re \equiv 4\delta\rho\overline{V}_z/\mu = 300$.

Table 2 Comparison of the data of Hasson and Perl [15] with predictions of the multispecies model for deposition flux and scale thickness at z = 1 m

	$N_{CaCO_3} \times 10^7$ (g/cm² s)			Thickness, τ (μm)		
Run no.	Exp.	Calc. 1	Calc. 2	Exp.	Calc. 1	Calc. 2
E-3	1.31	1.39	1.37	103	107	106
E-2	1.99	2.02	1.99	195	193	189
E-9	1.48	1.46	1.42	188	180	175
E-11	1.71	1.77	1.73	211	212	208
E-10	1.65	1.62	1.58	206	196	192
E-5	1.41	1.63	1.60	166	187	184
E-7	1.77	2.03	1.98	224	250	243
E-4	1.78	2.13	2.07	86	100	98
E-8	1.84	2.12	2.07	168	189	184

Calculation 1 uses the boundary condition $C_{CO_2} = 0$, whereas calculation 2 uses $\partial C_{CO_2}/\partial y = -k_{des} C_{CO_2,liq}/D_{CO_2}$.

Comparison of predicted with experimental values in Table 2 indicates an excellent agreement with a maximum error of about ±16%. It was found that the two boundary conditions given by Eq. (34) produced similar results. The model can predict local deposition with greater accuracy than Hasson's model; the later gives only average values. However, predicted inlet behavior is still inconsistent with experimental data. The assumption of a fully developed velocity profile at the entrance region may be one reason. The importance of pH and its effect on the CO_3^{2-} concentration might also be a reason. Another, though less likely, reason is that the model is in effect a

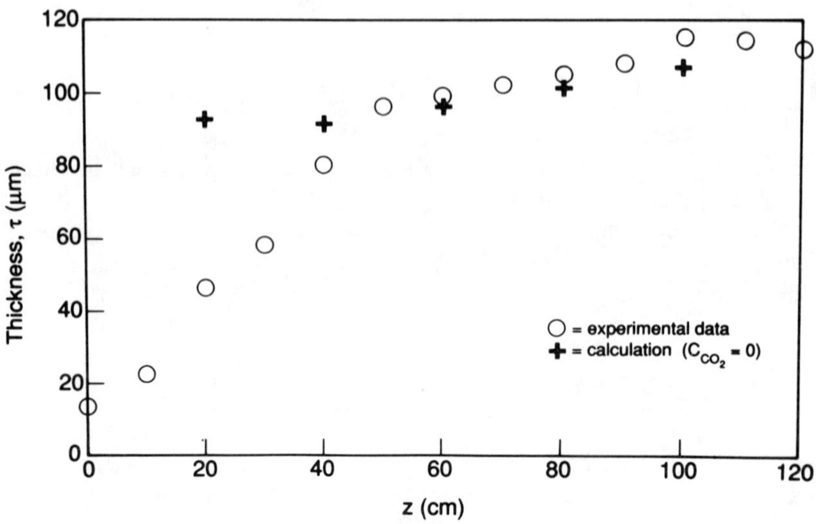

Figure 3 Experimental and calculated thickness profiles for Hansson's run E-3.

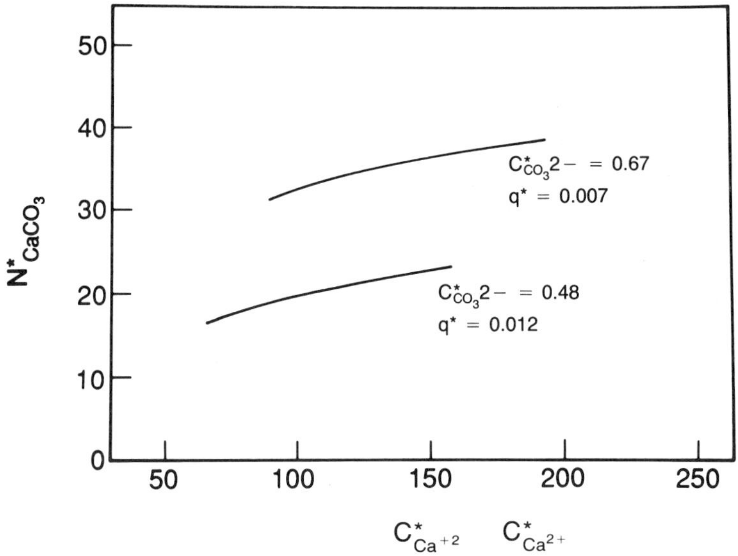

Figure 4 Nondimensional mass flux versus pertinent dimensionless operating parameters.

mass transport model. Processes such as nucleation, attachment, and removal, which have been assumed to be insignificant, could have some effects as well. They are subjects for future studies.

4 DEPOSITION OF $CaCO_3$ IN TURBULENT FLOW IN AN ANNULUS

With the encouraging success in analyzing the laminar falling film system, Chan and Ghassemi [19] proceeded to consider the deposition of $CaCO_3$ in turbulent shell-and-tube flows with constant heat flux, such as those in the experiments of Sheikholeslami and Watkinson [16]. The process variables are essentially the same as in the laminar case, except that the flow is turbulent and cylindrical coordinates must be used since calcium carbonate deposition occurs on the outside of the inner cylindrical surface in an annulus. The geometry of the problem is shown in Fig. 5. A constant flow of scale-producing water is maintained between the two concentric cylinders, while the inner cylinder is supplied with constant heat flux. The chemistry and equilibria of the system are similar to the laminar case and remain unchanged. Scale buildup on the surface of the inner tube causes the increase of fouling resistance. The prediction of the fouling resistance R_f as a function of time is desired, assuming that inlet temperature and composition are known. Since the deposition rate of $CaCO_3$ scale is related to the increase of R_f, the determination of the $CaCO_3$ deposition rate allows calculation of R_f.

4.1 Governing Transport Equations

4.1.1 Fluid Flow. Due to the turbulent nature of the flow in this problem, an analytical derivation of the velocity profile from the fundamental Navier-Stokes equation would be impractical. However, assuming a fully developed profile [$v_z = v_z(r)$], the mean turbulent velocity distribution in annuli, using a semiempirical approach, has been shown to be approximated by [23],

$$\overline{V}_{z,\max} - \overline{V}_z = \frac{1}{\kappa_1}\left(\frac{\tau_0}{\rho}\,\frac{\lambda^2 - \kappa^2}{\kappa}\right)^{1/2} \ln\frac{(\lambda - \kappa)R}{r - \kappa R} \qquad \text{for } r < \lambda R \qquad (37)$$

$$\overline{V}_{z,\max} - \overline{V}_z = \frac{1}{\kappa_1}\left[\frac{\tau_0}{\rho}(1 - \lambda^2)\right]^{1/2} \ln\frac{R - \lambda R}{R - r} \qquad \text{for } r > \lambda R \qquad (38)$$

where

$$\lambda^2 = \frac{1 - \kappa}{2\ln(1/\kappa)}$$

The shear stress at the wall, τ_0, can be obtained if the friction factor for the inner wall is known. This factor, f_1, for annuli has been correlated to the flow Reynolds number [26] by

$$f_1 = f_2\,\frac{(\lambda^2 - 2\kappa^2)}{\kappa(1 - \lambda^2)}$$

where f_2, the outer wall friction factor, is

$$f_2 = \frac{0.0791}{[\text{Re}(1 - \lambda^2)]^{1/4}}$$

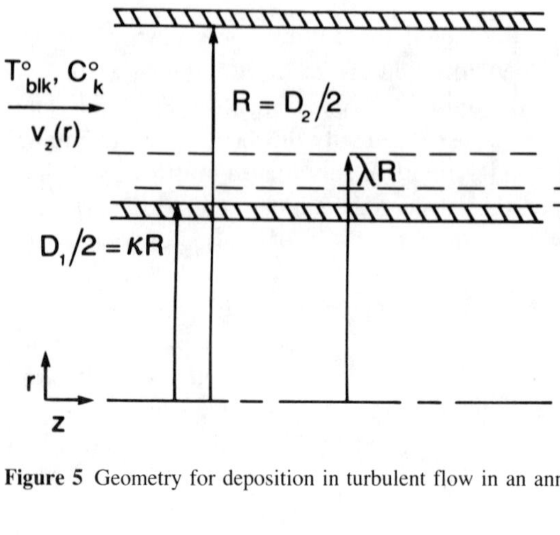

Figure 5 Geometry for deposition in turbulent flow in an annulus.

Then the wall shear stress may be calculated using

$$\tau_0 = f_1 \rho \langle \overline{v}_z \rangle^2 \tag{39}$$

Unfortunately, the velocity profile given above does not represent the profile for regions close to the annulus walls accurately. An accurate profile is needed for later calculation of the near-wall concentration profile. For the region adjacent to the walls, approximated by a boundary-layer thickness δ_t, one may use the following profile [23]:

$$v^+ = y^+ \qquad \text{for } r < \kappa R + \delta_t \text{ and } r > \kappa R - \delta_t \tag{40}$$

where

$$v^+ = \frac{\overline{v}_z / \langle \overline{v}_z \rangle}{f_1^{1/2}} \qquad \text{and} \qquad y^+ = \frac{y \langle \overline{v}_z \rangle f_1^{1/2} \rho}{\mu}$$

The layer thickness can be approximated as the critical thickness over a plate such that, as obtained from experiments,

$$\delta_t = \frac{10.8\mu}{\langle \overline{v}_z \rangle f_1^{1/2} \rho}$$

4.1.2 Heat Transfer. The development of expressions for bulk and surface temperatures from heat transfer consideration is similar to the laminar case. Only $\dot{m}_p$ should be redefined as the mass flow rate per unit of inner tube perimeter and a different heat transfer correlation referring to the inner tube should be used. For diameter ratios $d_1/d_2 > 0.2$, the following correlation for turbulent heat transfer in annuli is recommended [26]:

$$\text{Nu} = 0.02 \, \text{Re}^{0.8} \, \text{Pr}^{1/3} \left(\frac{d_2}{d_1} \right)^{0.53} \tag{41}$$

for calculation of the heat transfer coefficient. The Reynolds number for heat transfer in an annulus is given as

$$\text{Re} = \frac{d_e \dot{m}}{A_{\text{flow}} \mu}$$

where

$$d_e = \frac{d_2 - d_1}{d_1} \qquad \text{and} \qquad A_{\text{flow}} = \frac{\pi (d_2^2 - d_1^2)}{4} \tag{42}$$

4.1.3 Mass Transfer. For mass transfer in dilute solutions, the multicomponent equation for mass transport is still valid. For turbulent flows, however, a time-averaged equation must be used with mean quantities $\overline{C}_j$ as the dependent variables and with an effective turbulent diffusivity replacing the molecular diffusivity. The mean mass transport equation written for species j, where j can be Ca^{2+}, CO_3^{2-}, HCO_3^-, or CO_2, in cylindrical coordinates is

$$\overline{v}_z(r)\frac{\partial \overline{C}_j}{\partial z} = \frac{1}{r}\frac{\partial}{\partial r}\left(rD_{j,\mathrm{eff}}\frac{\partial \overline{C}_j}{\partial r}\right)$$

where

$$D_{j,\mathrm{eff}} = D_j + D_{j,\mathrm{turb}} \tag{43}$$

To determine the turbulent diffusivity, the mixing-length hypothesis is used. Assuming the turbulent Schmidt number to be unity, i.e., $\mathrm{Sc}_{\mathrm{turb}} = \nu_{\mathrm{turb}}/D_{\mathrm{turb}} = 1$, $D_{j,\mathrm{turb}}$ can be written as

$$D_{j,\mathrm{turb}} = l_m^2\left|\frac{d\overline{v}_z}{dr}\right| \tag{44}$$

in which l_m, the mixing length, is approximated by l_m in turbulent pipe flows [27]:

$$\frac{l_m}{R} = 0.14 - 0.08\left(1 - \frac{y}{R}\right)^2 - 0.06\left(1 - \frac{y}{R}\right)^4 \tag{45}$$

where y is the distance from the wall. Since the equation is being used for an annulus for which the axis of zero shear is not in the center, two separate regions must be considered—one from $r = \kappa R$ and λR and the other from $r = \lambda R$ to R. Equation (45) then becomes

$$r < \lambda R: \quad \frac{l_m}{R(\lambda - \kappa)} = 0.14 - 0.08H_1^2 - 0.06H_1^4 \qquad H_1 = 1 - \frac{r - \kappa R}{R(\lambda - \kappa)} \tag{46}$$

$$r > \lambda R: \quad \frac{1}{R(1 - \lambda)} = 0.14 - 0.08H_2^2 - 0.06H_2^4 \qquad H_2 = 1 - \frac{R - r}{R(1 - \lambda)} \tag{47}$$

The reaction-rate form of Eq. (26) is used, and only the Ca^{2+} and CO_3^{2-} species are considered. The reason for using only two species rather than three is that in the turbulent case, the composition of the inlet water is given as alkalinity and solution pH [16], from which the other concentrations can be calculated from equilibrium relations (assuming equilibrium conditions). This is obvious, in view of Eq. (9),

$$C_{CO_3^{2-}} = \frac{K(C_{HCO_3^-})^2}{C_{CO_2}}$$

and the equation from Langelier [28],

$$C_{\mathrm{AIKA}} + C_{H^+} = 2C_{CO_3^{2-}} + C_{HCO_3^-} + C_{OH^-}$$

From the above two equations, the concentration of HCO_3^- can be either eliminated or calculated, whereas for the laminar case all the relevant composition data are given for the experimental runs. The general initial and boundary conditions to Eq. (43) are then

$$\text{At } z = 0 \qquad \overline{C}_j = \overline{C}_j^0 \tag{48}$$

$$\text{At } r = \kappa R \qquad D_j \frac{\partial \overline{C}_j}{\partial r} = k_r(z)[\overline{C}_1 \, \overline{C}_2 - K_{sp}(z)] \tag{49}$$

$$\text{At } r = R \qquad \frac{\partial \overline{C}_j}{\partial r} = 0 \tag{50}$$

where $j = 1$ and 2 represent the Ca^{2+} and CO_3^{2-} species, respectively.

4.1.4 Solutions and Discussion. The calculation procedure used to solve for the concentration profiles of Ca^{2+} and CO_3^{2-} from Eq. (43) is the same as before, and the mass deposition can be calculated as

$$N_{CaCO_3} = M_{CaCO_3} D_j \left. \frac{\partial C_j}{\partial r} \right|_{r=R} \tag{51}$$

The relation between the mass of deposit and the fouling resistance, R_f, can be expressed as

$$\frac{dR_f}{dt} = \frac{N_{CaCO_3}}{\rho_f k_f} \tag{52}$$

from which the fouling resistance at any time can be calculated, assuming that the quantities on the right-hand side of Eq. (52) are time independent.

The average value of the calculated fouling resistance [19] for several flows was compared with data of Sheikholeslami and Watkinson [16]. The values of all physical properties and initial conditions used in the calculations for the turbulent model are given in Table 3. The values of E_a and A were obtained, as in the laminar case, by trial and error. The values for E_a/R_g and $\ln(A)$ used were 4800 K and 46 cm^4/mole S, respectively. It was found that the model is quite insensitive to the kinetic parameters (E_a and A) beyond these values. This is quite different from the laminar computations, which would increase without bound if the reaction kinetic parameters were continually increased.

The mass deposition flux averaged over the length of the surface, as well as the fouling resistance R_f, over a 70-h period were calculated for four of Sheikholeslami's experiments, and their values are given in Table 4. The experimental as well as the predicted behavior of fouling resistance versus time from both the present model and Hasson's model (see [15] and [16]) are all listed in Table 4 and compared in Figs. 6 and 7. The comparison shows that the multispecies transport model of Chan and Ghassemi [19] has improved over Hasson's model. The agreement between the predictions of the transport model and experimental data appears to be excellent for runs

Table 3 Physical properties and initial conditions used in calculations for deposition in turbulent flow

Run no.	$C^0_{Ca^{2+}}$ $(M \times 10^{-3})$	$C^0_{CO_3^-}$ $(M \times 10^3)$	Q_0 (J/s)	$\langle \bar{v}_z \rangle$ (m/s)
20	7.00	0.0090	11958	0.695
23	6.44	0.0080	11254	0.503
36	6.19	0.0092	13465	0.695
38	6.59	0.0082	9948	0.299
d_1	1.91×10^{-2} m		μ	0.001 kg/m s
d_2	3.70×10^{-2} m		ρ	1000 kg/m^3
T_{blk}	298 K		C_p	4186 J/kg K
ρ_f	2.65×10^3 kg/m^3		k	0.628 W/m K
k_f	2.2 W/m K		Section length 1.15 m	

with linear fouling behavior (see runs 20 and 36 in Fig. 6). But the agreement is not as good, especially during early stages of scaling, for the two runs (runs 23 and 38 shown in Fig. 7) with asymptotic behavior. However, Sheikholeslami and Watkinson [16] pointed out that most of their experiments produced linear plots of R_f versus time. They further stated that "asymptotic behavior was noted only in a few instances such as when the mild steel tube was used for the first time (run 23) or with the noncorroding copper tube at the lowest velocity (run 38)." They also continued to state that "for both cases, the initial rate was very much higher than usual, and in the former case

Table 4 Calculation of average deposition flux and fouling resistance for four experimental runs of Sheikoholeslami and Watkinson [16] over a 70-h period

	N_{CaCO_3} (kg/m^2 s $\times$ 10^5)	R_f (m^2 K/kW) at various times (h) 10	20	30	40	50	60	70
Run 20								
Experimental		0.01	0.02	0.03	0.04	0.05	0.06	0.07
Calc. (present)	0.167	0.010	0.021	0.031	0.041	0.052	0.062	0.072
Calc. (Hasson and Perl [15])	0.069	0.004	0.007	0.013	0.017	0.021	0.026	0.030
Run 23								
Experimental		0.08	0.12	0.15	0.17	0.18	0.18	0.19
Calc. (present)	0.368	0.023	0.045	0.068	0.091	0.114	0.136	0.159
Calc. (Hasson and Perl [15])	0.086	0.005	0.011	0.016	0.021	0.027	0.032	0.037
Run 36								
Experimental		0.02	0.05	0.07	0.09	0.12	0.14	0.16
Calc. (present)	0.444	0.027	0.055	0.082	0.110	0.137	0.165	0.192
Calc. (Hasson and Perl [15])	0.056	0.003	0.007	0.010	0.014	0.017	0.020	0.024
Run 38								
Experimental		0.05	0.07	0.07	0.07	0.07	0.07	0.07
Calc. (present)	0.268	0.017	0.033	0.050	0.066	0.083	0.099	0.116
Calc. (Hasson and Perl [15])	0.279	0.017	0.035	0.052	0.069	0.087	0.104	0.121

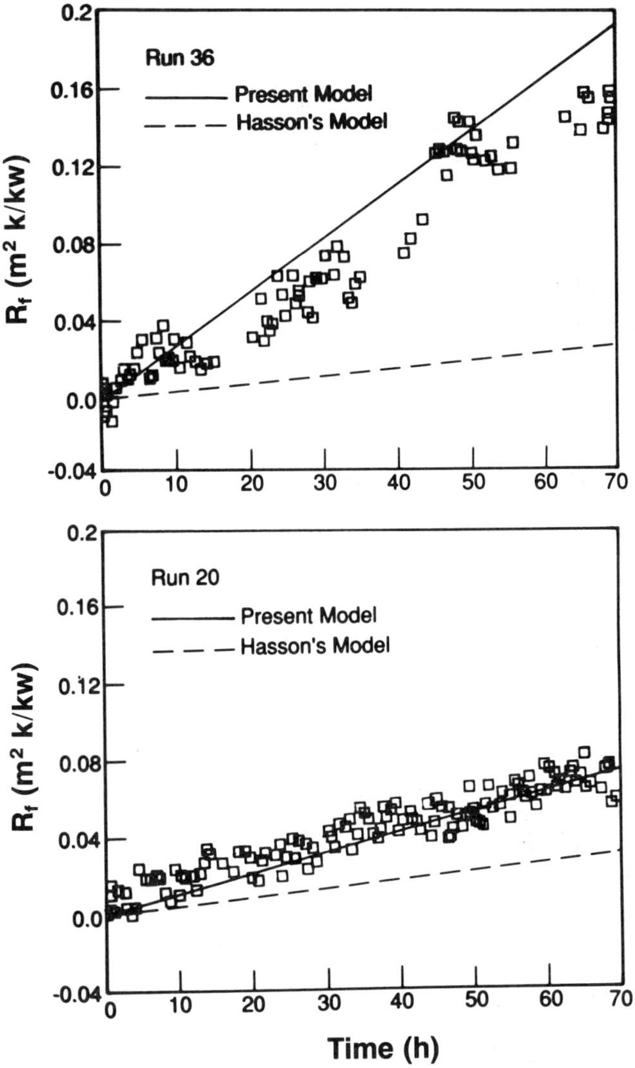

Figure 6 Comparison of the present and Hasson's model with fouling data of Sheikholeslami and Watkinson [16].

the ultimate scaling resistance was abnormally high. In a repeated run on the mild steel tube, scaling was linear, and less extensive.'' Therefore, for practical purposes, it was concluded that the prediction by the transport model appears to be satisfactory for most applications.

Like the previous laminar solution, the results from the turbulent model computations are also presented as plots of a nondimensional mass flux versus the dimensionless operating parameters of the inlet composition and flow Reynolds number, as

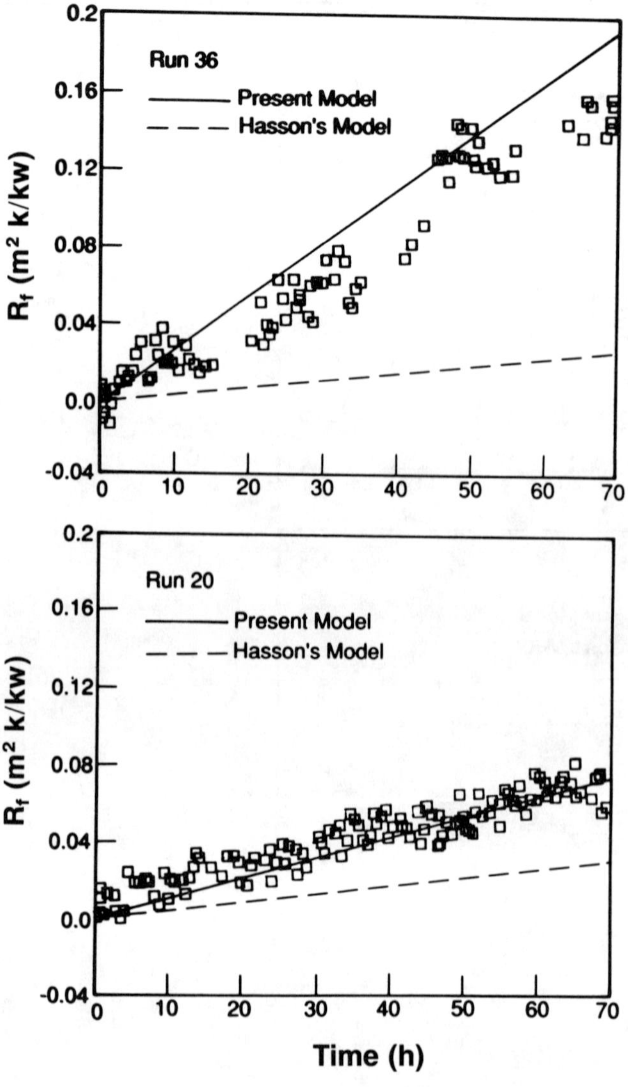

Figure 7 Comparison of the present and Hasson's model with fouling data of Sheikholeslami and Watkinson [16].

shown in Fig. 8. The plots are for a constant laminar Schmidt number value of 500. The nondimensional molar flux, species concentration, and Reynolds number were defined as

$$N^* = \frac{N_{\mathrm{CaCO_3}} d_r}{D K_{\mathrm{sp}}^{1/2}} \qquad C_j^* = \frac{C_j^0}{K_{\mathrm{sp}}^{1/2}} \qquad \mathrm{Re} = \frac{(d_2 - d_1)\langle \overline{v}_z \rangle \rho}{\mu} \tag{53}$$

where $d_r = (d_2^2 - d_1^2)/d_1$, the temperature dependent quantitites such as K_{sp} and D

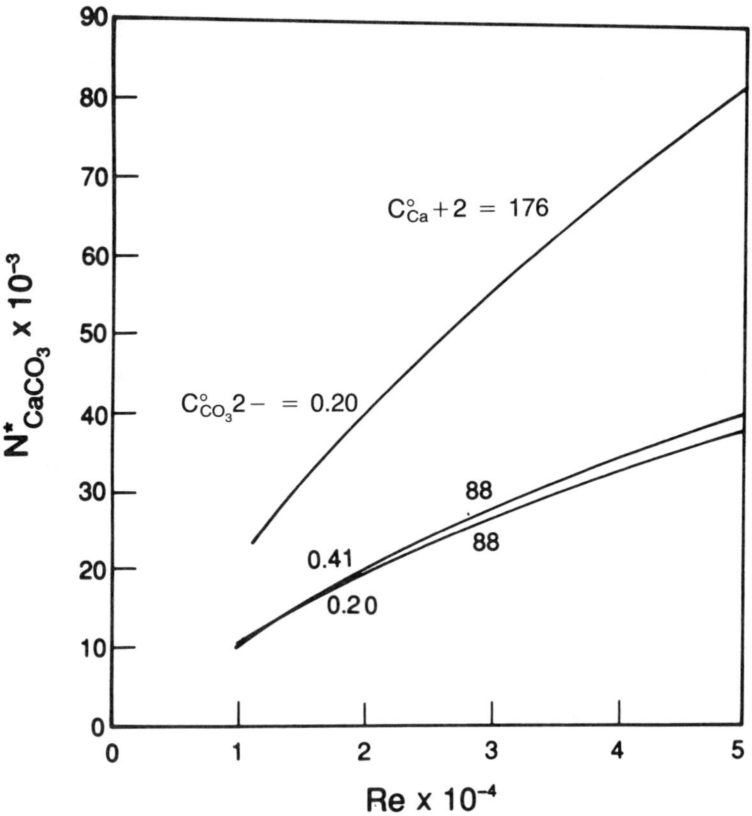

Figure 8 Nondimensional mass flux versus pertinent dimensionless operating parameters.

are evaluated at the inlet temperature, T^0_{blk}. It is clear that the calcium carbonate fouling increases significantly with Reynolds number and with the inlet Ca^{2+} ion concentrations but less with the inlet CO_3^{2-} ion concentrations.

5 SILICA SCALING

Unlike calcium carbonate, whose solubility decreases with increasing temperature, silica solubility increases with temperature. Thus calcium carbonate scale is often found on heated surfaces while silica scale is found on cooled surfaces. Like calcium carbonate scale, silica scale can also deposit on isothermal surfaces. For example, silica, being saturated in geothermal brines in the downhole environment in the monomeric state, can become supersaturated as the brine is cooled through a heat exchanger or when part of the brine is flashed into steam. The supersaturation causes precipitation of silica in an amorphous form on heat exchanger surfaces, separators, well lines, and discharge lines. Silica deposition kinetics is rather complex and has recently begun to attract fundamental research [29–35], and some attempts to model and to correlate experimental data have been made [36–44]. In the following, the related fundamentals in silica fouling are reviewed first, followed by presentation of the recently developed analytical model and experimental data on silica fouling.

5.1 The Solubility of Silica

Silica exists in both the crystalline and amorphous states [11, 45]. Because many possible ordered arrangements of tetrahedra joined by corners are possible, silica forms various crystal structures—polymorphs. The solubilities of the different polymorphs of silica in saturated water at various saturation temperatures are given in Fig. 9. The solubility of the polymorphs increases with increasing temperature up to 300°C; above this temperature the solubility decreases. It can be seen that amorphous silica is more soluble than quartz at all temperatures, but both exhibit similar temperature dependencies. The concentration of silica in underground geothermal brines is usually controlled by the solubility of quartz and downhole temperature, so that silica is generally present as quartz [11, 46]. However, under most conditions, it appears that silica

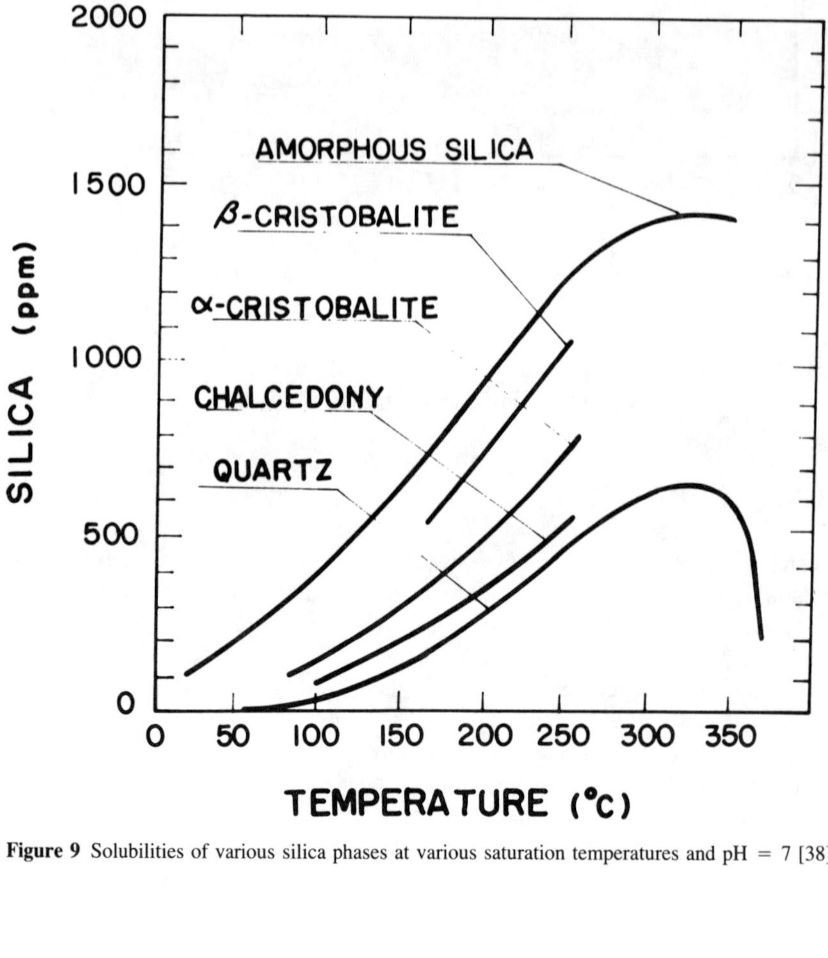

Figure 9 Solubilities of various silica phases at various saturation temperatures and pH = 7 [38].

precipitates out as relatively pure amorphous silica [11, 30–33], whatever the original silica source may be. The reason is that the kinetics of quartz crystallization below 300°C is much slower than that of amorphous silica [29].

Amorphous silica solubility is a weak function of pressure and therefore the pressure effect can be neglected for most practical applications [47].

Besides temperature, pH and salinity can have significant effects on the solubility of silica. Figure 10 shows that at low pH the solubility remains constant, but beyond a limit, the solubility increases drastically with pH value [35, 48].

When the salinity of brine is changed due to the addition of salts to the pure water, the magnitude of the activity of water decreases and so does the solubility of silica. Figure 11 shows that the solubility of amorphous silica decreases as the amount of additional NaCl increases at various temperatures.

The solubility of amorphous silica in aqueous sodium nitrate up to 300°C [49], and in sodium chloride, soldium sulfate, or magnesium sulfate solutions [50] up to 350°C, has been measured. Other investigators have also studied the solubility in 10 [51] and 11 [52] different salt solutions (NaCl, NaBr, NaI, LiCl, KCl, $NaNO_3$, NaAc, Na_2SO_4, $MgCl_2$, $CaCl_2$, and $SrCl_2$) at 25°C. For a given temperature and salt concentration the effect of different salts on solubility can be correlated by the hydration

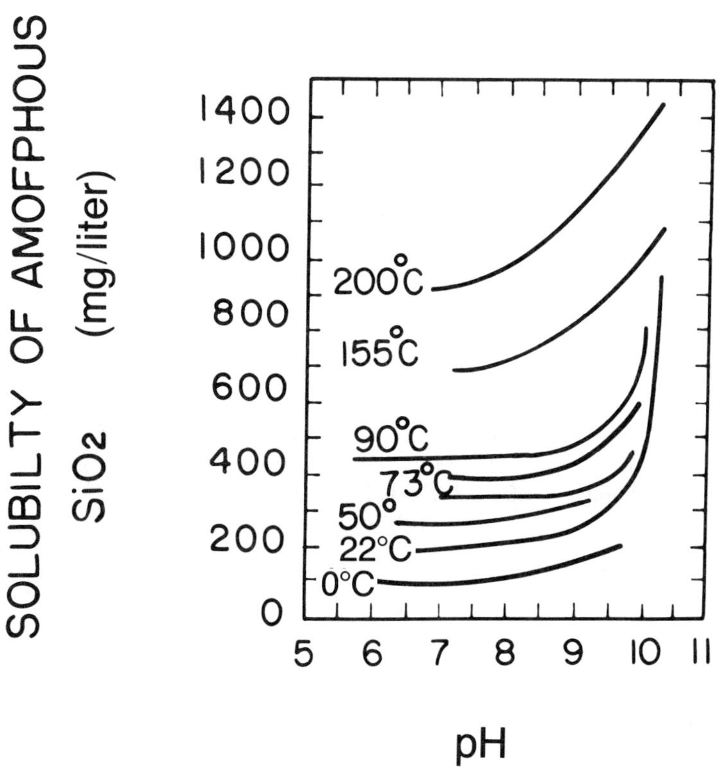

Figure 10 Solubility of amorphous silica as a function of pH at various temperatures [48].

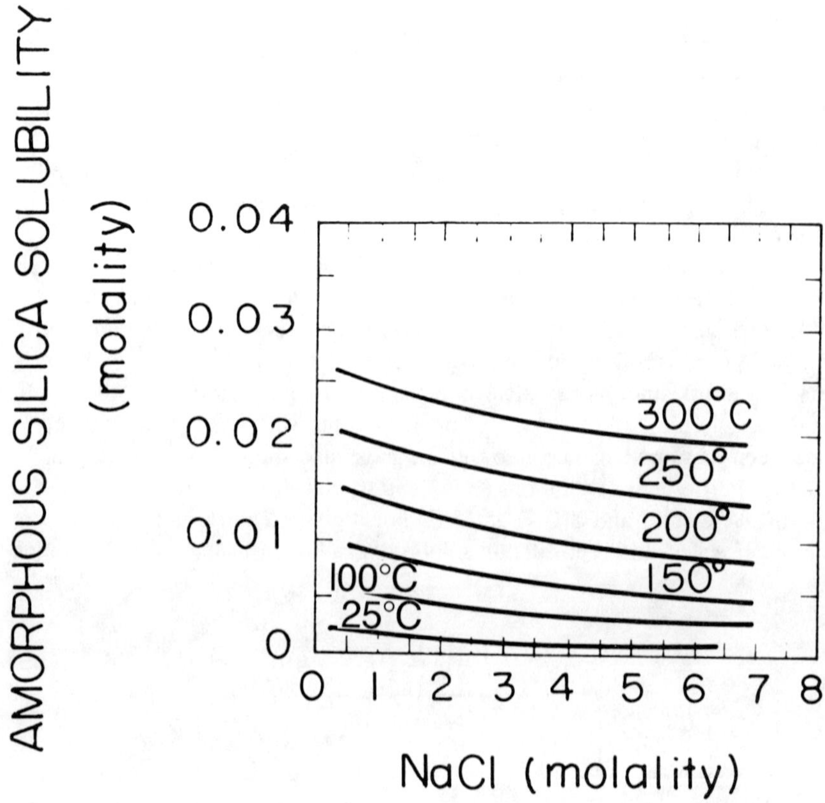

Figure 11 Molal solubility of amorphous silica in aqueous NaCl, solutions, 25–300°C [50].

number of salt cations. The order of decreasing effect of cation on the amorphous silica solubility is found as [51, 52]

$$Mg^{2+} > Ca^{2+} > Sr^{2+} > Li^{+} > Na^{+} > K^{+}$$

Similarly, the effects of the presence of sodium halide salts (NaCl, NaI, and NaBr) in brine solutions were studied. It was found that the gradients of silica concentration with respect to the salt concentration for sodium halide salts can be correlated in terms of the molecular weight of the anionic ions, such that the order of anionic effect can be stated as [52]

$$I^{-} > Br^{-} > Cl^{-}$$

and that the solubility of silica C_{SiO_2} (in ppm) in the sodium halide salt solution can be correlated as a function of salt concentration C_j (in molarity), temperature T, and the atomic weight of halide ions M_j as follows:

$$C_{SiO_2} = C_{op} - \frac{11.75[M_j + (378 - T)]}{281 - T} C_j$$

where C_{op} is the silica solubility (in ppm) in pure water,

$$\log C_{op} = 4.52 - \frac{731}{T}$$

5.2 Processes of Silica Polymerization and Physical Phenomena of Silica Scale Deposition

The high concentration of silica in most geothermal brines causes the brine solution to become supersaturated when extracted and cooled from the well. This supersaturation is thought to be the main factor in the deposition of silica scale in geothermal energy processes.

Usually, silica precipitates from geothermal brines as amorphous silica or amorphous silicates. The process of amorphous silica precipitation from a supersaturated bulk aqueous phase may consist of the following steps [30–33]:

1. Formation of silica polymers of less than critical nucleus size.
2. Nucleation of an amorphous silica phase in the form of colloidal particles that are large enough to grow spontaneously and without interruption, the so-called homogeneous nucleation.
3. Growth of the supercritical amorphous silica particles by further chemical deposition of silicic acid on their surfaces.
4. Coagulation or flocculation of colloidal particles to give a gel.
5. Cementation of the particles in the gel by chemical bonding and further deposition of silica between the particles.
6. Rarely, growth of a secondary phase in the interstices between the amorphous silica particles. Such secondary deposition of FeS and of calcite has been reported, but is not common.

The above homogeneous nucleation processes occur when the concentration of dissolved silica is highly supersaturated. This requires supersaturation by a factor of roughly 2.5 or more. When this condition is met, rapid polymerization occurs, and massive precipitation or scale deposition may follow.

If the concentration of dissolved silica is too low for rapid homogeneous nucleation to occur, relatively slow heterogeneous nucleation and the deposition of dissolved silica directly on solid surfaces become the dominant polymerization processes. The product of the latter process is a dense vitreous silica. At higher temperatures, this process may produce scale at a significant rate.

The polymerization of silica has been studied by some workers [30, 34, 53, 54], who employed silica derived from Na_3SiO_4 and thus the silica solution contained some significant impurities such as Na. Recently, pure silica from silica gel has been used [52]. A heated solution with saturated monomeric silica of a high grade of purity was brought to supersaturation by quenching the solution to a lower temperature. The rate of polymerization was determined by measuring the change of the concentration monomeric silica with time. The result showed that the induction time increases as the pH decreases. The reaction order was found to lie between 1 and 2 and the reaction constant was approximately a linear function of the pH value [52].

There are two different types of deposition of silica in the SiO_2-H_2O system, namely, the depositions from monomeric silicic acid and those from colloidal particles. Monomeric silicic acid forms an impervious glasslike film, whereas colloidal particles form a porous film, often white and opaque when dried [55]. The phenomena of both types of deposition are so complicated that, so far, neither has been modeled successfully to yield a general empirical equation for predicting silica deposition in the geothermal field. In the following, analytical models proposed by Chan and his co-workers [43, 44] to predict the deposition from monomeric silicic acid brine are presented.

6 SILICA DEPOSITION MODEL

The basis of the analytical model is a kinetic study by Bohlman et al. [29] on simulated geothermal brines in an isothermal system (isothermal in the sense that the supersaturated brine and the deposition surfaces are at the same temperature). They summarized the kinetic reaction rate of silica deposits by a correlation which shows that the apparent rate constant is 0.7, which is independent of system temperature over the range 60–100°C and that the apparent kinetic reaction order is 2 for hydroxide ion concentration and the supersaturation of $Si(OH)_4$. It is to be noted that the precise determination of reaction constant and reaction order is difficult and, not unexpectedly, somewhat different values are reported by Chan [52]. As pointed out by Fleming [56], reaction orders in the range of 1 to 4 have been reported.

Assuming that silica scale deposition is surface reaction rate controlled rather than diffusion controlled, the kinetic equation for an isothermal system [29] may be applicable to nonisothermal systems provided provisions are made for thermal-hydraulic effects. Therefore, it was proposed that the rate of change of fouling resistance (or the mass deposition flux, $dm_f/dt = (dR_f/dt)/\rho_f k_f$) be given as [44]

$$\frac{dR_f}{dt} = K_R \left[\frac{C_b - C_e(T_s)}{C_e(T_s)} \right]^P (C_{\mathrm{OH}^-})^q \tag{54}$$

where C_b is the bulk concentration of amorphous silica (SiO_2), C_e is the equilibrium solubility of amorphous silica evaluated at the scale-brine interface temperature T_s, and C_{OH^-} is the bulk molal concentration of OH^- ions in the brine. The reaction order of silica (SiO_2) can be taken as $p = 2$, the same as the isothermal system. This is because the fouling rate is expected to be about the same as long as the supersaturation of silica at the interface in the nonisothermal system is the same as that in the isothermal system. The other correlation constants, K_R and q, are obtained experimentally. As in isothermal systems, the supersaturation and pH effects are explicitly accounted for in Eq. (54). In addition, salinity effects due to the presence of impurities, and thermal-hydraulic effects due to heat transfer and fluid flow, are also accounted for as described below.

Since the surface kinetic reaction is assumed to be the controlling mechanism for silica deposition, the brine convective diffusion and the heat transfer from the hotter, supersaturated brine to the cooler deposition surface should not affect the deposition mechanism. They can, however, alter the scale-brine interface temperature T_s, which in turn has a direct bearing on the deposition rate because the equilibrium solubility, C_e, in Eq. (54) is a strong function of temperature. In other words, the model accounts for thermal-hydraulic effects by evaluating C_e at the interface temperature T_s, which is calculated from the heat transfer consideration. Normally, overall heat transfer coefficients, U_i, and the inlet and outlet temperatures of the brine and coolant, are measured and reported in the literature [36]. Then the mean interface temperature can be derived from the simple energy balance:

$$\dot{Q} = U_i A_i (\mathrm{LMTD}) = h_i A_i (T_m - T_s) \tag{55}$$

where h_i and A_i are, respectively, the brine-side heat transfer coefficient and surface area, LMTD is the log-mean temperature difference across a heat exchanger, and T_m is the mean mixed temperature of the brine $[T_m = (T_{i,\mathrm{in}} + T_{i,\mathrm{out}})/2]$. For pipe flows with a constant wall temperature, which is equivalent to a mean T_s, the use of simple temperature difference as in the third term of Eq. (55) is satisfactory.

Salinity effects due to the presence of impurities other than silica are included in the model by evaluating the equilibrium solubility of water, C_e, from

$$\frac{C_e(T_s)}{C_0(T_s)} = \left(\frac{a_{\mathrm{H_2O}}}{a_{\mathrm{H_2O},p}}\right)^2 \tag{56}$$

where C_0 is the equilibrium solubility of amorphous silica in pure water at the interface temperature [57], i.e.,

$$\log C_0(T_s) = -0.26 - \frac{731}{T_s} \tag{57}$$

The activity of pure water, $a_{\mathrm{H_2O},p}$, is one and the activity of water, $a_{\mathrm{H_2O}}$, in NaCl solution can be evaluated from [58]

$$\log a_{\mathrm{H_2O}} = -0.0157\phi I \tag{58}$$

where I is the ionic strength of all ion components in the brine,

$$I = \frac{1}{2} \sum_i C_i Z_i^2 \tag{59}$$

with C_i and Z_i standing for, respectively, the molality and ionic strength of ion i. The osmotic coefficient (ϕ) for water solutions of NaCl can be found in Lin and Lindsay [59], which provides a table of the coefficient as a function of temperature and salt (NaCl) concentration. The salt concentration can be replaced by the ionic strength of the brine if all the ions in the brine are assumed to have the same effect on the osmotic coefficient as NaCl salt.

To include the effect of salinity on the concentration of hydroxide ions, $C_{\mathrm{OH^-}}$ in Eq. (54) is calculated from

$$\log C_{OH^-} = \log Q_w + pH + \log \gamma_{H^+} \tag{60}$$

where Q_w is the dissociation quotient of water, which is a function of temperature and ionic strength [58]. The molal activity coefficient of H^+ ion is calculated from [60]:

$$\log \gamma_{H^+} = \frac{-C_1 Z_{H^+}^2 \sqrt{I}}{(1 + b_{H^+} C_2 \sqrt{I})} \tag{61}$$

with

$$b_{H^+} = 9 \text{ Å}$$

$$C_1 = \frac{1.82483 \times 10^{-6} \times \rho_b^{1/2}}{(\varepsilon T)^{3/2}} \quad (\text{molality})^{1/2}$$

$$C_2 = \frac{50.2916 \times \rho_b^{1/2}}{(\varepsilon T)^{1/2}}, \quad (\text{Å})^{-1} \, (\text{molality})^{-1/2}$$

$$Z_{H^+} = 1$$

where ε is the dielectric constant of pure water [61, 62].

The density of the brine solution is evaluated from

$$\rho_b = \rho_w + 0.0073 \text{ wt.} \, [1 + 1.6 \times 10^{-6} (T - 273)^2]$$

where ρ_w is the water density at T and wt is the concentration of salts in terms of the weight percentage of the total dissolved solids (TDS) in brine. Finally, T in the above equations is evaluated at the interface temperature. The accuracy of the model will be assessed later, when a comparison is made with experimental data.

7 MEASUREMENTS OF SILICA SCALE

Due to the inverse solubility, a low temperature and low pressure loop is sufficient to conduct tests for calcium carbonate scale. However, because of the progressive solubility nature of silica, a high temperature and high pressure loop is required to supply a heated supply of brine saturated with monomeric silica to a heat exchanger in which silica deposits. Since high temperature will accelerate corrosion due to the presence of salt in the brine, the loop design calls for the use of corrosion-resistant materials as well. Thus silica data is more difficult to obtain, and available data are still rather limited.

Five typical experimental data of Chan et al. [44] for silica scale are presented in Fig. 12. They are the overall transient fouling resistances formed on the inside surface of a titanium tube of a tube-and-shell design. Table 5 summarizes the average values of the monomeric and total silica concentrations in brines at the heat exchanger inlet during the courses of experiments. The thermal hydraulic conditions and chemistry for each run are given in Tables 6 and 7, respectively. Runs A and C used distilled, deionized water at a nominal pH of 6 and a variation of NaCl concentration. Runs D and DO used sodium acetate as a buffer to change the pH of the brine to above 7.

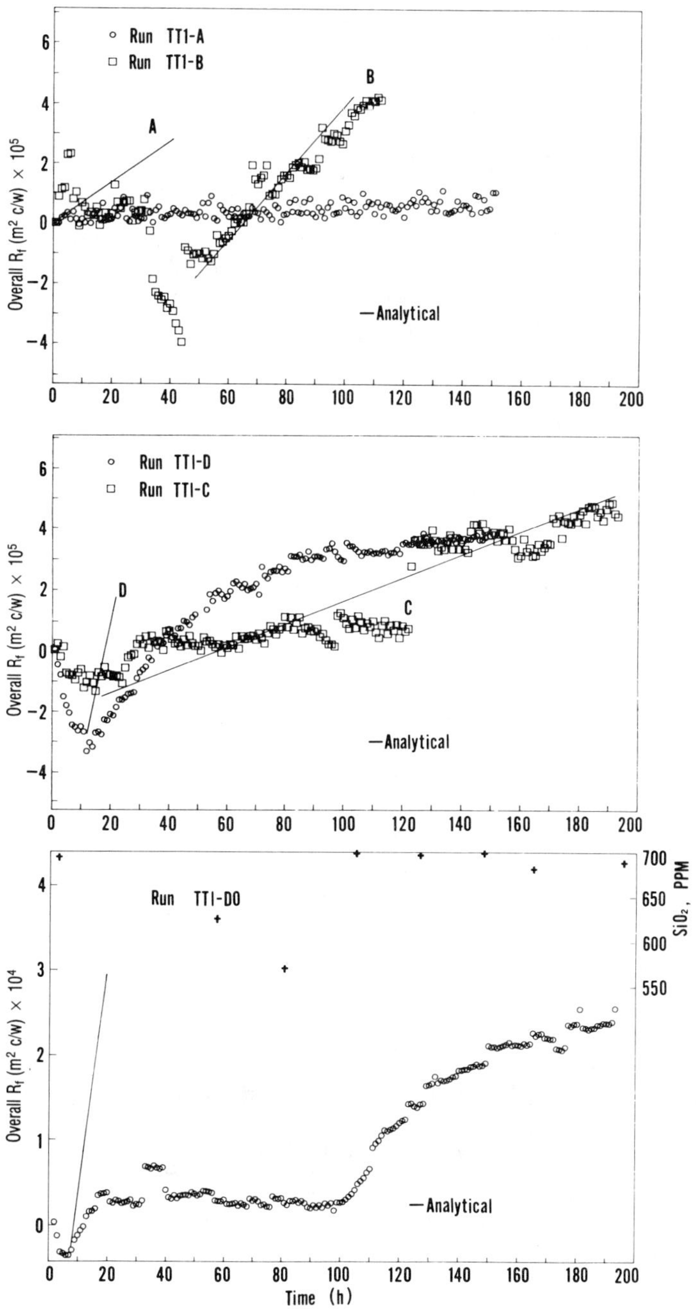

Figure 12 Overall thermal fouling resistance curves.

Table 5 Monomeric silica and total silica in brine drawn from brine tank

Run no.	TTI-A	TTI-B	TTI-C	TTI-D	TTI-DO
Monomeric acid, ppm S_iO_2	670	590	570	660	680
Total silica, ppm S_iO_2	587	—	560	662	668

The characteristic shapes of the fouling curves show an initial period in which the resistance drops for the runs with NaCl and sodium acetate (NaAc). With the exception of TT1B, experiments C, D, and DO show that the initial period is reduced as pH increases, in agreement with the batch type of experiments [30, 34]. The initial drop in resistance was reported in Gudmundsson and Bott [36] and by Bohlman et al. [42]. It is explained as an initial enhancement in heat transfer caused by scale-induced turbulence which is eventually overcome by the increasing thermal resistance of the deposit. Since the fouling resistance, R_f, is defined as the difference between the later and the initial total resistances, $R_f = 0$ initially. However, with heat transfer enhancement, the total resistance is reduced afterward, resulting in negative R_f values as shown in Fig. 12. All curves then proceed with increasing fouling functions, i.e., linear or asymptotic.

Examination of fouling surface in the heat exchanger by a boroscope revealed that the scale near the exit had a rough and rippled appearance transverse to the flow direction with high pointed ridges and valleys that left the tube surface exposed. The scale near the inlet, however, had a smoother layer which covered the entire surface, resulting in a higher thermal fouling resistance. This difference in physical characteristics of silica deposition makes silica modeling difficult.

A quantitative chemical analysis of the scale was made by x-ray energy spectrometry using a scanning electron microscope. The major component of the scale in all runs as shown in Table 8 is silica. This lends some support to the silica deposition model, which considers only the kinetics of silica at the brine-scale interface.

8 COMPARISON BETWEEN ANALYTICAL PREDICTIONS AND EXPERIMENTAL RESULTS

The analytical model given by Eq. (54) is used to correlate or to predict silica fouling from natural and synthetic brines. The data and experimental conditions of silica

Table 6 Thermal hydraulic conditions in fouling experiments

Run no.	Re_b	T_b(in)	T_b(out)	T_c(in)	T_c(out)	$\dot{V}_b$	$\dot{V}_c$	u_{0cl}	t_{tp}
TT1-A	73,130	438.6	396.6	344.5	345.6	0.94	169.5	2715.3	150
TT1-B	77,500	436.9	396.4	343.9	345.0	0.98	168.5	2709.1	133
TT1-C	75,790	438.2	397.4	344.3	345.4	0.95	167.8	2614.8	193
TT1-D	76,150	437.8	397.8	344.9	345.9	0.96	168.9	2608.0	148
TT1-DO	51,920	439.0	389.4	343.7	344.9	0.68	110.0	2257.1	193

T_b(out) and T_c(out) are based on the beginning value of run; Re_b is based on D_i, T_b, and water properties; U_0 is in units of w/m^2 C; t_{tp} = test period in h.

Table 7 Chemical composition in fouling experiments

Run no.	SiO_2[a]	NaCl[b]	NaAc	pH	TDS[c]
TT1-A	0.0111 (670)	0	0	5.9	670
TT1-B	0.0098 (590)	0.1 (5850)	0	6.8	6440
TT1-C	0.0095 (570)	1.0 (58500)	0	5.8	59070
TTI-D	0.0109 (660)	0	0.1 (13610)	7.3	14250
TT1-DO	0.0113 (680)	0	0.1 (13610)	7.4	14290

[a] Saturation concentration.
[b] Units of SiO_2, NaCl and NaAc concentrations in molarity and ppm.
[c] Total dissolved solids in ppm.

deposition rate on heat transfer surfaces at two geothermal fields in Iceland as reported in Gudmundsson and Bott [36] appear to be of sufficient detail that meaningful calculations can be made to incorporate supersaturation, pH factor, salinity, and thermal-hydraulic effects in the model.

Figure 13 shows that the experimental deposition rates of silica from both fields can be correlated surprisingly well by Eq. (54) with $K_R = 0.9954$ and $q = 0.862$. As discussed before, a second order of reaction for dissolved silica (i.e., $p = 2$) has been assumed. However, the comparison of the analytical calculations with the synthetic brine data is not totally satisfactory (see Fig. 12). In Fig. 12, the analytical lines are placed through the lowest value of the resistance curves. The fouling model is not expected to predict run A data, since no scale was found in this case. The comparison of run DO is not as poor as it appears since the flat portion of the data was caused by an experimental error and could be ignored. Better agreements are indicated for other runs. In any case, a further refinement of the analytical model is desirable.

Table 8 Scale deposition and composition

Run no.	TT1-A	TT1-B	TT1-C	TT1-D	TT1-DO
$m \times 10^3$	Neg.	23.2	18.2	66.2	171.2
$\dot{m} \times 10^5$	—	20.6	9.4	44.7	139.2
SiO_2[a]	—	87.6	82.3	84.9	97.1
Na_2O[b]	—	—	—	2.3	1.3
Cl[a]	—	—	<1	—	—
CaO[a]	—	<1	—	1.7	—
FeO[a]	—	4.7	9.3	5.6	1.6
ZNO[a]	—	6.2	6.3	5.6	—

[a] Composition in weight percent of total sample.
[b] Time = (193 h) − (70 h of zero fouling) = 123 h; m and $\dot{m}$ in kg/m^2 and kg/m^2 h, respectively.

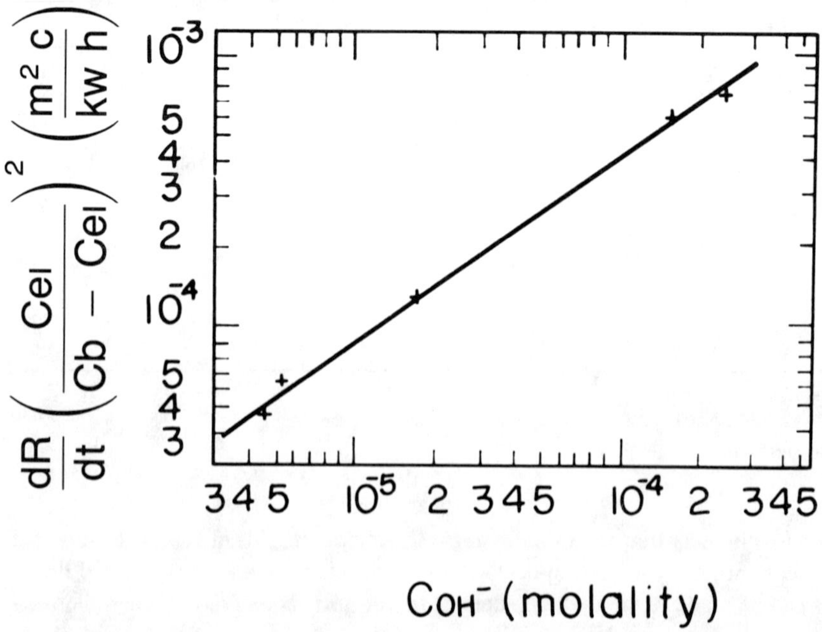

Figure 13 Correlation of silica deposition data from two geothermal fields.

9 IMPROVED SILICA FOULING MODEL

An attempt to refine the above model has been made, and a semianalytical model has been proposed [43] through an improvement in the deposition term and the addition of a removal term such that

$$\frac{dR_f}{dt} = \phi_d - \phi_r \tag{62}$$

The deposition term given in Eq. (54) has in a sense been split into two terms as follows:

$$\phi_d = \{K_{R1}[C_e(T_b) - C_e(T_s)]^P + K_{R2}[C_b - C_e(T_b)]^P\}(C_{\mathrm{OH}^-})^q \tag{63}$$

The first term, which is basically similar to that of Eq. (54), represents the contribution from the brine if the brine were at the saturation state with its dissolved monomeric silica reaching the solubility limit. The second term is to account for the supersaturation effect. Such a deposition term contains one more constant than the one in Eq. (54) but seems to be more general. For example, consider a heated saturated brine entering a heat exchanger. The second term is equal to zero at the inlet, and thus Eq. (63) reduces correctly to Eq. (54). At the other extreme, after the brine is discharged from the outlet to the downstream insulated pipe section, the first term automatically vanishes and the equation also reduces to the desired limit.

As for the removal term, it is commonly assumed to be a function of scale thickness and shear stress at the fluid deposit interface. A similar functional form is used,

$$\phi_r = k_3 R_f^a \left(\frac{\text{Re}}{10{,}000} \right)^b \tag{64}$$

where the fouling resistance R_f and Re take the roles of scale thickness and shear stress, respectively.

The correlation constants in ϕ_d and ϕ_r are [43] $k_{R1} = 1.65\ \text{m}^2\ \text{K/(w mol}^{1.2})$, $k_2 = 0.95\ \text{m}^2\ \text{K/(w mol}^{1.2})$, $k_3 = 2.34 \times 10^{-9}\ \text{m}^2\ \text{K/w}$, $a = 0.005$, $b = 4$, $p = 2$, and $q = 0.1$. Using those constants, the agreement with silica fouling data is improved considerably, including both the overall and local fouling resistances at various heat exchanger locations as illustrated in Chan et al. [43].

10 CONCLUSIONS

A review has been made of the fundamental information relevant to silica and calcium carbonate deposition from aqueous solutions in heat exchangers. Recent advances in analytical modeling and experimental measurements of both types of scale deposit have also been reviewed and discussed. The model for calcium carbonate fouling is a mechanistic one, based on conservation equations and surface reaction kinetics. It predicts well the overall fouling resistance averaged over the entire heat exchanger and the local fouling profile along the length of the heat exchanger. The silica deposition model is based on a kinetic-controlled mechanism with the monomer solubility evaluated at a scale-brine interface temperature governed by heat transfer. The comparison between the analytical prediction and silica data are encouraging but less satisfactory. This is due to the complexity of silica fouling, which is governed by a large number of variables, including solubility, polymerization, pH, salt contents, thermal-hydraulic conditions, etc. Nevertheless, progress has been made in understanding fouling in general and in predicting the growth of scale deposits with limited success.

ACKNOWLEDGMENTS

Support by the National Science Foundation under Grant No. CBT-8807406 is gratefully acknowledged. The author also wishes to thank many of his assistants in the Fouling Heat Transfer Laboratory at the University of Wisconsin—Milwaukee. They include H. Rau, C. BeBellis, K. F. Ghassemi, D. Alberts, D. Doers, D. Z. Zhou, and L. S. Lee, whose contributions formed the basis of this review.

REFERENCES

1. M. G. O'Callaghan, Fouling of Heat Transfer Equipment: Summary and Review, in *Heat Exchanger Sourcebook* (ed. J. W. Palen), Hemisphere, Washington, D.C., 1986.

2. J. Taborek, T. Aoki, R. B. Ritter, J. W. Palen, and J. G. Knudsen, Fouling: The Major Unresolved Problem in Heat Transfer, *Chem. Eng. Prog.*, vol. 68, no. 2, pp. 59–67.
3. E. F. C. Somerscales and J. G. Knudsen, (eds.), *Fouling of Heat Transfer Equipment*, Hemisphere, Washington, D.C., 1981.
4. R. W. Bryers, (ed.), *Fouling of Heat Exchanger Surface*, Proc. Engineering Foundation Conf., at White Haven, Pa., October 31–November 5, 1982.
5. L. F. Melo, T. R. Bott, and C. A. Bernardo, *Fouling Science and Technology*, NATO ASI Ser. E, vol. 145, Kluwer Academic Pub., Boston, 1987.
6. J. S. Gudmundsson and D. M. Thomas, (eds.), *Deposition of Solids in Geothermal Systems*, Geothermics, Int. J. Geothermal Res. and Its Application, vol. 18, no. 1/2, 1989.
7. N. Epstein, Fouling of Heat Exchangers, in *Heat Exchanger Sourcebook* (ed. J. W. Spalen), Hemisphere, Washington, D.C., 1986.
8. W. L. McCabe and C. S. Robinson, Evaporator Scale Formation, *Ind. Eng. Chem.*, vol. 16, pp. 478–479, 1924.
9. D. Q. Kern and R. E. Seaton, A Theoretical Analysis of Thermal Surface Fouling, *Br. Chem. Eng.*, vol. 4, pp. 258–262, 1959.
10. N. Epstein, Fouling: Technical Apsects, in *Fouling of Heat Transfer Equipment* (ed. E. F. C. Somerscales and J. G. Knudsen), Hemisphere, Washington, D.C., 1981.
11. E. F. Wahl, *Geothermal Energy Utilization*, John Wiley, New York, 1977.
12. J. Taborek, T. Aoki, R. B. Ritter, J. W. Palen, and J. G. Knudsen, Predictive Methods for Fouling Behavior, *Chem. Eng. Prog.*, vol. 68, no. 7, pp. 69–78, 1972.
13. J. W. Cleaver and B. Yates, The Effect of Re-entrainment on Particle Deposition, *Chem. Eng. Sci.*, vol. 31, pp. 835–845, 1976.
14. D. Hasson, M. Avriel, W. Resnick, T. Rozenman, and S. Windreich, Mechanism of Calcium Carbonate Deposition on Heat-Transfer Surfaces, *Ind. Eng. Chem. Fundamentals*, vol. 7, no. 1, pp. 59–65, 1968.
15. D. Hasson and I. Perl, Scale Deposition in a Laminar Falling-Film System, *Desalination*, vol. 37, pp. 279–292, 1981.
16. R. Sheikholeslami and A. P. Watkinson, Scaling of Plain and Externally Finned Heat Exchanger Tubes, *J. Heat Transfer*, vol. 108, pp. 147–152, 1986.
17. A. P. Watkinson and O. Martinez, Scaling of Heat Exchanger Tubes by Calcium Carbonate, *J. Heat Transfer*, pp. 504–508, 1975.
18. S. H. Chan and K. F. Ghassemi, Analytical Modeling of Calcium Carbonate Deposition for Laminar Falling Films and Turbulent Flow in Annuli: Part I—Formulation and Single-Species Model, HTD vol. 108, *Heat Transfer Equipment Fundamentals, Design, Applications and Operating Problems* (ed. R. S. Shah), ASME, New York, pp. 245–252, 1989. Also in *J. Heat Transfer* (in press).
19. S. H. Chan and K. F. Ghassemi, Analytical Model of Calcium Carbonate Deposition for Laminar Falling Films and Turbulent Flow in Annuli: Part II—Multi-Species Model, HTD vol. 108, *Heat Transfer Equipment Fundamentals, Design, Applications and Operating Problems* (ed. R. S. Shah), ASME, New York, pp. 253–260, 1989. Also in *J. Heat Transfer* (in press).
20. L. N. Plummer and E. Busenberg, The Solubilities of Calcite, Aragonite, and Vaterite in CO_2-H_2O Solutions Between 0 and 90°C, and an Evaluation of the Aqueous Model for the System $CaCO_3$-CO_2-H_2O, *Geochim. Cosmochim. Acta*, vol. 46, pp. 1011–1040, 1982.
21. H. Rau, Modeling of Calcium Carbonate and Silica Deposition in Geothermal Energy Systems, M.S. thesis, Milwaukee, Wisc., 1982.
22. H. N. S. Wiechers, P. Sturrock, and G. V. R. Marais, Calcium Carbonate Crystallization Kinetics, *Water Res.*, vol. 9, pp. 835–845, 1975.
23. R. B. Bird, W. E. Stewart, and E. N. Lightfoot, *Transport Phenomena*, John Wiley, New York, 1960.
24. K. Chun and R. Seban, Heat Transfer to Evaporating Liquid Films, *J. Heat Transfer*, vol. 91C, pp. 391–396, 1971.
25. G. H. Nancollas and M. M. Reddy, The Crystallization of Calcium Carbonate. II. Calcite Growth Mechanism, *J. Colloid Interface Sci.*, vol. 37, pp. 824–830, 1971.
26. R. H. Perry, D. W. Green, and J. O. Maloney (eds.), *Perry's Chemical Engineers' Handbook*, 6th ed., McGraw-Hill, New York, 1984.
27. H. Schlichting, *Boundary-Layer Theory*, McGraw-Hill, New York, 1955.

28. W. F. Langelier, Analytical Control of Anti-Corrosion Water Treatment, JAWWA, vol. 28, no. 10, pp. 1500–1521, 1936.
29. E. G. Bohlman, R. E. Mesmer, and P. Berlinski, Kinetics of Silica Deposition from Simulated Geothermal Brine, *J. Soc. Petroleum Eng.*, pp. 239–248, 1980.
30. O. Weres, A. W. Yee, and L. Tsao, Kinetic Equations and Type Curves for Predicting the Precipitation of Amorphous Silica from Geothermal Brines, SPE of AIME Int. Symp. Oilfield and Geothermal Chemistry, Houston, paper SPE 7888, January 22–24, 1979.
31. O. Weres, A. Yee, and L. Tsao, Kinetics of Silica Polymerization, Lawrence Berkeley Laboratory Report LBL-7033, Berkeley, Calif., 1980.
32. O. Weres, A. Yee, and L. Tsao, Kinetics of Silica Polymerization, *J. Colloid Interface Sci.*, vol. 84, no. 2, pp. 379–402, 1981.
33. O. Weres, A. Yee, and L. Tsao, Equations and Type Curves for Predicting the Polymerization of Amorphous Silica in Geothermal Brines, *Soc. Petroleum Eng. J.*, pp. 9–16, February 1982.
34. H. P. Rothbaum and A. G. Rohde, Kinetics of Silica Polymerization and Deposition from Dilute Solutions Between 5 and 180°C, *J. Colloid Interface Sci.*, vol. 71, no. 3, pp. 533–559, 1979.
35. H. L. Barnes, *Geochemistry of Hydrothermal Ore Deposits*, 2nd ed., John Wiley, New York, 1979.
36. J. S. Gudmundsson and T. R. Bott, Deposition of Silica from Geothermal Waters on Heat Transfer Surfaces, *Desalination*, vol. 28, pp. 125–145, 1979.
37. G. L. Lombard, Heat Exchanger Tests with Moderately Scaling Geothermal Brine, *AIChE Symp. Ser.*, vol. 74, no. 174, pp. 281–287, 1978.
38. R. O. Fournier, Silica in Thermal Waters: Laboratory and Field Investigations, in *Proc. Int. Symp. on Hydrogeochemistry and Biogeochemistry*, Tokyo, pp. 122–139, September 7–9, 1970.
39. D. E. Felsinger, D. K. Mulliner, and H. K. Bishop, Geothermal Field Tests Heat Exchanger Evaluation, San Diego Gas & Electric Co., San Diego, Calif., 1973.
40. Ben Holt Co., Geothermal Heat Exchanger Test at Heber, California, RP-846-1, EPRI ER-572, August 1978.
41. T. R. Bott and J. S. Gudmundsson, Rippled Silica Deposits in Heat Exchanger Tubes, *Proc. 6th Int. Heat Transfer Conf.*, Toronto, Ontario, August 7–11, 1978.
42. R. G. Bohlman, A. J. Shor, P. Berlinski, and R. R. Mesmer, Silica Scaling in Simulated Geothermal Brines, Oak Ridge National Laboratory, Oak Ridge, Tenn., 1981.
43. S. H. Chan, D. Z. Zhou, and K. F. Neusen, A Semi-Analytical Model of Silica Fouling, *Heat Transfer in Geophysical and Geothermal Systems*, HTD vol. 76, ASME, New York, pp. 45–51, 1987.
44. S. H. Chan, H. Rau, C. DeBellis, and K. F. Neusen, Silica Fouling of Heat Transfer Equipment—Experiments and Model, *J. Heat Transfer*, vol. 110, pp. 841–849, 1988.
45. K. B. Krauskopf, Dissolution and Precipitation of Silica at Low Temperature, *Geochim. Cosmochim. Acta*, vol. 10, pp. 1–26, 1956.
46. M. R. Hoffman, Brine Chemistry—Scaling and Corrosion—Geothermal Research Study in the Salton Sea Region of California, Environmental Quality Laboratory, California Institute of Technology, EQL Memorandum No. 14, 1975.
47. F. J. Millro, The Effect of Pressure on the Solubility of Minerals in Water and Seawater, *Geochim. Cosmochim. Acta*, vol. 46, pp. 11–22, 1982.
48. G. Okamota, T. Okura, and K. Goto, Properties of Silica in Water, *Geochim. Cosmochim. Acta*, vol. 12, pp. 123–132, 1957.
49. W. L. Marshall, Amorphous Silica Solubilities—I. Behavior in Aqueous Sodium Nitrate Solutions; 20–300°C, 0–6 Molal, *Geochim. Cosmochim. Acta*, vol. 44, pp. 907–913, 1980.
50. C. T. A. Chen and W. Marshall, Amorphous Silica Solubilities IV. Behavior in Pure Water and Aqueous Sodium Chloride, Sodium Sulfate, Magnesium Chloride, and Magnesium Sulfate Solutions up to 350°C, *Geochim. Cosmochim. Acta*, vol. 46, pp. 279–287, 1982.
51. W. L. Marshall and J. M. Warakomski, Amorphous Silica Solubilities—II, Effect of Aqueous Salt Solutions at 25°C, *Geochim. Cosmochim. Acta*, vol. 44, pp. 915–924, 1980.
52. S. H. Chan, K. F. Neusen, and C. T. Chang, The Solubility and Polymerization of Amorphous Silica in Geothermal Energy Applications, *Proc. 1987 ASME-JSME Thermal Engineering Joint Conf.*, vol. 3, pp. 103–108, 1987.

53. G. B. Alexander, The Polymerization of Monosilicic Acid, *J. Phys. Chem.*, vol. 76, pp. 2094–2096, 1954.
54. R. K. Iler, Polymerization of Polysilicia Acid Derived from 3.3 Ratio Sodium Silicate, *J. Phys. Chem.*, vol. 56, pp. 604–607, 1953.
55. R. K. Iler, *The Chemistry of Silica—Solubility, Polymerization, Colloid and Surface Properties, and Biochemistry*, John Wiley, New York, 1979.
56. B. A. Fleming, Polymerization Kinetics and Ionization Equilibria in Aqueous Silica Solutions, Ph.D. dissertation, Princeton University, Princeton, N.J., 1981.
57. R. P. Fournier and J. J. Rowe, The Solubility of Amorphous Silica in Water at High Temperatures and Pressure, *Am. Mineral*, vol. 52, pp. 1052–1056, 1977.
58. R. H. Busey and R. E. Mesmer, The Ionization of Water in NaCl Media to 300°C, *J. Solution Chem.*, vol. 5, no. 2, pp. 147–152, 1976.
59. C. Lin and W. T. Lindsay, Thermodynamics of Sodium Chloride Solutions at High Temperature, *J. Solution Chem.*, vol. 1, no. 1, pp. 45–69, 1972.
60. A. H. Truesdell and B. F. Jones, Water, A Computer Program for Calculating Chemical Equilibria of Natural Waters, *J. Res. U.S. Geol. Surv.*, vol. 2, no. 2, pp. 233–248, 1974.
61. C. G. Malmberg and A. A. Maryott, Dielectric Constant of Water from 0° to 100°C, *J. Res. Natl. Bur. Std.*, vol. 56, no. 1, pp. 1–8, 1956.
62. G. C. Akerlof and H. I. Ashry, The Dielectric Constant of Water at High Temperatures and in Equilibrium with Its Vapor, *J. Am. Chem. Soc.*, vol. 72, pp. 2044–2847, 1950.
63. D. Hasson, H. Sherman, and M. Biton, Prediction of $CaCO_3$ Scaling Rates, *Proc. 6th Int. Symp. Fresh Water from the Sea*, vol. 2, pp. 193–199, 1978.
64. R. K. Iler, *The Colloid Chemistry of Silica and Silicates*, Cornell University Press, Ithaca, N.Y., 1955.
65. E. F. Wahl, I. K. Yen, and W. J. Bartel, Silica Scale Control in Geothermal Brines, Office of Saline Water, U.S. Dept. of the Interior, Washington, D.C., #14-30-3-41, 1974.

INDEX